D0230817

for AQA

CHEMISTRY

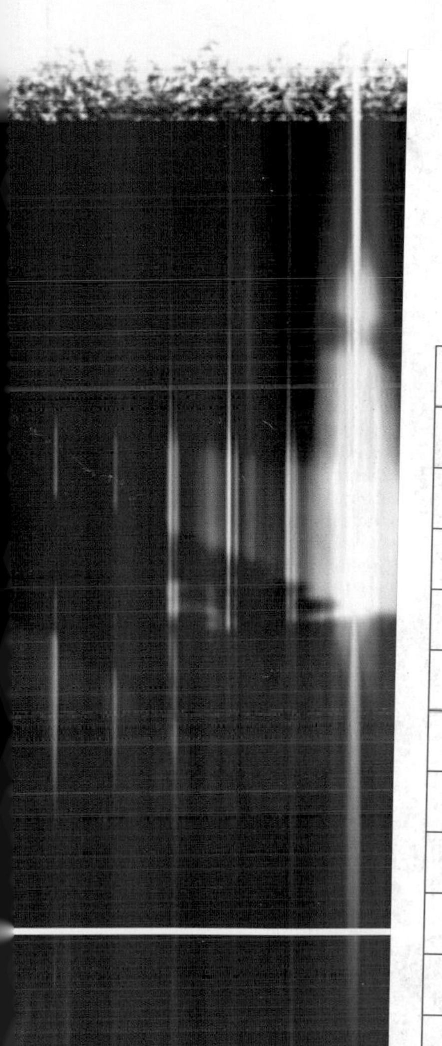

Ripon Grammar School

Book No. 139

Name	Form	Date Issued

gel Saunders
ela Saunders

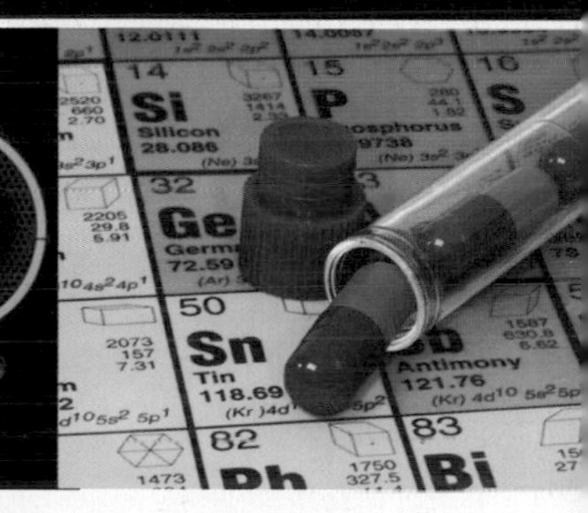

OXFORD
UNIVERSITY PRESS

Great Clarendon Street, Oxford OX2 6DP

Oxford University Press is a department of the University of Oxford.
It furthers the University's objective of excellence in research, scholarship, and education by publishing worldwide in

Oxford New York

Auckland Cape Town Dar es Salaam Hong Kong Karachi
Kuala Lumpur Madrid Melbourne Mexico City Nairobi
New Delhi Shanghai Taipei Toronto

With offices in

Argentina Austria Brazil Chile Czech Republic France Greece
Guatemala Hungary Italy Japan Poland Portugal Singapore
South Korea Switzerland Thailand Turkey Ukraine Vietnam

Oxford is a registered trade mark of Oxford University Press in the UK and in certain other countries

© Oxford University Press 2009

The moral rights of the author have been asserted

Database right Oxford University Press (maker)

First published 2009

Acknowledgements
The Press wishes to acknowledge the contribution of Dr Mike Clugston in the original preparation of some of the illustrations used herein.

We are grateful for permission to reproduce the following photographs;

Pages 14/15 Laguna Design/Science Photo Library; P17 Helene Rogers/Art Directors & Trip Photo Library; P20 Dr Ivan Polunin/NHPA; P21(T) Martyn F. Chillmaid / Science Photo Library; P21(L) Peter Alvey / Alamy; P24 Richard Megna/FPNY; P25 Martyn F Chilmaid/Science Photo Library; P26 Richard Megna/FPNY; P34 Hulton Archive/Stringer/Getty Images; P36 Martin Sookias; P37(L) CC Studio/Science Photo Library; P37R Andrew Lambert LGPL; P40 Andrew Lambert/LGPL; P41 Andrew Lambert/LGPL; P43 Edward Gill / Custom Medical Stock Photo/Science Photo Library; P44 Martin Shields/Alamy; P48 Charles D. Winters/Science Photo Library; P49(T) Sciencephotos/Alamy; P49(L) Sciencephotos/Alamy; P50 Martyn F. Chillmaid/Science Photo Library; P52 Jupiterimages; P53 Specta/Shutterstock; P59(T) James Steidl/Istockphoto; P59(L) Leonard Lessin/Science Photo Library; P60 Chili Sauce/Alamy; P62 Martin Asser Hansen; P63(T) Alfred Pasieka/Science Photo Library; P63(L) Sinclair Stammers/Science Photo Library; P65 Jupiterimages; P66 Andrew Lambert LGPL; P69 Aaron Wood/Shutterstock;P70 Martyn F. Chillmaid/Science Photo Library; P71 Jean-Louis Vosgien/Shutterstock;P72(T) SAV/Alamy; P72(L) Elke Dennis/123RF; P73 Tracy Martinez/Fotolia; P74 Sciencephotos/Alamy; P77 Andrew Lambert/LGPL; P80 A. Crump, Tdr, Who/Science Photo Library; P82 David Woodfall/NHPA; P85(T) Andrew Lambert Photography/Science Photo Library; P85(MB) Andrew Lambert Photography/Science Photo Library; P88 David J. Green/Alamy; P92 Jeremy Hofner/Panos Pictures; P93(T) Nordicphotos/Alamy; P98 Neville Elder/Corbis; P101 Andrew Lambert/LGPL; P105 (T)Grant Heilman Photography/Alamy; P105(L); Akira Suemori/Associated Press; P108 Colin Cuthbert/Science Photo Library; P110 Equinox Graphics/Science Photo Library: P113 Garo/Phanie/Rex Features; P119 Cordelia Molloy/Science Photo Library; P120 Construction Photography/Corbis; P121(T) Science Photo Library; P121(L) Chris Howes/Wild Places Photography/Alamy: P122 Istockphoto: P123(L) Mark Boulton/Alamy; P123R Mira/Alamy: P141Mauro Fermariello/Science Photo Library; P148 Maximilian Stock Ltd / Science Photo Library;P149 (T) Kit Kittle/Corbis; P149(L) Mauro Fermariello/Science Photo Library: P152/153 Andrew Syred/Science Photo Library; P157Shout/Alamy; P161 David R. Frazier Photolibrary, Inc.; P170 Janine Wiedel Photolibrary/Alamy; P174 Sharon Day/Big Stock Photo; P180(T) Andrew Lambert Photography/Science Photo Library; P180(M) Andrew Lambert Photography/Science Photo Library; P180(L) Andrew Lambert Photography/Science Photo Library; P181(L) Leslie Garland Picture Library/Alamy; P181R Pascal Goetgheluck/Science Photo Library; P185Workingwales/Alamy; P190 Andrew Lambert Photography/Science Photo Library; P200(T) Corbis; P200(L) Alamy; P203 Fine Art/Alamy; P204 Friedrich Saurer/Alamy: P205 Brooks Kraft/Corbis; P210 Andrew Lambert/LGPL; P211 Paul Silverman/NPNY; P212 Andrew Lambert/LGPL; P217(T) Document General Moters/Reuter R/Corbis Sygma;P217(L) Ace Stock Limited/Alamy; P219 Jerry Mason/Science Photo Library; P227 Phantatamix/Science Photo Library; P234 Andrew Lambert/Science Photo Library; P234 Andrew Lambert/Science Photo Library; P236 Andrew Lambert/Science Photo Library; P237(T)Martin F Chilmaid/Science Photo Library; P 237(L) Martin F Chilmaid/Science Photo Library; P239 Martin F Chilmaid/Science Photo Library; P243 Andrew Lambert/Science Photo Library; P244 Martin F Chilmaid/Science Photo Library; P246 Martin F Chilmaid/Science Photo Library; P247 Andrew Lambert/Science Photo Library; P249 Andrew Lambert/Science Photo Library; P251 Andrew Lambert/Science Photo Library; P252 Andrew Lambert/Science Photo Library; P261 Andrew Lambert/Science Photo Library.

In a few cases we have been unable to trace the copyright holder prior to publication. If notified the publishers will be pleased to amend the acknowledgements in any future edition.

All rights reserved. No part of this publication may be reproduced, stored in a retrieval system, or transmitted, in any form or by any means, without the prior permission in writing of Oxford University Press, or as expressly permitted by law, or under terms agreed with the appropriate reprographics rights organization. Enquiries concerning reproduction outside the scope of the above should be sent to the Rights Department, Oxford University Press, at the address above
You must not circulate this book in any other binding or cover and you must impose this same condition on any acquirer

British Library Cataloguing in Publication Data

Data available

ISBN: 978-0-19-915276-6

10 9 8 7 6 5 4 3 2 1

Printed in Singapore by KHL Printing Co Pte, Ltd.

Paper used in the production of this book is a natural, recyclable product made from wood grown in sustainable forests. The manufacturing process conforms to the environmental regulations of the country of origin.

Introduction

Chemistry is all around you. In a world without chemistry, there would be no advanced fuels, plastics, or medicines. There would be no flat screen colour displays for mobile phones. Millions of people might go hungry for lack of fertilizer for crops, and our water would not be safe to drink. Even something as simple as cooking a meal involves chemistry. But in the twenty-first century, chemists can design and control substances with astonishing precision. Chemists have a vital role to play in the modern world, and this is an exciting time to study chemistry.

Your A2 Chemistry studies will allow you to develop further your understanding of many important chemical ideas. You will discover how early discoveries became the shoulders on which modern chemistry stands, and you will analyse the present and look forward to the future. Chemistry is a refreshingly practical subject, and you will have the opportunity to enhance your laboratory skills during the course.

This book covers all the subject content you need for the AQA A2 Level Chemistry Specification. It aims to build on your AQA AS Chemistry studies, showing you how the knowledge and understanding of different aspects of chemistry can be woven together. Above all, *A2 Chemistry for AQA* aims to help you enjoy and take part fully in this exciting and challenging subject.

Nigel and Angela Saunders,
Harrogate, North Yorkshire 2008

Contents

* practical based material

Welcome: how to use *A2 Chemistry for AQA*

A2 Chemistry for AQA has been written specifically for students taking the new specification in GCE Advanced Level Chemistry for AQA, the Assessment and Qualifications Alliance. AQA is the largest of the three English examination boards.

The main part of the book covers all the subject content you will need to know for the assessments of Units 4 and 5. For easy access, the topics are covered in a series of double-page spreads that follow the order given in the specification. Each spread is self-contained and could be studied in a different order than the one in the book. But if you do this, you should check the prior knowledge needed to understand the concepts covered in the spread. Use the *Objectives* box in the left-hand margin to help you decide this.

It is assumed if you are taking A Level Chemistry and you are using this book, you will already have completed your AS Level in Chemistry using the AQA specification. Where calculations are needed, the mathematics involved is explained. Extension material is included to help you if you are seeking the highest grades. You will also find synoptic material there to help bring together concepts from AS Chemistry with those from A2 Chemistry, necessary to answer some of the trickier questions in the examinations.

Each main double-page spread contains a boxed list of prior knowledge and a list of outcomes which relate directly to the AQA specification.

- Text covering the subject content, with key terms shown in **bold** print and defined in the *Glossary* at the back of the book (it includes all the entries from the AS course too!)
- *Check your understanding* questions.
- The answers to calculations are printed at the end of the book; answers to all questions are found in the e-book and the teacher's resources.

Flexible use of the spreads means that you can easily follow your own route through the course. Unit 5 could easily be studied before Unit 4, or both could be studied at the same time.

How Science Works is an integral part of the new GCE A Level Chemistry specification, and is an integral part of this book. Many of the spreads contain a *Science@Work* section that deals with one or more of the related criteria. These are discussed more fully in the spread called *How Science Works* on page 10.

A2 Chemistry for AQA also includes six double-page spreads covering investigative and practical skills. They cover the tasks indicated in the specification, and are intended to help you prepare for the investigative skills assessments of Unit 6. These are discussed more fully on page 12 in the section called *Preparing for Assessment*. Also, there are five spreads of exam-style questions placed at the end of each major topic. There is also a selection of synoptic questions at the end of the book which need knowledge and understanding of more than one topic so that they can be answered fully.

AQA is the largest of the three English examination boards. It sets and marks examinations, including GCSEs and A Levels. AQA has a very useful website at **www.aqa.org.uk**. You can download examination papers and mark schemes from there to help you with your studies.

The Chemistry Specification

AQA's *AS and A Level Chemistry Specification* builds on its earlier chemistry course and aims to encourage you to:

- study chemistry in a modern context
- become enthusiastic about chemistry
- show that you can bring different ideas together
- develop your practical skills and data analysis skills
- appreciate how science works and its importance in the wider world

You will learn about the chemistry behind contemporary issues such as fuel cells, batteries, biodiesel, drug design, and biodegradable polymers. You will also look at some of the factors required for efficient industrial processes, necessary for sustainable development.

The A2 Chemistry course is divided into three Units. Units 4 and 5 cover chemistry knowledge and understanding and Unit 6 covers practical skills. But you can also expect to be assessed on your performance in class practicals that support the chemical ideas in Units 4 and 5.

Unit 4: Kinetics, equilibria, and organic chemistry

Unit 4 develops the concepts of physical chemistry and organic chemistry introduced in your AS Chemistry studies. These include:

- reaction kinetics
- reversible reactions and equilibria
- acids and bases
- nomenclature and isomerism in organic chemistry
- compounds containing the carbonyl group
- amines
- polymers
- organic synthesis and analysis

Unit 4 can be studied before Unit 5, but you may study Unit 5 first or even both Units together.

Unit 5: Energetics, redox, and inorganic chemistry

Unit 5 develops the concepts of inorganic chemistry introduced in your AS Chemistry studies, and more concepts in physical chemistry. These include:

- thermodynamics, including enthalpy and entropy
- periodicity, including the reactions of period 3 elements and their oxides
- redox equilibria
- electrochemical cells
- general properties of transition metals, including their use as catalysts and their reactions

Unit 5 can be studied after Unit 4, but you may study Unit 5 first or even both Units together.

Unit 6: Investigative and practical skills

Unit 6 assesses your practical skills and data analysis skills. There are two main parts for this:

- an assessment of your practical skills during several laboratory practicals
- a written test carried out under controlled conditions

Your teacher can choose one of two routes, depending on whether they intend to mark your work themselves, or have AQA examiners mark your work. Whichever route you do, your total marks are out of 50, and count for up to 60 UMS marks.

Centre-marked route (Route T)

This is the route where your teacher marks your work, and their marking is checked for accuracy by one of AQA's external moderators. There are two parts to this assessment.

1 An assessment of your practical skills during several laboratory practicals. You are awarded marks out of 2 for each assessed practical you do. You must do at least two assessed practicals from each of the three areas of chemistry. These are inorganic chemistry, physical chemistry, and organic chemistry. You will receive a total mark out of 12 for this part, called the **Practical Skills Assessment** or **PSA**.

2 At least one written test carried out under controlled conditions. This ensures that the answers are your own work, even if you collected the results as a class. It has two sections. Section

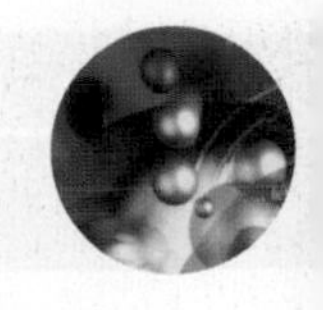

A consists of questions about your own data from an experiment. Section B consists of questions concerning a similar area of chemistry. Your teachers mark the papers and their marking is checked by a moderator appointed by AQA. You will receive a mark out of 38 for this part, called the **Investigative Skills Assignment** or **ISA**.

Externally-marked route (Route X)

This is the route where your work is marked by one of AQA's external examiners. There are two parts to this assessment.

1. An assessment of your practical skills during several laboratory practicals. You must do at least two assessed practicals from each of the three areas of chemistry, inorganic chemistry, physical chemistry, and organic chemistry. You will not receive a mark, but your teacher has to confirm on your *Candidate Record Form* that you have carried out each practical safely and skilfully. This part is called the **Practical Skills Verification** or **PSV**.
2. A written test carried out under controlled conditions. This is similar to the ISA from Route T but has an extra section. Sections A and B are similar to the sections in an ISA, but Section C asks you questions about how to carry out experiments correctly. You will receive a mark out of 50 for this part, called the **Externally Marked Practical Assignment** or **EMPA**.

The examinations

You can expect to see questions that address just Unit 4 or Unit 5. But you will also see questions that address other areas of the specification, too. These are called synoptic questions. You need to have a thorough understanding of the course to answer these effectively. The table summarizes how Units 4 and 5 are assessed.

Unit	Form of assessment	Duration (hours)	Type of questions	Percent of A2 marks
4	Written paper	1¾	Six to eight short questions, and two or three longer structured questions	40
5	Written paper	1¾	Five to seven short questions, and two or three longer structured questions	40

The total duration of the papers is three and a half hours. They contribute 80% of your A2 marks. The rest comes from the Unit 6 Investigative and Practical Skills Assessment. Your AS marks contribute half of your total A Level marks, and your A2 marks contribute the other half.

How Science Works, and you

It can be easy when you are learning new ideas in chemistry to forget that these ideas were discovered and developed by people. *How Science Works* seeks to explain how scientists carry out their investigations. It examines how their beliefs can influence their thinking and approach to their research, and shows how scientists contribute to decision-making in society. You can expect examination questions to assess your understanding of *How Science Works*, so you must be prepared for this. The specification identifies twelve key aspects of *How Science Works*, and *A2 Chemistry for AQA* contains numerous examples of these, particularly in the *Science@Work* sections.

Specification summary

A Theories, models, and ideas

Scientists use theories and models to explain their observations. These form the basis for scientific investigations. Progress is made when there is valid evidence to support a new theory. For example, experiments with cells confirm that electrons are transferred in redox reactions (Spread 16.03).

B Questions, problems, scientific arguments, and ideas

Scientists use their knowledge and understanding in their work. This includes when they make their observations, when they identify a scientific problem, and when they question scientific explanations. Progress is made when scientists contribute to new ideas and theories. For example, entropy can explain spontaneous reactions (Spread 14.03).

C Using appropriate methods

Scientists develop explanations or hypotheses from their observations. They can make predictions from their hypotheses, and these can be tested using carefully planned experiments. Data can be quickly collected, recorded, and analysed with the help of ICT. You learn about this aspect of *How Science Works* through your practical work in chemistry.

D Carrying out experiments and investigations

Scientists use a wide range of practical skills, including carrying out experiments and examining the data through graphs and statistical analysis. They need to choose the right equipment and record their results carefully. You learn about this aspect of *How Science Works* through your practical work in chemistry. For example, your Practical Skills Assessments cover this, whether assessed internally or externally.

E Analysing data

Scientists analyse observations and results. They look for patterns and correlations. They make informed decisions about what to do with anomalous results that fall outside the expected range. Data that matches scientific predictions increases the confidence that scientists have about their models. For example, it is recognized that entropy change is an important factor in determining the direction of spontaneous reactions (Spread 14.04).

F Evaluating methods and data

Scientists question the validity of new evidence and the conclusions drawn from it. Different research teams may come up with different results, even when using similar methods. In trying to resolve these differences, scientists may improve their methods or develop new hypotheses that can be tested. For example, thermochemical data from enthalpies of hydrogenation provides evidence consistent with the known structure of benzene (Spread 7.02).

G The nature of scientific knowledge

Scientific knowledge and understanding rarely stand still. Scientific explanations are based on experimental evidence, accepted by the scientific community. But new evidence may be found that needs a better explanation than the existing one, and scientific knowledge changes as a result. For example, the concept of Lewis bases was extended to include metal–aqua ions (Spread 20.01).

H Communicating information and ideas

Scientists share their findings, allowing other scientists to evaluate their work. It is important that precise scientific language is used to avoid confusion. You will need to develop your ability to use the appropriate words and phrases in your explanations and answers. For example, you should be able to use cell notation (Spread 16.03).

I Benefits and risks of science

Developments in science, medicine, and technology have improved the quality of life for most people. But there may also be risks involved with the methods that scientists develop and the way in which their research might be used. Scientists consider these benefits and risks in their work. For example, the benefits of using hydrogen–oxygen fuel cells has to be balanced with the hazards (Spread 16.10).

J Ethical and environmental issues

Scientific research is paid for through public funding or by private companies, and scientists must consider ethical and environmental issues in their research. Individual scientists have their own moral or religious beliefs, and they contribute to making decisions about what research should be allowed. For example, the relative biodegradability of polymers affects how they are disposed of or reused (Spread 10.05).

K The role of the scientific community

Scientific research is published in scientific journals. Before this happens, it is examined by other scientists in the same field to check its validity. This is called *peer review*. The scientific community is then able to study new research. This is important because it may be possible for the research and its conclusions to be influenced by the organization that funded the research. For example, the disadvantages of biofuels might be minimized by organizations that profit from their production (Spread 5.06).

L Society uses science to inform its decisions

Politicians and other decision-makers may use scientific findings to inform their decisions. Scientists may take part in making decisions, especially if the scientific evidence is incomplete. The final decisions are often influenced by the existing beliefs of the decision-makers, by special interest groups, by public opinion, and by the media. For example, thalidomide is a drug with optical isomers, one of which causes birth defects (Spread 4.04).

Chemistry resources on the Internet

There are many Internet resources that might support your chemistry studies. The US has a larger population than the UK, so a large proportion of sites is biased towards American courses. These sites can still be useful if you use them with care.

Many UK sites support the AQA course. Other sites' support might be helpful, but other courses are *not* the same. This might mean that the sites cover topics that you do not need. They might also miss out topics that you do need. Sometimes the chemistry vocabulary used is slightly different, too. Finally, there are some very good general chemistry websites that take care to explain things clearly to you.

There is a list of 13 useful chemistry websites here. It is not exhaustive and you may find others that you prefer. Remember that websites do move or disappear. You can use search engines to find other sites. Try key words and phrases such as 'chemistry revision', 'chemistry worksheets', and 'chemistry help'. Modify your search by adding 'UK', 'a2 level', or 'a level' to make it more specific to the UK. Remember that you can also search for particular topics.

Supporting AS/A2 Level Chemistry		
Site	**Content**	**Web address**
Creative Chemistry	worksheets, practical guides, revision notes, and interactive molecular models	www.creative-chemistry.org.uk/alevel
Knockhardy Publishing Science Notes	notes and PowerPoint® presentations	www.knockhardy.org.uk/sci.htm
Chemguide	many helpful pages	www.chemguide.co.uk
Chemistry in Perspective	web-based chemistry text book	www.chembook.co.uk
Rod's Pages	topic information, laboratory tips, and examination tips	www.rod.beavon.clara.net
S-Cool	topic guides, questions, and revision summaries	www.s-cool.co.uk
General Chemistry Online	general chemistry resources	http://antoine.frostburg.edu/chem/senese/101
Greener Industry	about sustainable industry	www.greener-industry.org
RSC Video Clips	video clips of chemistry experiments and reactions	www.chemsoc.org/networks/learnnet/videoclips.htm
WebElements™	chemistry information relating to the periodic table	www.webelements.com

Online revision quizzes and questions

These three sites have interactive or downloadable quizzes and questions to check your knowledge and understanding.

Site	Content	Web address
LearnNet	quizzes and revision	www.chemsoc.org/networks/learnnet/questions.htm
Revision Resources for A Level Chemistry	quizzes and revision	www.mp-docker.demon.co.uk
Timberlake's Chemistry	quizzes and revision	www.karentimberlake.com/quizzes.htm

Units 4 and 5

Units 4 and 5 are assessed by two examination papers, marked by external examiners rather than your teachers. The table summarizes the duration of these papers, and their contribution to your final marks.

Unit	Duration (hours)	Total marks on paper	UMS marks	Percent of A2 marks
4 (*Kinetics, equilibria, and organic chemistry*)	1¾	100	120	40
5 (*Energetics, redox, and inorganic chemistry*)	1¾	100	120	40

Papers 4 and 5 contribute 80% of your total A2 marks (and 40% of your total A Level marks).

The double-page spreads in *A2 Chemistry for AQA* cover all the content listed in the specification. One of your main tasks during your A2 Level course will be to learn and understand all the subject content. But you must be prepared to apply your knowledge and understanding to answer questions on any topic from your AS Level studies.

Unit 6

Unit 6 tests your investigative and practical skills. There are two routes, explained in detail on page 9. Whichever route you follow, Unit 6 is worth 50 raw marks and 60 UMS marks. It counts for 20% of the A2 marks (and so 10% of the full A Level).

A2 Chemistry for AQA contains spreads with guidelines for carrying out practical work. These are based on the tasks and suggested contexts in the A2 Chemistry Specifications. The double-page spreads involved are:

Unit	Spread	Task(s)	Page
4	1.05	The iodine clock experiment	24
4	2.04	Finding the value of K_c	32
4	3.08	Investigating pH changes	48
4	6.05	Making aspirin	84
5	17.03	Analysing iron	212
5	21.02	Preparing inorganic compounds	254
5	22.02	Some copper chemistry	260

Towards effective study

Use the Objectives

Use the Objectives for each spread to help you organize your learning. They relate directly to the subject content in the AQA specification. It tells you what you should be able to do when you have finished studying each topic. Before an assessment in January or June, check that you can recall and show an understanding of the scientific knowledge in the Unit or Units you will be sitting. You will also be expected to be able to apply your knowledge and understanding in unfamiliar contexts, including the ideas behind *How Science Works*.

Use the *Check your understanding* questions

These questions are designed to help you check that you understand and can recall key ideas covered in each spread. Immediately after you have finished a spread, read the questions and write down your answers. Check through the text to make sure that your answers are correct. If you cannot find the answer, or if you do not understand the question, make a note of it and discuss it with your lecturer or teacher. You may also check the answers in the support material available separately from Oxford University Press.

Use the Glossary

One of the challenges of studying chemistry is getting to grips with lots of specialist words and phrases. It can be a bit daunting when first confronted with a lot of new words. It is a good idea to begin to build up your chemical vocabulary right from the start of your studies. You will find it much easier to explain chemical ideas if you have a good grasp of chemical vocabulary. For each topic, you might like to put together your own list of key words and phrases, and use the Glossary or a chemistry dictionary to define them. Your examiners will be looking to see that you can use specialist vocabulary when appropriate.

Use the exam-style questions

There are five spreads of more challenging questions together with a selection of synoptic questions with which to practise further.

Make concept maps

Concept maps (sometimes called mind maps or patterned notes) are a useful way of organizing your ideas for revision. It is often best to keep them simple.

1 Choose a theme, and then identify related concepts as key words or phrases. Write simple definitions for each one. The Glossary will help here. Then start looking for connections between the concepts. Some connections will be obvious, but you might need to look at the relevant spread again to see all the connections.

2 Link concepts with lines or arrows. Where appropriate, add simple joining phrases between them to make the connections clear. Some phrases you could use are: *such as*, *necessary for*, *leads to*, *produces*, *and*, *or*, *during*.

3 Draw a box around each concept to separate it from the connecting phrases.

You might need to use a concept more than once, but concept maps are usually clearer and more useful if concepts are used only once. You might find it helpful to use different colours to group similar concepts together.

Use past examination papers

The AQA website provides links to specimen question papers, past papers, and mark schemes. These are free and are a great way to check your revision progress. Download and print out the papers. Have a go at answering the questions, then use the mark scheme to check your answers. It might take several hours the first time you try a question paper. But stick at it: you will get faster and more accurate with practice. You can also download the Examiners' Reports for past examinations. These tell you how many marks you need for each A Level grade.

Discuss your work

Talk to other students about the chemistry you are studying, if you can. Form a small group that meets to discuss chemistry topics. There are thousands of other students studying your course. These are the people you are competing against, not your friends and classmates. Encourage and help each other to develop a better understanding of chemistry. Get together, order a pizza, and make your revision fun!

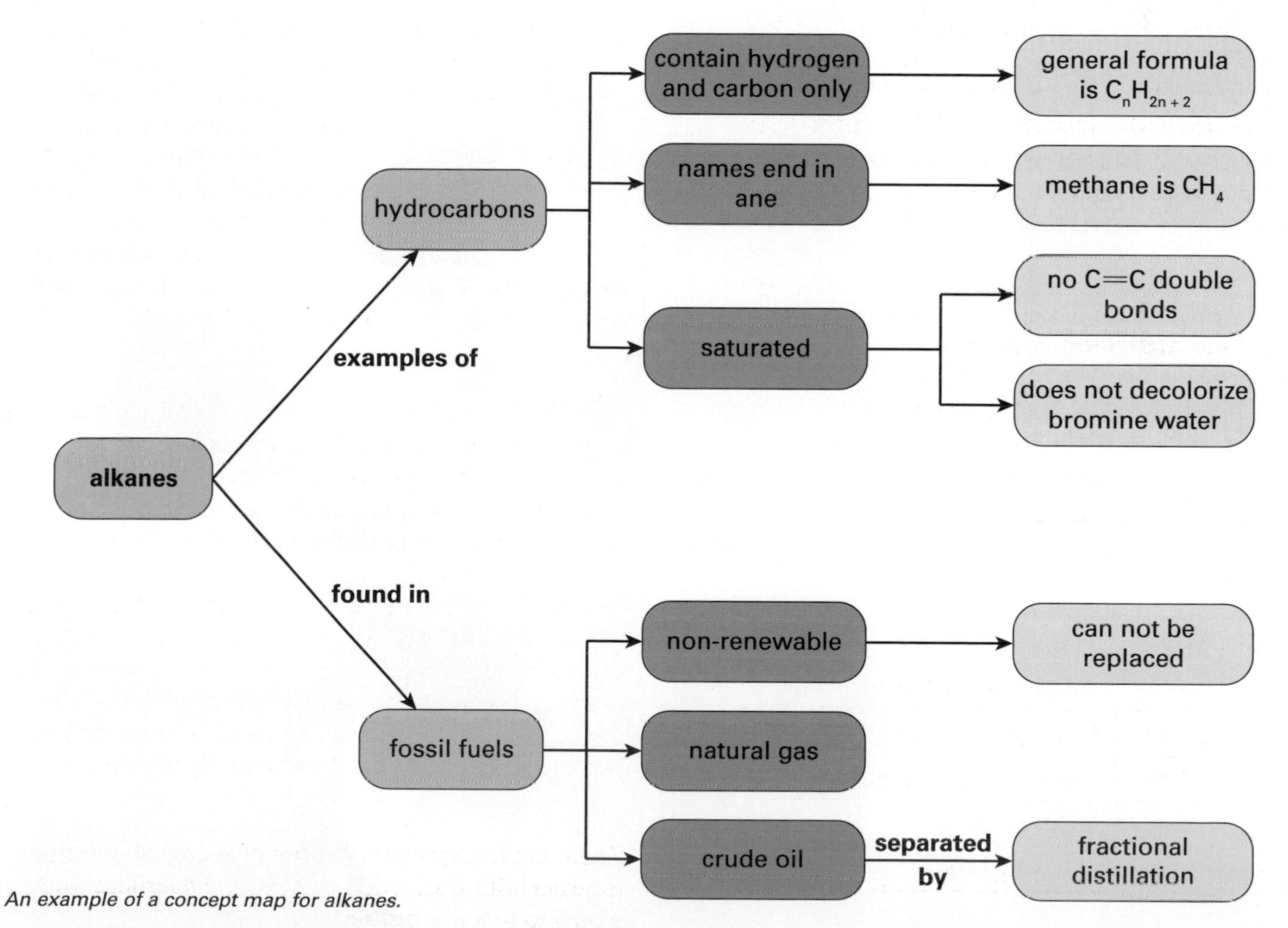

An example of a concept map for alkanes.

Unit 4: *Kinetics, equilibria, and organic chemistry* develops the concepts of physical chemistry introduced at AS Level. You find out how to treat equilibria and rates of reaction quantitatively, rather than just qualitatively. You also discover how to calculate the pH values of acids, bases, and buffer solutions. The changes in pH observed during titrations are explained, too.

Your study of organic chemistry is extended to include more compounds. These include those containing the carbonyl group, aromatic compounds, amines, amino acids, and polymers. In the last section, you find out how spectroscopic techniques are used to determine the molecular formulae and structures of organic compounds. The emphasis here is on solving problems, rather than on the theory of how spectroscopy works.

Kinetics, equilibria, and organic chemistry

A computer-generated representation of the benzene molecule, C_6H_6.

1.01 Order of reaction

OUTCOMES

already from AS Level, you understand

- that reactions can only occur when collisions take place between particles with sufficient energy
- the qualitative effect of changes in concentration on rate of reaction

and after this spread you should understand

- how analysing the results from rates experiments can reveal the order of reaction with respect to a reactant

For a reaction to happen, the reactant particles must collide with each other with sufficient energy. The minimum amount of energy that colliding particles need to cause a reaction is called the **activation energy**. A reaction will not happen if the particles have less energy than this. One way to increase the rate of collision is to increase the concentration of the reactants. But does this always lead to an increase in the rate of reaction?

First order

The equation represents the reaction between hydrogen and iodine:

$$H_2(g) + I_2(g) \rightarrow 2HI(g)$$

It is possible to measure the initial rate of reaction at a constant temperature. The table shows the results of three experiments in which the initial concentrations of the two reactants were varied.

Experiment	$[H_2(g)]$ ($mol\ dm^{-3}$)	$[I_2(g)]$ ($mol\ dm^{-3}$)	Initial rate ($mol\ dm^{-3}\ s^{-1}$)
1	1.0×10^{-3}	1.0×10^{-2}	2.0×10^{-6}
2	2.0×10^{-3}	1.0×10^{-2}	4.0×10^{-6}
3	2.0×10^{-3}	2.0×10^{-2}	8.0×10^{-6}

The square brackets [ ] round $H_2(g)$ and $I_2(g)$ represent their concentrations in $mol\ dm^{-3}$.

If you compare experiments 1 and 2, you will see that the initial concentration of iodine stays the same. But the initial concentration of hydrogen is doubled, and so is the initial rate of reaction. This means that the reaction is **first order** with respect to hydrogen. If the initial concentration of hydrogen were to be increased four times, the initial rate of reaction would also be increased four times.

If you compare experiments 2 and 3, you will see that the initial concentration of hydrogen stays the same. But the initial concentration of iodine is doubled, and so is the initial rate of reaction. This means that the reaction is also first order with respect to iodine.

Other orders

Other orders of reaction are possible, other than first order.

Zero order

Propanone reacts with iodine to form iodopropanone and hydrogen iodide:

$$CH_3COCH_3(aq) + I_2(aq) \rightarrow CH_3COCH_2I(aq) + HI(aq)$$

The reaction is **zero order** with respect to iodine. Changing the initial concentration of iodine has no effect on the initial rate of reaction.

Second order

Nitrogen monoxide and oxygen react to form nitrogen dioxide:

$$2NO(g) + O_2(g) \rightarrow 2NO_2(g)$$

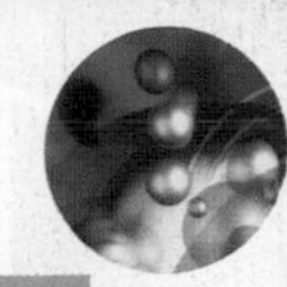

The reaction is **second order** with respect to nitrogen monoxide. Doubling the initial concentration of nitrogen monoxide increases the initial rate of reaction by four times.

Overall order of reaction

If the orders of reaction with respect to each reactant are known, the **overall order of reaction** can be worked out. This is the sum of the individual orders involved. For example, the overall order of reaction for the reaction between hydrogen and iodine is second order (1 + 1).

Nitrogen oxides or NO_x are produced by factories, power stations, and vehicle engines. They give smog its brown appearance.

Orders at A Level

At A Level, you will only meet reactions that are zero, first, or second order with respect to individual reactants. But you could meet reactions where the overall order is greater. For example, in the reaction A + B → C, the reaction might be first order with respect to A and second order with respect to B. The overall order of reaction would be third order.

Determining order

The **order of reaction** cannot be worked out by looking at the balanced equation. Instead, it is worked out by inspecting a table of experimental data or by plotting an appropriate graph.

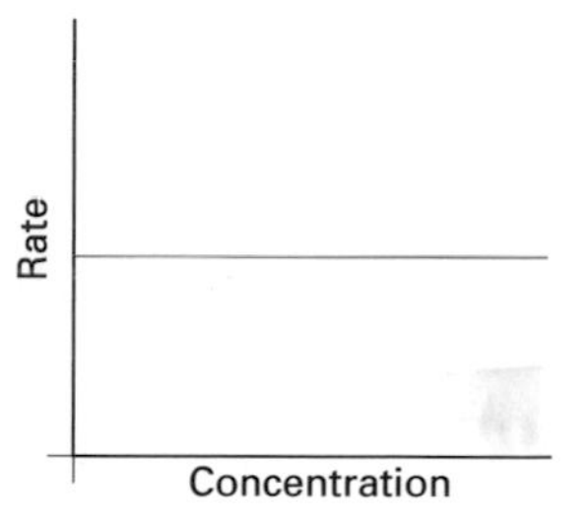

A graph of initial rate against concentration of a reactant, where the reaction is zero order with respect to that reactant. The line is horizontal.

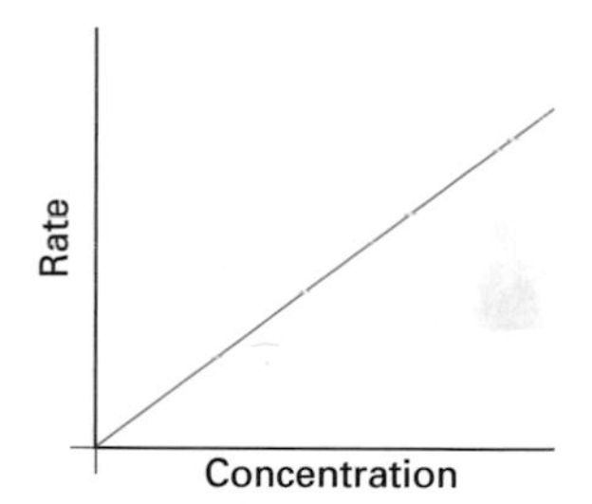

A graph of initial rate against concentration of a reactant, where the reaction is first order with respect to that reactant. The line is straight with a positive gradient.

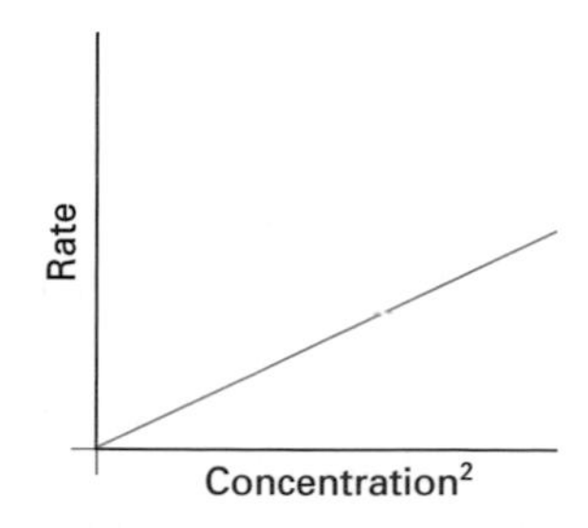

A graph of initial rate against the square of the concentration of a reactant, where the reaction is second order with respect to that reactant. The line is straight with a positive gradient.

Check your understanding

1 The initial concentration of a certain reactant is increased ten times.
   a What is the effect on the initial rate of reaction when the order of reaction with respect to that reactant is: **i** zero order; **ii** first order; **iii** second order?
   b Sketch appropriate graphs to illustrate each of these orders.

2 Iodate(V) ions and sulfate(IV) ions react in the presence of an acid catalyst:

$$2IO_3^-(aq) + 5SO_3^{2-}(aq) + 2H^+(aq) \rightarrow I_2(aq) + 5SO_4^{2-}(aq) + H_2O(l)$$

The reaction is second order with respect to iodate(V) ions, zero order with respect to sulfate(IV) ions, and first order with respect to hydrogen ions. What is the overall order of reaction?

3 Bromide ions and bromate(VI) ions react in the presence of an acid catalyst:

$$5Br^-(aq) + BrO_3^-(aq) + 6H^+(aq) \rightarrow 3Br_2(aq) + 3H_2O(l)$$

Here are the results of some rate experiments.

Experiment	$[Br^-]$ (mol dm^{-3})	$[BrO_3^-]$ (mol dm^{-3})	$[H^+]$ (mol dm^{-3})	Initial rate (mol dm^{-3} s^{-1})
1	0.10	0.10	0.10	1.2×10^{-3}
2	0.10	0.20	0.10	2.4×10^{-3}
3	0.20	0.20	0.10	4.8×10^{-3}
4	0.10	0.10	0.20	4.8×10^{-3}

   a Find the orders with respect to each of the three reactants.
   b What is the overall order of reaction?

1.02 The rate equation

OUTCOMES

already from A2 Level, you understand

- how analysing the results from rates experiments can reveal the order of reaction with respect to a reactant

and after this spread you should

- be able to derive the rate equation for a reaction from data relating initial rate to the concentrations of the different reactants
- understand and be able to use rate equations of the form:

  rate = $k[A]^m[B]^n$

If you know the individual orders of reaction for a reaction, you can write an expression that links the initial rate of reaction with the concentrations of the reactants. Such an expression is called a **rate equation**.

Deriving the rate equation

Consider the reaction between iodate(V) ions and sulfate(IV) ions in the presence of an acid catalyst:

$$2IO_3^-(aq) + 5SO_3^{2-}(aq) + 2H^+(aq) \rightarrow I_2(aq) + 5SO_4^{2-}(aq) + H_2O(l)$$

The reaction is second order with respect to iodate(V) ions. This means that the initial rate of reaction is proportional to the square of the initial concentration of iodate(V) ions. The relationship can be shown using this expression:

$$\text{rate} \propto [IO_3^-]^2$$

The symbol $\propto$ means 'proportional to'. It can be replaced by an equals sign if a constant of proportionality, k, is introduced:

$$\text{rate} = k[IO_3^-]^2$$

In the context of rates of reaction, the constant of proportionality is called the **rate constant**. Different reactions have different values for the rate constant.

The reaction is first order with respect to hydrogen ions:

$$\text{rate} = k[H^+]$$

The reaction is zero order with respect to sulfate(IV) ions:

$$\text{rate} = k[SO_3^{2-}]^0 = k$$

The three expressions can be combined to give the overall rate equation for the reaction:

$$\text{rate} = k[IO_3^-]^2[H^+]$$

Notice that the concentration of sulfate (IV) ions does not appear in the overall rate equation. This is because the reaction is zero order with respect to sulfate(IV) ions.

Working out the rate constant

The value of the rate constant can be calculated using experimental data. Consider the reaction between hydrogen and iodine:

$$H_2(g) + I_2(g) \rightarrow 2HI(g)$$

Here are the results of three experiments in which the initial concentrations of the two reactants were varied.

Experiment	$[H_2(g)]$ (mol dm^{-3})	$[I_2(g)]$ (mol dm^{-3})	Initial rate (mol dm^{-3} s^{-1})
1	1.0×10^{-3}	1.0×10^{-2}	2.0×10^{-6}
2	2.0×10^{-3}	1.0×10^{-2}	4.0×10^{-6}
3	2.0×10^{-3}	2.0×10^{-2}	8.0×10^{-6}

A Level rate equations

In general, the rate equation for a reaction can be written in the form:

rate = $k[A]^m[B]^n$

where m and n are the orders of reaction with respect to reactants A and B

At A Level, m and n will only be 0, 1, or 2.

The value of the rate constant is different at different temperatures and for different reactions, but it is always shown as k.

The reaction is first order with respect to hydrogen and first order with respect to iodine. This is the corresponding rate equation:

$$\text{rate} = k[H_2][I_2]$$

It can be rewritten to find the rate constant, k:

$$k = \frac{\text{rate}}{[H_2][I_2]}$$

The value of k can be found using any one of the three experiments. For example, using the data from experiment 2:

$$k = \frac{\text{rate}}{[H_2][I_2]} = \frac{4.0 \times 10^{-6}}{(2.0 \times 10^{-3})\,(1.0 \times 10^{-2})} = 0.20\ \text{mol}^{-1}\,\text{dm}^3\,\text{s}^{-1}$$

Once the value of k has been found, the complete rate equation can be written:

$$\text{rate} = 0.2[H_2][I_2]$$

The units of the rate constant

The units for the rate constant k depend on the rate equation. This is how you can work them out quickly:

1. Add together all the powers in the rate equation.
2. Subtract the answer to step 1 from the number 1.
3. Raise the unit, mol dm^{-3}, to the power given at step 2.
4. Write the unit, s^{-1}, on the end of the unit found at step 3.

Worked example

What are the units for k *given the rate equation below?*

$$\text{rate} = k[IO_3^-]^2[H^+]$$

1. Add the powers: $2 + 1 = 3$
2. Subtract answer from 1: $1 - 3 = -2$
3. $(\text{mol dm}^{-3})^{-2} = \text{mol}^{-2}\,\text{dm}^6$
4. Units for k are mol^{-2} dm^6 s^{-1}

Units for k from the overall order

The table summarizes the units for the rate constant, based on the overall order of reaction. Notice that there is a pattern.

Overall order	Units for k
first	s^{-1}
second	mol^{-1} dm^3 s^{-1}
third	mol^{-2} dm^6 s^{-1}
fourth	mol^{-3} dm^9 s^{-1}

Check your understanding

1. The reaction between NO and O_2 is second order with respect to NO and first order with respect to O_2.

Experiment	[NO] (mol dm^{-3})	[O_2] (mol dm^{-3})	Initial rate (mol dm^{-3} s^{-1})
1	2.0×10^{-3}	5.0×10^{-2}	2.0×10^{-6}
2	4.0×10^{-3}	6.0×10^{-2}	*to calculate*

   a Write the rate equation for the reaction.
   b Use the results from experiment 1 to calculate the value of the rate constant k, and give its units.
   c Calculate the initial rate of reaction for experiment 2.

2. Propanone reacts with iodine in acid solution:

$$CH_3COCH_3(aq) + I_2(aq) \rightarrow CH_3COCH_2I(aq) + HI(aq)$$

The rate equation is:

$$\text{rate} = k[CH_3COCH_3][H^+]$$

The initial rate of reaction is 1.0×10^{-4} mol dm^{-3} s^{-1} when the initial concentration of propanone is 1.25 mol dm^{-3} and the initial concentration of hydrogen ions is 0.08 mol dm^{-3}.

   a Calculate the value of the rate constant k, and give its units.
   b Calculate the initial rate of reaction when each reactant is added to an equal volume of water before being mixed together.
   c How can you tell that the hydrogen ions act as a catalyst?

1.03 The rate constant

OUTCOMES

already from AS Level, you

- have a qualitative understanding of the Maxwell–Boltzmann distribution of molecular energies in gases
- can draw and interpret distribution curves for different temperatures
- understand the qualitative effect of temperature changes on the rate of reaction

already from A2 Level, you

- can derive the rate equation for a reaction from data relating initial rate to the concentrations of the different reactants
- understand and can use rate equations of the form:

$$\text{rate} = k[A]^m[B]^n$$

and after this spread you should

- be able to explain the qualitative effect of changes in temperature on the rate constant k

Arrhenius plots

The Swedish chemist Svante Arrhenius discovered a link between the rate constant k and the activation energy E_a. A graph of $\log_e(k)$ against 1/temperature produces a straight line. Its gradient equals $-E_a/R$, where R is the gas constant.

Tropical fireflies flash more rapidly on warm nights. The change in flash rate with temperature leads to an Arrhenius plot with an E_a of about 50 kJ mol^{-1}.

The Maxwell–Boltzmann distribution

The Maxwell–Boltzmann distribution curve shows how the energies of the molecules in a sample of gas are distributed. The diagram shows the distribution for a sample of gas at two different temperatures, T_1 and T_2. T_2 is higher than T_1.

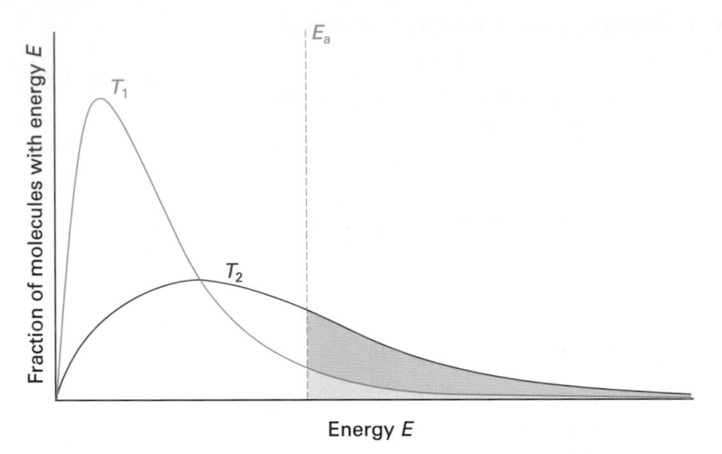

The Maxwell–Boltzmann distribution for a sample of gas at two different temperatures. The shaded areas represent the fraction of molecules with the activation energy E_a or more.

These are the key features of the curves.

- The area under the curve is proportional to the total number of molecules in the sample.
- Since these curves represent the same sample of gas at two temperatures, the area under each curve is the same.
- Only very small fractions of the molecules have very low or very high energies.
- The mean energy does not occur at the peak, but to the right of it.

The activation energy, E_a, is shown in the diagram. At T_1, the lower temperature, only the fraction of molecules shown by the blue area has sufficient energy to react. At T_2, the higher temperature, the fraction represented by the blue and red areas combined has enough energy to react.

Effect of temperature on the rate constant

An increase in temperature increases the rate constant. At high temperatures, particles move more quickly than they do at low temperatures. They are more likely to collide and, when they do, more of them have the activation energy or greater. So the higher the temperature, the greater the rate of reaction. If the rate of reaction increases, the rate constant does, too.

The sulfur clock reaction

When dilute hydrochloric acid is added to aqueous sodium thiosulfate, a yellow precipitate of sulfur forms. This gradually obscures the view of a cross on a piece of paper underneath the reaction vessel. The time taken is recorded, and the relative rate is calculated from 1/time. The overall order for this 'sulfur clock' reaction is first order, so the rate constant is proportional to the relative rate. The graph shows how the rate constant changes with temperature.

The beaker on the left contains dilute hydrochloric acid and aqueous sodium thiosulfate. They have only just started reacting, so you can still see the blue cross. On the right, enough sulfur has formed to obscure the cross.

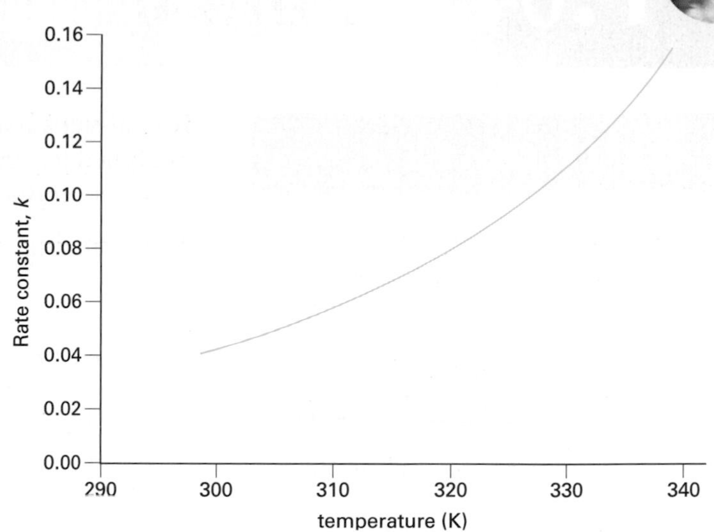

A graph of k *against temperature for the sulfur clock experiment.*

Some exceptions

The rate constant is only constant at a constant temperature. For this reason, it is also called the *rate coefficient*. It increases as the temperature increases, but there are some exceptions. The rate constants for a small number of reactions involving nitrogen monoxide decrease instead. Several factors lead to this observation.

Catalysts also increase the rate constant. Enzymes are biological catalysts. They are proteins that are damaged or *denatured* above their optimum temperature. So the rate constant for an enzyme-catalysed reaction decreases when the temperature increases above the optimum temperature.

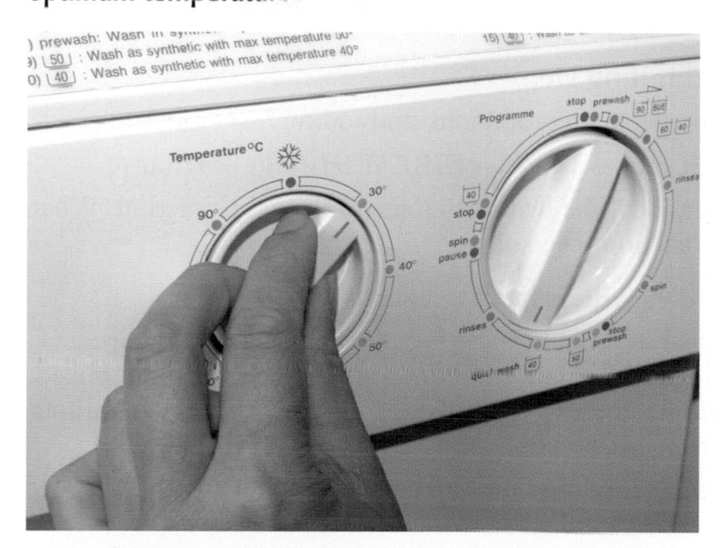

Washing clothes at 30°C helps to reduce energy consumption, and it also prevents the enzymes in biological detergents from being denatured.

Check your understanding

1 a Explain why an increase in temperature leads to an increase in the rate of reaction.

b What happens to the rate constant when the temperature changes?

c What is unusual about the rate constants of enzyme-catalysed reactions?

1.04 Rate limiting steps

OUTCOMES

already from AS Level, you understand

- some reaction mechanisms

already from A2 Level, you understand

- how analysing the results from rates experiments can reveal the order of reaction with respect to a reactant

and after this spread you should understand

- that the orders of reactions with respect to reactants can be used to provide information about the rate determining/limiting step of a reaction

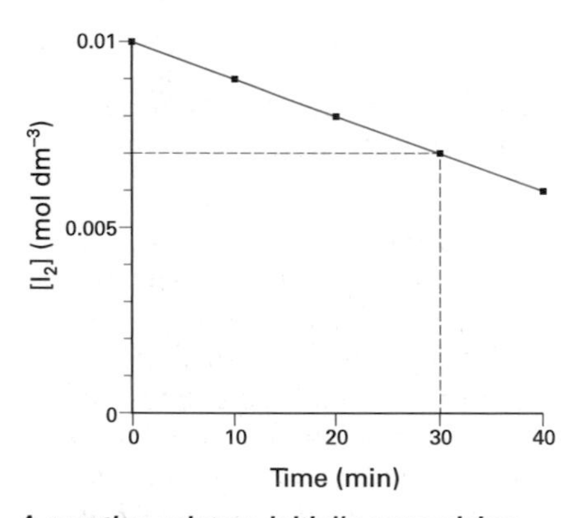

A reaction mixture initially comprising 0.25 mol dm^{-3} propanone, 0.01 mol dm^{-3} iodine, and 0.25 mol dm^{-3} sulfuric acid was studied for 40 minutes. This graph shows how the concentration of iodine decreased with time.

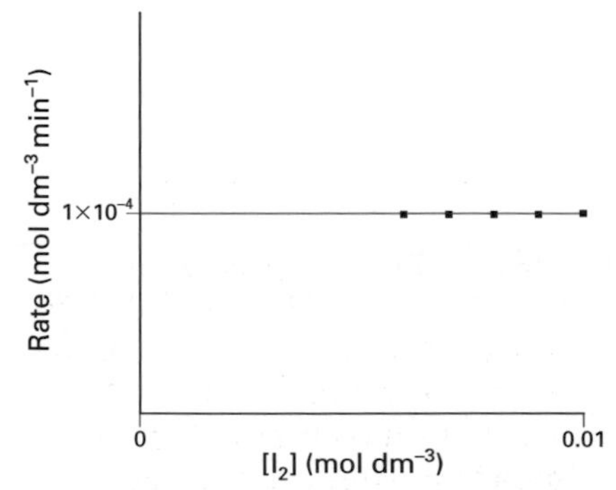

A graph of rate of reaction against iodine concentration, using data from the results in the previous graph.

Several steps may be involved in a reaction. They can be identified and combined to form a **reaction mechanism**. Orders of reaction provide information that helps to identify the steps in reaction mechanisms.

Rate limiting steps

Imagine a production line in a cake factory. Plain cakes go in at one end of the production line and decorated cakes in boxes leave at the other end. There are three steps: plain cakes are put onto the conveyor belt, decorated, and then boxed.

The worker who puts the plain cakes onto the conveyor belt can do this quickly. So can the person who puts decorated cakes into their boxes. It is the middle worker who slows the whole process down because cake decorating can't be rushed. This step is the **rate limiting step**, also called the **rate determining step**. The rate of the whole process depends on the rate of this step.

Decorating the cake limits the overall rate of the cake production line.

Identifying a rate limiting step

Consider the reaction between propanone and iodine:

$$CH_3COOCH_3(aq) + I_2(aq) \rightarrow CH_3COCH_2I(aq) + HI(aq)$$

When dilute acid is added, the hydrogen ions catalyse the reaction. What are the steps in the reaction mechanism?

Determining the orders with respect to each reactant

The order with respect to iodine can be found using a simple experiment. Known concentrations of propanone, iodine, and sulfuric acid are mixed together. The remaining amount of iodine is found by titration at various times from the start.

The mean rate at each sampling time can be calculated using this expression:

$$\text{rate} = \frac{\text{change in iodine concentration}}{\text{time taken for the change}}$$

For example, at 30 min the iodine concentration had decreased from 0.010 mol dm^{-3} to 0.007 mol dm^{-3}.

$$\text{rate} = \frac{0.010 - 0.007}{30} = \frac{0.003}{30} = 1 \times 10^{-4}\ \text{mol dm}^{-3}\ \text{min}^{-1}$$

The mean rate at each time point can be calculated in a similar way, and plotted on a graph of rate against iodine concentration.

The graph of rate against iodine concentration gives a horizontal line, so the reaction is zero order with respect to iodine. The rate of reaction does not depend on iodine concentration. Other studies show that the reaction is first order with respect to propanone and hydrogen ions.

The rate limiting step

The reaction is zero order with respect to iodine, so iodine is not involved in the rate limiting step. The discovery that the reaction is first order with respect to both propanone and hydrogen ions suggests that they are involved in the rate limiting step. The rate of reaction depends on the concentration of these two reactants, and it is likely that they react together in this step.

The reaction mechanism

The iodination of propanone takes place in two steps.

- The first step is the slow, rate limiting step. Propanone is changed into a more reactive form, called an enol form.
- In the second step, the reactive enol form of propanone reacts rapidly with iodine.

The suggested two-step reaction mechanism for the iodination of propanone.

Check your understanding

1 2-bromo-2-methylpropane, $(CH_3)_3CBr$, reacts with hydroxide ions to form 2-methylpropan-2-ol, $(CH_3)_3COH$.
The rate equation is: rate = $k[(CH_3)_3CBr]$.
Explain which of these two steps is likely to be the rate limiting step.

$$(CH_3)_3CBr \rightarrow (CH_3)_3C^+ + Br^-$$
$$(CH_3)_3C^+ + OH^- \rightarrow (CH_3)_3COH$$

2 Nitrogen monoxide and oxygen react together to produce nitrogen dioxide.
The rate equation is: rate = $k[NO]^2[O_2]$.
Explain which of these two steps is likely to be the rate limiting step.

$$NO + NO \rightarrow N_2O_2$$
$$N_2O_2 + O_2 \rightarrow 2NO_2$$

Second order steps

When a reaction is second order with respect to a particular reactant, a small increase in reactant concentration gives a large increase in reaction rate. So a step involving a second order reactant is less likely to be the rate limiting step than one involving a first order reactant.

Activation energy and the rate limiting step

If a step has a high activation energy, only a small fraction of molecules in the reaction mixture will have enough energy to react. So the rate of that step is low. If a step has a low activation energy, a larger fraction of molecules will have enough energy to react, so the rate of that step is high.

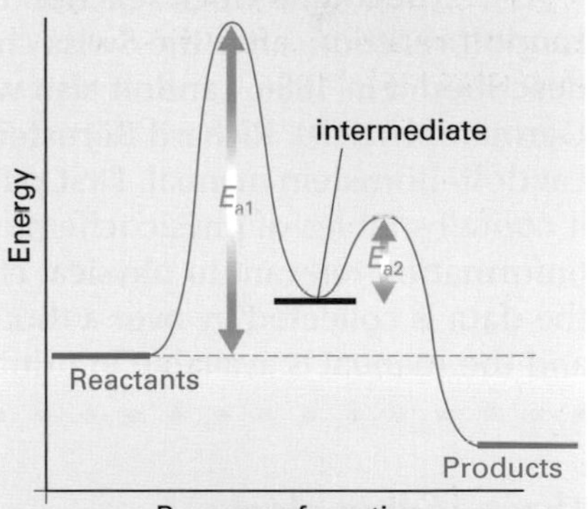

The rate limiting step has a higher activation energy than the other step.

Intermediates, such as the enol form of propanone, often have an unusual structure or bonding. They are at a higher energy level than the reactants or products.

1.05 The iodine clock experiment

OUTCOMES

already from A2 Level, you understand

- how analyzing the results from rates experiments can reveal the order of reaction with respect to a reactant

and after this spread you should

- know how to carry out a kinetic study to determine the order of a reaction

One of the Investigative and Practical Skills tasks for A2 Physical Chemistry is to carry out a kinetic study to determine the order of a reaction. The 'iodine clock' experiment is often used. There are several versions but they usually involve these steps:

- Iodine or iodate(V) ions are reduced to form iodide ions.
- The iodide ions immediately react with a limited amount of a oxidizing agent.
- Free iodine accumulates rapidly when all the reducing agent is used up.

Starch suspension is added to the reaction mixture. It forms a blue-black complex with iodine when the iodine accumulates. This happens suddenly and at various times, depending on the amount of reducing agent present.

Hans Heinrich Landolt (1831–1910)

The iodine clock reaction is also called the Landolt reaction, after the Swiss chemist who first described it in 1886. Landolt also worked with a German physicist, Richard Börnstein, to produce the Landolt–Börnstein manual. First published in 1883, it contains tables of physicochemical data (information relevant to physical chemistry). Today the data is collected by over a thousand scientists, and the manual is available in print and online.

Three iodine clocks

Here are three different versions of the iodine clock experiment.

Iodine and peroxodisulfate ions

Iodide ions are oxidized by peroxodisulfate ions $S_2O_8^{2-}$ to form iodine and sulfate(VI) ions:

$$2I^-(aq) + S_2O_8^{2-}(aq) \rightarrow I_2(aq) + 2SO_4^{2-}(aq)$$

A fixed amount of aqueous thiosulfate ions is added to reduce the iodine as it forms:

$$I_2(aq) + 2S_2O_3^{2-}(aq) \rightarrow 2I^-(aq) + S_4O_6^{2-}(aq)$$

Iodine and hydrogen peroxide

Iodide ions are oxidized by hydrogen peroxide, H_2O_2, to form iodine and water:

$$2I^-(aq) + H_2O_2(aq) + 2H^+(aq) \rightarrow I_2(aq) + 2H_2O(aq)$$

Sulfuric acid is added to supply the hydrogen ions needed. Again, a fixed amount of aqueous thiosulfate ions is added to reduce the iodine as it forms.

Iodate(V) and sulfate(IV) ions

Iodate(V) ions are reduced by sulfate(IV) ions, SO_3^{2-}, to form iodine and sulfate(VI) ions:

$$2IO_3^-(aq) + 5SO_3^{2-}(aq) + 2H^+(a) \rightarrow$$
$$I_2(aq) + 5SO_4^{2-}(aq) + H_2O(l)$$

Again, sulfuric acid is added to supply the hydrogen ions needed. But this time, the sulfate(IV) ions are the fixed amount of reducing agent:

$$I_2(aq) + SO_3^{2-}(aq) + H_2O(l) \rightarrow$$
$$2I^-(aq) + 2H^+ + SO_4^{2-}(aq)$$

When the reactants are first mixed, the reaction mixture is clear and colourless.

Once the fixed amount of reducing agent has been used up, the reaction mixture very quickly turns blue-black.

An outline method

This example involves the iodine and hydrogen peroxide experiment. The order of reaction with respect to iodide ions is being determined.

Prepare different reaction mixtures

Set up different mixtures in which only the volume of potassium iodide varies. For example, add the volumes shown in the table to three conical flasks.

Experiment	volume (cm^3)			
	0.10 mol dm^{-3} KI	water	0.05 mol dm^{-3} $Na_2S_2O_3$	starch suspension
A	5	20	5	1
B	15	10	5	1
C	25	0	5	1

Enough water is added to make the total volume constant.

Measuring volumes

The percentage error in a measurement depends upon the precision of the measuring device and how much is being measured. For example, a 25 cm^3 measuring cylinder is usually precise to $\pm$ 0.5 cm^3. So the error in measuring 25 cm^3 using it is:

$$\% \text{ error} = \frac{0.5}{25} \times 100 = 2\%$$

But if it is used to measure 5 cm^3, the error increases to 10%. It would be better to use a pipette for small volumes, as these are more accurate.

Using a pipette and pipette filler may be a little tricky, but it is an accurate way to measure and deliver liquids.

Carry out the experiments

Add 25 cm^3 of dilute sulfuric acid and 10 cm^3 of 0.10 mol dm^{-3} hydrogen peroxide to a beaker. Swirl one of the conical flasks to get the liquid moving, quickly add the contents of the beaker and start timing. Stop timing as soon as the colour appears. Repeat for the other two mixtures.

Analyse and evaluate the results

Here are some typical results obtained from the experiment described.

Experiment	Volume of 0.10 mol dm^{-3} KI (cm^3)	Initial concentration of $I^-(aq)$ (mol dm^{-3})	Time of reaction (s)	$\frac{1}{\text{time}}$ (s^{-1})
A	5	7.58×10^{-3}	126	7.94×10^{-3}
B	15	2.27×10^{-2}	45	2.22×10^{-2}
C	25	3.79×10^{-2}	27	3.70×10^{-2}

The rate is proportional to 1/time. A graph of 1/time against the initial concentration of $I^-(aq)$ produces a straight line. The gradient is 0.96, so the rate is likely to be first order with respect to iodide ions. The percentage error in this result would be:

$$\% \text{ error} = \frac{\text{experimental value} - \text{accepted value}}{\text{accepted value}} \times 100$$

$$\% \text{ error} = \frac{0.96 - 1}{1} \times 100 = -4\%$$

This shows that the experimental value is less than the accepted value.

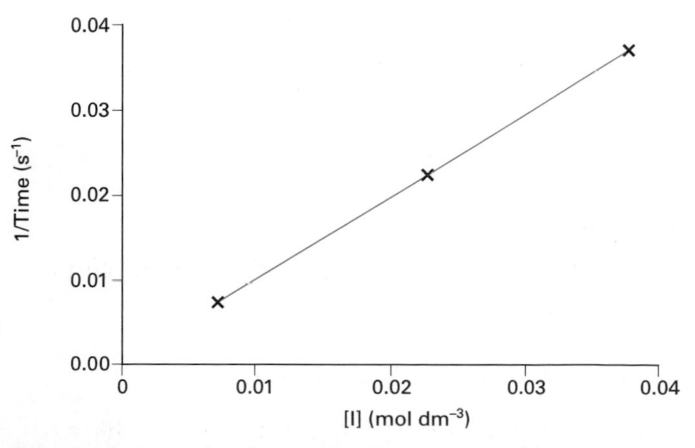

This graph shows that the reaction is first order with respect to iodide ions.

2.01 Reversible reactions and equilibria

OUTCOMES

already from AS Level, you

- know that many chemical reactions are reversible
- understand that for a reaction in equilibrium, although the concentrations of reactants and products remain constant, both the forward and reverse reactions are still proceeding at equal rates
- can use Le Chatelier's principle to predict the effects of changes in temperature, pressure, and concentration on the position of equilibrium in homogeneous reactions
- know that a catalyst does not affect the position of equilibrium
- can apply these concepts to given chemical processes
- can predict qualitatively the effect of temperature on the position of equilibrium from the sign of ΔH for the forward reaction
- understand why a compromise temperature and pressure may be used in industrial processes

and after this spread you should

- have familiarized yourself again with the key concepts of reversible reactions and equilibria from *AS Level Chemistry Unit 2*

Many chemical reactions are **reversible** and all have a tendency to come to **equilibrium**. But the **position of equilibrium** depends on the nature of the reaction itself and the conditions. There are two possibilities when the position of equilibrium lies far to one side.

- If mostly reactants remain, very little reaction has happened.
- If mostly products remain, the reaction has gone almost to completion.

In this section we look at reactions in which significant proportions of both reactants and products remain.

Dynamic equilibrium

At equilibrium, the concentrations of each reactant and product stay constant. This is not because the **forward reaction** and **reverse reaction** have stopped. Instead, both reactions are still happening, and they are happening at the same rate. This is why the equilibrium is described as **dynamic**. The products are being converted into reactants as quickly as reactants are being converted into products. For example, nitrogen dioxide molecules react with each other to form dinitrogen tetraoxide:

$$2NO_2(g) \rightleftharpoons N_2O_4(g)$$

Remember that the $\rightleftharpoons$ symbol means 'reversible'. At equilibrium, NO_2 is being converted into N_2O_4 at the same rate as N_2O_4 is being converted into NO_2.

Le Chatelier's Principle

Le Chatelier's Principle states that

- If a small change is applied to a system at equilibrium, the position of equilibrium moves in such a way as to minimize the effect of the change.

Pressure change

In a reaction involving gases, increasing the pressure moves the position of equilibrium in the direction of the fewest moles of gas. For example, consider this reaction:

$$\underset{\text{red-brown}}{2NO_2(g)} \rightleftharpoons \underset{\text{colourless}}{N_2O_4(g)}$$

Increasing the pressure on a mixture of these two gases causes the position of equilibrium to move to the right.

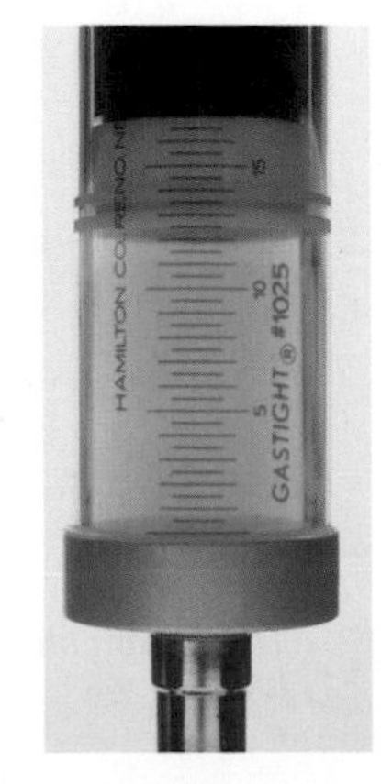

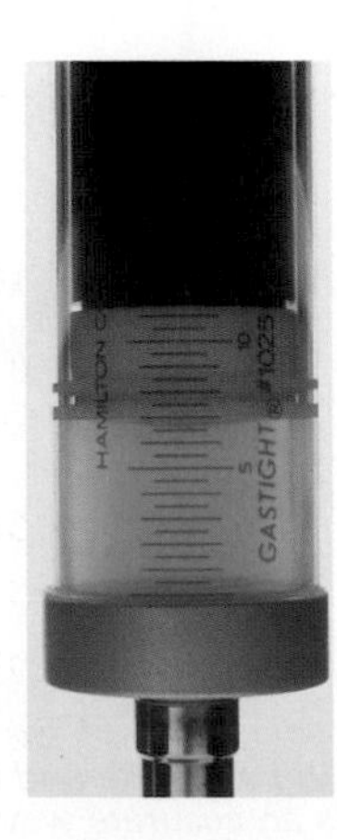

A mixture of NO_2 and N_2O_4 is at equilibrium (left). When the plunger is depressed (centre), the pressure is increased. The mixture is darker because it is more concentrated. The mixture reaches equilibrium again after a while (right). It is lighter in colour because it contains a lower proportion of NO_2.

If the number of moles of gas is the same on both sides, changing the pressure has no effect on the position of equilibrium. For example, it happens in this reversible reaction:

$$H_2(g) + I_2(g) \rightleftharpoons 2HI(g)$$

Concentration change

Increasing the concentration of a substance on one side will cause the position of equilibrium to move towards the opposite side. For example, consider this reaction:

$$\underset{\text{yellow}}{2CrO_4^{2-}(aq)} + 2H^+(aq) \rightleftharpoons \underset{\text{orange}}{Cr_2O_7^{2-}(aq)} + H_2O(l)$$

If acid is added, the concentration of $H^+(aq)$ increases. The position of equilibrium moves to the right. If an alkali is added, $H^+(aq)$ ions react with $OH^-(aq)$ ions from the alkali and are removed. The position of equilibrium moves to the left. These movements are observed as colour changes.

The beaker on the left contains yellow $CrO_4^{2-}(aq)$ ions. The beaker on the right shows what happens when sulfuric acid is added. The position of equilibrium moves to the right, producing orange $Cr_2O_7^{2-}(aq)$ ions.

Temperature change

Increasing the temperature will cause the position of equilibrium to move in the direction of the endothermic process. For example, consider again this reaction:

$$2NO_2(g) \underset{\Delta H = +24\ \text{kJ mol}^{-1}}{\overset{\Delta H = -24\ \text{kJ mol}^{-1}}{\rightleftharpoons}} N_2O_4(g)$$

The forward reaction is exothermic and the reverse reaction is endothermic. If the temperature is increased, the position of equilibrium moves to the left. If it is decreased, the position of equilibrium moves to the right.

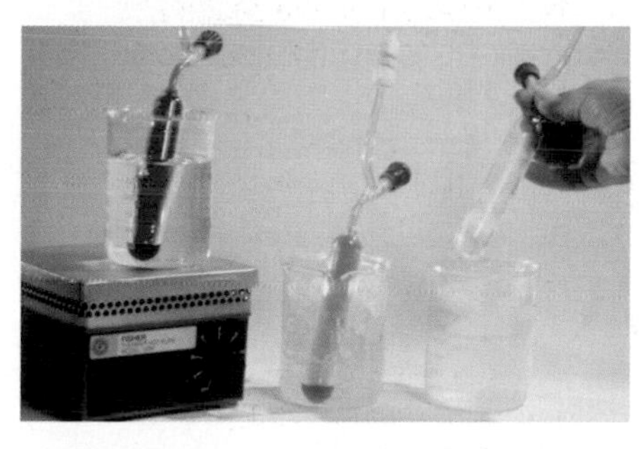

At high temperatures (left) there is a high proportion of red-brown NO_2 at equilibrium. At 0°C (centre) there is much less NO_2, and in a freezing mixture (right) the equilibrium mixture is nearly all colourless N_2O_4.

Effect of catalysts

Catalysts change the rate of the forward and reverse reactions by the same ratio. For example, if the rate of the forward reaction doubles, the rate of the reverse reaction doubles, too. So catalysts do not change the position of equilibrium: they just reduce the time taken to reach equilibrium. The relative concentrations of the reactants and products at equilibrium are the same with or without a catalyst.

Check your understanding

1 **a** What does the term *dynamic equilibrium* mean?

b State Le Chatelier's Principle.

2 In the Haber process, nitrogen and hydrogen react together to produce ammonia: $N_2(g) + 3H_2(g) \rightleftharpoons 2NH_3(g)$. The forward reaction is exothermic. Explain the effects of the following on the equilibrium yield of ammonia:

a increasing the pressure

b increasing the temperature

c using an iron catalyst

2.02 The equilibrium constant

OUTCOMES

already from AS Level, you understand

- key concepts of reversible reactions and equilibria, including Le Chatelier's Principle

and after this spread you should

- know that K_c is the equilibrium constant calculated from equilibrium concentrations for a system at constant temperature
- be able to construct an expression for K_c for a homogeneous system in equilibrium
- be able to perform calculations involving K_c

The position of equilibrium is usually described as being 'to the left' or 'to the right', or moving to the left or the right. This is a qualitative description rather than a quantitative one, because it does not involve numbers. The **equilibrium constant** is a value that describes the position of equilibrium for a reaction at a given temperature. It is calculated from the equilibrium concentrations of the reactants and products.

The equilibrium constant, K_c

The c in K_c shows that the value has been calculated using the equilibrium concentrations measured in mol dm^{-3}. Look at this general equation for a reversible reaction:

$$aA + bB \rightleftharpoons cC + dD$$

This is how you would write the corresponding expression for the reaction:

$$K_c = \frac{[C]^c [D]^d}{[A]^a [B]^b}$$

Remember that [D] means the concentration of substance D in mol dm^{-3}.

Writing expressions for K_c

Worked example 1

Dichromate(VI) ions react with hydroxide ions in the following way:

$$Cr_2O_7^{2-}(aq) + 2OH^-(aq) \rightleftharpoons 2CrO_4^{2-}(aq) + H_2O(l)$$

Write the expression for K_c for this reaction.

$$K_c = \frac{[CrO_4^{2-}]^2[H_2O]}{[Cr_2O_7^{2-}][OH^-]^2}$$

$[OH^-]$ and $[CrO_4^{2-}]$ are raised to the power of 2 because there are 2 mol of each of them in the balanced equation. Similarly, $[Cr_2O_7^{2-}]$ and $[H_2O]$ are raised to the power of 1, but note that you do not write the number 1.

Worked example 2

Write the expression for K_c for the reaction between nitrogen and hydrogen to form ammonia:

$$N_2(g) + 3H_2(g) \rightleftharpoons 2NH_3(g).$$

$$K_c = \frac{[NH_3]^2}{[N_2][H_2]^3}$$

Working out the units for K_c

The units for K_c depend on each particular expression. Some values for K_c have no units at all. This is how you can work out the units quickly:

1 Add together all the powers on the top of the expression.

2 Add together all the powers on the bottom of the expression.

3 Subtract the answer to step 2 from the answer to step 1.

4 Raise the unit, mol dm^{-3}, to the power given at step 3.

Check your understanding

1 For each of the following reactions, write the expression for K_c and give its unit:

a $2SO_2 + O_2 \rightleftharpoons 2SO_3$

b $PCl_5 \rightleftharpoons PCl_3 + Cl_2$

c $H_2 + I_2 \rightleftharpoons 2HI$

2 Hydrogen reacts with iodine to produce hydrogen iodide. At equilibrium, there are 0.025 mol of H_2, 0.20 mol of I_2, and 2.0 mol of HI. Calculate the value of K_c.

3 1.5 mol of nitrogen reacts with 1.8 mol of hydrogen in a 3.0 dm^3 flask. At equilibrium there are 0.45 mol of ammonia. Calculate the value of K_c and give its units.

Worked example

What are the units for K_c, given the expression below?

$$K_c = \frac{[NH_3]^2}{[N_2][H_2]^3}$$

1 On the top: 2
2 On the bottom: 1 + 3 = 4
3 2 − 4 = −2
4 Units for K_c are $(mol\ dm^{-3})^{-2}$, which is $mol^{-2}\ dm^6$

Stoichiometry

In the example, 1 mol of N_2 is needed to form 2 mol of NH_3. This is why the change is $0.12 \times \frac{1}{2} = 0.06$ mol. In the same way, 3 mol of H_2 is needed to form 2 mol of NH_3. This is why the change is $0.12 \times \frac{3}{2} = 0.18$ mol.

Calculations using K_c

It is relatively easy to calculate K_c if you are given the equilibrium concentrations of all the components. But often you only know the starting concentrations of the reactants, and the equilibrium concentration of one of the reactants or products. You have to work out the equilibrium concentrations first, using the balanced equation to help.

K_c and the reverse process

The value of K_c for the reverse process is the reciprocal of the original K_c value. In the example, the value of K_c for the reaction $N_2 + 3H_2 \rightleftharpoons 2NH_3$ is $0.11\ mol^{-2}\ dm^6$. The value of K_c for the reaction $2NH_3 \rightleftharpoons N_2 + 3H_2$ is $1 \div 0.11 = 9.1\ mol^2\ dm^{-6}$. The units change, too.

Worked example 1

1.0 mol of nitrogen reacts with 1.0 mol of hydrogen in a 2.0 dm³ flask. At equilibrium there are 0.12 mol of ammonia. Calculate the value of K_c and give its units.

equation	N_2 +	$3H_2$ $\rightleftharpoons$	$2NH_3$
at start	1.0 mol	1.0 mol	0 mol
change	$-(0.12 \times \frac{1}{2}) =$ −0.06 mol	$-(0.12 \times \frac{3}{2}) =$ −0.18 mol	+ 0.12 mol
at equilibrium	1.0 − 0.06 = 0.94 mol	1.0 − 0.18 = 0.82 mol	0 + 0.12 = 0.12 mol
equilibrium concentration	0.94 ÷ 2 = 0.47 mol dm^{-3}	0.82 ÷ 2 = 0.41 mol dm^{-3}	0.12 ÷ 2 = 0.06 mol dm^{-3}

$$K_c = \frac{[NH_3]^2}{[N_2]H_2]^3} = \frac{(0.06)^2}{0.47 \times (0.41)^3} = \frac{0.0036}{0.47 \times 0.0689} = 0.11\ mol^{-2}\ dm^6$$

K_c and the position of equilibrium

When the value of K_c is large, the position of equilibrium lies to the right. When the value of K_c is small, the position of equilibrium lies to the left.

You can use moles instead of concentration if the value of K_c has no units.

Worked example 2

Ethanol and ethanoic acid react together to form ethyl ethanoate and water. At the start, 0.50 mol of ethanol and 1.0 mol of ethanoic acid react together. At equilibrium there is 0.42 mol of ethyl ethanoate. Calculate the value of K_c and give its units.

equation	CH_3CH_2OH +	CH_3COOH $\rightleftharpoons$	$CH_3COOCH_2CH_3$ +	H_2O
at start	0.50 mol	1.00 mol	0 mol	0 mol
change	− 0.42 mol	− 0.42 mol	+ 0.42 mol	+ 0.42 mol
at equilibrium	0.50 − 0.42 = 0.08 mol	1.00 − 0.42 = 0.58 mol	0 + 0.42 = 0.42 mol	0 + 0.42 = 0.42 mol

$$K_c = \frac{[CH_3COOCH_2CH_3]H_2O]}{[CH_3CH_2OH][CH_3COOH]} = \frac{0.42 \times 0.42}{0.08 \times 0.58} = 3.8 \text{ (no units)}$$

2.03 Changing K_c

OUTCOMES

already from A2 Level, you

- know that K_c is the equilibrium constant calculated from equilibrium concentrations for a system at constant temperature
- can construct an expression for K_c for a homogeneous system in equilibrium
- can perform calculations involving K_c

and after this spread you should

- understand that the value of the equilibrium constant is not affected by either changes in concentration or the addition of a catalyst
- be able to predict the effects of changes of temperature on the value of the equilibrium constant

Max Bodenstein (1871–1942)

Max Bodenstein carried out a series of classic experiments on equilibria involving hydrogen iodide, which he published in 1899. He kept sealed tubes of hydrogen iodide at different temperatures until their contents reached equilibrium, and then cooled them rapidly to 'freeze' the position of equilibrium. Bodenstein analysed the amount of hydrogen iodide present and then calculated the equilibrium concentrations of hydrogen and iodine. In this way he was able to support the law of chemical equilibrium proposed between 1864 and 1879 by two Norwegian scientists, Cato Guldberg and Peter Waage.

You know two ways to describe an equilibrium. In a quantitative description you would use the equilibrium constant, K_c. In a qualitative description you would say that the position of the equilibrium lies to the left or to the right. It is important that you can recognize the conditions under which these stay the same or change.

Effect of catalysts

Catalysts change the rate of the forward and reverse reactions by the same ratio. For example, if the rate of the forward reaction doubles, the rate of the reverse reaction doubles, too. The relative concentrations of the reactants and products at equilibrium are the same with or without a catalyst. So catalysts do not change the value of K_c or the position of equilibrium: they just reduce the time taken to reach equilibrium.

Concentration change

When the concentration of one of the species present at equilibrium is changed, the value of K_c stays the same. But the position of equilibrium changes. It moves in the direction that minimizes the effect of the change, which keeps the value of K_c the same as before. Consider this reaction:

$$H_2 + I_2 \rightleftharpoons 2HI \qquad K_c = \frac{[HI]^2}{[H_2][I_2]} \text{ (no units)}$$

Suppose that at equilibrium $[H_2] = 1.0\ \text{mol dm}^{-3}$, $[I_2] = 1.0\ \text{mol dm}^{-3}$, and $[HI] = 8.0\ \text{mol dm}^{-3}$. (Note that these concentrations are much higher than would likely in practice, but the numbers used make the mathematics easier to follow.)

$$K_c = \frac{[HI]^2}{[H_2][I_2]} = \frac{(8.0)^2}{1.0 \times 1.0} = 64$$

Then, extra HI is added to increase the concentration to 10.5 mol dm^{-3}. The steps below show what happens afterwards to the concentrations of each species, assuming that the temperature stays constant.

	$[H_2]$ (mol dm^{-3})	$[I_2]$ (mol dm^{-3})	[HI] (mol dm^{-3})
equilibrium at start	1.0	1.0	8.0
HI added			+ 2.5
change because some HI reacts	+ (0.5 ÷ 2) = + 0.25	+ (0.5 ÷ 2) = + 0.25	− 0.5
equilibrium at end	1.0 + 0.25 = 1.25	1.0 + 0.25 = 1.25	8.0 + 2.5 − 0.5 = 10

K_c stays the same:

$$K_c = \frac{[HI]^2}{[H_2][I_2]} = \frac{(10)^2}{1.25 \times 1.25} = 64$$

Temperature change

A change in temperature will change the position of equilibrium, and it will change the value of K_c. Increasing the temperature will cause the position of equilibrium to move in the direction of the endothermic process.

Consider the manufacture of ammonia by the Haber process:

$$N_2 + 3H_2 \underset{\Delta H^\ominus_{298} = +92.4 \text{ kJ mol}^{-1}}{\overset{\Delta H^\ominus_{298} = -92.4 \text{ kJ mol}^{-1}}{\rightleftharpoons}} 2NH_3$$

If the temperature is increased, the position of equilibrium moves to the left in the direction of the endothermic process. The relative proportions of nitrogen and hydrogen are increased, so the value of K_c decreases.

Consider the manufacture of hydrogen by the reaction of steam with coke:

$$H_2O + C \underset{\Delta H^\ominus_{298} = -131 \text{ kJ mol}^{-1}}{\overset{\Delta H^\ominus_{298} = +131 \text{ kJ mol}^{-1}}{\rightleftharpoons}} H_2 + CO$$

If the temperature is increased, the position of equilibrium moves to the right in the direction of the endothermic process. The relative proportions of carbon monoxide and hydrogen are increased, so the value of K_c increases.

Exothermic or endothermic?

A process can be identified as exothermic or endothermic by examining the way in which the value of K_c changes with temperature.

- If the process is exothermic, K_c decreases as the temperature increases.
- If the process is endothermic, K_c increases as the temperature increases.

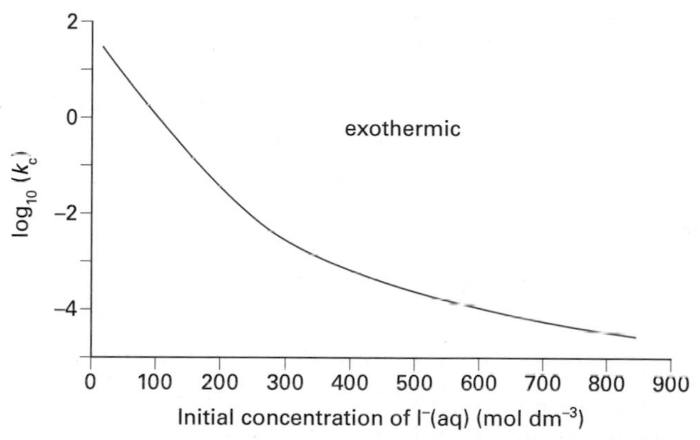

The manufacture of ammonia from nitrogen and hydrogen is an exothermic process. The value of K_c decreases as the temperature increases.

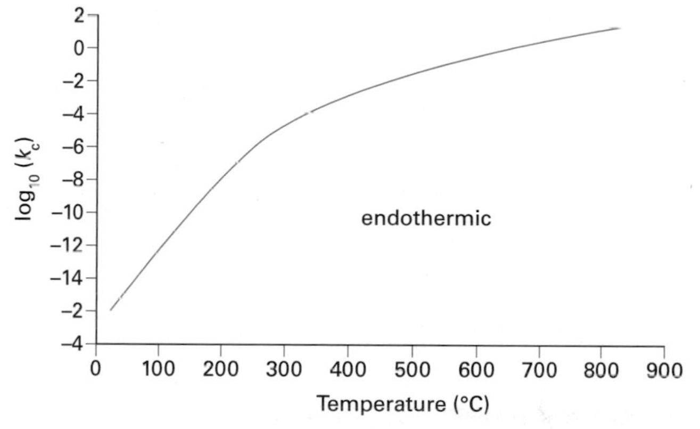

The manufacture of hydrogen from steam and coke is an endothermic process. The value of K_c increases as the temperature increases.

Check your understanding

1 For a reaction in equilibrium, under what conditions are the following changed?
  a position of equilibrium b equilibrium constant, K_c

2 The forward reaction in this reaction is exothermic: $H_2 + I_2 \rightleftharpoons 2HI$. If the temperature is increased, what effect will there be on:
  a the position of equilibrium
  b the value of the equilibrium constant, K_c?

3 The value of K_c for this reaction increases as the temperature is increased: $N_2O_4 \rightleftharpoons 2NO_2$.
  a What does this tell you about the forward reaction?
  b What can be concluded about the reverse reaction?
  c What change, if any, would there be in the value of K_c if a catalyst were added? Explain your answer.

2.04 Finding the value of K_c

OUTCOMES

already from AS Level, you can

- make up a volumetric solution
- carry out a simple acid–base titration

already from A2 Level, you

- can perform calculations involving K_c

and after this spread you should

- be able to carry out an experiment to determine an equilibrium constant

One of the Investigative and Practical Skills tasks for A2 Physical Chemistry is the determination of an equilibrium constant. The equilibrium involving ethanol, ethanoic acid, ethyl ethanoate, and water is often used for this purpose:

$$\underset{\text{ethanoic acid}}{CH_3COOH(l)} + \underset{\text{ethanol}}{CH_3CH_2OH(l)} \rightleftharpoons \underset{\text{ethyl ethanoate}}{CH_3COOCH_2CH_3(l)} + \underset{\text{water}}{H_2O(l)}$$

To carry out such an investigation successfully, you will need to use these techniques skilfully:

- transferring known volumes of liquid
- weighing
- acid–base titration

Experiment outline

Ethyl ethanoate and water, or ethanol and ethanoic acid, are mixed together in a container. Dilute hydrochloric acid is added to act as a catalyst. A control consisting of the same volume of dilute hydrochloric acid is also prepared. The containers are sealed, and left aside at a constant temperature.

The containers are then unsealed, and their contents titrated with a standard solution of aqueous sodium hydroxide. The titres are used in the calculation of K_c. But they cannot be used directly. The starting masses of the reactants must also be known so that their equilibrium amounts can be calculated.

Setting up

It is assumed here that the starting mixture comprises ethyl ethanoate, water, and hydrochloric acid. The mixture is in a conical flask with a stopper.

Sample method

1 Weigh the conical flask and record its mass.
2 Use a pipette and pipette filler to add 5.0 cm^3 of 3.00 mol dm^{-3} hydrochloric acid, and weigh the flask again.
3 Add 4.0 cm^3 of ethyl ethanoate and weigh the flask again.
4 Add 1.0 cm^3 of de-ionized water and weigh the flask again.
5 Prepare and weigh a control consisting of just 5.0 cm^3 of dilute hydrochloric acid.

Contents	Reaction mixture	Control
Volume of HCl(aq) added (cm^3)	5.0	5.0
Volume of ethyl ethanoate added (cm^3)	4.0	na
Volume of water added (cm^3)	1.0	na
Mass of empty flask (g)	126.01	123.24
Mass of flask with HCl(aq) (g)	131.25	128.68
Mass of HCl(aq) added (g)	131.25 − 126.01 = 5.24	128.68 − 123.24 = 5.24
Mass of flask with HCl(aq) and ethyl ethanoate (g)	134.84	na
Mass of ethyl ethanoate added (g)	134.84 − 131.25 = 3.59	na
Final mass of flask (g)	135.84	na
Mass of water added (g)	135.84 − 134.84 = 1.00	na

Some typical results obtained when setting up the investigation.

Seal the flasks and leave them to stand for several days. You might use a thermostatically controlled water bath to maintain a constant temperature. Or you might simply leave them at room temperature.

The titration

Unseal each flask and immediately titrate the contents with 1.00 mol dm^{-3} sodium hydroxide. Phenolphthalein is a suitable indicator (see Spread 3.07 for more information). Sometimes the equilibrium position is 'frozen' by adding excess water to slow down the rate of reaction. This is particularly useful if the equilibrium is established at a high temperature but the titration is carried out at room temperature.

The calculation

The calculation can be complex, so you may be given a spreadsheet to do it for you. But you must be prepared to calculate a K_c value yourself from experimental results.

Amount of ethanoic acid

The titres for the control and reaction mixture are used to calculate the total amount of acid present in each flask. Remember that the equilibrium mixture contains the hydrochloric acid catalyst and any ethanoic acid formed. The amount of ethanoic acid in the equilibrium mixture is found by subtracting the amount of hydrochloric acid present in the control from the total amount of acid in the equilibrium mixture.

Amount of ethanol

The amount of ethanol is equal to the amount of ethanoic acid formed.

Amount of ethyl ethanoate

The amount of ethyl ethanoate at the start is found from the mass used and its M_r. The amount of ethyl ethanoate that reacted is the same as the amount of ethanoic acid at equilibrium. So at equilibrium, the amount of ethyl ethanoate is the starting amount of ethyl ethanoate minus the amount of ethanoic acid formed.

Amount of water

Water was added at the start and some was added in the dilute hydrochloric acid. Here is how you find the mass of water added in the acid:

1. Use the amount of HCl found by titration and its M_r to calculate the mass of HCl added.
2. Subtract the mass found in step 1 from the measured mass of hydrochloric acid added.

The mass of water at equilibrium is the total mass at the start minus the amount of ethanoic acid at equilibrium. The amount of water is then found using its M_r.

Ethanoic acid

From titration:

Amount of HCl(aq) = 0.0150 mol

Total amount of acid = 0.0440 mol

Amount of ethanoic acid = 0.0440 − 0.0150
= 0.0290 mol

Ethanol

Amount of ethanol = 0.0290 mol

Ethyl ethanoate

Amount of ethyl ethanoate at start = mass ÷ M_r
= 3.59 ÷ 88.0 = 0.0408 mol

Amount of ethyl ethanoate at equilibrium
= 0.0408 − 0.0290 = 0.0118 mol

Water

Amount of HCl from titration = 0.0150 mol

Mass of HCl = amount × M_r = 0.0150 × 36.5
= 0.55 g

Mass of water in HCl(aq) = mass of HCl(aq) − mass of HCl = 5.24 − 0.55 = 4.69 g

Total mass of water at start = 1.00 + 4.69 = 5.69 g

Amount of water at start = mass ÷ M_r
= 5.69 ÷ 18 = 0.316 mol

Amount of water at equilibrium = 0.316 − 0.0290
= 0.287 mol

Equilibrium constant

$$K_c = \frac{[CH_3COOCH_2CH_3][H_2O]}{[CH_3CH_2OH][CH_3COOH]} = \frac{0.0118 \times 0.287}{0.0290 \times 0.0290}$$
= 4.03 (no units)

A typical calculation using the results opposite.

3.01 Brønsted–Lowry acids and bases

OUTCOMES

already from GCSE, you know that

- $H^+(aq)$ ions make solutions acidic and $OH^-(aq)$ ions make solutions alkaline
- in neutralization reactions, hydrogen ions react with hydroxide ions to produce water

and after this spread you should know that

- an acid is a proton donor
- a base is a proton acceptor
- acid–base equilibria involve the transfer of protons

Svante Arrhenius, a Swedish scientist, introduced his theory about acids and bases in 1884. He suggested that acids ionize in water to produce hydrogen ions, $H^+(aq)$, and that bases ionize in water to produce hydroxide ions, $OH^-(aq)$. These ions react with each other when acids are neutralized by bases, forming water:

$$H^+(aq) + OH^-(aq) \rightarrow H_2O(l)$$

Arrhenius' theory was improved upon in 1923 by two scientists working independently of each other, Johannes Brønsted from Denmark and T. Martin Lowry from England.

Science @ Work

Svante Arrhenius (1859–1927)

Starting as a young university student, Arrhenius developed ideas about salts in solution that were different from the accepted ideas of his time. He suggested that salts dissociate in water to form ions, and that this happens even if no electricity is passed through the solution. He also suggested that when an acid and a base react with each other, water is the primary product and the salt is the secondary product. His evidence for this was that the molar enthalpy change for neutralization is the same whatever combination of strong acid and strong base is used. But at the time it was generally believed that the salt was the primary product.

The work needed for Arrhenius to be awarded a doctorate, his thesis, was only just passed by the professors at his university. But eventually his ideas were accepted, and he received the 1903 Nobel Prize for Chemistry.

Arrhenius was a busy scientist. One of his other discoveries was that carbon dioxide in the atmosphere could cause a greenhouse effect.

Acids and bases

Remember that a hydrogen atom consists of a proton and an electron. A hydrogen atom loses its electron when it forms a hydrogen ion. So the hydrogen ion H^+ is identical to the proton. According to the Brønsted–Lowry theory of acids and bases:

- A **Brønsted–Lowry acid** is a proton donor.
- A **Brønsted–Lowry base** is a proton acceptor.

Acids lose hydrogen ions and bases gain them. When an acid is in aqueous solution, it donates protons to water molecules. In general, using HA to represent the acid:

$$HA(aq) + H_2O(l) \rightarrow H_3O^+(aq) + A^-(aq)$$

The anion A^- depends upon the acid used. For example, A^- would be Cl^- if hydrochloric acid were used, and SO_4^{2-} if sulfuric acid were used:

$$HCl(aq) + H_2O(l) \rightarrow H_3O^+(aq) + Cl^-(aq)$$

$$H_2SO_4(aq) + 2H_2O(l) \rightarrow 2H_3O^+(aq) + SO_4^{2-}(aq)$$

Notice that water is acting as a Brønsted–Lowry base here because it is accepting protons from the acids.

A base in aqueous solution accepts protons from water molecules. For example, for aqueous ammonia:

$$NH_3(aq) + H_2O(l) \rightarrow NH_4^+(aq) + OH^-(aq)$$

Notice that water is acting as a Brønsted–Lowry acid here because it is donating protons to the base. Water can act as both an acid and a base: it is **amphoteric**.

Acid–base equilibria

In the Brønsted–Lowry theory, a substance can only act as an acid when there is a base present. A substance can only act as a base when there is an acid present. Look at this neutralization reaction between ethanoic acid and aqueous ammonia:

$$CH_3COOH(aq) + NH_3(aq) \rightleftharpoons CH_3COO^-(aq) + NH_4^+(aq)$$

It is an equilibrium reaction involving the transfer of protons from ethanoic acid to ammonia. Ethanoic acid is a Brønsted–Lowry acid and ammonia is a Brønsted–Lowry base.

Conjugate acid–base pairs

The ammonium ion and ammonia are a conjugate acid–base pair. Ammonium ions form when ammonia accepts protons. They can donate protons, so the ammonium ion is the **conjugate acid** of ammonia.

Ethanoic acid and the ethanoate ion are a **conjugate acid–base pair.** Ethanoate ions form when ethanoic acid donates protons. They can accept protons, so the ethanoate ion is the **conjugate base** of ethanoic acid.

The oxonium ion

The aqueous hydrogen ion is usually shown as $H^+(aq)$. But it has a very high charge density, so it attracts a lone pair of electrons in a water molecule. The water molecule forms a co-ordinate bond with the hydrogen ion, producing the oxonium ion, $H_3O^+(aq)$. The neutralization reaction is sometimes shown using the oxonium ion instead of the aqueous hydrogen ion:

$$H_3O^+(aq) + OH^-(aq) \rightarrow 2H_2O(l)$$

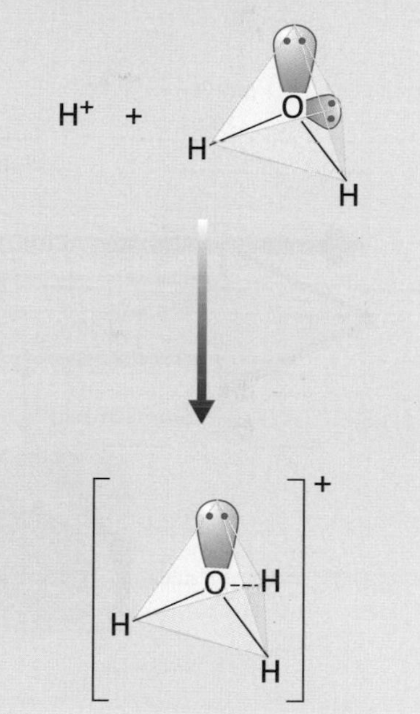

The oxonium ion forms when a proton from an acid bonds with a water molecule.

Check your understanding

1 Define the terms *Brønsted–Lowry acid* and *Brønsted–Lowry base.*

2 For each of the following, give the formula of its conjugate base:
 a CH_3COOH **b** H_2O **c** HCl

3 For each of the following, give the formula of its conjugate acid:
 a NH_3 **b** H_2O **c** OH^-

4 Identify the acids and bases in this reaction:

$$HNO_3 + H_2SO_4 \rightleftharpoons H_2NO_3^+ + HSO_4^-$$

3.02 The pH of strong acids

OUTCOMES

already from GCSE, you know that

- the pH scale is a measure of the acidity or alkalinity of a solution
- acids have a pH less than 7

already from A2 Level, you know that

- an acid is a proton donor

and after this spread you should

- know that pH = $-\log_{10}[H^+]$, where [ ] represents the concentration in mol dm^{-3}
- be able to calculate the pH of a solution of a strong acid from its concentration
- be able to convert pH into concentration

At the start of the twentieth century, a German chemist called Hans Friedenthal studied different acids. He observed the effects of the acids on over a dozen indicators, and calculated their aqueous hydrogen ion concentrations. In 1904 he suggested that acids could be organized on a scale based on these concentrations. But the aqueous hydrogen ion concentrations were often very small and the range of values was enormous. For example, they could be as low as 1×10^{-14} mol dm^{-3} and at least as high as 1 mol dm^{-3}. Friedenthal's idea needed an improvement to make the scale more manageable.

The pH scale

Søren Sørensen, a Danish chemist, provided the improvement needed when he proposed the **pH scale** in 1909. You should already be familiar with this scale from your previous studies. Its main features are:

- the typical range of pH numbers is 0 to 14
- **acidic** solutions have pH numbers less than 7
- **neutral** solutions have a pH of 7 at 25°C
- **basic** solutions have pH numbers greater than 7

You can have negative pH numbers and pH numbers greater than 14.

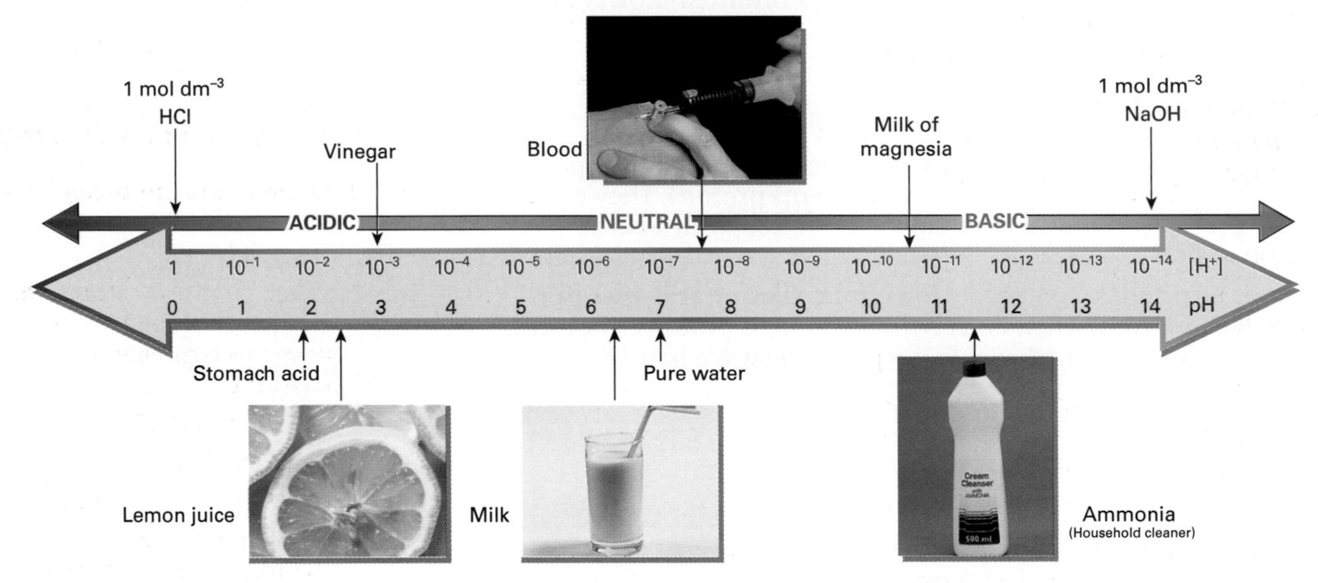

The pH scale and aqueous hydrogen ion concentration.

Sørensen decided to use a scale based on logarithms. This meant that the enormous range in aqueous hydrogen ion concentrations could be turned into something more manageable. Here is the expression needed to calculate a pH number:

$$\text{pH} = -\log_{10}[H^+]$$

$[H^+]$ means the concentration of hydrogen ions in mol dm^{-3}. The use of the negative sign means that high concentrations of hydrogen ions give low pH numbers. Remember that 1.0 mol dm^{-3} can also be written as 1.0 M.

$[H^+]$ to pH calculations

Worked example

Calculate the pH of human blood, if its aqueous hydrogen ion concentration is 3.89×10^{-8} mol dm^{-3}

$$pH = -\log_{10}[H^+]$$
$$pH = -\log_{10}(3.89 \times 10^{-8}) = 7.41$$

pH to [H1] calculations

Here is the expression you need to calculate the aqueous hydrogen ion concentration from a pH number:

$$[H^+] = 10^{-pH}$$

Worked example

A certain solution has a pH of 3.20. What is its aqueous hydrogen ion concentration?

$$[H^+] = 10^{-pH}$$
$$[H^+] = 10^{-3.20} = 6.31 \times 10^{-4}\ mol\ dm^{-3}$$

A common mistake is to forget to reverse the sign of the pH before finding its antilogarithm. You can check your answer by working out the pH from your calculated hydrogen ion concentration, and then comparing it with the pH you were given in the question.

Strong acids

A **strong acid** is fully dissociated or ionized in solution. For example, hydrochloric acid is a strong acid:

$$HCl(aq) \rightarrow H^+(aq) + Cl^-(aq)$$

Its pH depends only on its concentration. For example, 2.0 M hydrochloric acid has an aqueous hydrogen ion concentration of 2.0 mol dm^{-3}. Its pH is:

$$pH = -\log_{10}(2.0) = -0.30$$

Hydrochloric acid is a **monoprotic acid** because each molecule can donate just one proton. Sulfuric acid is a **diprotic acid** because each molecule can donate two protons:

$$H_2SO_4(aq) \rightarrow 2H^+(aq) + SO_4^{2-}(aq)$$

The aqueous hydrogen ion concentration in sulfuric acid is twice the concentration of the acid. For example, 2.0 M sulfuric acid has an aqueous hydrogen ion concentration of 4.0 mol dm^{-3}. Its pH is:

$$pH = -\log_{10}(4.0) = -0.60$$

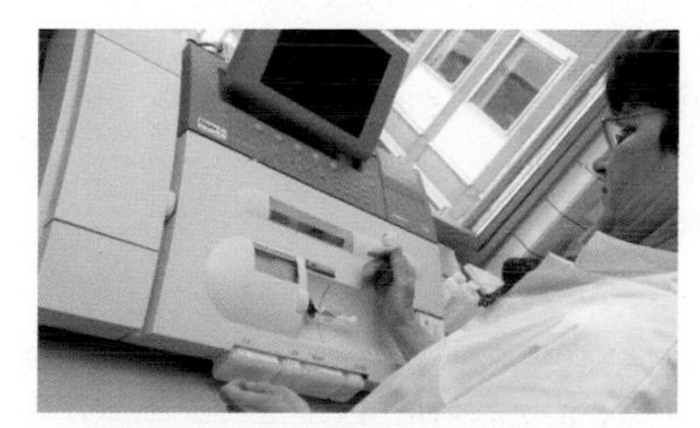

Machines like this can measure the pH of the blood, and the blood concentrations of gases, ions, glucose, and haemoglobin.

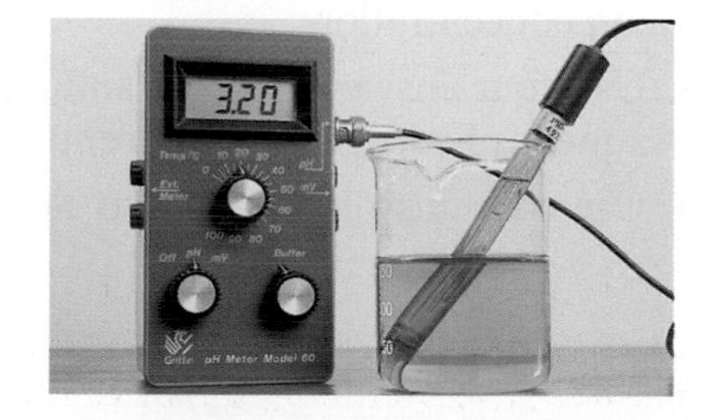

A pH meter gives a precise measurement of the pH of a sample.

Using logarithms

Sørensen used logarithms to base ten, written as $\log_{10}$. Your calculator will probably have a button labelled *'log'* to calculate these numbers. Check that you can use your calculator correctly. For example, $\log_{10}(5 \times 10^{-2}) = -1.3$ and $\log_{10}(2 \times 10^5) = 5.3$.

You need antilogarithms to calculate an aqueous hydrogen ion concentration from a pH number. Calculators usually have the symbol *'10^x'* just above the *'log'* key, accessed using a *'shift'* button. Again, check that you can use your calculator correctly. For example, $10^2 = 100$ and $10^{-3} = 1 \times 10^{-3}$.

Check your understanding

1. Define the term *strong* when applied to acids, and the term *pH*.
2. Calculate the pH numbers corresponding to these aqueous hydrogen ion concentrations:
   - **a** 0.05 mol dm^{-3}
   - **b** 6.5 mol dm^{-3}
   - **c** 2.0×10^{-4} mol dm^{-3}
3. Calculate the aqueous hydrogen ion concentrations corresponding to these pH numbers:
   - **a** 2.50
   - **b** 7.00
   - **c** 11.3
4. Calculate the pH of the following strong acids:
   - **a** 0.100 mol dm^{-3} HCl
   - **b** 0.750 mol dm^{-3} H_2SO_4
   - **c** 0.040 mol dm^{-3} H_3PO_4 (Hint: phosphoric acid is a triprotic acid)

3.03 The pH of strong bases

OUTCOMES

already from GCSE, you know that

- basic solutions have a pH more than 7
- neutral solutions have a pH of 7

already from A2 Level, you

- know that a base is a proton acceptor
- know that $pH = -\log_{10}[H^+]$
- can calculate the pH of a solution of a strong acid from its concentration

and after this spread you should

- know that water is weakly dissociated
- know that $K_w = [H^+][OH^-]$
- be able to calculate the pH of a strong base from its concentration

The concentration of water

It is possible to calculate the concentration of water using the density of water and its relative molecular mass. The density of water at 25°C is 997 g dm^{-3} and its M_r is 18.0, so the concentration of water is $997 \div 18.0 = 55.4$ mol dm^{-3}. This is very high indeed compared to the typical range of concentrations of aqueous hydrogen ions and hydroxide ions.

Just like a strong acid, a **strong base** is fully dissociated in solution. Sodium hydroxide, for example, is a strong base:

$$NaOH(aq) \rightarrow Na^+(aq) + OH^-(aq)$$

To calculate the pH of a strong base in solution, you need to know its aqueous hydrogen ion concentration. But all you know at this stage is that the concentration of hydroxide ions, OH^-, is equal to the concentration of the base. You need some way to link hydroxide ion concentration with hydrogen ion concentration. This is where knowledge of the ionic product of water helps.

The ionic product of water, K_w

Water is amphoteric. It can act as an acid and as a base. One water molecule can donate a proton to another water molecule:

$$H_2O(l) + H_2O(l) \rightarrow H_3O^+(aq) + OH^-(aq)$$

The oxonium ion, H_3O^+, is a strong acid and the hydroxide ion is a strong base. The ions react together to form water and a dynamic equilibrium is formed. This can be shown as:

$$2H_2O(l) \rightleftharpoons H_3O^+(aq) + OH^-(aq)$$

The equation can be simplified to show the hydrogen ion instead of the oxonium ion:

$$H_2O(l) \rightleftharpoons H^+(aq) + OH^-(aq)$$

It is possible to write an expression for the equilibrium constant, K_c, for this equilibrium:

$$K_c = \frac{[H^+(aq)][OH^-(aq)]}{[H_2O(l)]}$$

The concentration of water is very large and water is only weakly dissociated. So $[H_2O(l)]$ can be left out of the expression as it is essentially constant. The new expression for the equilibrium constant is called the **ionic product of water**. It has the symbol K_w:

$$K_w = [H^+][OH^-]$$

This is the expression you need to help you to calculate the pH of a strong base in solution. At 25°C, $K_w = 1.0 \times 10^{-14}$ mol^2 dm^{-6}.

Calculating the pH of a strong base

There are three steps for calculating the pH of a strong base:

1. Use the concentration of the base to calculate the aqueous hydroxide ion concentration.
2. Use your answer to step 1 and the value of K_w to calculate the aqueous hydrogen ion concentration.
3. Use your answer to step 2 to calculate the pH.

Worked example

What is the pH of 0.5 mol dm^{-3} sodium hydroxide solution?

1 $[OH^-] = 0.5$ mol dm^{-3}

2 $K_w = [H^+][OH^-]$. It can be rearranged like this: $[H^+] = \frac{K_w}{[OH^-]}$

$[H^+] = \frac{1.0 \times 10^{-14}}{0.5} = 2.0 \times 10^{-14}$ mol dm^{-3}

3 $pH = -\log_{10}[H^+]$

$pH = -\log_{10}(2.0 \times 10^{-14}) = 13.7$

The pH of water

The ionic product of water can be used to calculate the pH of water. In neutral water, $[H^+] = [OH^-]$. This means the expression for K_w can be rewritten like this:

$$K_w = [H^+][H^+] = [H^+]^2$$

It can be rearranged to find the aqueous hydrogen ion concentration in neutral water:

$$[H^+] = \sqrt{K_w}$$

At 25°C, $[H^+] = \sqrt{1.0 \times 10^{-14}} = 1.0 \times 10^{-7}$ mol dm^{-3}

So at 25°C, $pH = -\log_{10}(1.0 \times 10^{-7}) = 7.0$

The dissociation of water is an endothermic process. This means that the value of K_w increases as the temperature increases. As a result, the pH of neutral water decreases as the temperature increases: at 0°C it is 7.47 but at 100°C it is 6.14. The pH of neutral water is only 7 at 25°C.

When you calculate the pH of a strong base, assume that the temperature is 25°C and that $K_w = 1.0 \times 10^{-14}$ mol^2 dm^{-6} unless you are told otherwise.

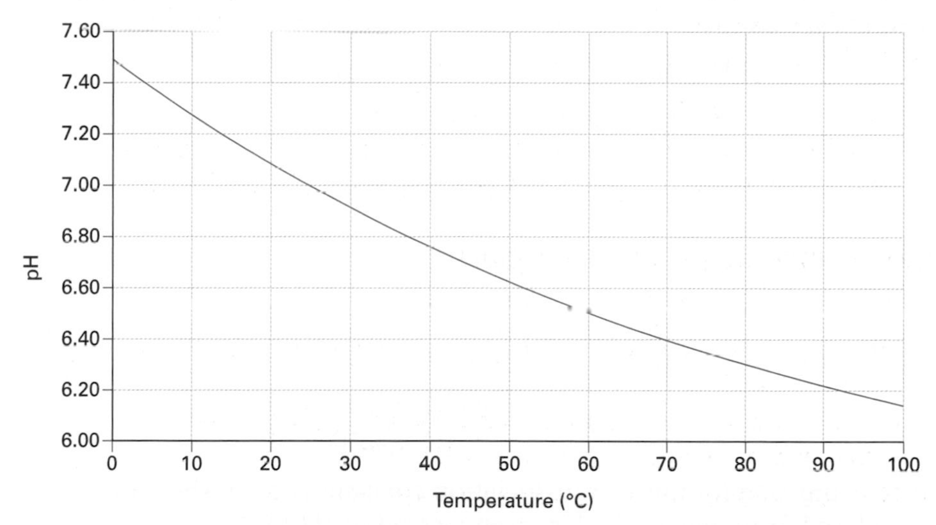

The pH of neutral water decreases as the temperature increases. At 70°C, for example, a solution would be acidic if its pH were less than 7.40.

Check your understanding

1 Define the term *strong* when applied to bases, and the term *ionic product of water*.

2 Calculate the pH of the following strong bases in solution:

a 0.100 mol dm^{-3} NaOH

b 0.750 mol dm^{-3} KOH

c 2.00 mol dm^{-3} NaOH

3 At 40°C, $K_w = 2.92 \times 10^{-14}$ mol^2 dm^{-6}.

a Calculate the pH of neutral water at this temperature.

b Calculate the pH of 0.100 mol dm^{-3} NaOH at this temperature.

3.04 Weak acids and bases

OUTCOMES

already from A2 Level, you know that

- an acid is a proton donor
- a base is a proton acceptor
- water is weakly dissociated
- pH = $-\log_{10}[H^+]$

and after this spread you should

- know that weak acids and weak bases dissociate only slightly in aqueous solution
- be able to construct an expression, with units, for the dissociation constant K_a for a weak acid
- know that $pK_a = -\log_{10}K_a$

Strong acids and strong bases are fully dissociated in aqueous solution, but **weak acids** and **weak bases** are only partially dissociated in aqueous solution. It is important to understand the difference between 'concentration' and 'strength' in acids and bases.

- The concentration or molarity of an acid or base is a measure of its amount in aqueous solution.
- The strength of an acid or base depends on the extent to which it is dissociated in aqueous solution.

For example, hydrochloric acid is a strong acid and ethanoic acid is a weak acid. When the two acids have the same molarity, hydrochloric acid has a greater aqueous hydrogen ion concentration than ethanoic acid. So the pH of the hydrochloric acid will be lower than the pH of the ethanoic acid.

1.0 mol dm^{-3} ethanoic acid has a pH of 2.4, while 1.0 mol dm^{-3} hydrochloric acid has a pH of 0. Magnesium ribbon reacts more vigorously with the hydrochloric acid on the right than it does with the ethanoic acid on the left.

The acid dissociation constant, K_a

When a weak acid HA dissolves in water, it partially dissociates and an equilibrium forms:

$$HA(aq) + H_2O(l) \rightleftharpoons H_3O^+(aq) + A^-(aq)$$

The expression for the corresponding equilibrium constant, K_c, is:

$$K_c = \frac{[H_3O^+(aq)][A^-(aq)]}{[HA(aq)][H_2O(l)]}$$

The concentration of water is very large compared to the other species present. So $[H_2O(l)]$ can be left out of the expression as it is essentially constant. The new expression for the equilibrium constant is called the **acid dissociation constant**. It has the symbol K_a:

$$K_a = \frac{[H_3O^+ (aq)][A^-(aq)]}{[HA(aq)]}$$

The expression for the acid dissociation constant is often shown in a simplified form, where $[H^+]$ is given instead of $[H_3O^+]$:

$$K_a = \frac{[H^+][A^-]}{[HA]}$$

This corresponds to:

$$HA(aq) \rightleftharpoons H^+(aq) + A^-(aq)$$

Worked example

Ethanoic acid, CH_3COOH, is a weak acid. It produces aqueous ethanoate ions, $CH_3COO^-(aq)$. Write an equation, with state symbols, for the formation of ethanoate ions and oxonium ions in aqueous ethanoic acid. Write an expression for the acid dissociation constant, K_a.

Equation: $CH_3COOH(aq) + H_2O(l) \rightleftharpoons CH_3COO^-(aq) + H_3O^+(aq)$

Expression: $K_a = \dfrac{[H^+][CH_3COO^-]}{[CH_3COOH]}$

Both beakers contain 0.1 mol dm^{-3} acid. The left hand one contains ethanoic acid, which is only partially dissociated. It has a relatively low concentration of aqueous ions, so it is a poor conductor of electricity (shown by the small current). The right hand beaker contains hydrochloric acid, which is fully dissociated. It has a relatively high concentration of aqueous ions, so it is a good conductor of electricity (shown by the high current).

K_a and pK_a

The range of values for K_a is immense. For example, the K_a value of silicic acid H_2SiO_3 is just 1.3×10^{-10} mol dm^{-3}, but the K_a value of trichloroethanoic acid CCl_3COOH is as high as 0.23 mol dm^{-3}. A similar problem existed at the beginning of the last century with Hans Friedenthal's idea to organize acids on a scale based on their aqueous hydrogen ion concentrations. You will recall that it was solved by Søren Sørensen when he proposed the pH scale, which is based on logarithms. In the same way, K_a values can be expressed in terms of pK_a values:

$$pK_a = -\log_{10}K_a$$

The weakest weak acids have the smallest K_a values and the largest pK_a values. The table shows some weak acids, organized in order of decreasing strength.

acid	equilibrium	K_a at 298 K (mol dm^{-3})	pK_a at 298 K
trichloroethanoic acid	$CCl_3COOH \rightleftharpoons H^+ + CCl_3COO^-$	2.3×10^{-1}	0.63
ethanoic acid	$CH_3COOH \rightleftharpoons H^+ + CH_3COO^-$	1.7×10^{-5}	4.8
chloric(I) acid	$HClO \rightleftharpoons H^+ + ClO^-$	3.7×10^{-8}	7.4
silicic acid	$H_2SiO_3 \rightleftharpoons H^+ + HSiO_3^-$	1.3×10^{-10}	9.9

Trichloroethanoic acid is the strongest acid in the table and silicic acid is the weakest.

Check your understanding

1 Define the term *weak* when applied to acids and bases.

2 Hydrocyanic acid, HCN, is a weak acid. It produces aqueous cyanide ions, $CN^-(aq)$, in aqueous solution.

  a Write an equation, with state symbols, for the formation of cyanide ions and H_3O^+ ions in aqueous hydrocyanic acid.

  b Write an expression for the acid dissociation constant, K_a.

  c The K_a value is 4.9×10^{-10} mol dm^{-3}. Calculate the corresponding pK_a value.

  d Explain whether hydrocyanic acid is weaker or stronger than chloric(I) acid.

3.05 The pH of weak acids

OUTCOMES

already from A2 Level, you

- know that weak acids dissociate only slightly in aqueous solution
- can construct an expression for the dissociation constant, K_a, for a weak acid
- know that pH = $-\log_{10}[H^+]$

and after this spread you should

- be able to perform calculations relating the pH of a weak acid to the dissociation constant, K_a, and the concentration

Strong acids are fully dissociated in aqueous solution. When you calculate the pH of a strong monoprotic acid, you can assume that the aqueous hydrogen ion concentration is the same as the concentration of the acid. But you cannot make the same assumption for weak acids, because they are only partially dissociated in aqueous solution. To calculate the pH of a weak acid, you must also take into account the acid dissociation constant, K_a.

K_a and pH for weak acids

This is the expression for K_a for the equilibrium,

$$HA(aq) \rightleftharpoons H^+(aq) + A^-(aq):$$

$$K_a = \frac{[H^+][A^-]}{[HA]}$$

For a weak acid in aqueous solution, $[H^+] = [A^-]$, so the expression can be rewritten as:

$$K_a = \frac{[H^+]^2}{[HA]}$$

It can be rearranged to find $[H^+]$:

$$[H^+] = \sqrt{K_a \times [HA]}$$

If we assume that the amount of dissociation is very small, [HA], the equilibrium concentration of acid, will be essentially the same as the quoted concentration of the acid. So the expression can be rewritten like this:

$$[H^+] = \sqrt{K_a \times [acid]}$$

Calculating the pH of a weak acid

To calculate the pH of a weak acid you need to know its K_a (or pK_a) value and its concentration. There are two steps involved.

1. Use the K_a value and the concentration of the acid to calculate the aqueous hydrogen ion concentration.
2. Use your answer to step 1 to calculate the pH.

Worked example

What is the pH of 0.01 mol dm^{-3} acetylsalicylic acid (aspirin), if the K_a value is 3.3×10^{-4} mol dm^{-3} at 25°C?

$[H^+] = \sqrt{K_a \times [acid]}$

$[H^+] = \sqrt{3.3 \times 10^{-4} \times 0.01} = \sqrt{3.3 \times 10^{-6}} = 1.82 \times 10^{-3}$ mol dm^{-3}

pH $= -\log_{10}[H^+]$

pH $= -\log_{10}(1.82 \times 10^{-3}) = 2.7$

Remember that if you are asked to calculate the pH of an acid and you are unsure whether it is strong or weak, you will be given the K_a value if it is a weak acid.

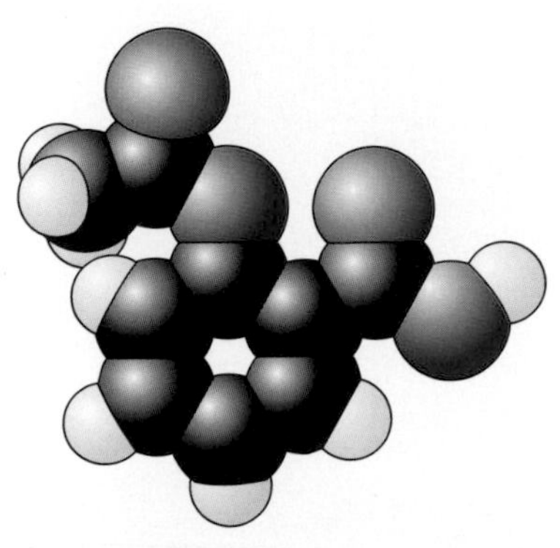

Aspirin or acetylsalicylic acid is an analgesic (pain-relieving drug). It is a weak monoprotic acid.

pK_a and pH

It is relatively easy to calculate the pH of a weak acid if you know its pK_a:

$$pH = \frac{1}{2}pK_a - \frac{1}{2}\log_{10}[acid]$$

For example, the pK_a for benzoic acid is 4.2 at 25°C. What is the pH of 0.010 mol dm^{-3} benzoic acid?

$$pH = \frac{1}{2}pK_a - \frac{1}{2}\log_{10}[acid]$$

$$pH = \frac{4.2}{2} - \frac{\log_{10}(0.010)}{2} = 2.1 - \frac{(-2.0)}{2} = 3.1$$

People sometimes try to treat the painful symptoms of toothache by keeping an aspirin tablet against the diseased tooth. Unfortunately the acidity of the aspirin damages the inside of the mouth, which can take days to heal.

Calculating the K_a of a weak acid

The K_a of a weak acid can be calculated from its pH and concentration:

1 Use the pH to calculate the aqueous hydrogen ion concentration.

2 Use your answer to step 1 and the acid concentration to calculate K_a.

Worked example

0.025 mol dm^{-3} benzoic acid has a pH of 2.90 at 25°C. Calculate the K_a value for benzoic acid.

$pH = -\log_{10}[H^+]$, so $[H^+] = 10^{-pH}$

$[H^+] = 10^{-pH} = 10^{-2.90} = 1.26 \times 10^{-3}$ mol dm^{-3}

$$[H^+] = \sqrt{K_a \times [acid]}, \text{ so } K_a = \frac{[H^+]^2}{[acid]}$$

$$K_a = \frac{[H^+]^2}{[acid]} = \frac{(1.26 \times 10^{-3})^2}{0.025} = \frac{1.5876 \times 10^{-6}}{0.025} = 6.35 \times 10^{-5} \text{ mol dm}^{-3}$$

Check your understanding

1 Write the expression that links aqueous hydrogen ion concentration, acid dissociation constant, and concentration of weak acid.

Use data in this table to help you answer the questions below.

weak acid	K_a at 25°C (mol dm^{-3})
hydrofluoric acid	5.6×10^{-4}
ethanoic acid	1.7×10^{-5}
ammonium ion	5.6×10^{-10}

2 a Calculate the pH of 0.10 mol dm^{-3} hydrofluoric acid at 25°C.

b Calculate the pH of 0.10 mol dm^{-3} hydrochloric acid at 25°C.

c Explain why the two different acids have a different pH at the same concentration and temperature.

3 a Calculate the pK_a of ethanoic acid to two decimal places.

b Use your answer to part a to calculate the pH of 0.25 mol dm^{-3} ethanoic acid at 25°C.

4 Calculate the pH of 0.50 mol dm^{-3} aqueous ammonium chloride at 25°C.

3.06 pH curves

OUTCOMES

already from AS Level, you can

- carry out a simple titration

and after this spread you should understand

- the typical shape of pH curves for acid–base titrations in all combinations of weak and strong monoprotic acids and bases

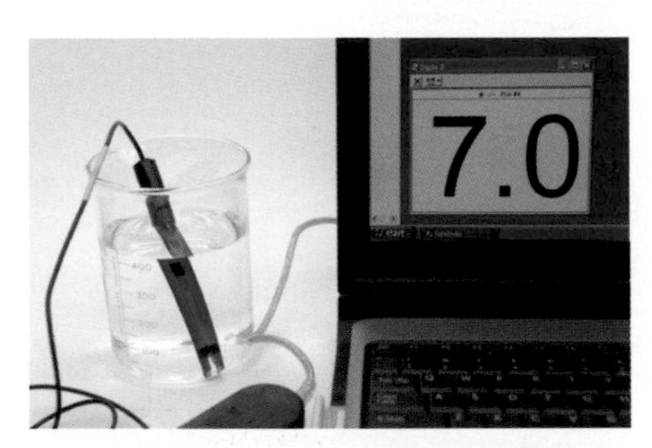

The pH of an aqueous solution can be measured by a pH probe connected to a computer.

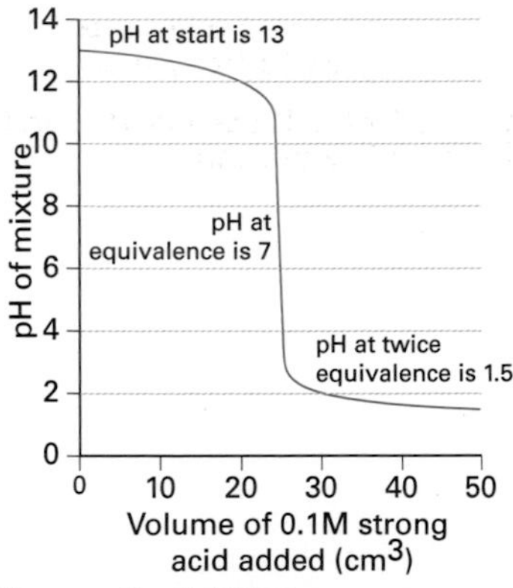

A pH curve for 0.1 M strong monoprotic acid added to 25 cm^3 of 0.1 M strong base.

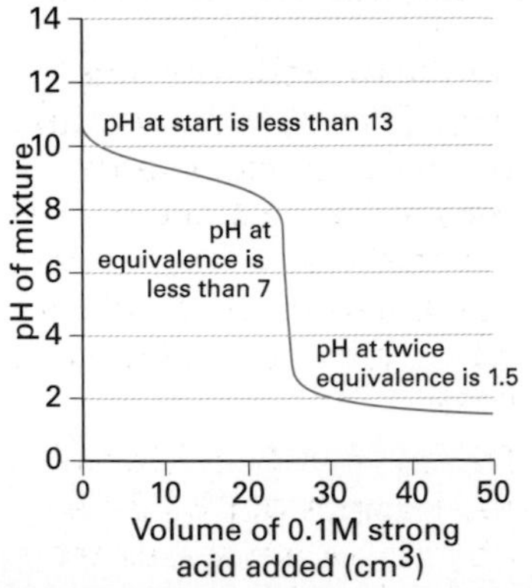

A pH curve for 0.1 M strong monoprotic acid added to 25 cm^3 of 0.1 M weak base.

Aqueous acids and bases neutralize each other when they are mixed together. Acid–base **titrations** are used to find out the volumes needed to obtain a neutral solution. The course of the titration can be monitored using a pH meter. The graph of pH against volume of acid or base added is called a **pH curve**.

Combinations of acid and base

There are four possible combinations, assuming that an aqueous acid is delivered from a burette to an aqueous base in a conical flask:

- strong acid into strong base
- strong acid into weak base
- weak acid into strong base
- weak acid into weak base

These produce different pH curves. You need to recognize and understand each curve, and be able to estimate the pH at the **equivalence point**.

When a titration involves monoprotic acids and bases, the equivalence point occurs when there are **equimolar** amounts of acid and base present (the number of moles of acid and base are the same). For monoprotic acids and bases at the same concentration, this occurs when equal volumes are mixed. This is the same whether the acids and bases involved are strong or weak.

Strong acid into strong base

The base is in large excess at the start of the titration. Only a small proportion is neutralized on each addition of acid, so the pH decreases slightly each time. Just before the equivalence point, which happens at pH 7.0, a large proportion of the remaining base is neutralized by each addition of acid. At the equivalence point, the next addition of acid causes a large fall in pH, as it is being added to a neutral solution. As more acid is added, the pH once more decreases slightly each time.

Strong acid into weak base

At the very start of the titration, the pH decreases rapidly on each addition of acid. But the curve becomes less steep as more acid is added. This is because a **buffer solution** is being formed, consisting of the weak base and the salt of its conjugate acid. You will find out about buffers in Spread 3.10 – they resist changes in pH when small amounts of acid or base are added.

Near the equivalence point, which happens below pH 7.0, each addition of acid causes a large fall in pH. As more acid is added, the shape of the curve is similar to the one for strong acid into strong base.

Weak acid into strong base

At the start of the titration, the shape of the curve is similar to the one for strong acid into strong base. On each side of the equivalence point, which happens above pH 7.0, each addition of acid causes a large fall in pH. As more acid is added, the curve becomes less steep as more acid is added. This is because a buffer solution is being formed, consisting of the weak acid and the salt of its conjugate base.

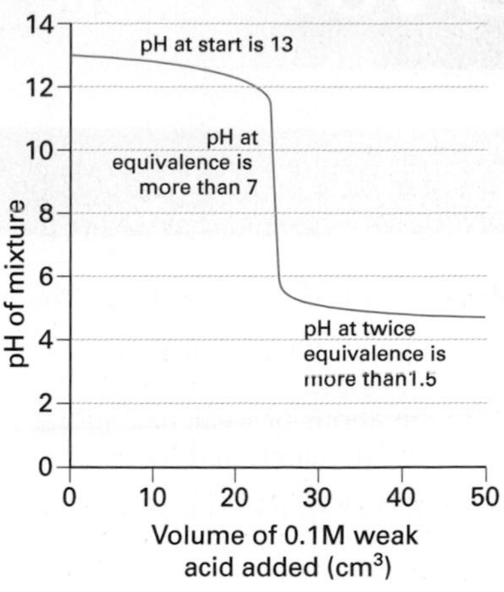

A pH curve for 0.1 M weak acid added to 25 cm^3 of 0.1 M strong base.

Weak acid into weak base

At the start of the titration, the shape of the curve is similar to the one for strong acid into weak base. There is often no obvious change in pH at the equivalence point. This happens at approximately pH 7.0 and depends on the relative strength of the acid and base. As more acid is added, the shape of the curve is similar to the one for weak acid into strong base.

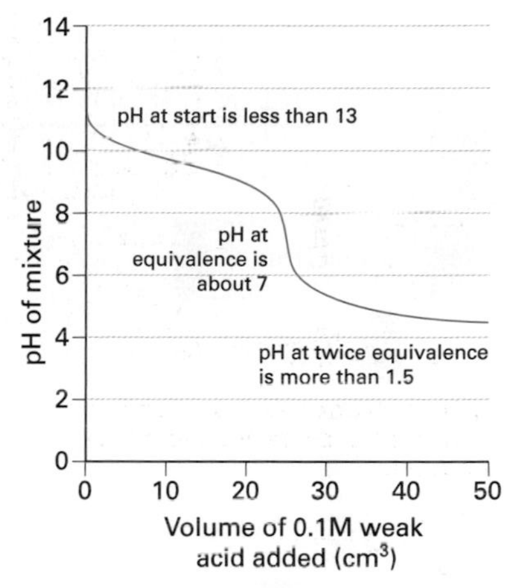

A pH curve for 0.1 M weak acid added to 25 cm^3 of 0.1 M weak base.

Adding a base to an acid

The pH curves obtained when a base in a burette is added to an acid in a conical flask are very similar to the ones described here for acid into base. They are the same shapes but upside down.

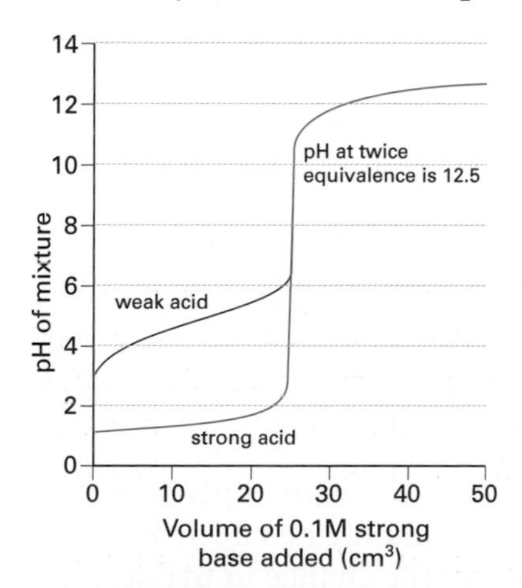

Typical pH curves obtained when 0.1 M strong base is added to 25 cm^3 of 0.1 M weak or strong acid.

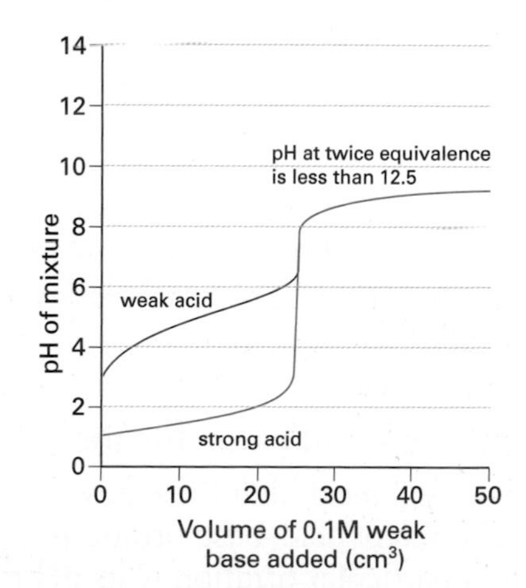

Typical pH curves obtained when 0.1 M weak base is added to 25 cm^3 0.1 M weak or strong acid.

Check your understanding

1. Summarize the starting pH, pH at equivalence, and pH at twice equivalence (double the volume needed for equivalence) for all four combinations of titrations where an acid is added to a base.
2. Describe and explain the four shapes of pH curve obtained when bases are added to acids.

3.07 Indicators

OUTCOMES

already from A2 Level, you understand

- the typical shape of pH curves for acid–base titrations in all combinations of weak and strong monoprotic acids and bases

and after this spread you should be able to

- use pH curves to select an appropriate indicator

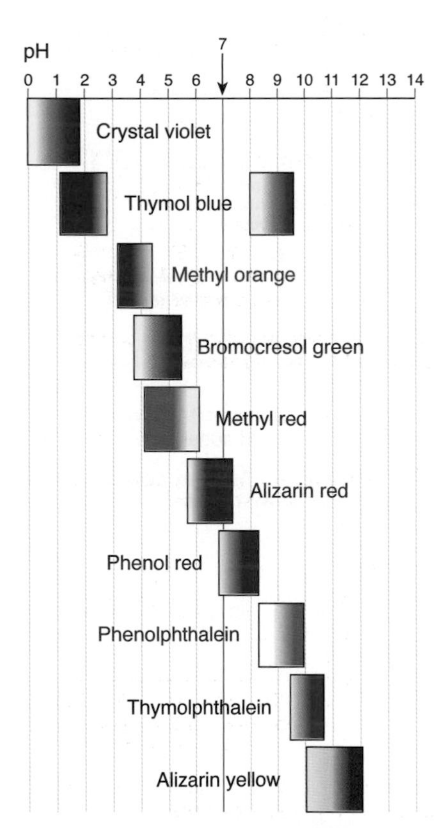

Some common acid–base indicators and the pH ranges over which they change colour.

Acid–base **indicators** are weak organic acids or bases. They have different colours in aqueous solution, depending on the pH of the solution. There are many such indicators. Universal indicator is commonly used in schools. It is a mixture of four different indicators, thymol blue, methyl red, bromothymol blue, and phenolphthalein. Universal indicator is suitable for determining the approximate pH of a solution, but its large range of colours makes it unsuitable for titrations. Just one indicator is usually used in a titration, and different indicators are appropriate for different titrations.

Changing colour

Phenolphthalein is one of the ingredients of universal indicator. It is useful for many titrations. It is a weak organic acid that is colourless in acidic solution and pink in basic solution. If phenolphthalein is shown as HIn:

$$\underset{\text{colourless}}{HIn(aq)} \rightleftharpoons H^+(aq) + \underset{\text{pink}}{In^-(aq)}$$

In acidic solution, the presence of excess $H^+(aq)$ ions causes the position of equilibrium to move to the left, producing the colourless HIn species. In basic solution, excess $OH^-(aq)$ ions react with $H^+(aq)$ ions and remove them as water. The position of equilibrium moves to the right, producing the pink In^- species. Similar equilibria exist for other acid–base indicators.

Phenolphthalein changes colour between pH 8.3 and pH 10.0. Other indicators change colour over difference pH ranges. For example, methyl orange is also commonly used in titrations. It is red in acidic solution and yellow in basic solution. It changes colour between pH 3.1 and pH 4.4.

The pK_a of indicators

An expression for the acid dissociation constant of an indicator can be written like this:

$$K_a = \frac{[H^+][In^-]}{[HIn]}$$

Halfway through a colour change, $[In^-] = [HIn]$, so $K_a = [H^+]$. This means that at that point, pK_a = pH. The colour change for an indicator happens approximately one pH unit either side of its pK_a. For phenolphthalein, its pK_a is 9.3 and its colour changes between pH 10.0 and pH 8.3.

The right indicator for the job

You will recall that the equivalence point occurs at different pH values in different acid–base titrations. An indicator will only be effective in a particular titration if its pH range matches the change in pH at equivalence.

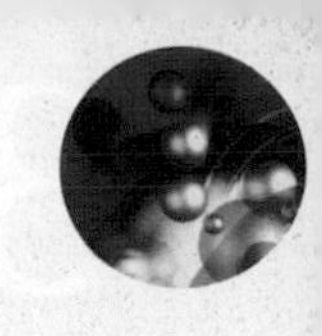

Strong acid into strong base: When 0.1 M strong acid is added to 25 cm^3 of strong base, there is a sharp drop in pH close to the equivalence point. The pH is 11.0 when 24.5 cm^3 of acid has been added, but it drops to just 3.0 when the next 1.0 cm^3 of acid is added. Phenolphthalein and methyl orange would both be suitable for this titration. Phenolphthalein will become colourless at pH 8.3, which is very close to the equivalence point at pH 7.0. Methyl orange will produce an orange tinge at pH 4.4, which is again very close to the equivalence point.

Strong acid into weak base: The equivalence point occurs below pH 7.0, so phenolphthalein would not be suitable. Methyl orange is the better choice because it changes colour between pH 3.1 and pH 4.4.

Weak acid into strong base: The equivalence point occurs above pH 7.0, so methyl orange would not be suitable. Phenolphthalein is the better choice because it changes colour between pH 8.3 and pH 10.0.

Weak acid into weak base: The equivalence point occurs at about pH 7.0 but no indicator would be suitable. This is because there is only a gradual change in pH at equivalence. A pH meter would be needed to follow the titration in this situation.

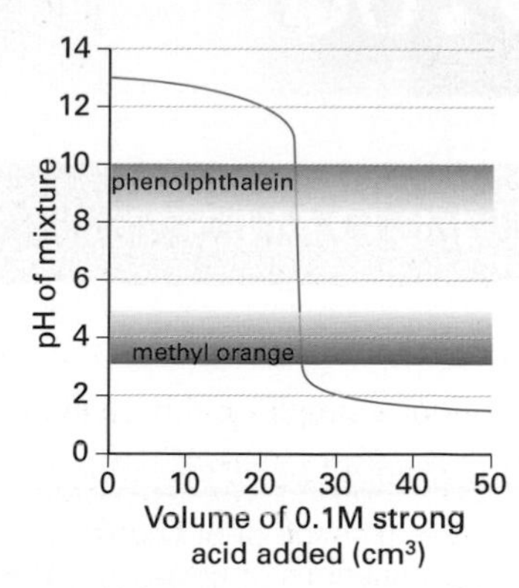

Phenolphthalein and methyl orange are both suitable for strong acid–strong base titrations.

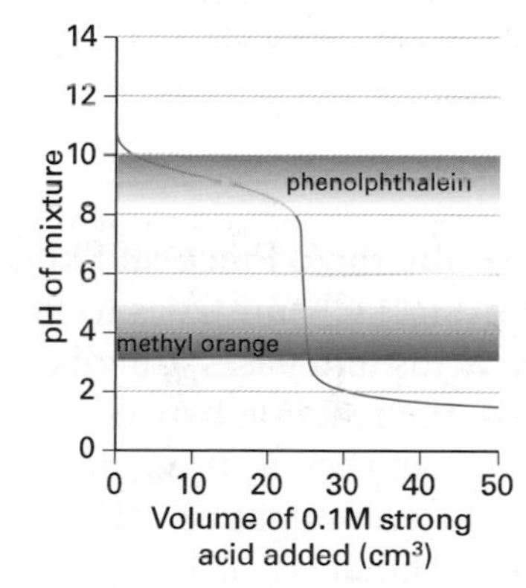

Methyl orange is suitable for strong acid–weak base titrations.

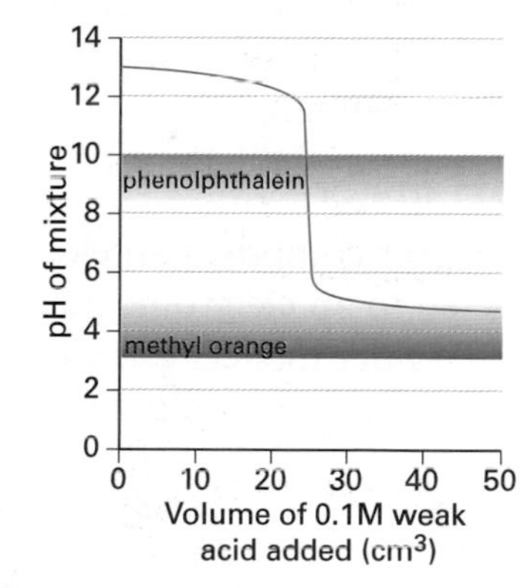

Phenolphthalein is suitable for weak acid–strong base titrations.

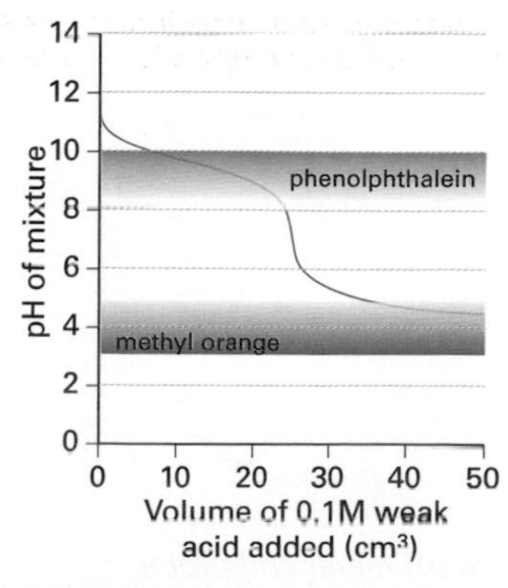

Phenolphthalein and methyl orange are not suitable indicators for weak acid–weak base titrations.

Fading phenolphthalein

The pink colour of phenolphthalein gradually fades in strongly basic solutions. This is because a colourless species forms. This can sometimes be confusing in titrations where the conical flask contains a strong base and phenolphthalein.

Check your understanding

1 The list shows some common acid–base indicators and the pH ranges over which they change colour.

thymol blue: 1.2–2.8	**bromophenol blue: 3.0–4.6**
methyl red: 4.2–6.3	**bromothymol blue: 6.0–7.6**
thymol blue: 8.0–9.6	**cresolphthalein: 8.2–9.8**
thymolphthalein: 9.3–10.5	

a What is unusual about thymol blue in the table?

b Why would cresolphthalein be a suitable indicator where strong acid is added to strong base, but not where strong acid is added to weak base?

c Identify a suitable indicator from the table for a titration where a weak acid is added to a strong base.

2 The pK_a for methyl orange is 3.7. Use this value to estimate the pH range over which it will change colour.

3.08 Investigating pH changes

OUTCOMES

already from AS Level, you can

- carry out a simple acid-base titration

already from A2 Level, you understand

- the typical shape of pH curves for acid–base titrations in all combinations of weak and strong monoprotic acids and bases

and after this spread you should know

- how to investigate the pH changes that occur when a weak acid reacts with a strong base or when a strong acid reacts with a weak base

One of the three Practical Skills Assessments for A2 Physical Chemistry is to investigate pH changes when acids and bases are mixed. This PSA can involve the reaction between a weak acid and a strong base, or between a strong acid and a weak base. For example, you might be asked to determine the pH curve for ethanoic acid reacting with aqueous sodium hydroxide, or for hydrochloric acid reacting with aqueous ammonia. In each case, you will need to use the normal apparatus for a titration, and use a pH meter competently.

The pH meter

A **pH meter** comprises an electrode or pH probe connected to an electronic meter that displays the reading. Older models use a needle and dial, but modern models have digital displays. You might use a simple pH stick in which the probe and meter are combined into a single device. Or you might use a probe connected to a separate meter, datalogger, or computer by a cable. Whatever type of pH meter you use, it will need to be calibrated before use.

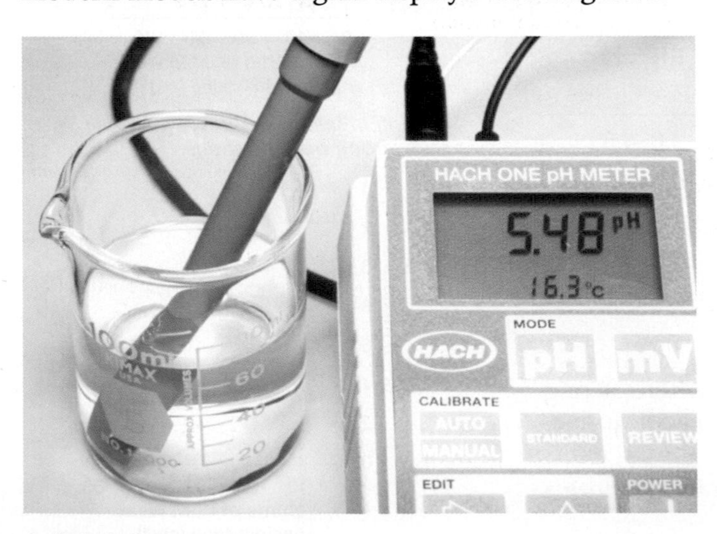

A digital pH meter and probe.

Calibration

The accuracy of a pH meter gradually decreases with storage and use, so it is wise to calibrate the pH meter before you use it. This is done with the help of pH calibration buffers. These are usually at pH 4, pH 7, or pH 10. The idea is to make sure that the pH meter gives the correct reading in these buffers so you can be sure your experimental results are accurate. Some models automatically detect which buffer the pH probe is in, but on others you may have to press a button, or turn a screw or knob. Here is an outline method for calibrating a pH meter:

1. Turn the pH meter on.
2. Rinse the probe with deionized water and gently blot excess liquid from the outside with tissue.
3. Place the tip of the probe in pH 7 buffer and adjust the reading to 7.
4. Remove the probe.
5. Repeat steps 2 to 4 with the other buffers, adjusting the reading to the pH of the buffer each time.

It is very important that the probe is not allowed to dry out during your experiment. The glass bulb of the probe is delicate and usually protected by a plastic cover. Resist any temptation to poke about and fiddle with it, and handle the probe carefully.

Determining a pH curve

Here are the basic steps involved in determining a pH curve.

- Calibrate the pH meter.
- Clean and set up a burette containing the acid.
- Use a transfer pipette and pipette filler to transfer 25 cm^3 of aqueous base to a conical flask.
- Carefully clamp the pH probe so that its tip is in the aqueous base.
- Run the acid into the aqueous base and take regular pH readings until 50 cm^3 of acid has been added.

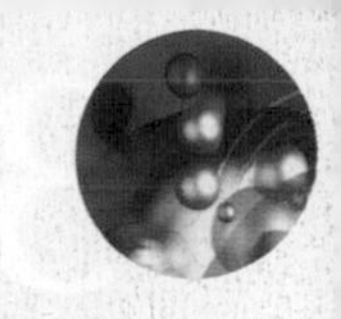

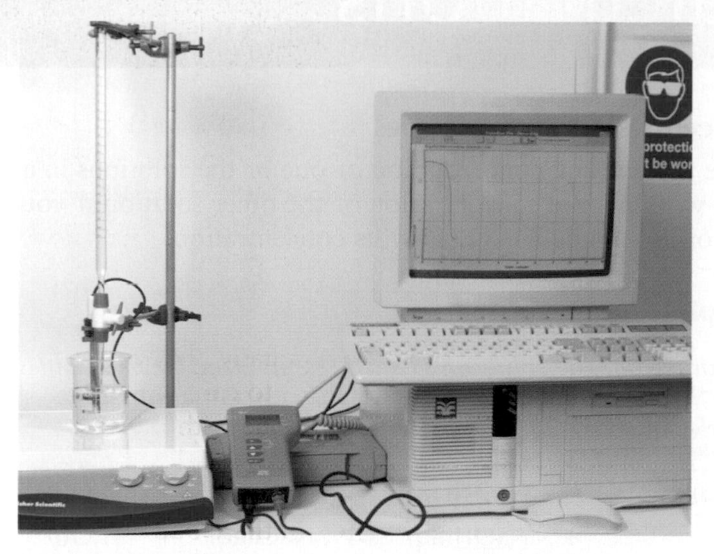

Determining a pH curve.

If your pH probe is connected to a datalogger or computer, you will be able to record the pH automatically during the experiment. This is the most convenient way to determine a pH curve, as many readings can be collected easily while the acid drains from the burette.

If your pH probe is not connected like this, you will need to add a small amount of acid from the burette, record the pH reading, and repeat until the end of the experiment. This shows how useful datalogging can be in science: if you intend to add 50 cm^3 of acid from the burette in 0.5 cm^3 amounts, you will have 101 readings to take by hand (including the one from the start).

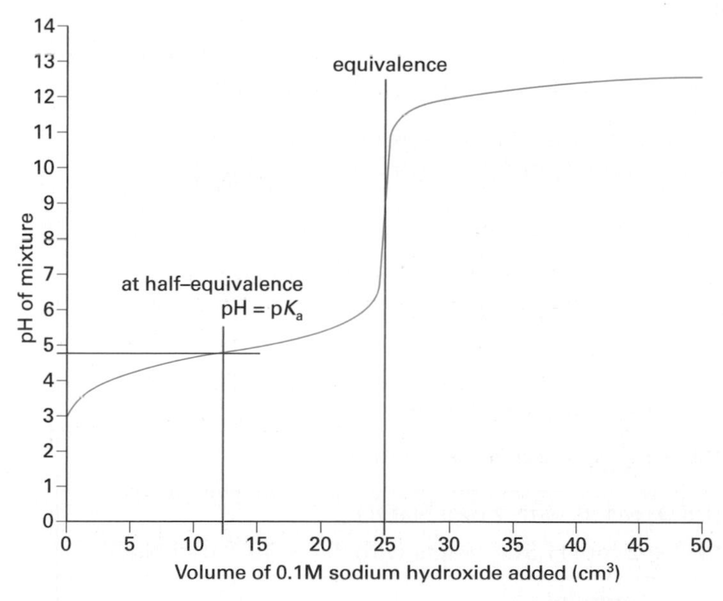

This is a pH curve obtained when 0.1 mol dm^{-3} sodium hydroxide is added to 0.1 mol dm^{-3} ethanoic acid. Such a curve can be used to determine the pK_a of the weak acid. At half-equivalence, pH = pK_a.

Magnetic stirrers

It is important to mix the acid and base before taking a pH measurement. This can be done by hand with a stirring rod, taking care not to bash the pH probe. It can also be done automatically with a magnetic stirrer, which is a machine that contains a rotating magnet under a level surface. A beaker or conical flask containing the reaction mixture is placed on top, and a small plastic-coated magnetic follower or 'flea' is put inside. This spins around and stirs the mixture. It is important to adjust the speed of rotation so that the follower does not spin too quickly. If it does, it can leap suddenly to one side, which may damage the container or the pH probe.

A magnetic stirrer takes the effort out of continually stirring a reaction mixture.

3.09 Titration calculations

OUTCOMES

already from AS Level, you can calculate

- concentrations and volumes for reactions in solutions

already from A2 Level, you can calculate the pH of

- a solution of a strong acid from its concentration
- a strong base from its concentration
- a weak acid using its dissociation constant, K_a, and its concentration

and after this spread you should be able to

- perform calculations for the titrations of monoprotic and diprotic acids with sodium hydroxide

You have taken great care to achieve a sharp end-point in your titration, so make sure you can carry out calculations based on your results.

Calculating concentrations and volumes

You will know the volume and concentration of one of the solutions in a titration. You can work out the concentration of the other solution if you know its volume, or its volume if you know its concentration.

Worked example

25.0 cm³ of sodium hydroxide solution is exactly neutralized by 24.50 cm³ of 0.050 mol dm⁻³ hydrochloric acid. What was the concentration of the sodium hydroxide solution?

Step 1	$HCl(aq) + NaOH(aq) \rightarrow NaCl(aq) + H_2O(l)$
Step 2	$\underline{HCl}(aq) + \underline{NaOH}(aq) \rightarrow NaCl(aq) + H_2O(l)$
Step 3	$24.50\ cm^3$ (HCl), $0.05\ mol\ dm^{-3}$ (HCl); $25.0\ cm^3$ (NaOH)
Step 4	moles = concentration × volume $n = 0.050 \times \frac{24.50}{1000} = 1.225 \times 10^{-3}$ mol of HCl
Step 5	1 mol HCl reacts with 1 mol NaOH 1.225×10^{-3} mol HCl reacts with 1.225×10^{-3} mol NaOH
Step 6	$\text{concentration} = \frac{\text{amount}}{\text{volume}}$ $c = \frac{1.225 \times 10^{-3}}{25.0} \times 1000 = 0.049\ mol\ dm^{-3}$

The method is just the same if the acid is diprotic, such as sulfuric acid. But at step 5, remember that one mole of acid will react with two moles of sodium hydroxide.

Worked example

What volume of 0.125 mol dm⁻³ sulfuric acid is needed to exactly neutralize 25.00 cm³ of 0.100 mol dm⁻³ sodium hydroxide?

Step 1	$H_2SO_4(aq) + 2NaOH(aq) \rightarrow Na_2SO_4(aq) + 2H_2O(l)$
Step 2	$\underline{H_2SO_4}(aq) + \underline{2NaOH}(aq) \rightarrow Na_2SO_4(aq) + 2H_2O(l)$
Step 3	$0.125\ mol\ dm^{-3}$ (H_2SO_4); $25.00\ cm^3$, $0.100\ mol\ dm^{-3}$ (NaOH)
Step 4	moles = concentration × volume $n = 0.100 \times \frac{25.00}{1000} = 2.5 \times 10^{-3}$ mol of NaOH
Step 5	1 mol H_2SO_4 reacts with 2 mol NaOH $2.5 \times 10^{-3} \div 2$ mol H_2SO_4 reacts with 2.5×10^{-3} mol NaOH
Step 6	$\text{volume} = \frac{\text{amount}}{\text{concentration}}$ $\text{volume} = \frac{1.25 \times 10^{-3}}{0.125} \times 1000 = 10.0\ cm^3$

Check your understanding

1 a 25.0 cm³ of 0.180 mol dm⁻³ sodium hydroxide is exactly neutralized by 15.0 cm³ of hydrochloric acid. What is the concentration of the acid?

b What volume of 0.200 mol dm⁻³ sulfuric acid is needed to exactly neutralize 25.0 cm³ of 0.500 mol dm⁻³ sodium hydroxide?

Calculating pH values in titrations

The pH of a reaction mixture can be measured during a titration using a pH meter. But it can also be calculated.

Strong acids and sodium hydroxide

Step 1 Calculate the number of moles of H^+ ions and OH^- ions in each solution.

Step 2 Work out which ion is in excess.

Step 3 Calculate the total volume of the mixture.

Step 4 Use your answers to steps 1 to 3 to calculate the concentration of the excess ion. If the OH^- ions are in excess, use $[H^+] = K_w/[OH^-]$ to calculate $[H^+]$.

Step 5 Calculate the pH from $pH = -\log_{10}[H^+]$.

Worked example

40.0 cm³ of 0.200 mol dm⁻³ hydrochloric acid is added to 10.0 cm³ of 0.150 mol dm⁻³ sodium hydroxide. Calculate the pH of the solution formed.

Step 1	Amount of H^+ from acid $= \frac{40.0}{1000} \times 0.200 = 8.0 \times 10^{-3}$ mol
	Amount of OH^- from base $= \frac{10.0}{1000} \times 0.150 = 1.5 \times 10^{-3}$ mol
Step 2	Excess $H^+ = (8.0 \times 10^{-3}) - (1.5 \times 10^{-3}) = 6.5 \times 10^{-3}$ mol
Step 3	Total volume of mixture $= 40.0 + 10.0 = 50.0$ cm³
Step 4	Concentration of excess $H^+ = \frac{6.5 \times 10^{-3}}{50.0} \times 1000 = 0.13$ mol dm⁻³
Step 5	pH of solution $= -\log_{10}(0.13) = 0.886$

Weak acids and sodium hydroxide

Step 1 Calculate the number of moles of weak acid and OH^- ions in each solution.

Step 2 Work out which is in excess (acid or OH^- ions).

Step 3 Calculate the total volume of the mixture.

If the OH^- ions are in excess, continue with steps 4 and 5 as described (left). But if the acid is in excess, calculate the pH as you do for a weak acid (Spread 3.05).

Worked example

25.0 cm³ of 0.100 mol dm⁻³ sodium hydroxide is added to 10.0 cm³ of 0.600 mol dm⁻³ ethanoic acid. Calculate the pH of the solution formed. K_a for ethanoic acid is 1.74×10^{-5} mol dm⁻³ at 298 K.

Step 1	Amount of acid $= \frac{10.0}{1000} \times 0.600 = 6.0 \times 10^{-3}$ mol
	Amount of OH^- from base $= \frac{25.0}{1000} \times 0.100 = 2.5 \times 10^{-3}$ mol
Step 2	Excess acid $= (6.0 \times 10^{-3}) - (2.5 \times 10^{-3}) = 3.5 \times 10^{-3}$ mol
Step 3	Total volume of mixture $= 25.0 + 10.0 = 35.0$ cm³

The weak acid is in excess, so calculate its pH:

$$[\text{acid}] = \frac{3.5 \times 10^{-3}}{35.0} \times 1000 = 0.100 \text{ mol dm}^{-3}$$

$$[H^+] = \sqrt{K_a \times [\text{acid}]}$$

$$[H^+] = \sqrt{1.74 \times 10^{-5} \times 0.100} = \sqrt{1.74 \times 10^{-6}} \text{ mol dm}^{-3}$$

$$pH = -\log_{10}(1.32 \times 10^{-3}) = 2.88$$

Check your understanding

2 Calculate the final pH when the following solutions are mixed:

a 25.0 cm³ of 0.100 mol dm⁻³ HCl and 15.0 cm³ of 0.100 mol dm⁻³ NaOH.

b 25.0 cm³ of 0.100 mol dm⁻³ H_2SO_4 and 15.0 cm³ of 0.200 mol dm⁻³ NaOH.

c 15.0 cm³ of 0.100 mol dm⁻³ HCl and 35.0 cm³ of 0.120 mol dm⁻³ NaOH.

3 Calculate the final pH when the following solutions are mixed (K_a for ethanoic acid is 1.74×10^{-5} mol dm⁻³ at 298 K).

a 25.0 cm³ of 0.100 mol dm⁻³ ethanoic acid and 5.00 cm³ of 0.200 mol dm⁻³ NaOH.

b 25.0 cm³ of 0.100 mol dm⁻³ ethanoic acid and 25.0 cm³ of 0.200 mol dm⁻³ NaOH.

3.10 Buffers

OUTCOMES

already from A2 Level, you

- know that weak acids and weak bases dissociate only slightly in aqueous solution
- can construct an expression, with units, for the dissociation constant K_a for a weak acid
- know that $pH = -\log_{10}[H^+]$, where [ ] represents the concentration in mol dm^{-3}

and after this spread you should

- know some applications of buffer solutions
- be able to explain qualitatively the action of acidic and basic buffers
- be able to calculate the pH of acidic buffer solutions

Avoid a bad hair day with the help of acidic buffers in pH balanced shampoos.

A buffer solution resists changes in pH when small amounts of acid or base are added to it, or when it is diluted. Note the use of the words 'resists' and 'small'. The pH of a buffer will still change if an acid or base is added to it, but much less than if it were not a buffer solution. And if large amounts of acid or base are added to a buffer solution, its pH will change a lot.

Applications of buffers

Buffer solutions are used in the laboratory to calibrate pH meters. They are very important in biochemical research. Many enzymes are easily denatured by extremes of pH once they are extracted from living cells. Buffer solutions help to maintain them at the correct pH and ionic concentration. In industry, buffer solutions achieve the correct conditions for dyeing fabrics.

Shampoos contain acidic buffers. The surface of a hair has overlapping scales. The scales stand up in basic solutions, making hair look rough and dull. They lie flat in slightly acidic solutions, making hair look smooth and shiny. The buffer solution in a shampoo keeps it slightly acidic, even when it is mixed with water and dirt.

Biological buffers

Buffer solutions are common in nature. The pH of your blood is maintained between 7.35 and 7.45 by a buffer in the plasma. This involves carbonic acid, H_2CO_3, which forms when carbon dioxide dissolves in water, and the hydrogencarbonate ion HCO_3^-:

$$H_2CO_3(aq) \rightleftharpoons H^+(aq) + HCO_3^-(aq)$$

If a small amount of acid enters the bloodstream, its H^+ ions combine with hydrogencarbonate ions to form undissociated carbonic acid, and the position of equilibrium moves to the left. The concentration of hydrogen ions stays almost the same, maintaining the pH in the correct range.

How buffers work

There are two types of buffer:

- acidic buffers, which contain a weak acid and a salt of its conjugate base
- basic buffers, which contain a weak base and a salt of its conjugate acid

An acidic buffer can be made by mixing ethanoic acid with aqueous sodium ethanoate. An equilibrium forms:

$$CH_3COOH(aq) + H_2O(l) \rightleftharpoons CH_3COO^-(aq) + H_3O^+(aq)$$

The ethanoic acid itself is partially dissociated and the majority of the ethanoate ions come from the salt.

If a small amount of an acid is added to the buffer, its H^+ ions combine with ethanoate ions to form undissociated ethanoic acid. The position of equilibrium moves to the left. If a small amount of a base is added, its OH^- ions combine with H_3O^+ ions to form water. More ethanoic acid dissociates and the position of equilibrium moves to the right.

A basic buffer can be made by mixing aqueous ammonia with aqueous ammonium chloride. An equilibrium forms between the ammonia and ammonium ions:

$$NH_3(aq) + H^+(aq) \rightleftharpoons NH_4^+(aq)$$

If a small amount of an acid is added to the buffer, its H^+ ions combine with ammonia to form ammonium ions. The position of equilibrium moves to the right. If a small amount of a base is added, its OH^- ions combine with ammonium ions to form ammonia and water, and the position of equilibrium moves to the left.

Calculating the pH of an acidic buffer

The hydrogen ion concentration of an acidic buffer is found using this expression:

$$[H^+] = K_a \times \frac{[acid]}{[salt]}$$

The expression $pH = -\log_{10}[H^+]$ is then used to calculate the pH of the buffer solution.

If the pK_a of the acid is given instead, the pH of the buffer solution can be calculated using this expression:

$$pH = pK_a \times \log_{10}\left(\frac{[salt]}{[acid]}\right)$$

Notice that the two terms, [salt] and [acid], are in opposite positions in the two expressions, so take care.

The acidity of the oceans is maintained by the relative concentrations of carbon dioxide, hydrogencarbonate ions, and carbonate ions. The oceans absorb excess carbon dioxide from burning fossil fuels, and the pH of sea water is gradually falling. This causes difficulty for coral and other marine invertebrates because their shells contain carbonates. These react with acids, making the shells weaker.

Worked example

Calculate the pH of a buffer solution containing 0.20 mol dm^{-3} ethanoic acid and 0.10 mol dm^{-3} sodium ethanoate. The value of K_a for ethanoic acid is 1.74×10^{-5} mol dm^{-3} at 298 K.

$$[H^+] = K_a \times \frac{[acid]}{[salt]} = 1.74 \times 10^{-5} \times \frac{0.20}{0.10} = 3.48 \times 10^{-5} \text{ mol dm}^{-3}$$

$$pH = -\log_{10}[H^+] = -\log_{10}(3.48 \times 10^{-5}) = 4.46$$

Check your understanding

1 a What is a buffer solution?

  b Give two examples of applications of buffer solutions.

2 a Methanoic acid is a weak acid. What could you add to it to make an acidic buffer solution?

  b Explain how this buffer would be able to resist a decrease in pH when a small amount of strong acid is added to it.

3 Calculate the pH of a buffer solution containing 0.15 mol dm^{-3} ethanoic acid and 0.20 mol dm^{-3} sodium ethanoate. The value of K_a for ethanoic acid is 1.74×10^{-5} mol dm^{-3} at 298 K.

Questions: Chs 1–3

1 a A chemical reaction is first order with respect to compound **L** and second order with respect to compound **M**.

- **i** Write the rate equation for this reaction.
- **ii** What is the overall order of this reaction?
- **iii** By what factor will the rate increase if the concentrations of **L** and **M** are both doubled?

[4]

b The table below shows the initial concentrations of two compounds, **P** and **Q**, and the initial rate of the reaction that takes place between them at constant temperature.

experiment	[P] (mol dm^{-3})	[Q] (mol dm^{-3})	initial rate (mol dm^{-3} s^{-1})
1	0.3	0.3	7.875×10^{-4}
2	0.6	0.6	3.150×10^{-3}
3	1.2	0.6	1.260×10^{-2}

- **i** Determine the overall order of the reaction between **P** and **Q**. Explain how you reached your conclusion.
- **ii** Determine the order of reaction with respect to compound **Q**. Explain how you reached your conclusion.
- **iii** Write the rate equation for the overall reaction.
- **iv** Calculate the value of the rate constant, stating its units. [7]

[Total 11 marks]

2 a A large excess of calcium carbonate was added to 100 cm^3 of 0.2 M hydrochloric acid. After the reaction had ended, 240 cm^3 of carbon dioxide had been formed. In three further experiments, extra substances were added to the original mixture as shown in the table below. Copy and complete the table to show the total volume of carbon dioxide formed in each experiment and the qualitative effect of these additions on the initial rate of reaction compared to the original experiment.

substances added to an excess of calcium carbonate and 100 cm^3 of 0.2 M hydrochloric acid	volume of carbon dioxide (cm^3)	effect on initial rate of reaction
100 cm^3 water		
10 g calcium carbonate		
50 cm^3 0.2 M hydrochloric acid		

[6]

b The rate of reaction between compounds **W** and **X** was studied at a fixed temperature and some results obtained are shown in the table below.

experiment	initial concentration of W (mol dm^{-3})	initial concentration of X (mol dm^{-3})	initial rate (mol dm^{-3} s^{-1})
1	0.40	0.20	5.0×10^{-5}
2	0.60	0.20	7.5×10^{-5}
3	0.80	0.10	5.0×10^{-5}
4	0.30	0.15	to be calculated

Use the data in the table to deduce the order of reaction with respect to compound **W** and the order of reaction with respect to compound **X**. Hence calculate the initial rate of reaction in experiment 4. [4]

c The rate equation for a reaction between substances **Y** and **Z** is:

$$\text{rate} = k[\mathbf{Y}]^2[\mathbf{Z}]^2$$

The initial rate is found to be 4.8×10^{-3} mol dm^{-3} s^{-1} when the initial concentration of **Y** is 0.20 dm^{-3} and the initial concentration of **Z** is 0.40 mol dm^{-3}.

- **i** Calculate the value of the rate constant, k, at this temperature and deduce its units.
- **ii** Sketch a graph to show how the value of k varies as temperature is increased over a large range. [4]

[Total 14 marks]

3 Tetrafluoroethene, C_2F_4, is obtained from chlorodifluoromethane, $CHClF_2$:

$$2CHClF_2(g) \rightarrow C_2F_4(g) + 2HCl(g)$$

$$\Delta H^{\ominus} = +128 \text{ kJ mol}^{-1}$$

a A 1.2 mol sample of $CHClF_2$ is put in a 20.0 dm^3 container and heated. When equilibrium is reached, the mixture contains 0.40 mol of $CHClF_2$.

- **i** Calculate the number of moles of C_2F_4 and the number of moles of HCl present at equilibrium.
- **ii** Write an expression for K_c for the equilibrium.
- **iii** Calculate a value for K_c and give its units. [6]

b
- **i** State how the temperature should be changed at constant pressure to increase the equilibrium yield of C_2F_4.
- **ii** State how the total pressure should be changed at constant temperature to increase the equilibrium yield of C_2F_4. [2]

[Total 8 marks]

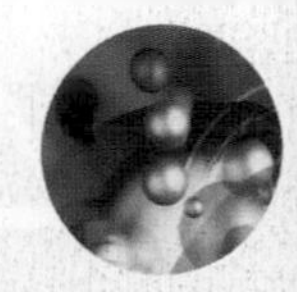

4 At 25°C, the constant K_w has the value 1.00×10^{-14} mol^2 dm^{-6}.

a Define the term K_w. [1]

b Define the term pH. [1]

c Calculate the pH at 25°C of 1.00 M HNO_3. [1]

d Calculate the pH at 25°C of 2.50 M KOH. [2]

e Calculate the pH at 25°C of the solution that results from mixing 19.0 cm^3 of 2.00 M HNO_3 with 16.0 cm^3 of 2.50 M KOH. [6]

[Total 11 marks]

5 The value of the acid dissociation constant, K_a, for the weak acid HA, at 298 K, is 1.45×10^{-4} mol dm^{-3}.

a Write an expression for the term K_a for the weak acid HA. [1]

b Calculate the pH of a 0.450 mol dm^{-3} solution of HA at 298 K. [4]

c A mixture of the acid HA and the sodium salt of this acid, NaA, can be used to prepare a buffer solution.

i State and explain the effect on the pH of this buffer solution when a small amount of dilute hydrochloric acid is added.

ii The concentration of HA in a buffer solution is 0.450 mol dm^{-3}. Calculate the concentration of A^- in this buffer solution when the pH is 3.59. [6]

[Total 11 marks]

6 a Three titration curves labelled **A**, **B**, and **C** are shown below. All solutions have a concentration of 0.1 mol dm^{-3}.

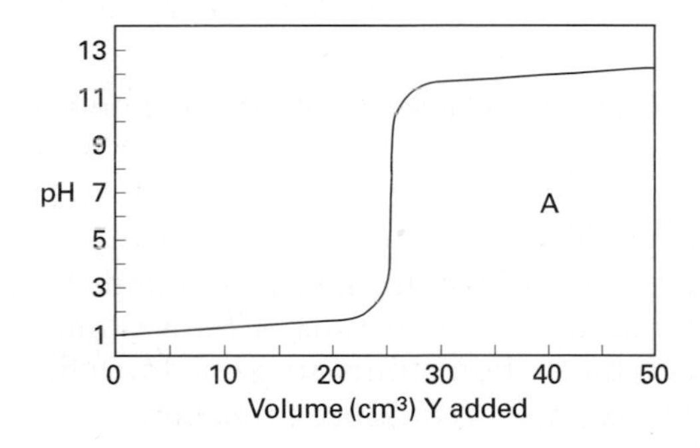

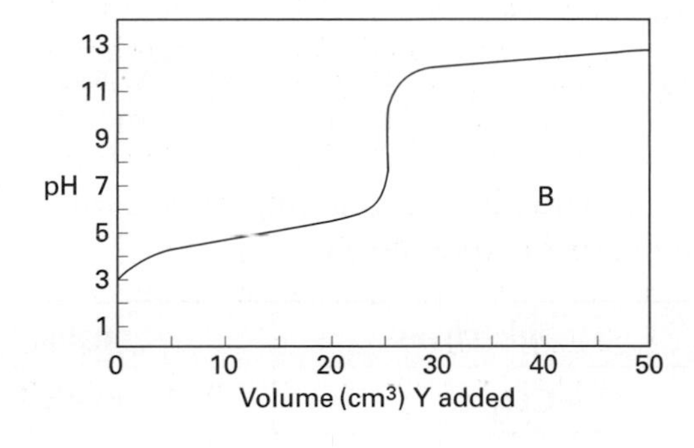

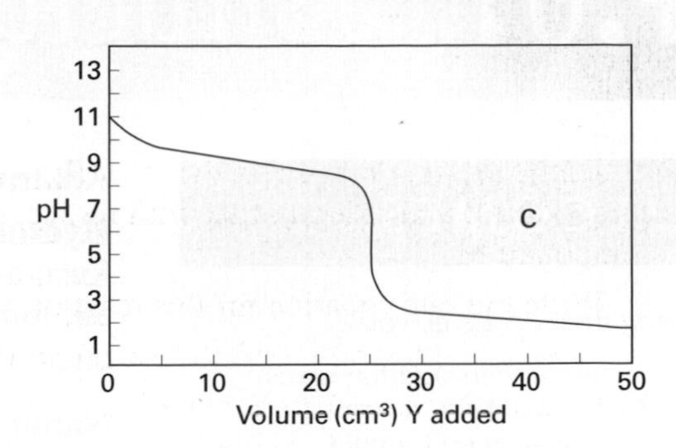

Select from **A**, **B**, and **C** the curve that would be produced by the addition of:

i ammonia to 25 cm^3 of nitric acid

ii ethanoic acid to 25 cm^3 of potassium hydroxide

iii potassium hydroxide to 25 cm^3 of hydrochloric acid [3]

b A table of acid–base indicators and the pH ranges over which they change colour is shown below.

indicator	pH range
Thymol blue	1.2–2.8
Congo red	3.0–5.0
Methyl red	4.2–6.3
Cresolphthalein	8.2–9.8
Thymolphthalein	9.3–10.5

Select from the table an indicator which could be used in the titration which produces curve **A** but not in the titration which produces curve **B**. [1]

c **i** Define the term pH.

ii A solution of sodium hydroxide has a pH of 11.90 at 25°C. Calculate the concentration of sodium hydroxide in the solution. [4]

d The acid dissociation constant, K_a, for ethanoic acid has the value of 1.70×10^{-5} mol dm^{-3} at 25°C.

$$K_a = \frac{[H^+][CH_3COO^-]}{[CH_3COOH]}$$

i Calculate the pH of a 0.117 mol dm^{-3} aqueous solution of ethanoic acid. Give your answer to two decimal places.

ii Calculate the pH of a mixture formed by adding 25 cm^3 of a 0.117 mol dm^{-3} aqueous solution of sodium ethanoate to 25 cm^3 of a 0.117 mol dm^{-3} aqueous solution of ethanoic acid. Give your answer to two decimal places. [5]

[Total 13 marks]

4.01 Structural isomers

OUTCOMES

already from AS Level, you

- know and understand the terms molecular formula, structural formula, displayed formula, homologous series, and functional group
- can apply IUPAC rules for nomenclature to alkanes, alkenes, and haloalkanes with chains of up to six carbon atoms
- know and understand the meaning of the term structural isomerism
- can draw the structures of chain and position isomers

and after this spread you should

- have familiarized yourself again with the key concepts of naming and representing organic compounds from *AS Level Chemistry*
- know and understand the meaning of the term structural isomerism
- be able to draw the structural formulae and displayed formulae of isomers

```
C — C — C — C

    ↓

  |   |   |   |
— C — C — C — C —
  |   |   |   |

    ↓

    H   H   H   H
    |   |   |   |
H — C — C — C — C — H
    |   |   |   |
    H   H   H   H
```

How to draw the displayed formula of butane, C_4H_{10}.

```
    H   H   H
    |   |   |
H — C — C — C — H
    |   |   |
    H   |   H
    H — C — H
        |
        H
```

Methylpropane is a chain isomer of butane.

Alkanes

Hydrocarbons consist of hydrogen and carbon atoms only. The **alkanes** form a **homologous series** of hydrocarbons in which the carbon atoms are joined to each other by single covalent bonds. Alkanes have similar chemical properties and the same general formula, C_nH_{2n+2}.

Naming unbranched alkanes

The names of alkanes tell you how many carbon atoms they contain. For unbranched alkanes, this information is found in the first part of the name, the *prefix*.

prefix	number of C atoms	prefix	number of C atoms
meth	1	but	4
eth	2	pent	5
prop	3	hex	6

The names of alkanes end in *ane*. So hexane is an alkane containing six carbon atoms. Its **molecular formula** is C_6H_{14}, following the general formula for alkanes.

Structural formulae

A **structural formula** shows the number and type of each atom in a molecule, and how they are joined together. Shortened structural formulae are the easiest to type, but they do not show the bonds. Ethane is shown as CH_3CH_3 and propane as $CH_3CH_2CH_3$. Butane can be shown as $CH_3CH_2CH_2CH_3$ but this can be simplified to $CH_3(CH_2)_2CH_3$.

Displayed formulae

A **displayed formula** is difficult to type but it shows the bonds clearly as straight lines.

Structural isomers of alkanes

Structural isomerism occurs when two or more compounds have the same molecular formula, but different structures. **Chain isomers** are structural isomers in which the carbon atoms are joined together in different arrangements, for example to produce branches.

Naming branched alkanes

The longest chain of carbon atoms is the **main chain**. The branches are **side chains** attached to the main chain. The alkane is named after its main chain, and the position and type of each side chain. For example, methylpropane comprises a main chain containing three carbon atoms, with a side chain containing one carbon atom.

number of C atoms	side chain	name
1	$-CH_3$	methyl
2	$-CH_2CH_3$	ethyl
3	$-CH_2CH_2CH_3$	propyl

Side chains are named after the number of carbon atoms they contain.

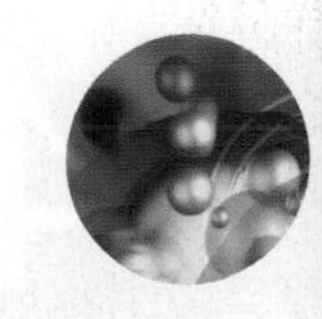

Numbers are used if the side chains could be attached to different carbon atoms on the main chain. The numbering is done so that you get the lowest total number. Side chains with different numbers of carbon atoms are written in alphabetical order.

Each side chain has its own number, and each number is separated from the next one by a comma. A number and a word are separated by a dash. If you have two identical side chains, the prefix *di* is put in front of their name (three = tri, four = tetra).

```
   H   H   H   H   H   H
   |   |   |   |   |   |
H— C — C — C — C — C — C —H
   |   |   |   |   |   |
   H  CH3 CH3 CH2  H   H
               |
              CH3
```

4-ethyl-2,3-dimethylhexane.

Alkenes

The **alkenes** are hydrocarbons containing the **functional group** >C=C<. They have the general formula C_nH_{2n}. Alkenes are named in a similar way to alkanes but their names end in *ene*. For example, ethene is $CH_2{=}CH_2$ and propene is $CH_3CH{=}CH_2$.

Position isomers

Position isomerism occurs when compounds have the same molecular formula and functional group, but different structures because the functional group is on different parts of the same carbon chain. Alkenes with four or more carbon atoms may have position isomers. The position of the double bond is numbered. For example, but-1-ene and but-2-ene are position isomers of butene.

If the alkene is branched, you identify the longest chain containing the double bond. You name this first, then base the numbering of the side chains on that name.

```
   H   H   H   H   H   H
   |   |   |   |   |   |
H— C — C — C — C = C — C —H
   |   |   |           |
   H  CH3 CH3          H
```

This is 4,5-dimethylhex-2-ene rather than 2,3-dimethylhex-4-ene.

Haloalkanes

Haloalkanes are alkanes in which at least one hydrogen atom is replaced by a halogen atom. The halogen atom is the functional group. A haloalkane is named after the alkane on which it is based and the halogen atoms it contains. For example, CH_3F is fluoromethane, CH_3Cl is chloromethane, CH_3Br is bromomethane, and CH_3I is iodomethane.

Position isomers

Position isomers of haloalkanes are named in a similar way to chain isomers of alkanes. For example, $CH_3CH_2CH_2Br$ is 1-bromopropane and $CH_3CHBrCH_3$ is 2-bromopropane. If the haloalkane contains different halogen atoms, the halogens are named in alphabetical order. When the main chain is branched, it is numbered so that the halogen atom has the lowest possible number.

```
    H   F   H   H
    |   |   |   |
Cl— C — C — C — C —Br
    |   |   |   |
    H  CH3  Br  I
```

1,2-dibromo-4-chloro-3-fluoro-1-iodo-3-methylbutane.

Check your understanding

1 For each of the following terms, explain what they mean and give a suitable named example:

homologous series, molecular formula, structural formula, structural isomerism, chain isomerism, position isomerism

2 Draw the displayed formulae for the following compounds:

a 2,2-dimethylpropane
b 4-ethyl-2,2,4-trimethylhexane
c 2-bromopropane
d 1-bromo-2,3-dichloro-3-fluoro-2-iodobutane
e 2-chloro-2-methylpropane
f 1,3-dibromobut-1-ene

4.02 Stereoisomers

OUTCOMES

already from AS Level, you

- know that the alkenes can exhibit *E-Z* isomerism
- can draw the structures of *E* and *Z* isomers
- understand that *E-Z* isomers exist owing to restricted rotation about the C=C bond

and after this spread you should

- have familiarized yourself again with the key concepts of *E-Z* isomerism from *AS Level Chemistry*
- know that *E-Z* isomerism and optical isomerism are forms of stereoisomerism

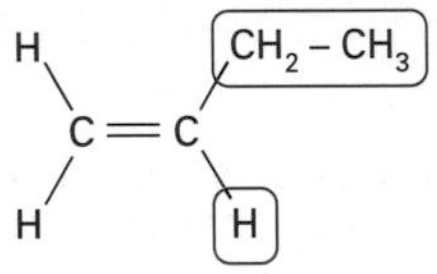

But-1-ene does not have E-Z *isomers.*

Structural isomers have the same molecular formula but different structural formulae. In contrast, **stereoisomers** have the same molecular formula and the same structural formula. The atoms in stereoisomers are arranged differently in space. Stereoisomers include ***E-Z*** **isomers** and **optical isomers**.

E-Z isomers

Atoms and groups are usually free to rotate around a C—C bond. But there is only restricted rotation around a C=C bond. This makes it possible for stereoisomers, called *E-Z* isomers, to exist in alkenes.

These two conditions must apply for a compound to have *E-Z* isomers:

- It must have a C=C bond.
- Both the carbon atoms involved in this bond must have different groups attached to them.

For example, butene exists as two position isomers, but-1-ene and but-2-ene. But-1-ene does not have *E-Z* isomers, because one of the carbon atoms involved in the C=C bond has two hydrogen atoms attached to it.

But-2-ene does have *E-Z* isomers. Both of the carbon atoms involved in the C=C bond have two different groups attached to them.

(*E*)-but-2-ene (*Z*)-but-2-ene

E-Z *isomers of but-2-ene.*

Naming *E–Z* isomers

The *E* comes from the German word *entgegen*, which means *opposite*. The *Z* comes from the German word *zusammen*, which means *together*. To decide whether an alkene is the *E* isomer or the *Z* isomer, you need to work out the priority of the groups attached to the carbon atoms in the double bond. Priority increases in the order $H < CH_3 < C_2H_5$. You then work out whether the highest priority group on each carbon atom is above or below the double bond:

- If one of the highest priority groups is above the double bond and the other is below, it is an *E* isomer.
- If both of the highest priority groups are above the double bond, or both are below, it is the *Z* isomer.

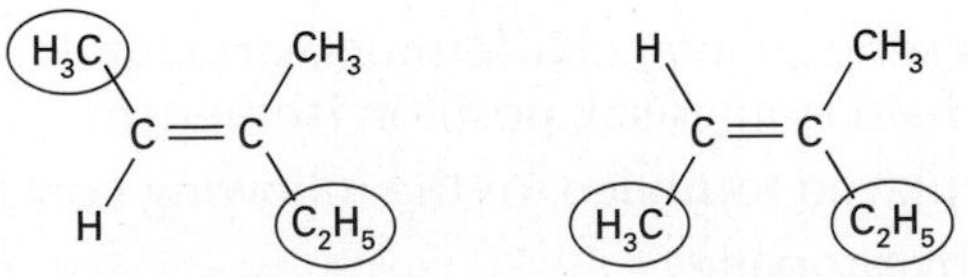

(*E*)-3-methylpent-2-ene (*Z*)-3-methylpent-2-ene

The two E-Z *isomers of 3-methylpent-2-ene. The highest priority groups on each side have been circled for clarity. The ones in the* E *isomer are on '**e**ither half' and the ones on the* Z *isomer are on the '**z**ame half' of the double bond.*

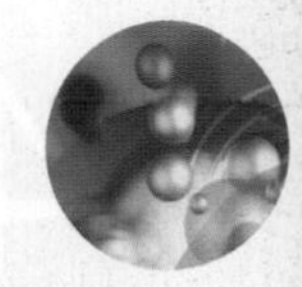

Haloalkenes

The order of priority in *E-Z* isomers is decided by the *Cahn–Ingold–Prelog priority rules*. Note that you do not need to know these for A Level. They identify priority according to the atomic numbers of the atoms, or groups of atoms, attached to the carbon atoms involved in the double bond. This allows haloalkenes to be named, too. For example, consider 1-bromo-1-chloro-2-fluoro-2-iodoethene. The bromine atom (Z = 35) takes priority over the chlorine atom (Z = 17) on C_1. The iodine atom (Z = 53) takes priority over the fluorine atom (Z = 9) on C_2.

(*E*)-1-bromo-1-chloro-2-fluoro-2-iodoethene

(*Z*)-1-bromo-1-chloro-2-fluoro-2-iodoethene

The two E-Z isomers of 1-bromo-1-chloro-2-fluoro-2-iodoethene. The highest priority groups on each side have been circled.

Trans fats

The carbon chains in unsaturated fats contain C=C bonds, so *E-Z* isomers are possible. But they are usually known by the older *cis–trans* naming system. The *cis* isomers are *Z* isomers and the *trans* isomers are *E* isomers.

Vegetable oils contain unsaturated fats but they are too runny to make margarine, cakes, and pastry. Food manufacturers can partially hydrogenate the vegetable oils by reacting them with hydrogen in the presence of a nickel catalyst. This converts some of the double bonds into single bonds. The oils become more saturated and their melting point increases. But the process also converts *cis* bonds into *trans* bonds, producing trans fats. These are associated with an increased risk of coronary heart disease, and many UK food manufacturers phased them out in 2007.

Trans fats were commonly used in processed foods such as these.

Optical isomers

Optical isomers are stereoisomers that exist as mirror images of each other. They do not need to have double bonds in them, unlike *E-Z* isomers. You will find out more about these stereoisomers in the next spread.

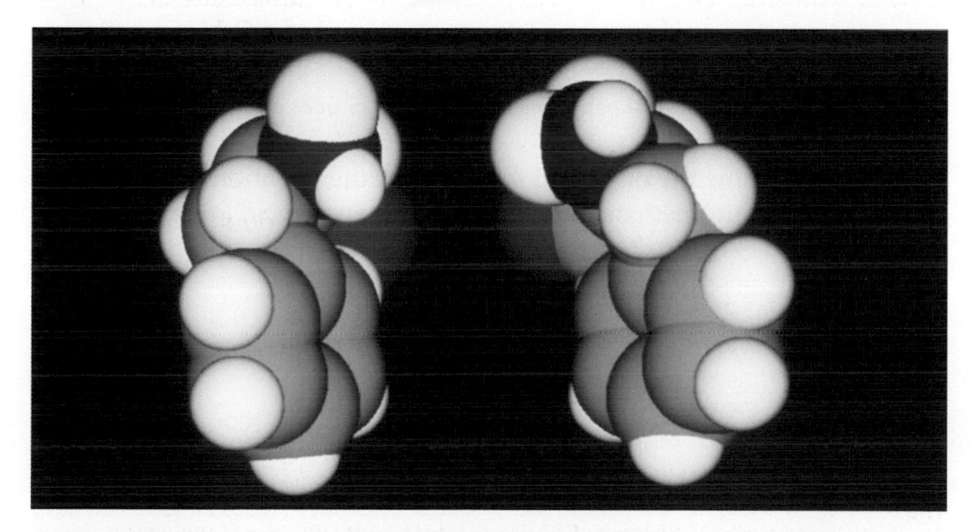

Computer graphics of the two optical isomers of phenylalanine, an amino acid.

Check your understanding

1 Explain why pent-2-ene has *E-Z* isomers but pent-1-ene does not.

2 Show the displayed formulae for the *E-Z* isomers of pent-2-ene.

3 Is this an *E* isomer or a *Z* isomer? Explain how you know.

H, CH_3, H_3C, CH_2CH_3 (C=C)

4.03 Optical isomers

OUTCOMES

already from A2 Level, you

- can draw the structural formulae and displayed formulae of isomers
- know that *E-Z* isomerism and optical isomerism are forms of stereoisomerism

and after this spread you should

- know that an asymmetric carbon atom is chiral and gives rise to optical isomers
- know that optical isomers exist as non-superimposable mirror images and differ only in their effect on plane-polarized light
- understand the meaning of the terms enantiomer and racemate
- understand why racemates are formed

Chiral centres

Human hands are chiral objects. An asymmetric carbon atom is a **chiral centre**, pronounced 'ky-ral'. This name comes from the Greek word for *hand*.

A pair of gloves comprises two gloves that are non-superimposable mirror images of each other.

Optical isomers occur when a compound contains a carbon atom with four different atoms, or groups of atoms, joined to it. This is an **asymmetric carbon atom**. Its presence means that the compound can have two optical isomers, called **enantiomers**. These are mirror images of each other, and they are **non-superimposable**. No matter how you turn the two enantiomers around, you cannot get all the atoms and groups attached to the asymmetric carbon atom to match up.

Enantiomers

Picture a carbon atom with four different atoms joined to it. For example, this happens in bromo(chloro)fluoro(iodo)methane. The diagram shows its two enantiomers. Notice that they are non-superimposable mirror images. You can match up the carbon atom and two of the halogen atoms, but not the other two halogen atoms.

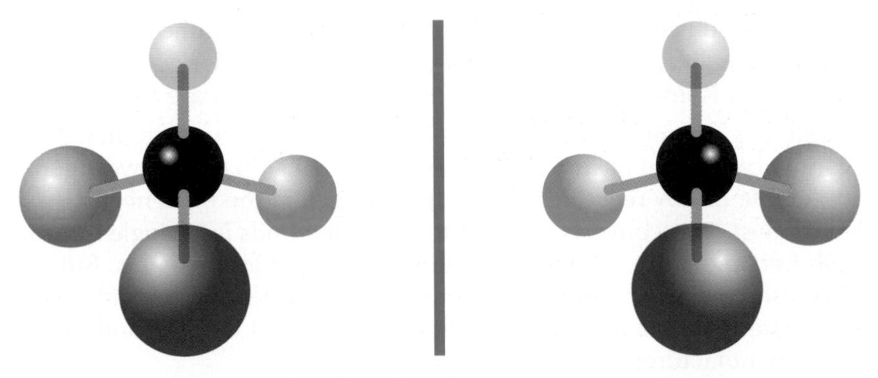

The enantiomers of bromo(chloro)fluoro(iodo)methane are mirror images.

Representing enantiomers

Enantiomers extend in three dimensions, so their bonds are drawn to give a sense of depth.

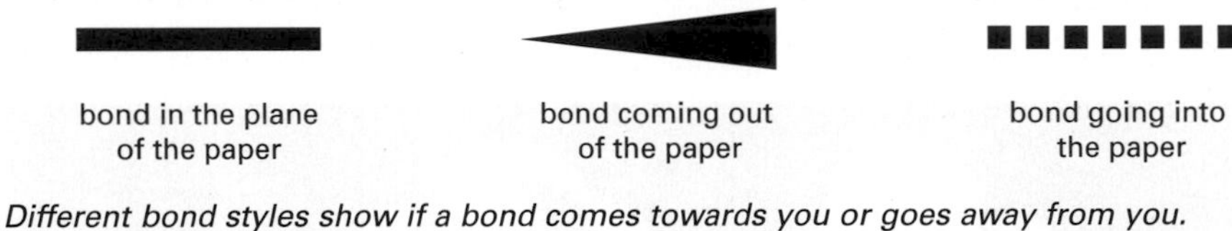

Different bond styles show if a bond comes towards you or goes away from you.

The diagrams show the displayed formulae of the enantiomers of bromo (chloro)fluoro(iodo)methane and 2-aminopropanoic acid (alanine, an amino acid).

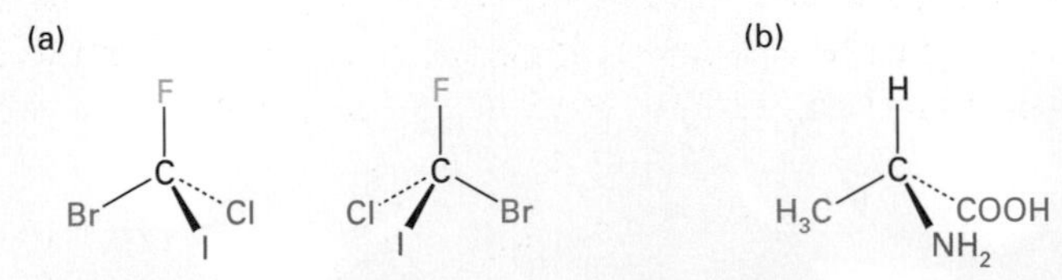

Enantiomers of (a) bromo(chloro)fluoro(iodo)methane and (b) alanine.

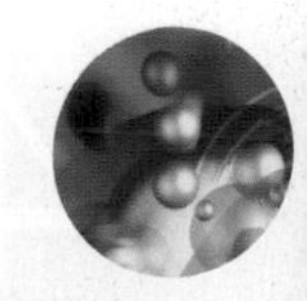

Showing optical activity

Enantiomers of a compound have identical chemical properties, unless they are reacting with another chiral compound. Optical activity is shown using a **polarimeter**.

Light waves can vibrate in an infinite number of directions or planes. Polaroid filters only let rays that vibrate in a particular plane to pass through. Sunglasses often have Polaroid filters to block the glare from bright sunlight. The light that emerges from a Polaroid filter is **plane-polarized light**. Optical isomers can rotate the plane of this light.

A simple polarimeter comprises a light source, two Polaroid filters, and a glass cylinder. Light passes upwards through the first filter, the cylinder, then through the second filter and into the observer's eye. The filter nearest the eye is rotated until the two filters are 'crossed' and no light emerges. A solution of one of the enantiomers is poured into the cylinder. Light emerges again because optical isomers can rotate the plane of plane-polarized light. The filter is then rotated again until light no longer emerges, and the angle needed is measured.

Naming enantiomers

Enantiomers can be named according to the direction in which they rotate plane-polarized light. One enantiomer rotates it in one direction, and the other enantiomer rotates it in the opposite direction. Looking through a polarimeter towards the light source, the enantiomer that rotates light anticlockwise, or to the left, is the **laevorotatory** or (−) enantiomer. The enantiomer that rotates light clockwise, or to the right, is the **dextrorotatory** or (+) enantiomer.

Racemates

A mixture containing equal amounts of the two enantiomers is called a **racemic mixture** or **racemate**. A racemate does not show optical activity, because the rotating effect of one enantiomer is cancelled by the opposite effect from the other enantiomer.

Racemates typically occur when chemicals are synthesized in the laboratory. The reaction between ethanal and hydrogen cyanide provides a good example.

$$CH_3-C(=O)H + HCN \longrightarrow CH_3-C(OH)(CN)-H$$

The carbonyl group >C=O in ethanal is planar. The cyanide ion CN^- can attack from above or below the molecule, so a racemate of 2-hydroxypropanenitrile is formed. If only the (−) or (+) enantiomer were formed, it would show that the cyanide ion could only attack in one direction.

Science @ Work: Factors affecting the angle of rotation

The angle of rotation for an enantiomer depends on its concentration, temperature, and depth, and on the wavelength of light used. The concentration of the enantiomer can be measured using a polarimeter. If the other factors are kept constant, the angle of rotation is proportional to the concentration.

Science @ Work: Other ways to name enantiomers

Enantiomers can also be named according to their structure. For example, D and L isomers are named by comparing them to 2,3-dihydroxypropanal (glyceraldehyde). *S* and *R* isomers are named following more of the Cahn–Ingold–Prelog priority rules.

Check your understanding

1. Explain how optical isomerism arises.
2. Explain the terms *enantiomer* and *racemate*.
3. Explain why 2-chlorobutane shows optical activity but 1-chlorobutane does not.
4. Show the displayed formulae for the two enantiomers of 2-hydroxypropanoic acid, $CH_3CH(OH)COOH$.

4.04 Stereochemistry and drugs

OUTCOMES

already from A2 Level, you

- know that an asymmetric carbon atom is chiral and gives rise to optical isomers
- know that optical isomers exist as non-superimposable mirror images and differ only in their effect on plane-polarized light
- understand the meaning of the terms enantiomer and racemate
- understand why racemates are formed

and after this spread you should

- appreciate that drug action may be determined by the stereochemistry of the molecule
- appreciate that different optical isomers may have very different effects

When a compound with a chiral centre is synthesized in the laboratory, the product is usually a racemate. But the same compound synthesized by living organisms consists of just one enantiomer. This is because only one enantiomer of a reactant can bind to an enzyme's active site, the place where an enzyme-catalyzed reaction happens. And enzymes can only make one enantiomer of a product. For example, (+)glucose can be digested but (−)glucose cannot. Some enantiomers can be distinguished by their taste or smell:

- One enantiomer of asparagine tastes sweet and the other tastes bitter.
- One enantiomer of carvone tastes of spearmint and the other of caraway.
- One enantiomer of limonene smells of oranges and the other of lemons.

The enantiomers of medical drugs can be the difference between a toxic substance and one with beneficial effects on the body. For example, D-penicillamine is used to treat rheumatoid arthritis, but L-penicillamine is toxic. When penicillamine was first introduced in the US, the synthetic racemate was used. It caused optic nerve damage and the drug was initially withdrawn. The penicillamine introduced in the UK was derived from penicillin. Only the D enantiomer was present and so the toxic effects were not seen.

```
        CH3  NH2    OH
        |    |     /
  HS —  C —  C — C
        |    |     \\
        CH3  H      O
```

Penicillamine. The chiral centre is coloured red.

Louis Pasteur (1822–1895)

The Pasteur Institute in Paris was opened in 1888. Today, it is one of the world's leading scientific research institutes.

Louis Pasteur was the French scientist who invented pasteurization, the process used to kill bacteria in milk so that it keeps longer. He also proved that micro-organisms cause disease, and he developed and used vaccines against rabies and anthrax. In 1848 Pasteur became the first person to isolate enantiomers.

Pasteur studied 2,3-dihydroxybutanedioic acid or tartaric acid. This is formed during the fermentation of grapes, and it can be synthesized in the laboratory. The natural form rotates the plane of polarized light but the synthetic form does not. Pasteur discovered that crystals of sodium ammonium tartrate had two mirror image shapes. He was able to pick out the two different types of crystal and dissolve them separately in water. The two solutions rotated the plane of polarized light in opposite directions. Pasteur converted them to tartaric acid and discovered that one form was dextrorotatory, like natural tartaric acid. The other form was laevorotatory and had not been described before.

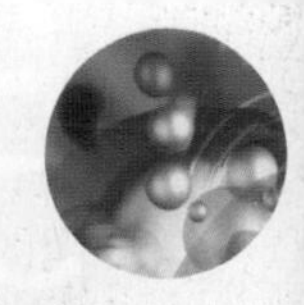

Thalidomide

In the late 1950s, thalidomide was introduced as a drug to treat morning sickness during pregnancy. But evidence emerged that it was teratogenic and so caused serious birth defects. Thousands of babies were born between 1956 and 1962 with deformities such as very short limbs.

Thalidomide is chiral and was available as the racemate. One enantiomer relieves morning sickness and is not teratogenic, but the other enantiomer is teratogenic. Unfortunately, even if the safe enantiomer had been synthesized, it would not have helped. This is because the body converts one enantiomer into the other.

Thalidomide was withdrawn from general use in 1962 and the regulations regarding drug testing were made more stringent. The compound is available today, but this time to treat leprosy. And it is not prescribed to pregnant women.

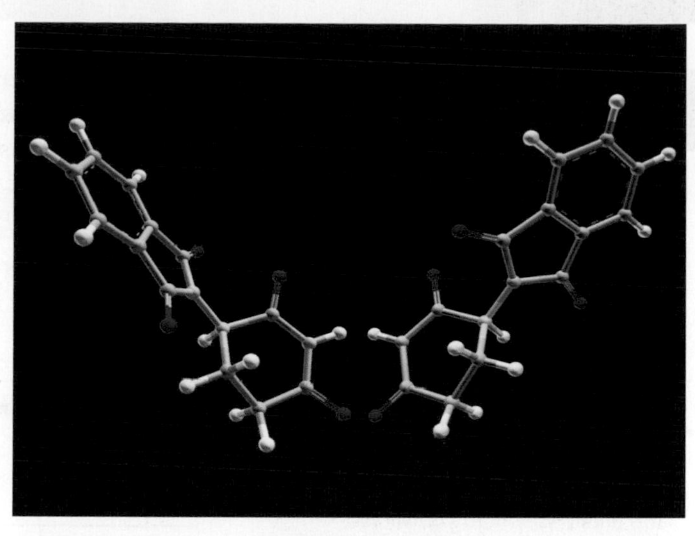

The two enantiomers of thalidomide. The chiral centre is coloured yellow.

Praziquantel

Schistosomiasis is a tropical disease caused by a parasitic worm. The parasite lives in the blood vessels of the small intestine, where it feeds on red blood cells. Over 200 million people are infected around the world. Schistosomiasis is spread through contact with contaminated water such as irrigation channels, rivers, and lakes. Praziquantel kills the worms, with few side effects for the patient. It is available as the racemate but only one enantiomer actually kills the worms. Research is being carried out to synthesize the active enantiomer at a cost that is competitive with the racemate.

Many drugs have undergone this 'chiral switch'. The use of single enantiomer drugs reduces the risk of side effects and lowers the dose needed. It may be easier for patients to take a drug if its tablets are smaller as a result.

A pair of Schistosoma mansoni *worms, about 1 cm long. The male is the larger of the two. The female can lay around 300 eggs a day.*

Check your understanding

1 Outline some effects on the body of different enantiomers.

2 Compare the three drugs penicillamine, thalidomide, and praziquantel. To what extent does the existence of enantiomers of these drugs cause problems for patients? How have chemists attempted to avoid these problems?

5.01 The carbonyl group

OUTCOMES

already from AS Level, you

- understand that primary alcohols can be oxidized to aldehydes and carboxylic acids
- understand that secondary alcohols can be oxidized to ketones
- can use a simple chemical test to distinguish between aldehydes and ketones (e.g. Fehling's solution or Tollens' reagent)

and after this spread you should

- be able to apply IUPAC rules for naming aldehydes, ketones, and carboxylic acids
- know that aldehydes are readily oxidized to carboxylic acids and that this forms the basis of a simple chemical test to distinguish between aldehydes and ketones (e.g. Fehling's solution and Tollens' reagent)

The **carbonyl** functional group is >C=O. It is found in a wide range of organic compounds, including **aldehydes**, **ketones**, and **carboxylic acids**. There are other carbonyl compounds, too, which you will study later.

compound	example		
aldehyde	propanal	CH_3CH_2CHO	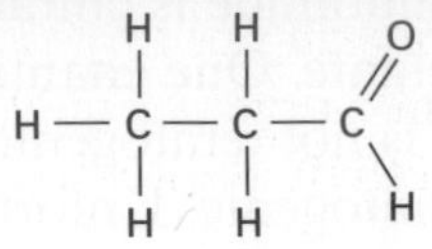
ketone	propanone	CH_3COCH_3	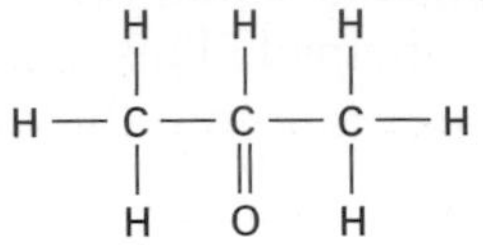
carboxylic acid	propanoic acid	CH_3CH_2COOH	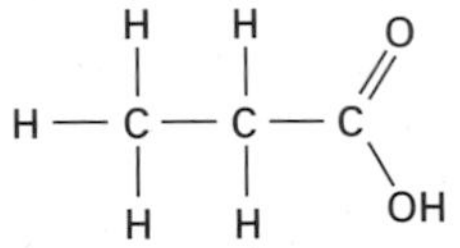
acyl chloride	propanoyl chloride	CH_3CH_2COCl	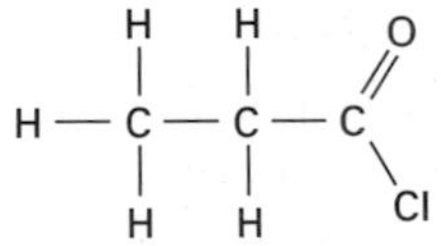
amide	propanamide	$CH_3CH_2CONH_2$	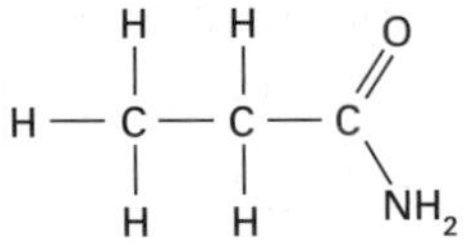
ester	methyl propanoate	$CH_3CH_2COOCH_3$	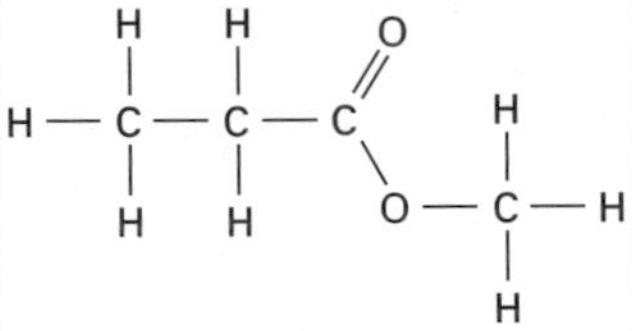
acid anhydride	propanoic anhydride	$(CH_3CH_2CO)_2O$	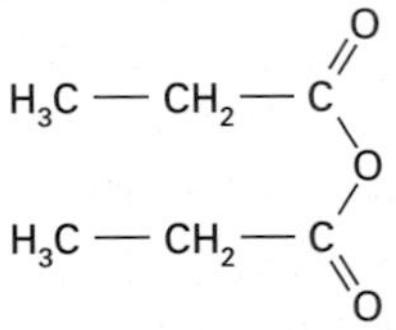

These compounds (above) all contain the carbonyl group, coloured red here.

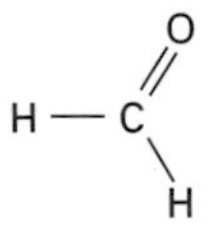

methanal 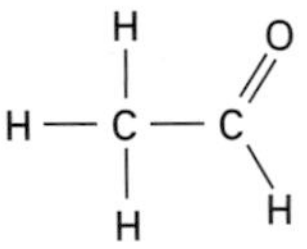 ethanal

The displayed formulae of methanal and ethanal. Methanal is also referred to by its older, non-systematic name of formaldehyde.

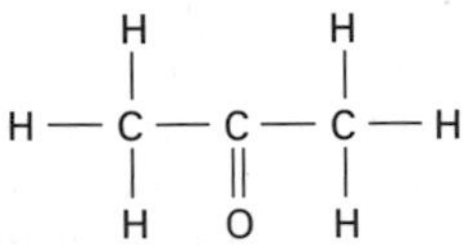

The displayed formula of propanone.

Naming aldehydes

Aldehydes are named after the number of carbon atoms they contain, including the carbon atom in the carbonyl group, and their names end in *al*. The carbonyl group in aldehydes is shown as –CHO in shortened structural formulae. Methanal is HCHO and ethanal is CH_3CHO.

Naming ketones

Ketones are named after the number of carbon atoms they contain, including the carbon atom in the carbonyl group, and their names end in *one*. The carbonyl group in ketones is shown as –CO in shortened structural formulae. Propanone is CH_3COCH_3 and butanone is $CH_3CH_2COCH_3$.

Position isomers

Ketones with five or more carbon atoms have position isomers. For example, pentanone has two position isomers. Pentan-2-one is $CH_3(CH_2)_2COCH_3$ and pentan-3-one is $CH_3CH_2COCH_2CH_3$.

pentan-2-one

pentan-3-one

The displayed formulae of pentan-2-one and pentan-3-one. The ends of the names rhyme with 'stone'.

Chain isomers

Ketones with five or more carbon atoms have chain isomers. You name the main chain containing the carbonyl group first, then base the numbering of any side chains on this name. For example, the branched chain isomer of pentanone, $CH_3CH(CH_3)COCH_3$, is called 3-methylbutan-2-one and not 2-methylbutan-3-one.

Naming carboxylic acids

The carbonyl group in carboxylic acids is attached to a carbon atom that also has a **hydroxyl group**, –OH. It forms a new functional group, the **carboxyl group**, which is shown as –COOH. Carboxylic acids are named after the number of carbon atoms they contain, including the carbon atom in the carboxyl group, and their names end in *oic acid*. Methanoic acid is HCOOH and ethanoic acid is CH_3COOH.

methanoic acid

ethanoic acid

The displayed formulae of methanoic acid and ethanoic acid.

Distinguishing between aldehydes and ketones

Aldehydes are readily **oxidized** to carboxylic acids but ketones resist oxidation. For example, ethanal is oxidized to form ethanoic acid:

$CH_3CHO + [O] \rightarrow CH_3COOH$ [O] represents the **oxidizing agent**

This observation is the basis of two simple laboratory tests to distinguish between aldehydes and ketones.

Tollens' reagent

Aldehydes produce a *silver mirror* with **Tollens' reagent** but ketones do not. The test is carried out in this way. A small volume of aqueous sodium hydroxide is added to some aqueous silver nitrate. A precipitate forms, which is changed to aqueous $[Ag(NH_3)_2]^+$ ions by adding some aqueous ammonia. The test substance is added and the mixture is warmed. Aldehydes reduce the $[Ag(NH_3)_2]^+$ ions to silver, and become oxidized to carboxylic acids in the process.

Fehling's solution

Fehling's solution is a blue solution containing a complex of Cu^{2+} ions. When heated with Fehling's solution, aldehydes reduce it to form a brick-red precipitate of copper(I) oxide, Cu_2O. Ketones do not react with Fehling's solution, so no change is observed.

test substance	Tollens' reagent	Fehling's solution
aldehyde	silver mirror forms	blue solution changes to red precipitate
ketone	no change	no change

A summary of the changes seen with Tollens' reagent and Fehling's solution.

Propanone is also referred to by its older, non-systematic name of acetone. It is a solvent in nail polish remover.

Check your understanding

1 Name these compounds:
- a $CH_3(CH_2)_4CHO$
- b $CH_3CO(CH_2)_3CH_3$
- c $CH_3(CH_2)_2COOH$
- d $CH_3CH(CH_3)CH_2CHO$
- e $CH_3COCH_2CH(CH_3)_2$

2 a Describe a simple laboratory test to distinguish between propanal and propanone.
- b Using [O] to represent the oxidizing agent, write an equation to show the oxidation of 3-methylbutanal. Name the product formed.

5.02 Alcohols from aldehydes and ketones

OUTCOMES

already from AS Level, you understand

- that primary alcohols can be oxidized to aldehydes
- that secondary alcohols can be oxidized to ketones
- that haloalkanes are susceptible to nucleophilic attack
- the mechanism of nucleophilic substitution in primary haloalkanes

and after this spread you should

- know that aldehydes can be reduced to primary alcohols, and ketones to secondary alcohols, using reducing agents such as $NaBH_4$
- be able to write equations for the reduction of aldehydes and ketones, showing [H] as the reducing agent
- understand the mechanism of the nucleophilic addition reaction involving aldehydes and ketones, and H^-

Nucleophiles

A nucleophile is a species with a lone pair of electrons available to form a co-ordinate bond. It is attracted to regions of positive charge, such as the electron-deficient carbon atom in the carbonyl group.

Sodium tetrahydridoborate. It is a white crystalline solid that is used in aqueous solution or ethanolic solution.

You discovered in Unit 2 of your AS Chemistry studies that **primary alcohols** can be oxidized to aldehydes, and **secondary alcohols** to ketones. The oxidizing agent or **oxidant** is often potassium dichromate(VI), $K_2Cr_2O_7$, acidified with sulfuric acid. It can be represented as [O] in equations:

$$\underset{\text{propan-1-ol}}{CH_3CH_2CH_2OH} + [O] \rightarrow \underset{\text{propanal}}{CH_3CH_2CHO} + H_2O$$

$$\underset{\text{propan-2-ol}}{CH_3CH(OH)CH_3} + [O] \rightarrow \underset{\text{propanone}}{CH_3COCH_3} + H_2O$$

The colour changes during the reaction from orange to green as dichromate(VI) ions are reduced to blue-green chromium(III) ions.

The process can be reversed, with aldehydes being reduced to primary alcohols, and ketones being reduced to secondary alcohols.

Oxidation and reduction

There are several ways to describe oxidation and reduction, shown in the table.

	oxidation	reduction
electrons	lost	gained
oxygen atoms	gained	lost
hydrogen atoms	lost	gained

Reducing aldehydes and ketones

Sodium tetrahydridoborate $NaBH_4$ is commonly used to reduce carbonyl compounds such as aldehydes and ketones. Lithium tetrahydridoaluminate $LiAlH_4$ is a more powerful **reducing agent** or **reductant** that may be used instead. Both reductants produce the hydride ion H^-. This acts as a **nucleophile** and attacks the carbon atom in the carbonyl group of aldehydes and ketones.

The reductant is shown as [H] in equations, just as an oxidant is shown as [O].

Aldehydes and primary alcohols

This is the general equation for the reduction of aldehydes.

$$RCHO + 2[H] \rightarrow RCH_2OH$$

For example, propanal is reduced to propan-1-ol, a primary alcohol:

$$CH_3CH_2CHO + 2[H] \rightarrow CH_3CH_2CH_2OH$$

Ketones and secondary alcohols

This is the general equation for the reduction of ketones.

$$R_1COR_2 + 2[H] \rightarrow R_1CH(OH)R_2$$

For example, propanone is reduced to propan-2-ol, a secondary alcohol:

$$CH_3COCH_3 + 2[H] \rightarrow CH_3CH(OH)CH_3$$

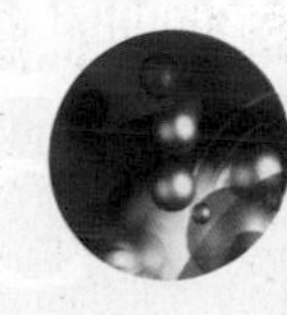

Nucleophilic addition

Oxygen is more **electronegative** than carbon, so the C=O bond is **polar**. The carbon atom has a partial positive charge δ+ and the oxygen atom has a partial negative charge δ−. The δ+ carbon atom can be attacked by nucleophiles such as the hydride ion H^-. The reaction is a **nucleophilic addition reaction** in which the nucleophile forms a co-ordinate bond with the carbon atom, and the C=O bond becomes a C—O bond. Note that the nucleophilic reactions you met in Unit 2 were nucleophilic substitution reactions in haloalkanes.

Herc is the general reaction mechanism for the reduction of aldehydes and ketones. Remember that a curly arrow shows the movement of a pair of electrons, such as a lonc pair of electrons or the pair of electrons in a covalent bond.

In the first step, the hydride ion is attracted to thc electron-deficient carbon atom and forms a co-ordinate bond. Notice that you should draw a lone pair of electrons in the nucleophile :H⁻. The curly arrow is drawn from one side of the lone pair of electrons and points towards the electron-deficient carbon atom.

In the second step, the intermediatc produced in the first stcp gains a proton from water, the solvent.

The complete mechanism is drawn like this. In this example, the carbonyl compound is an aldehyde, propanal.

propanal → propan-1-ol

The complete reaction mechanism for the reduction of propanal, with ticks to show the features an examiner is likely to look for in an examination.

Here is the complete mechanism for the reduction of a ketone, propanone.

propanone → propan-2-ol

The complete reaction mechanism for the reduction of propanone.

Check your understanding

1 a What is a nucleophile?
  b Explain why carbonyl compounds are susceptible to nucleophilic attack.

2 Write equations for the reduction of the following compounds, showing the reducing agent as [H]. In each case, name the product and state whether it is a primary alcohol or a secondary alcohol.
  a methanal
  b butanal
  c butanone

3 Draw the mechanism for the reaction that occurs when ethanal is reduced by sodium tetrahydridoborate.

5.03 Hydroxynitriles from aldehydes and ketones

OUTCOMES

already from A2 Level, you

- understand the mechanism of the nucleophilic addition reaction involving aldehydes and ketones, and H^-
- know that an asymmetric carbon atom is chiral and gives rise to optical isomers

and after this spread you should

- understand the mechanism of the reaction of carbonyl compounds with HCN as a further example of nucleophilic addition, producing hydroxynitriles
- appreciate the hazards of synthesis using HCN or KCN

Naming nitriles and hydroxynitriles

Nitriles are named after the number of carbon atoms they contain, including the carbon atom in the nitrile group, and their names end in *nitrile*. The nitrile group is shown as –CN in shortened structural formulae. Notice that the letter e is included in the name, so it is propanenitrile, not propanitrile or propannitrile.

Hydroxynitriles are first named after the main chain containing the nitrile group. The hydroxyl group –OH is then indicated in the name as *hydroxy*. If position isomers are possible, a number shows the carbon atom to which the hydroxyl group is attached.

Aldehydes and ketones react with hydrogen cyanide, HCN, to produce **hydroxynitriles**. For example:

$$\underset{\text{ethanal}}{CH_3CHO} + HCN \rightarrow \underset{\text{2-hydroxypropanenitrile}}{CH_3CH(OH)CN}$$

The mechanism is another example of nucleophilic addition, and the nucleophile is the cyanide ion CN^-. The production of hydroxynitriles from aldehydes and ketones is interesting because an extra carbon atom is added to the carbon chain during the reaction, and hydroxynitriles may have optical isomers.

Nucleophilic addition

Here is the general reaction mechanism for the reaction of aldehydes and ketones with hydrogen cyanide. It is identical to the reaction mechanism for the reduction of aldehydes and ketones, except that the nucleophile is CN^- rather than H^-.

In the first step, the cyanide ion is attracted to the electron-deficient carbon atom and forms a co-ordinate bond.

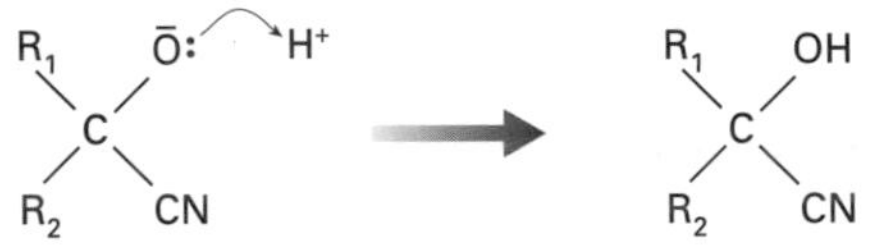

In the second step, the intermediate produced in the first step gains a proton from the solvent or hydrogen cyanide.

The complete mechanism is drawn like this. In this example, the carbonyl compound is an aldehyde, ethanal.

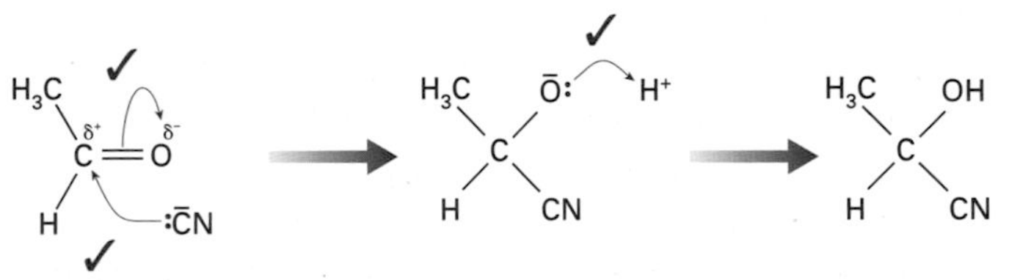

ethanal

2-hydroxypropanenitrile

The complete reaction mechanism for the reaction of ethanal with hydrogen cyanide, with ticks to show the features an examiner is likely to look for in an examination.

Toxic cyanide

Hydrogen cyanide, HCN, is a colourless gas with a faint smell like almonds. Potassium cyanide, KCN, is a white powder. It also smells like almonds, because it reacts with moisture in the air to release hydrogen cyanide gas.

$$KCN + H_2O \rightarrow HCN + KOH$$

Hydrogen cyanide and its salts are toxic by swallowing or inhalation. Cyanide inhibits reactions needed for cells to respire. People exposed to cyanide can suffer convulsions, loss of consciousness, and respiratory failure leading to death. Survivors of serious cyanide poisoning may develop heart and brain damage.

An aqueous solution of potassium cyanide with some sulfuric acid is usually used in reactions instead of hydrogen cyanide. The solution is adjusted to around pH 4 before use. CLEAPSS® is the organization that advises local authorities and schools on everything to do with practical science. In its *Hazcards®*, CLEAPSS does not advise the use of hydrogen cyanide or potassium cyanide in schools. Any proposed use would require a special risk assessment.

The most infamous use of hydrogen cyanide was by the Nazis during World War II. Zyklon B, *an insecticide, releases the gas on contact with the air. It was used to murder millions of people in concentration camps, including this one at Auschwitz in Poland.*

Hydrolysis of the nitrile group

One more carbon atom is added to the carbon chain when hydrogen cyanide reacts with an aldehyde or ketone. For example, ethanal with two carbon atoms becomes 2-hydroxypropanenitrile with three carbon atoms.

```
        OH
        |
  H  —  C  —  CN
        |
        H
```

Hydroxyethanenitrile does not have an asymmetric carbon atom.

The nitrile group is hydrolyzed in acidic conditions to become a carboxyl group. For example:

$$CH_3CH(OH)CN + 2H_2O + H^+ \rightarrow CH_3CH(OH)COOH + NH_4^+$$

2-hydroxypropanenitrile → 2-hydroxypropanoic acid

Optical isomers

Ethanal reacts with hydrogen cyanide to form 2-hydroxypropanenitrile. This has an asymmetric carbon atom, so it has optical isomers. 2-hydroxypropanenitrile can be hydrolyzed to form 2-hydroxypropanoic acid (lactic acid), which also has an asymmetric carbon atom. Note that not all hydroxynitriles have optical isomers. Methanal reacts with hydrogen cyanide to produce hydroxyethanenitrile, which does not have an asymmetric carbon atom.

```
        OH                      OH
        |                       |
 H3C  — C — CN           H3C  — C — COOH
        |                       |
        H                       H
```

2-hydroxypropanenitrile　　2-hydroxypropanoic acid

These two compounds each have an asymmetric carbon atom.

Check your understanding

1 What are the hazards of using hydrogen cyanide and potassium cyanide in laboratory syntheses?

2 a Draw the mechanism for the reaction between hydrogen cyanide and propanone.
  b Name the product of the reaction.
  c Does the product show optical isomerism? Explain your answer.

3 a Draw the mechanism for the reaction between hydrogen cyanide and butanone.
  b Name the product of the reaction.
  c Does the product show optical isomerism? Explain your answer.

5.04 Esters from carboxylic acids and alcohols

OUTCOMES

already from A2 Level, you

- know that aldehydes are readily oxidized to carboxylic acids
- can perform calculations relating the pH of a weak acid to the dissociation constant, K_a

and after this spread you should know that

- carboxylic acids are weak acids but will liberate carbon dioxide from carbonates
- carboxylic acids and alcohols react, in the presence of a strong acid catalyst, to give esters

Ethanoic acid reacts with sodium carbonate, releasing carbon dioxide gas.

Carboxylic acids contain the carboxyl group $-COOH$. They are synthesized in the laboratory by oxidizing primary alcohols or aldehydes using acidified potassium dichromate(VI). For example:

$$\underset{\text{ethanol}}{CH_3CH_2OH} + [O] \rightarrow \underset{\text{ethanal}}{CH_3CHO} + H_2O$$

In excess oxidizing agent, the aldehyde is oxidized further:

$$\underset{\text{ethanal}}{CH_3CHO} + [O] \rightarrow \underset{\text{ethanoic acid}}{CH_3COOH}$$

The overall reaction is:

$$\underset{\text{ethanol}}{CH_3CH_2OH} + 2[O] \rightarrow \underset{\text{ethanoic acid}}{CH_3COOH} + H_2O$$

Weak acids

Carboxylic acids are weak acids. The strength of an acid can be shown by its pK_a value: the stronger the acid, the lower its pK_a. The pK_a of ethanoic acid is 4.8 at 25°C. The pH of 0.1 mol dm^{-3} ethanoic acid is 2.9 but the pH of 0.1 mol dm^{-3} hydrochloric acid is 1. Carboxylic acids react with carbonates to release carbon dioxide gas.

$$\underset{\text{ethanoic acid}}{2CH_3COOH(aq)} + Na_2CO_3(s) \rightarrow \underset{\text{sodium ethanoate}}{2CH_3COO^-Na^+(aq)} + H_2O(l) + CO_2(g)$$

This gives a simple test for the presence of carboxylic acids. Fizzing or **effervescence** is seen when a carboxylic acid is added to sodium carbonate. For example, this can reveal if an aldehyde, produced when an alcohol is oxidized, has oxidized further to form a carboxylic acid.

Carboxylate ions

Carboxylic acids form **carboxylate ions** when they ionize. For example, ethanoic acid CH_3COOH forms ethanoate ions CH_3COO^-. The negative charge is **delocalized** so that both C—O bonds are equivalent. The formula for a carboxylate ion may be written as RCO_2^- rather than $RCOO^-$ to show this.

The length of the C—O bond in the ethanoate ion is halfway between the lengths of the C—O and C=O bonds in ethanoic acid.

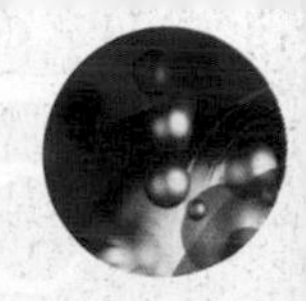

Making esters

Carboxylic acids react with alcohols to produce **esters**. Concentrated sulfuric acid or hydrochloric acid is added as a catalyst. Here is the general equation for the reaction, called **esterification**:

$$R_1COOH + R_2OH \rightarrow R_1COOR_2 + H_2O$$

The formulae are coloured so you can see which parts of the carboxylic acid and alcohol contribute to the ester.

Different combinations of carboxylic acid and alcohol produce different esters:

$HCOOH$	+	CH_3CH_2OH	→	$HCOOCH_2CH_3$	+	H_2O
methanoic acid	+	ethanol	→	ethyl methanoate	+	water
CH_3COOH	+	CH_3OH	→	CH_3COOCH_3	+	H_2O
ethanoic acid	+	methanol	→	methyl ethanoate	+	water

Ethanoic acid and ethanol react to produce ethyl ethanoate and water. In this diagram, the atoms that produce the water are coloured.

Naming esters

Esters are named after the alcohol and carboxylic acid that would form them. The first part of the name comes from the alcohol. For example, *methyl* comes from methanol and *ethyl* from ethanol. The second part of the name comes from the carboxylic acid. For example, *methanoate* comes from methanoic acid and *ethanoate* from ethanoic acid. You can recognize the carboxylic acid part from the structural formula, as it is the part that includes COO. Here are two examples:

- $HCOOCH_2CH_3$ is ethyl methanoate.

- CH_3COOCH_3 is methyl ethanoate.

Rhubarb, rhubarb

Rhubarb stalks make tasty fillings for fruit pies, but rhubarb leaves contain potentially hazardous concentrations of oxalic acid. This is a toxic substance that causes stomach irritation and kidney problems in high doses. Oxalic acid is ethanedioic acid. It contains two carboxyl groups and is stronger than ethanoic acid. It reacts with calcium in the teeth and causes the 'rough' feeling when you eat rhubarb.

ethanedioic acid

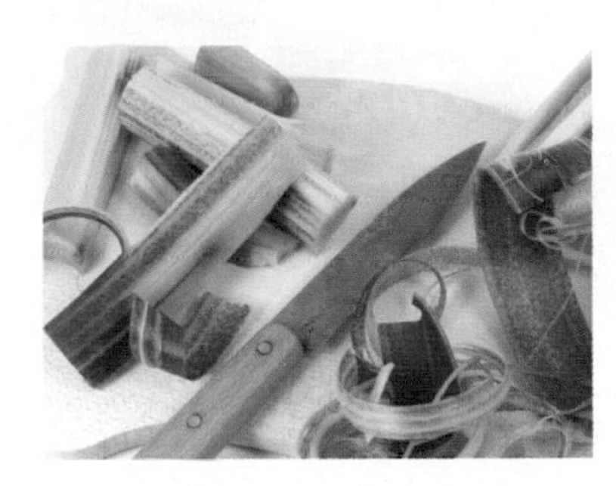

Rhubarb contains ethanedioic acid.

Check your understanding

1 For each of the following esters, name the carboxylic acids and alcohols needed to make them, write an equation for the esterification reaction involved, and draw their displayed formulae.

- **a** ethyl propanoate
- **b** butyl methanoate
- **c** methylethyl ethanoate
- **d** methyl 2-methylbutanoate

5.05 Uses of esters

OUTCOMES

already from A2 Level, you know

- that carboxylic acids and alcohols react, in the presence of a strong acid catalyst, to give esters

and after this spread you should know

- that esters can have pleasant smells
- the common uses of esters (e.g. in solvents, plasticizers, perfumes, and food flavourings)

Ethyl ethanoate is the solvent in the glue for modelling kits.

Taste and smell

Around three-quarters of our sensation of taste actually comes from the smell or aroma of food. Many esters are volatile. They easily evaporate in the mouth and reach the olfactory receptor cells in the nose. This is why food does not taste very good when you have a cold with a blocked nose.

Professional cooks prefer to use natural vanilla in their food.

Esters are used as solvents for varnishes, perfumes, and glues. For example, polystyrene cement, used for gluing modelling kits together, consists of polystyrene dissolved in ethyl ethanoate. The ester evaporates to leave solid polystyrene behind, which joins the two parts. If you have ever made a model aircraft, you may have discovered that the cement also dissolves the surface of model parts if it is spilled.

Esters have pleasant 'fruity' smells, and are found naturally in fruits, nuts, and flowers. They are also used as flavourings in food.

Food flavourings

Different esters have different distinctive smells and flavours. The table shows some esters and their flavours, but there are many others. Mixtures of different esters can produce more complex smells and flavours.

ester	flavour
butyl butanoate	pineapple
ethyl methanoate	raspberry
ethyl pentanoate	apple
3-methylbutyl ethanoate	pear
pentyl butanoate	strawberry

Vanilla

Vanilla is a flavouring used in foods such as ice cream, yogurt, and chocolate. Natural vanilla flavouring is an extract from the vanilla bean. It contains many chemicals that contribute to its flavour, but methyl vanillin, 'vanillin', is the main one. Natural vanilla is extracted using a sequence of processes, which takes several months to complete. Vanillin is also synthesized artificially from 2-methoxyphenol. This can be extracted from lignin, a waste product of the paper industry, or synthesized from crude oil products.

2-methoxyphenol | methyl vanillin | ethyl vanillin

Methyl vanillin is synthesized from 2-methoxyphenol. Ethyl vanillin is more expensive to synthesize but has a stronger flavour.

The cost of manufacturing esters is much less than the cost of extracting natural flavours. For example, artificial vanilla flavour is several times cheaper than natural vanilla. But if you have very sensitive taste buds, you may be able to tell the difference between natural and artificial flavours.

Plasticizers

Polymers are useful materials but the polymer itself may not have all the desired properties for its intended use. **Plasticizers** are added to polymers to let the polymer molecules slide over each other more easily. This makes the polymer softer and more flexible. The majority of plasticizers are esters, in particular phthalate esters. For example, di-2-ethylhexyl phthalate (DEHP) accounts for around 20% of the plasticizer used in Europe.

Poly(chloroethene) or PVC is a hard, tough polymer. Unplasticized PVC or uPVC is used to make window frames, pipes, and gutters. PVC containing plasticizers is soft and flexible. It is suitable for raincoats, floor coverings, car dashboards, and electrical insulation. Over time, the plasticizer may gradually evaporate, causing the polymer to become hard and inflexible as it ages.

Phthalates and toys

Animal studies have shown that high doses of phthalate plasticizers cause liver damage and birth defects. Breakdown products of phthalates are also found in the urine of people exposed to these plasticizers. Scientists carried out a risk assessment for the European Union. It was decided that phthalate plasticizers do not pose a risk to health. But the European Parliament took a cautious view and banned phthalates from children's toys, especially ones that are intended to be chewed. Phthalates continue to be used in medical applications, such as tubes and bags, because it was decided that any possible risk to health was outweighed by their advantages.

The European Union banned phthalate plasticizers from children's teething rings and other plastic toys meant to go in the mouth.

Check your understanding

1 Apart from food flavourings, describe two uses of esters. Give an example of each use.

2 Suggest an advantage and a disadvantage of using esters for artificial flavours.

3 a Outline how you could make an ester with the aroma of strawberries.

b How would you modify your method to make an ester with the aroma of pineapple instead?

5.06 Oils, fats, and biodiesel

OUTCOMES

already from A2 Level, you know

- that carboxylic acids and alcohols react, in the presence of a strong acid catalyst, to give esters

and after this spread you should

- know that esters can be hydrolyzed
- know that vegetable oils and animal fats are esters of propane-1,2,3-triol (glycerol)
- understand that vegetable oils and animal fats can be hydrolyzed to give soap, glycerol, and long-chain carboxylic (fatty) acids
- know that biodiesel is a mixture of methyl esters of long-chain carboxylic acids

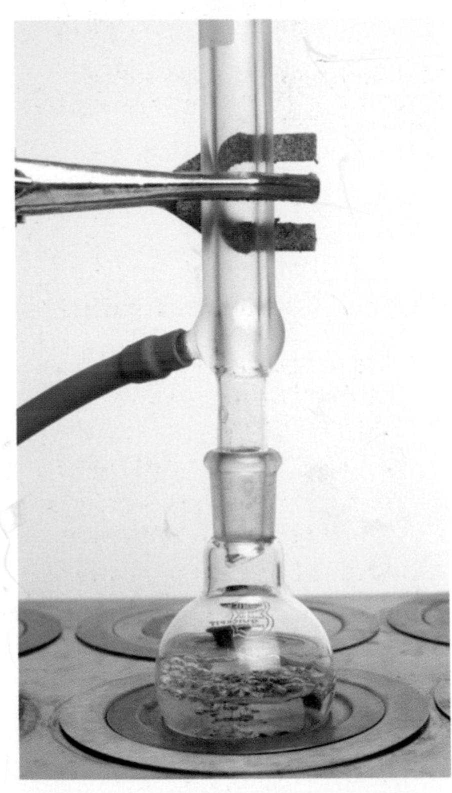

When reactants are heated under reflux conditions, their vapours continually cool in the condenser and drip back into the reaction mixture. This allows the reaction to continue at high temperatures for a long time.

Carboxylic acids react with alcohols to produce esters and water. For example:

$$\underset{\text{ethanoic acid}}{CH_3COOH(aq)} + \underset{\text{methanol}}{CH_3OH(aq)} \rightarrow \underset{\text{methyl ethanoate}}{CH_3COOCH_3(aq)} + \underset{\text{water}}{H_2O(l)}$$

This process can be reversed in a reaction called **ester hydrolysis**. The products of ester hydrolysis are useful for manufacturing soap.

Ester hydrolysis

Esters can be hydrolyzed in acidic or basic solutions. An alcohol is produced whichever method is used, but only acid-catalysed hydrolysis produces a carboxylic acid. Base-catalysed hydrolysis produces a salt of the carboxylic acid instead.

Acid-catalysed hydrolysis

Esters are hydrolyzed by heating them under reflux conditions in the presence of dilute hydrochloric acid or dilute sulfuric acid. Here is the general equation for the reaction:

$$R_1COOR_2(aq) + H_2O(l) \rightarrow R_1COOH(aq) + R_2OH(aq)$$

For example:

$$\underset{\text{methyl ethanoate}}{CH_3COOCH_3(aq)} + \underset{\text{water}}{H_2O(l)} \rightarrow \underset{\text{ethanoic acid}}{CH_3COOH(aq)} + \underset{\text{methanol}}{CH_3OH(aq)}$$

Base-catalysed hydrolysis

Esters are hydrolyzed by heating them under reflux conditions in the presence of a dilute aqueous strong base, such as sodium hydroxide. Here is the general equation for the reaction:

$$R_1COOR_2(aq) + NaOH(aq) \rightarrow R_1COO^-Na^+(aq) + R_2OH(aq)$$

For example:

$$\underset{\text{methyl ethanoate}}{CH_3COOCH_3(aq)} + NaOH(aq) \rightarrow \underset{\text{sodium ethanoate}}{CH_3COO^-Na^+(aq)} + \underset{\text{methanol}}{CH_3OH(aq)}$$

Base-catalysed ester hydrolysis is the method of choice because it is quicker than acid-catalysed ester hydrolysis.

Carboxylic acids from their salts

Base-catalysed ester hydrolysis produces salts of carboxylic acids. These can be heated with excess dilute hydrochloric acid or dilute sulfuric acid to produce the carboxylic acids themselves. For example:

$$\underset{\text{sodium ethanoate}}{CH_3COO^-Na^+(aq)} + HCl(aq) \rightarrow \underset{\text{ethanoic acid}}{CH_3COOH(aq)} + NaCl(aq)$$

Vegetable oils and animal fats

Vegetable oils and animal fats are **triacylglycerols**, which are often called **triglycerides**. They are **triesters** of

- propane-1,2,3-triol (glycerol)
- long-chain carboxylic acids (fatty acids)

The alkyl chains in vegetable oils are usually unsaturated, and the alkyl chains in animal fats are usually saturated.

Triacylglycerols are hydrolysed by heating them with aqueous sodium hydroxide or aqueous potassium hydroxide. This produces glycerol, a colourless viscous liquid with a sweet taste. It is an ingredient of processed foods, cosmetics, and toothpaste. Hydrolysis of triacylglycerols also produces the sodium salts or potassium salts of long-chain carboxylic acids, used in soap.

Ester hydrolysis of oils and fats is also called **saponification**, after the Latin word for soap. Sodium hydroxide produces hard soap, and potassium hydroxide produces soft soap and liquid soap.

$$\underset{\text{triacylglycerol}}{\begin{array}{l} H-\overset{H}{\overset{|}{C}}-O-\overset{O}{\overset{\|}{C}}-R \\ H-C-O-\overset{O}{\overset{\|}{C}}-R \\ H-\underset{H}{\underset{|}{C}}-O-\overset{O}{\overset{\|}{C}}-R \end{array}} + 3NaOH \longrightarrow \underset{\text{propane-1,2,3-triol}}{\begin{array}{l} H-\overset{H}{\overset{|}{C}}-OH \\ H-C-OH \\ H-\underset{H}{\underset{|}{C}}-OH \end{array}} + \underset{\text{sodium carboxylate}}{3\,R-\overset{O}{\overset{\|}{C}}-O^-Na^+}$$

A general saponification reaction using sodium hydroxide. Note that for each mole of triacylglycerol, three moles of sodium hydroxide are needed and three moles of sodium carboxylate are produced. The group labelled R represents the alkyl chain. These groups can be identical or different from each other.

Biodiesel

Biodiesel is a fuel usually made from vegetable oils, but it can also be made from animal fats and waste oil. Most biodiesel is made by heating vegetable oils with methanol and sodium hydroxide. A reaction called **base-catalysed transesterification** takes place. The products are a mixture of methyl esters of long-chain carboxylic acids, and propane-1,2,3-triol. The biodiesel forms a layer on top of the denser propane-1,2,3-triol. The oil from rapeseed is reacted with methanol to produce rape methyl ester or RME, a biodiesel with properties almost identical to diesel from crude oil.

$$\underset{\text{triacylglycerol}}{\begin{array}{l} H-\overset{H}{\overset{|}{C}}-O-\overset{O}{\overset{\|}{C}}-R \\ H-C-O-\overset{O}{\overset{\|}{C}}-R \\ H-\underset{H}{\underset{|}{C}}-O-\overset{O}{\overset{\|}{C}}-R \end{array}} + \underset{\text{methanol}}{3\,H_3C-OH} \longrightarrow \underset{\text{propane-1,2,3-triol}}{\begin{array}{l} H-\overset{H}{\overset{|}{C}}-OH \\ H-C-OH \\ H-\underset{H}{\underset{|}{C}}-OH \end{array}} + \underset{\text{methyl ester}}{3\,R-\overset{O}{\overset{\|}{C}}-O-CH_3}$$

Transesterification of vegetable oils with methanol produces a mixture of methyl esters of long-chain carboxylic acids, used as biodiesel.

Science @ Work

How soaps work

The alkyl chain or 'tail' of a soap molecule is non-polar, so it dissolves in oil and grease. The 'head' is charged, so it dissolves in water. Soap molecules surround particles of oil and grease, and carry them away in water during cleaning.

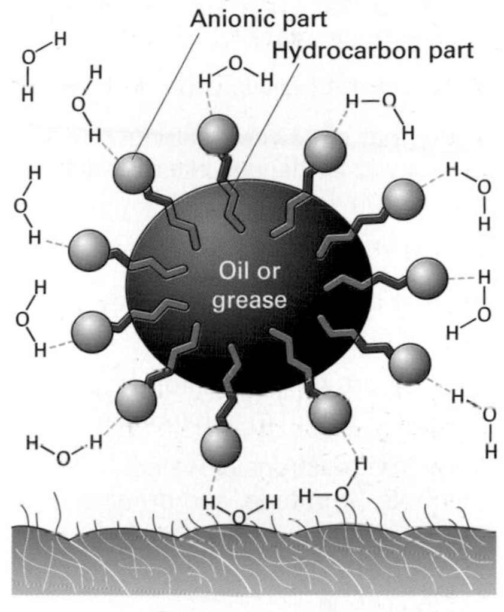

The anionic 'head' of a soap molecule dissolves in water and the non-polar 'tail' dissolves in oil and grease.

Check your understanding

1. Write equations for the ester hydrolysis of ethyl methanoate by:
   a. acid-catalysed hydrolysis
   b. base-catalysed hydrolysis using sodium hydroxide
2. Stearic acid is octadecanoic acid, $CH_3(CH_2)_{16}COOH$.
   a. Show the structure of the triester formed by stearic acid and propane-1,2,3-triol.
   b. Show how this ester can be hydrolysed using sodium hydroxide.
   c. Give a possible use of the sodium stearate formed in part b.
3. What is biodiesel and how is it made?

6.01 Acyl chlorides and acid anhydrides

OUTCOMES

already from AS Level, you understand

- that haloalkanes contain polar bonds
- that haloalkanes are susceptible to nucleophilic attack

already from A2 Level, you

- know that aldehydes, ketones, carboxylic acids, and esters contain the carbonyl group

and after this spread you should

- be able to name acyl chlorides and acid anhydrides
- understand why acyl chlorides are more reactive than chloroalkanes
- know the reactions of water, alcohols, ammonia, and primary amines with acyl chlorides and acid anhydrides

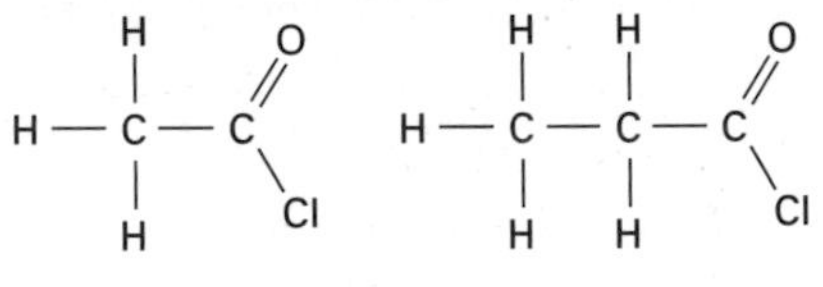

ethanoyl chloride propanoyl chloride

two acyl chlorides

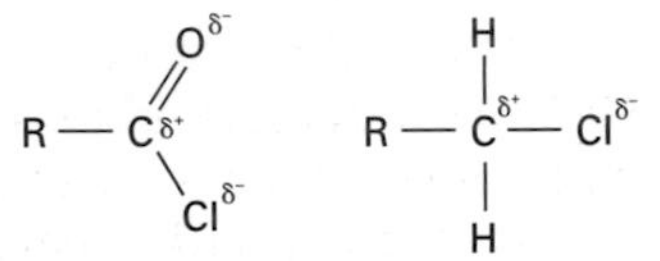

The functional groups of acyl chlorides and chloroalkanes.

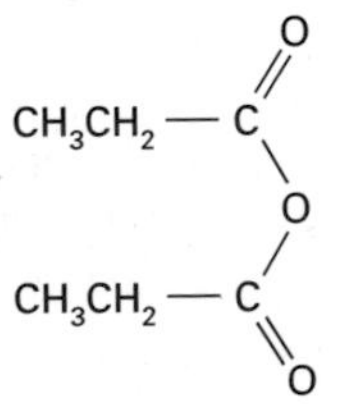

propanoic anhydride $(CH_3CH_2CO)_2O$

Acyl chlorides and **acid anhydrides** contain the carbonyl group >C=O. They are reactive compounds that are useful in organic syntheses.

Acyl chlorides

Acyl chlorides contain the functional group −COCl. They are synthesized in the laboratory by reacting carboxylic acids with phosphorus(V) chloride. For example:

$$\underset{\text{ethanoic acid}}{CH_3COOH} + PCl_5 \rightarrow \underset{\text{ethanoyl chloride}}{CH_3COCl} + POCl_3 + HCl$$

Acyl chlorides are sometimes called **acid chlorides**.

Naming acyl chlorides

Acyl chlorides are named after the number of carbon atoms they contain, including the carbon atom in the functional group, and their names end in *oyl chloride*. The functional group is shown as −COCl in shortened structural formulae. For example, CH_3COCl is ethanoyl chloride and CH_3CH_2COCl is propanoyl chloride. Note that ethanoyl is pronounced 'ethan-o-ile' not 'ethanoil'.

Reactivity of acyl chlorides

Acyl chlorides and chloroalkanes both contain the C—Cl bond. Chloroalkanes can take part in **nucleophilic substitution reactions**. So can acyl chlorides, but they are more reactive.

Chlorine is more electronegative than carbon, so it withdraws electron density towards itself in a covalent bond. This means that the carbon atom in the C—Cl bond in acyl chlorides and chloroalkanes has a partial positive charge $\delta+$. But this charge is greater in acyl chlorides because oxygen, another electronegative atom, is also attached to the carbon atom. As a result, acyl chlorides are more susceptible to nucleophilic attack than chloroalkanes.

Acid anhydrides

Acid anhydrides can be synthesized in the laboratory by dehydrating a carboxylic acid. For example:

$$\underset{\text{ethanoic acid}}{2CH_3COOH} \xrightarrow{P_4O_{10}} \underset{\text{ethanoic anhydride}}{(CH_3CO)_2O} + H_2O$$

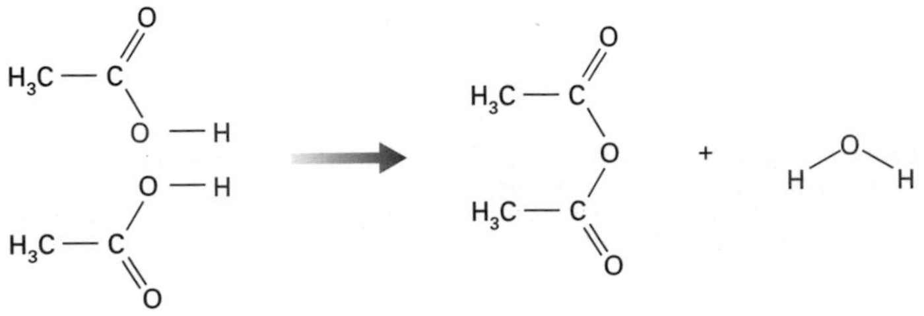

The formation of ethanoic anhydride from ethanoic acid.

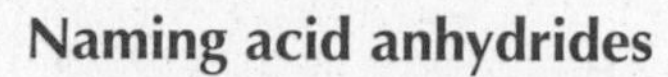

Naming acid anhydrides

Acid anhydrides are named after the carboxylic acid that formed them. For example, ethanoic acid forms ethanoic anhydride, and propanoic acid forms propanoic anhydride.

Asymmetric acid anhydrides

Acid anhydrides can also be synthesized by **refluxing** an acyl chloride with the sodium salt of a carboxylic acid. It is possible to produce asymmetric acid anhydrides this way. For example:

$$\underset{\text{ethanoyl chloride}}{CH_3COCl} + \underset{\text{sodium methanoate}}{HCOO^-Na^+} \rightarrow \underset{\text{ethanoic methanoic anhydride}}{(CH_3CO)O(OCH_3)} + NaCl$$

Reactivity of acid anhydrides

Acid anhydrides are often preferred in syntheses because they are cheaper than acyl chlorides, less easily hydrolysed, and have less violent reactions. Acyl chlorides release corrosive fumes of hydrogen chloride when they react but acid anhydrides release a carboxylic acid, which is less hazardous.

Ethanoyl chloride is readily hydrolysed by water. Fumes of hydrogen chloride gas are given off when a drop of water on a glass rod is put into ethanoyl chloride vapour.

Reactions of acyl chlorides and acid anhydrides

Acyl chlorides and acid anhydrides produce the same primary product when they react with water, alcohols, ammonia, or primary amines. They react with

- water to produce carboxylic acids
- alcohols to produce esters
- ammonia to produce amides
- primary amines to produce *N*-substituted acid amides

The secondary product is hydrogen chloride in reactions with acyl chlorides, and a carboxylic acid in reactions with acid anhydrides.

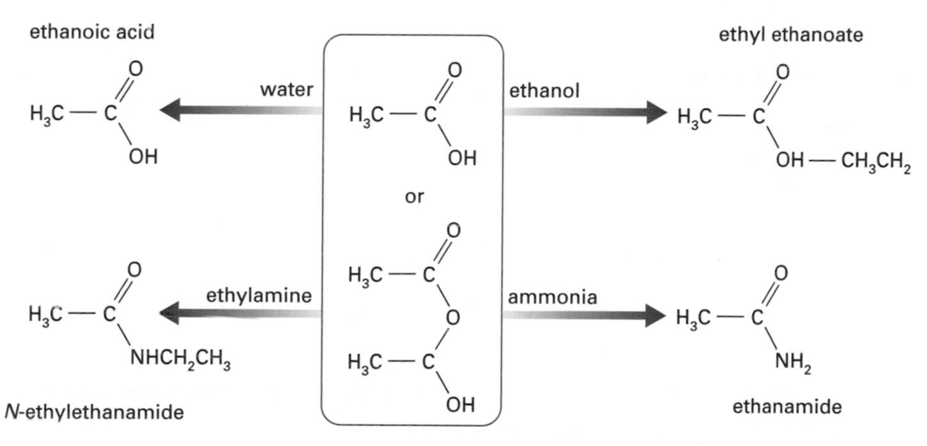

The products of the reactions of water, alcohols, ammonia, and primary amines with ethanoyl chloride or ethanoic anhydride.

Check your understanding

1 Explain why acyl chlorides are susceptible to nucleophilic attack.

2 Name these compounds:
  a $HCOCl$
  b $CH_3(CH_2)_2COCl$
  c $(CH_3CO)_2O$

3 Carboxylic acids are formed when acyl chlorides or acid anhydrides react with water.
  a What organic compounds are formed when acyl chlorides react with i alcohols, ii ammonia, iii primary amines?
  b Apart from carboxylic acids, what organic compounds are formed when acid anhydrides react with i alcohols, ii ammonia, iii primary amines?
  c Outline why the use of acid anhydrides in organic syntheses may be preferred to the use of acyl chlorides.

6.02 Carboxylic acids and esters from acyl chlorides

OUTCOMES

already from A2 Level, you

- can name acyl chlorides and acid anhydrides
- understand why acyl chlorides are more reactive than chloroalkanes
- know the reactions of water, alcohols, ammonia, and primary amines with acyl chlorides and acid anhydrides

and after this spread you should

- understand the mechanism of nucleophilic addition–elimination reactions between acyl chlorides and water or alcohols

Acyl chlorides react with water to produce carboxylic acids, and with alcohols to produce esters. The overall reaction results in the chlorine atom being swapped for a different group, and it takes place in two steps:

1. an addition step, in which the nucleophile attacks the $\delta+$ carbon atom in the —COCl functional group
2. an elimination step, in which a chloride ion and a hydrogen ion are removed

The reaction mechanism involved is called **nucleophilic addition–elimination**.

Acyl chlorides and water

Acyl chlorides react with water to produce carboxylic acids and hydrogen chloride. For example:

$$CH_3COCl + H_2O \rightarrow CH_3COOH + HCl$$

ethanoyl chloride → ethanoic acid

The addition step

The oxygen atom in a water molecule has two lone pairs of electrons, which can form co-ordinate bonds. Water can act as a nucleophile and attack the $\delta+$ carbon atom in the —COCl group of acyl chlorides.

- A co-ordinate bond forms between the oxygen atom in a water molecule and the $\delta+$ carbon atom.
- A pair of electrons in the C═O bond is transferred to the oxygen atom in the —COCl group.

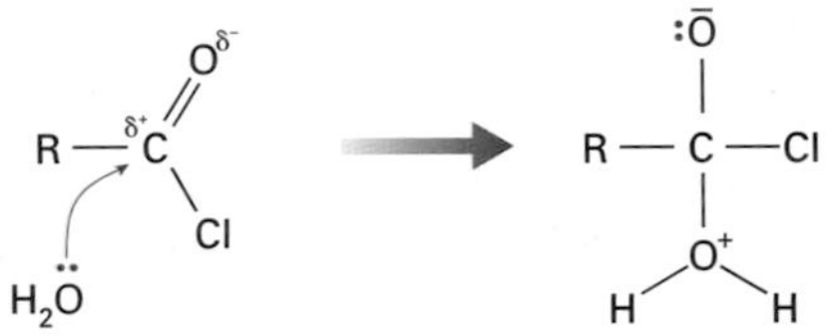

The addition step for an acyl chloride, RCOCl, reacting with water. Note that you should show a lone pair of electrons in the water molecule and on the negatively charged oxygen atom. Do not forget to show the negative and positive charges on the intermediate structure.

The elimination step

Three processes happen in the elimination step.

- The pair of electrons on the negatively charged oxygen atom forms a bond between carbon and oxygen again, re-forming the C═O bond.
- A chloride ion is eliminated.
- The bond between a hydrogen atom and the positively charged oxygen atom breaks, and a hydrogen ion is lost.

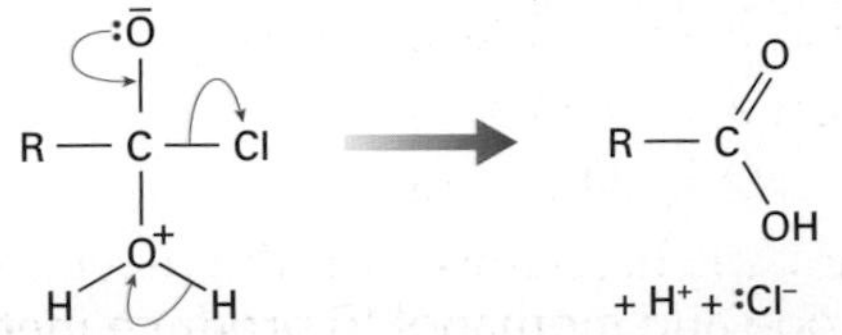

The elimination step for an acyl chloride, RCOCl, reacting with water.

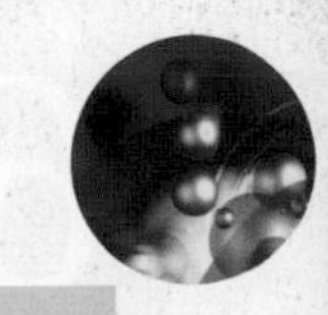

Here is the full reaction mechanism. In this example, ethanoyl chloride reacts with water to produce ethanoic acid and hydrogen chloride.

✓ (for three arrows)

$R-C^{\delta+}(=O^{\delta-})Cl$ + $H_2\ddot{O}$ → $R-C(-\ddot{O}^-)(-Cl)(-O^+H_2)$ → $H_3C-C(=O)OH$ + HCl

✓ (for structure)

The full nucleophilic addition–elimination reaction mechanism, with ticks to show the features an examiner is likely to look for in an examination.

Acyl chlorides and alcohols

Acyl chlorides react with alcohols to produce esters and hydrogen chloride. For example:

$$CH_3COCl + CH_3OH \rightarrow CH_3COOCH_3 + HCl$$

ethanoyl chloride | methanol | methyl ethanoate

The reaction mechanism is exactly the same as the one for the reaction with water. It is easy to see how this works if the alcohol is shown like this.

H–O–H H_3C–O–H

In the reaction mechanism, show the alcohol molecule as a water molecule with one of its hydrogen atoms replaced by the carbon chain.

Here is the full reaction mechanism for the reaction between ethanoyl chloride and methanol.

$H_3C-C^{\delta+}(=O^{\delta-})Cl$ + $CH_3\ddot{O}H$ → $H_3C-C(-\ddot{O}^-)(-Cl)(-O^+(H)CH_3)$ → $H_3C-C(=O)-O-CH_3$ + HCl

The reaction mechanism for the production of methyl ethanoate from ethanoyl chloride and methanol.

Acylation

An **acyl group** has the general formula RC=O. This is why compounds with the general formula RCOCl are called acyl chlorides. When an acyl group is introduced into a compound, the reaction is called **acylation**. Carboxylic acids are formed when water is acylated, and esters are formed when alcohols are acylated.

Check your understanding

1 Write balanced equations for the reactions between:
  - a methanoyl chloride and water
  - b 2-methylpropanoyl chloride and water
  - c butanoyl chloride and ethanol
  - d ethanoyl chloride and butanol

2 Name and outline the mechanism for the reaction between propanoyl chloride and methanol to produce methyl propanoate.

6.03 Amides from acyl chlorides

OUTCOMES

already from A2 Level, you

- know the reactions of water, alcohols, ammonia, and primary amines with acyl chlorides and acid anhydrides
- understand the mechanism of nucleophilic addition–elimination reactions between acyl chlorides and water or alcohols

and after this spread you should

- understand the mechanism of nucleophilic addition–elimination reactions between acyl chlorides and ammonia or primary amines

Pyrazinamide

Pyrazinamide is one of the drugs used to treat tuberculosis, also known as TB. This is an infectious disease in which the lungs are damaged. It can damage other parts of the body, too, such as the central nervous system. Tuberculosis can lead to death if untreated. Pyrazinamide stops the growth of *Mycobacterium tuberculosis* bacteria, the pathogen that causes tuberculosis.

Displayed formula of pyrazinamide.

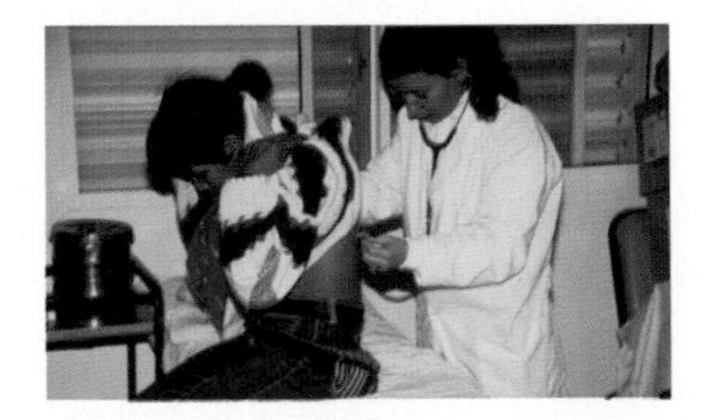

A doctor using a stethoscope to listen to the lungs of a woman with tuberculosis. It is important that patients complete their drug treatment so that all the M. tuberculosis *bacteria in the body are killed.*

Acyl chlorides react with ammonia to produce **amides**, and with primary amines to produce **N-substituted amides**. The reaction mechanism involved is nucleophilic addition–elimination. It is identical to the mechanism involved in the reactions between acyl chlorides and water to produce carboxylic acids, and between acyl chlorides and alcohols to produce esters.

Acyl chlorides and ammonia

Acyl chlorides react with ammonia to produce amides and hydrogen chloride. For example:

$$CH_3COCl + NH_3 \rightarrow CH_3CONH_2 + HCl$$

ethanoyl chloride → ethanamide

Ammonia can act as a nucleophile because it has a lone pair of electrons on its nitrogen atom. The reaction mechanism is the same as the one for the reaction with water. Here is the full reaction mechanism for the reaction between ethanoyl chloride and ammonia.

The reaction mechanism for the production of ethanamide from ethanoyl chloride and ammonia.

The hydrogen ion may be removed by a chloride ion, forming HCl. It may also be removed by an ammonia molecule, forming the ammonium ion NH_4^+. In this situation, the ammonia molecule is acting as a Brønsted–Lowry base, rather than as a nucleophile. The ammonium ion will react with a chloride ion to form ammonium chloride, NH_4Cl. So you may see the equation for the reaction between an acyl chloride and excess ammonia written like this:

$$CH_3COCl + 2NH_3 \rightarrow CH_3CONH_2 + NH_4Cl$$

ethanoyl chloride → ethanamide

Acyl chlorides and primary amines

Acyl chlorides react with primary amines to produce N-substituted amides and hydrogen chloride. For example:

$$CH_3COCl + CH_3NH_2 \rightarrow CH_3CONHCH_3 + HCl$$

ethanoyl chloride — methanamine — *N*-methylethanamide

A primary amine can act as a nucleophile because it has a lone pair of electrons on its nitrogen atom. The reaction mechanism is the same as the one for the reaction with ammonia. Here is the full reaction mechanism for the reaction between ethanoyl chloride and methanamine.

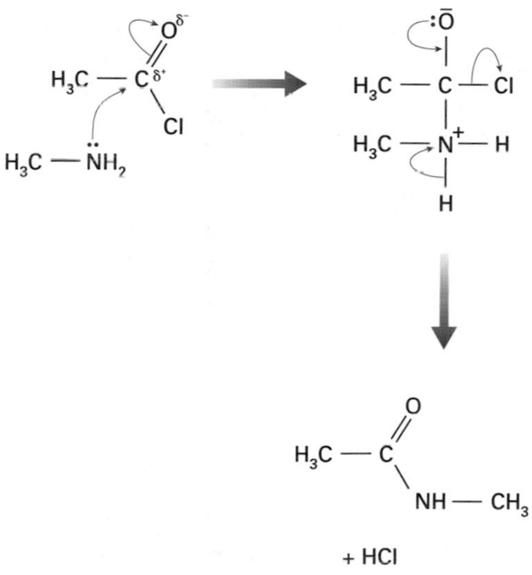

The reaction mechanism for the production of N-*methylethanamide from ethanoyl chloride and methanamine.*

Naming amides

Amides are named after the number of carbon atoms they contain, including the carbon atom in the functional group, and their names end in *amide*. The functional group is shown as $-CONH_2$ in shortened structural formulae. For example, $HCONH_2$ is methanamide and $CH_3CH_2CONH_2$ is propanamide. Amides are also called **acid amides**.

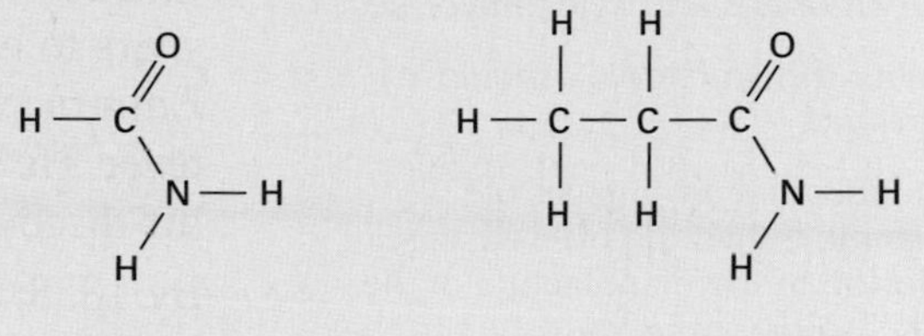

methanamide — propanamide

Displayed formulae of methanamide and propanamide.

Naming primary amines

Primary amines contain the **amino group** $-NH_2$. They are named after the number of carbon atoms they contain, and their names end in *amine*. For example, CH_3NH_2 is methanamine and $CH_3CH_2NH_2$ is ethylamine.

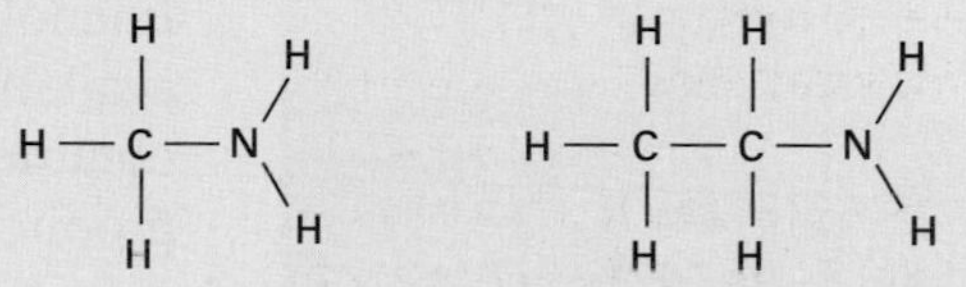

methanamine — ethanamine

Displayed formulae of methanamine and ethanamine. (also called methylamine and ethylamine)

Check your understanding

1 Write balanced equations for the reactions between:
  a propanoyl chloride and ammonia
  b methanoyl chloride and ethanamine (ethylamine)

2 a Name and outline the mechanism for the reaction between propanoyl chloride and ethanamine (ethylamine).
  b Name the organic product made in the reaction.

Naming N-substituted amides

N-substituted amides are named as an amide with one of the hydrogen atoms in the $-CONH_2$ replaced by an alkyl group. For example, $CH_3CH_2CONH_2$ is propanamide, but $CH_3CH_2CONHCH_3$ is *N*-methylpropanamide.

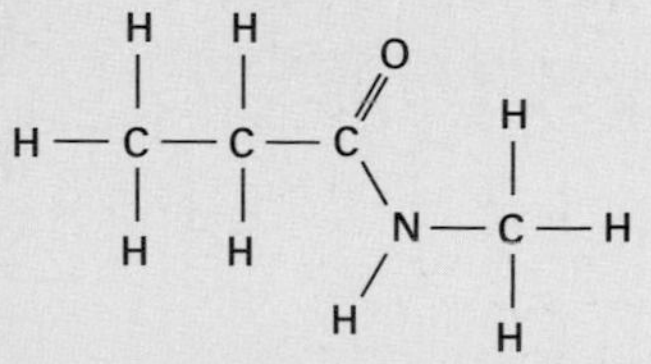

Displayed formula of N-methylpropanamide.

6.04 Aspirin

OUTCOMES

already from A2 Level, you know

- the reactions of water, alcohols, ammonia, and primary amines with acyl chlorides and acid anhydrides

and after this spread you should understand

- the industrial advantages of ethanoic anhydride over ethanoyl chloride in the manufacture of the drug aspirin

The bark of the English willow tree, Salix alba, *contains salicylic acid.*

Ancient remedies relied on extracts from animals and plants. Naturally occurring substances are the starting point for medical drugs in the modern world, too. Aspirin is one such drug. It is widely used to treat pain, fevers, and inflammation.

From willow to aspirin

Extracts from the bark of certain trees have been used for thousands of years to ease aches and pains. In the eighteenth century, the Reverend Edmund Stone noticed that an extract from the willow tree bark reduced fever. He successfully tested his powder on other people and published his discovery in a letter to the Royal Society in 1763.

By 1838, the active ingredient of willow tree bark had been isolated. It is salicylic acid or 2-hydroxybenzoic acid. The use of natural sources for medicines has disadvantages compared to synthesizing the chemicals artificially. For example:

- The source may be rare, difficult to cultivate, or only available at certain times of the year.
- The concentration of the active ingredient may vary from source to source.
- The extract may be contaminated with harmful substances.

It would be better if salicylic acid could be synthesized from another, more common substance.

Synthesizing salicylic acid

Phenol has a very similar structure to salicylic acid. It was readily available in the nineteenth century. At that time, gas for lighting was obtained by the destructive distillation of coal. Phenol was one of the by-products made at the gas works.

The German chemist Hermann Kolbe discovered how to convert phenol into salicylic acid in 1860. His *Kolbe process* involves converting phenol into its sodium salt by reacting it with sodium hydroxide, then heating the salt with carbon dioxide under pressure. Salicylic acid is then made by reacting the product with sulfuric acid.

OH

NaOH

CO_2

O

O^-Na^+

C

O^-Na^+

H_2SO_4

O

OH

C

OH

phenol

2-hydroxybenzoic acid (salicylic acid)

Phenol can be converted into salicylic acid by the Kolbe process.

Problems with salicylic acid

Synthetic salicylic acid was widely used in the second half of the nineteenth century as an **analgesic** or painkiller. Unfortunately, salicylic acid causes painful side effects, including stomach pains. The father of another German chemist, Felix Hoffmann, suffered from rheumatism.

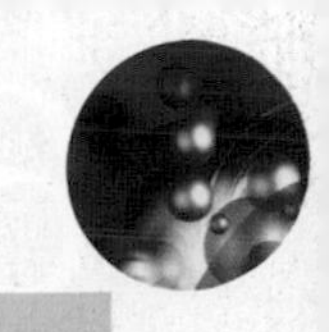

The younger Hoffmann aimed to modify salicylic acid so that it was still effective but without side effects. He synthesized various chemicals based on salicylic acid, and tried them out on his father. Luckily for both of them, acetylsalicylic acid was found to answer the problem.

Acetylsalicylic acid

Acetylsalicylic acid is 2-ethanoyloxybenzoic acid. Hofmann synthesized it in 1898 by reacting 2-hydroxybenzoic acid with ethanoic anhydride. This acylated the hydroxyl group and produced a chemical that still has an analgesic effect, but with fewer side effects. It was named aspirin and it went on sale in 1899.

2-hydroxybenzoic acid (salicylic acid) + ethanoic anhydride → 2-ethanoyloxybenzoic acid (aspirin) + ethanoic acid

Salicylic acid is converted into acetylsalicylic acid using ethanoic anhydride.

The manufacture of aspirin

Aspirin could be synthesized using ethanoyl chloride, rather than ethanoic anhydride. Ethanoyl chloride is more reactive than ethanoic anhydride, so it might be a better choice. But the reaction involving ethanoic anhydride is safer and easier to control, and there are other advantages to using ethanoic anhydride including:

- It is cheaper than ethanoyl chloride.
- It is less susceptible to hydrolysis than ethanoyl chloride.
- It is less corrosive than ethanoyl chloride, because it produces ethanoic acid rather than fumes of hydrogen chloride.

Paracetamol

Paracetamol is another common analgesic. It is manufactured by acylating 4-aminophenol rather than 2-hydroxybenzoic acid.

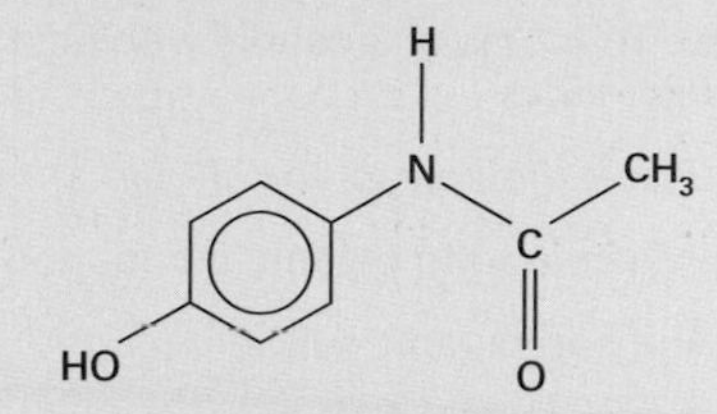

Paracetamol, N-(4-hydroxyphenyl) ethanamide

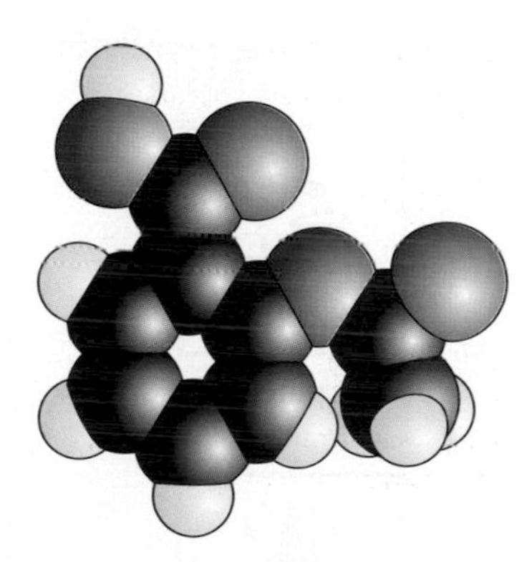

The global consumption of aspirin is about 35 000 tonnes per year, equivalent to about 100 billion tablets.

Check your understanding

1. Give two disadvantages of using medicines prepared from natural sources.
2. Why was phenol, one of the materials needed to manufacture aspirin, widely available in the nineteenth century?
3. Explain the industrial advantages of using ethanoic anhydride over ethanoyl chloride in the manufacture of aspirin.

6.05 Making aspirin

OUTCOMES

already from A2 Level, you know

- the reactions of alcohols with acyl chlorides and acid anhydrides

and after this spread you should know how to

- prepare a solid organic compound
- purify an organic solid
- test the purity of an organic solid

Keeping safe

WEAR EYE PROTECTION

If you see this sign you must wear eye protection, but eye protection is always wise in chemistry experiments.

Hazard warning symbols

CORROSIVE

TOXIC

FLAMMABLE

HARMFUL

OXIDIZING

IRRITANT

These are some of the common hazard symbols seen in laboratories.

There are three Investigative and Practical Skills tasks for A2 Organic Chemistry. You are expected to prepare a solid organic compound, purify it, and test its purity. For example, you might be asked to make a solid organic compound, such as aspirin. Whatever compound you are asked to work with, you will be given instructions to follow carefully and skilfully. Procedures involving aspirin are outlined in this spread as examples only.

Preparing a solid organic compound

Take care to observe any hazard warning symbols. It is particularly important to wear eye protection, and you may need to wear gloves and work in a fume cupboard. Here is an outline method for making aspirin.

Typical reagents needed

- 2-hydroxybenzoic acid (salicylic acid) IRRITANT
- Ethanoic anhydride CORROSIVE
- Concentrated phosphoric acid CORROSIVE

Making crude aspirin

1. Place 2 g of 2-hydroxybenzoic acid in a 100 cm^3 conical flask.
2. Add 4 cm^3 of ethanoic anhydride to the conical flask, then 5 drops of concentrated phosphoric acid.
3. Heat the flask in a hot water bath for about 10 minutes, swirling occasionally.
4. Remove the flask from the hot water bath and let it cool.
5. Carefully add 20 cm^3 of ice-cold water to the flask, starting cautiously a few drops at a time. Stir with a glass rod to start the precipitation process, if needed. Stand the flask in an iced water bath until the precipitation of crystals seems to have finished.

Special precautions

An organic synthesis may need special precautions. For example, the apparatus must be completely dry when ethanoic anhydride is used. If it is damp, the ethanoic anhydride will hydrolyse and the preparation will fail.

Reagents such as ethanoic anhydride may react vigorously or give off harmful fumes. Reactions should be carried out in a fume cupboard to avoid inhaling such fumes. Water baths avoid the danger

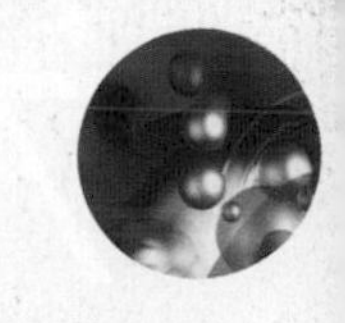

presented by the naked flame of a Bunsen burner. They can be as simple as a large beaker filled using a kettle.

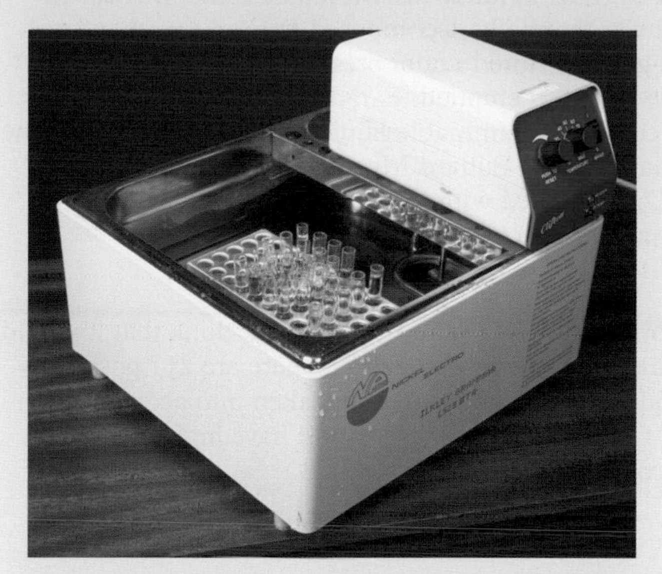

Thermostatically controlled electric water baths like this one are a convenient way to maintain reactants at a constant temperature.

Purifying an organic solid

Washing the crude product

Filtration under reduced pressure or **vacuum filtration** quickly separates a solid from a liquid. The **Buchner flask** has thick walls to withstand reduced pressure inside, and a side arm to connect it to a pump to draw air out. The **Buchner funnel** fits into the neck of the flask, making an airtight fit. It has a flat perforated surface inside for the filter paper to rest on.

With the pump working, the crude aspirin suspension is carefully poured into the middle of the filter paper. The liquid is rapidly drawn through and the solid is filtered from the liquid. The aspirin can easily be washed with chilled water while it is in the Buchner funnel.

The apparatus needed for vacuum filtration. It is a good idea to dampen the filter paper before adding your product, otherwise the paper might float off the perforated surface in the funnel.

Recrystallization

A solid product can be purified by **recrystallization**. The impure product is removed from the filter paper and dried. It is added to a test tube and dissolved in the *minimum volume* of hot solvent needed. Hot ethyl ethanoate can be used to dissolve aspirin. Insoluble impurities can be filtered off but this must be done while everything is hot.

The hot saturated solution of aspirin is allowed to cool *slowly* to room temperature, then finally in an ice bath. This allows pure crystals to separate out. Soluble impurities may be present, but these can be removed by vacuum filtration again. This also dries the product. A warm oven can be used, too.

Testing the purity of an organic solid

A **melting point determination** is often carried out to test the purity of an organic solid. The purer a solid is, the closer its melting point will be to the known melting point. An impure solid will melt over a range of temperatures but a pure solid will have a sharply defined melting point.

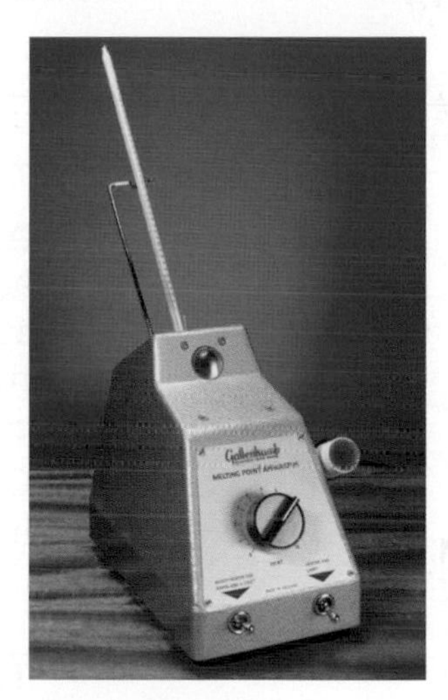

A melting point apparatus. The capillary tube holding the sample fits into the top near the thermometer and is observed through the glass window.

If you are lucky you will have sealed glass capillary tubes to hand. If not, you need to seal one end in a Bunsen burner flame. Take care not to heat the tube until it bends.

Add the product to a depth of about 5 mm and put the tube into the melting point apparatus. Set the temperature to increase slowly, as you will get inaccurate results if it rises too quickly. Reduce the rate of heating near the known melting point of the product to get a sharp end point. Note the temperature at which the solid melts. For aspirin, this should be between 138°C and140°C.

7.01 Benzene structure

OUTCOMES

already from AS Level, you know that

- alkanes are saturated hydrocarbons
- alkenes are unsaturated hydrocarbons
- bonding in alkenes involves a double covalent bond
- bromine can be used to test for unsaturation

and after this spread you should

- understand the nature of the bonding in a benzene ring, limited to planar structure and bond length intermediate between single and double

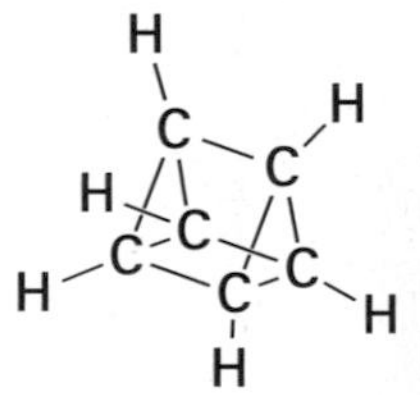

prismane

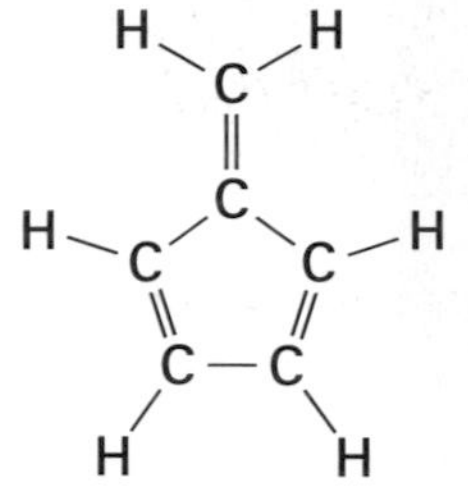

fulvene

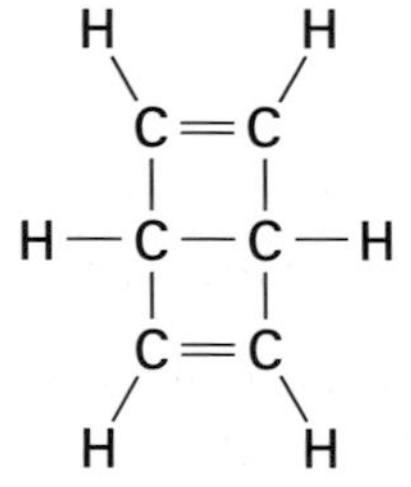

Dewar benzene

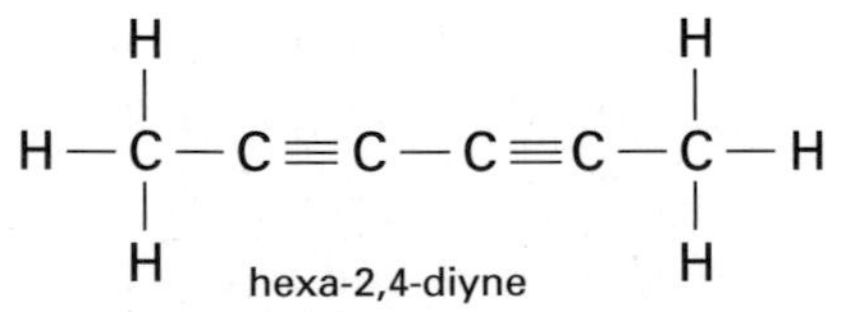

hexa-2,4-diyne

Some isomers of C_6H_6.

The English scientist Michael Faraday discovered benzene in 1825. In those days, whale oil was thermally decomposed to produce ethyne gas for street lamps. The gas was stored under pressure, and a runny oil was left behind as the cylinders emptied. Faraday distilled this oil and obtained benzene, a colourless flammable liquid with a sweet smell. Nine years later, the German chemist Eilhard Mitscherlich worked out its molecular formula, C_6H_6. Armed with its molecular formula, chemists tried to work out benzene's structure.

Isomers of C_6H_6

Given a molecular formula for a substance, you might think that it would be a simple task to find its structure. Amazingly there are 217 possible isomers of C_6H_6. And if stereoisomers are also taken into account, this number rises to 328. Luckily most of these would have highly strained covalent bonds and could not exist.

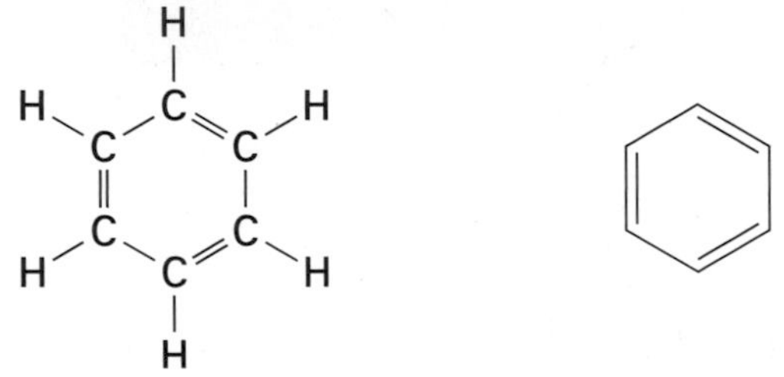

The Kekulé structure of benzene, shown as a displayed formula and as a skeletal formula.

August Kekulé was a German chemist who considered the structure of benzene. He suggested in 1864 that benzene contains a ring of carbon atoms, arranged in a hexagon. The following year he proposed that the ring contains alternating single and double carbon–carbon bonds. This is called the **Kekulé structure** of benzene.

One piece of evidence in support of this structure is that benzene can be hydrogenated to form cyclohexane, another **cyclic** compound:

$$C_6H_6(g) + 3H_2(g) \xrightarrow{\text{Pt catalyst}} C_6H_{12}(g)$$

This suggests that benzene could contain three C=C bonds, with its carbon atoms arranged in a ring.

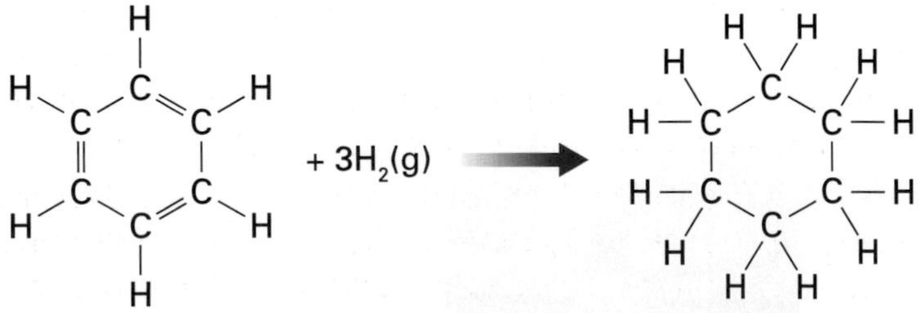

$+ 3H_2(g)$

Benzene can be hydrogenated to form cyclohexane.

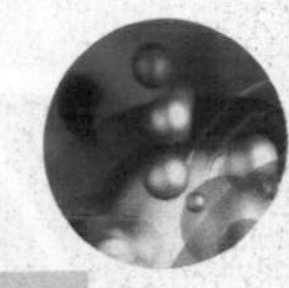

Problems with the Kekulé structure

Too few isomers of dichlorobenzene

Benzene reacts with chlorine in the presence of iron(III) chloride as a catalyst. Kekulé predicted the existence of four isomers of dichlorobenzene $C_6H_4Cl_2$. But only three of these exist. The chlorine atoms in 1,2-dichlorobenzene and 1,6-dichlorobenzene are on adjacent carbon atoms, which are joined either by a double bond or by a single bond. These two isomers have not been made.

1,6-dichlorobenzene ✗ 1,2-dichlorobenzene ✔

1,4-dichlorobenzene ✔ 1,3-dichlorobenzene ✔

The four predicted isomers of dichlorobenzene. 1,5-dichlorobenzene is identical to 1,3-dichlorobenzene, so it is not a fifth isomer. 1,6-dichlorobenzene and 1,2-dichlorobenzene do not exist as separate isomers.

Substitution reactions not addition reactions

Alkenes contain C=C bonds and they readily undergo **electrophilic addition** reactions. For example, ethene reacts with bromine to produce 1,2-dibromoethane. The Kekulé structure for benzene contains three C=C bonds, so benzene should be highly unsaturated and readily undergo addition reactions, too. But this does not happen. Benzene only reacts with bromine in the presence of iron(III) bromide, $FeBr_3$, as a catalyst. When it does, a substitution reaction happens instead, producing bromobenzene, C_6H_5Br.

The modern structure for benzene

Kekulé's structure for benzene could not account for all its chemical properties. Benzene is now known to consist of a **planar** ring of six carbon atoms, just as Kekulé suggested. But there are no double carbon–carbon bonds. Instead, the carbon atoms are joined by single bonds, with **delocalized electrons** above and below the plane of the ring.

The mean bond length of the C—C bond is 0.154 nm and that of the C=C bond is 0.134 nm. In fact, all six carbon–carbon bonds in benzene are 0.140 nm long. This is intermediate between the lengths of single and double carbon–carbon bonds. All the bond angles in benzene are 120°.

Kekulé's solution

In 1872, Kekulé proposed a solution to the problem presented by the missing isomer of dichlorobenzene. He suggested that single bonds convert into double bonds, and back again, continuously and very rapidly. As a result it would be impossible to detect or isolate 1,2-dichlorobenzene and 1,6-dichlorobenzene.

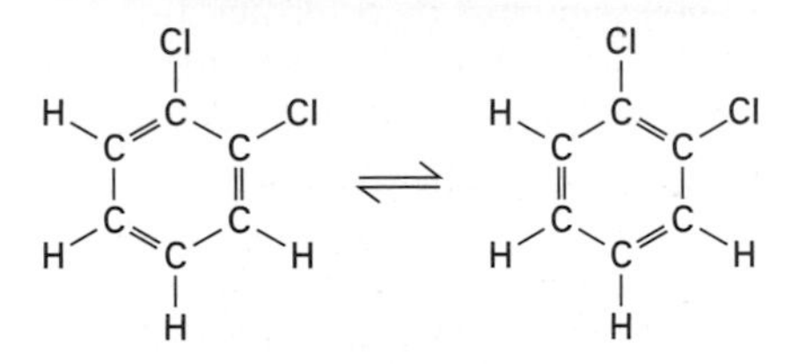

The proposed rapid interconversion of two isomers of dichlorobenzene.

Benzene has two 'rings' of delocalized electrons.

Check your understanding

1. a Describe the appearance of benzene.
   b Draw the skeletal formula for the Kekulé structure of benzene, and give its molecular formula.
   c Outline the way in which Kekulé modified his ideas in response to experimental data.
   d Describe the modern structure for benzene.

7.02 Stability of benzene

OUTCOMES

already from A2 Level, you understand

- the nature of the bonding in a benzene ring, limited to planar structure and bond length intermediate between single and double

and after this spread you should

- understand that delocalization confers stability to the benzene molecule
- be able to use thermochemical evidence from enthalpies of hydrogenation to show that delocalization confers stability to the benzene molecule

Arenes

Benzene is a colourless flammable liquid with a sweet smell. It is also highly toxic, and repeated exposure to benzene can cause leukaemia. As a result, benzene is banned from use in schools. But it is the parent molecule of all **arenes**, many of which are not hazardous. Arenes contain one or more benzene rings. They used to be called **aromatic compounds** because they have characteristic smells.

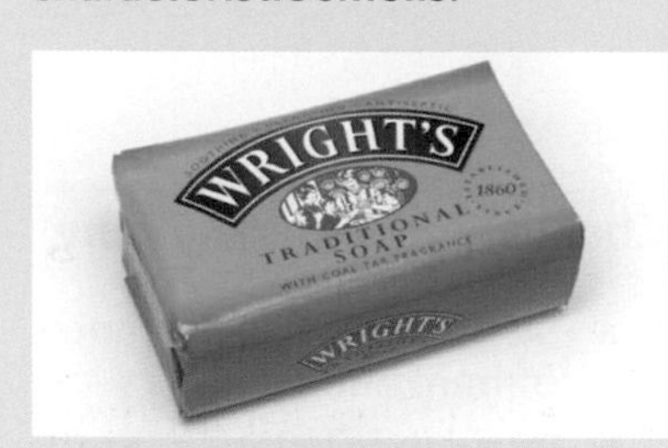

Traditional coal tar soap has a characteristic smell because of the arenes it contains.

The Kekulé structure for benzene has six carbon atoms joined in a ring by alternating single and double carbon–carbon bonds. If the Kekulé structure were correct, benzene would be cyclohexa-1,3,5-triene. It is possible to study whether benzene behaves as this compound would be expected to behave.

As discussed in the previous spread, the bond lengths of all six carbon–carbon bonds in benzene are identical. They are intermediate in length between single and double carbon–carbon bonds. All the bond angles are 120° and the molecule is planar. Benzene is also much less reactive than expected, assuming that it contains three double carbon–carbon bonds. This suggests that benzene is more stable than the Kekulé structure would predict.

Thermochemical evidence for stability

Cyclohexene has a ring of six carbon atoms with five single carbon–carbon bonds and one double bond. It reacts with hydrogen to produce cyclohexane:

cyclohexene $+ H_2 \longrightarrow$ cyclohexane

$\Delta H_r^{\ominus} = -120\ kJ\ mol^{-1}$

The Kekulé structure for benzene contains three double bonds, not just one. So the predicted enthalpy of hydrogenation of the Kekulé structure is three times that of the hydrogenation of cyclohexene.

Kekulé structure $+ 3H_2 \longrightarrow$ cyclohexane

$\Delta H_r^{\ominus} = 3 \times (-120)\ kJ\ mol^{-1}$
$= -360\ kJ\ mol^{-1}$ (predicted)

The experimental value for the enthalpy of hydrogenation of benzene is not -360 kJ mol^{-1}, as predicted. It is only –208 kJ mol^{-1}. This shows that benzene is more stable than predicted. It happens because benzene contains delocalized electrons. The difference between the predicted and experimental values for the enthalpy of hydrogenation of benzene, 152 kJ mol^{-1}, is called the **delocalization enthalpy**.

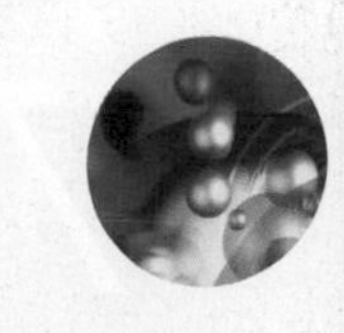

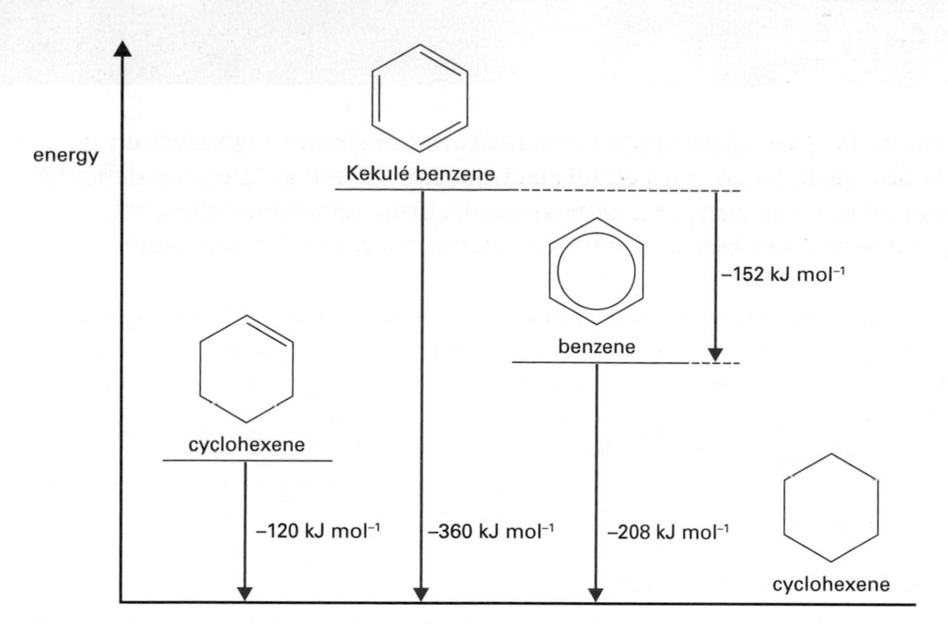

Benzene is at a lower energy level and so is more stable than expected from the Kekulé structure.

Delocalization in the benzene ring

The structure of benzene involves six carbon atoms arranged in a planar hexagonal structure. The electron configuration of carbon is $1s^2\ 2s^2\ 2p^2$. Each carbon atom has four electrons available to form covalent bonds: two s electrons and two p electrons. Three of these electrons form the main structure of benzene, leaving one p electron from each carbon atom.

These six electrons are not associated with any particular carbon atom, but instead are delocalized in ring-shaped molecular orbitals above and below the hydrocarbon skeleton. The skeletal structure of benzene is shown as a hexagon with a circle inside it, which represents the delocalization present.

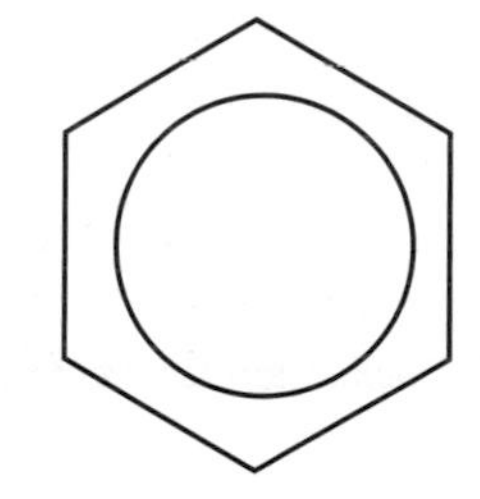

Fused rings

Many arenes contain two or more benzene rings fused together. The delocalization in these compounds extends across all the rings. It is usual to represent their skeletal structure using the Kekulé structure.

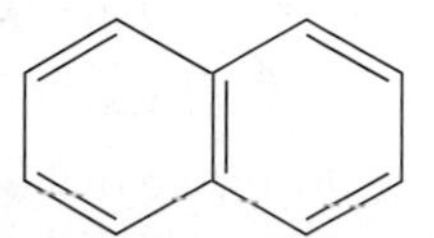

The skeletal structure of naphthalene, an arene used in mothballs to keep moths away from stored clothing. The delocalization extends across both rings.

Check your understanding

1 a Explain what the term *delocalized* means.
  b Describe the accepted structure of benzene.
  c Explain how enthalpy of hydrogenation values can be used to provide evidence for delocalization in benzene.

7.03 Nitro compounds from benzene

OUTCOMES

already from AS Level, you

- understand the mechanism of electrophilic addition of alkenes with HBr, H_2SO_4, and Br_2

already from A2 Level, you

- understand the nature of the bonding in a benzene ring, limited to planar structure and bond length intermediate between single and double
- understand that delocalization confers stability to the benzene molecule
- can use thermochemical evidence from enthalpies of hydrogenation to illustrate this principle

and after this spread you should

- understand that electrophilic attack in arenes results in substitution
- understand the mechanism of nitration, including the generation of the nitronium ion

The delocalized electrons in benzene form a region of high electron density that can be attacked by **electrophiles**. These are electron-deficient species that can accept a lone pair of electrons. In alkenes, the C=C bond is attacked by electrophiles in addition reactions. For example, ethene reacts with bromine to produce 1,2-dibromoethane. But benzene undergoes **electrophilic substitution** reactions instead. This is because an addition reaction would cause a significant loss of delocalization, which would be energetically unfavourable.

Electrophilic substitution

Electrophilic substitution in benzene has two main steps, an addition step followed by an elimination step. Here is the general reaction mechanism for electrophilic substitution reactions in benzene. Remember that a curly arrow shows the movement of a pair of electrons.

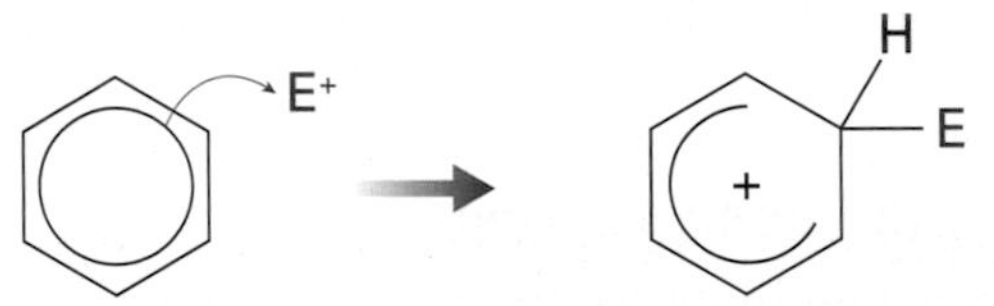

E^+ → H, E, +

In the first step, the electrophile E^+ is attracted to the delocalized electron cloud. The curly arrow is drawn from the circle and points towards the electrophile. A co-ordinate bond forms between one of the carbon atoms in the benzene ring and the electrophile. An unstable positively charged intermediate is formed.

Note that the partial circle should not extend any further than the two carbon atoms next to the one involved in the new bond. Remember that a hydrogen atom is attached to each carbon atom in benzene, so one has not appeared from nowhere.

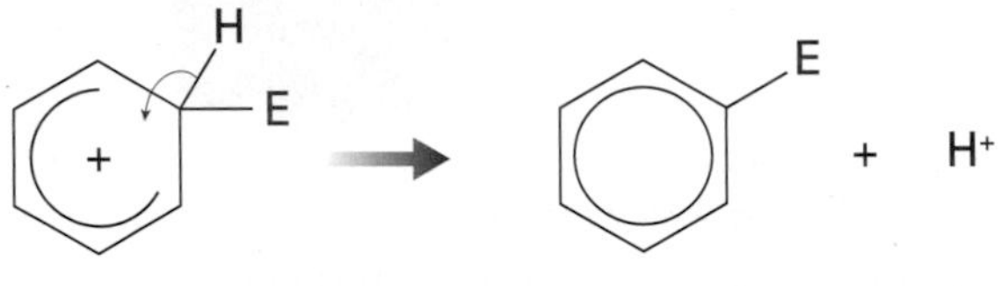

H, E, + → E + H^+

In the second step, the C—H bond breaks. Its pair of electrons restores the stable delocalized electron structure, and a hydrogen ion is eliminated.

The complete mechanism is drawn like this.

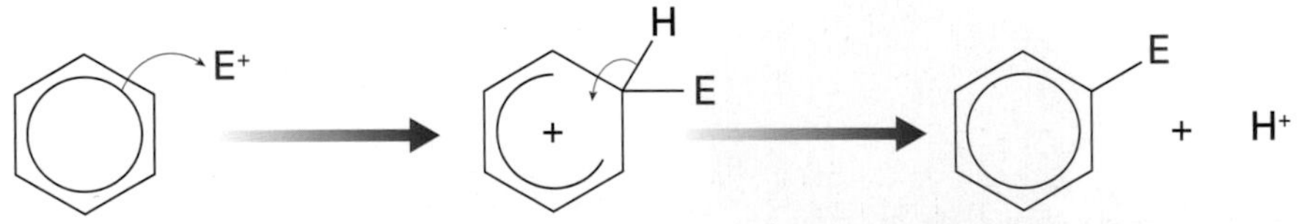

E^+ → H, E, + → E + H^+

The reaction mechanism for electrophilic substitution in benzene.

Nitration of benzene

The **nitro group** is shown as $—NO_2$. Substituting a nitro group into benzene is called **nitration**. Benzene is converted into nitrobenzene by nitration. Nitrobenzene is an oily yellow liquid with a smell of bitter almonds. Here is the overall equation for the nitration of benzene:

$$C_6H_6 + NO_2^+ \rightarrow C_6H_5NO_2 + H^+$$

NO_2^+ is the **nitronium ion**. It is an electrophile. The nitronium ion is generated in a reaction mixture called a **nitrating mixture**.

Generating the nitronium ion

The nitrating mixture is a mixture of concentrated nitric acid and concentrated sulfuric acid. Here is the overall equation for the reaction that produces the nitronium ion:

$$HNO_3 + 2H_2SO_4 \rightarrow NO_2^+ + H_3O^+ + 2HSO_4^-$$

The reaction takes place in two steps.

Step 1 Sulfuric acid is a stronger acid than nitric, so it protonates the nitric acid:

$$H_2SO_4 + HNO_3 \rightarrow H_2NO_3^+ + HSO_4^-$$

Step 2 The protonated nitric acid breaks down to form the nitronium ion:

$$H_2NO_3^+ + H_2SO_4 \rightarrow NO_2^+ + H_3O^+ + HSO_4^-$$

The reaction mechanism

When you draw the reaction mechanism for the nitration of benzene, remember that the nitronium ion is the electrophile, not the nitric acid.

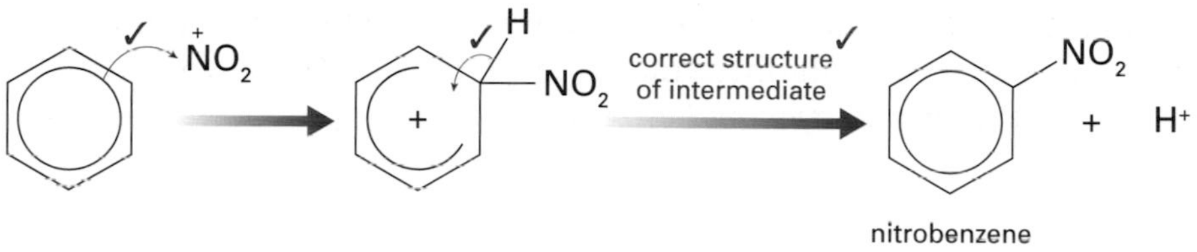

The reaction mechanism for the nitration of benzene, with ticks to show the features an examiner is likely to look for in an examination.

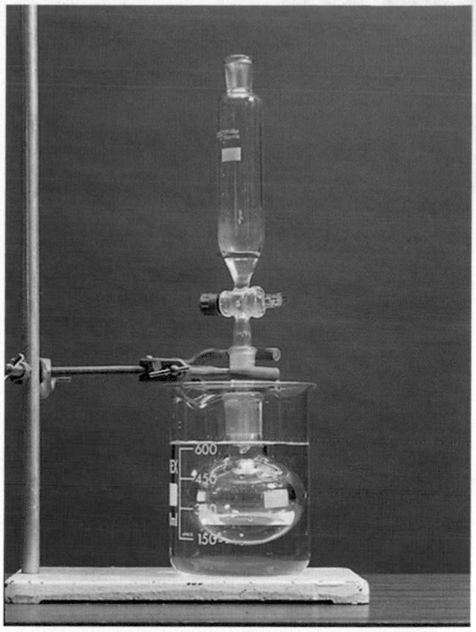

The nitrating mixture of concentrated nitric acid and concentrated sulfuric acid is prepared in the flask. Benzene is slowly run into the flask from the funnel while the flask is cooled in water. The benzene is converted into nitrobenzene when the flask is heated to 50°C in a water bath.

Sulfuric acid as a catalyst

Concentrated sulfuric acid, H_2SO_4, acts as a catalyst in the nitration of benzene. It is regenerated when hydrogen ions H^+, which are eliminated from the positively charged intermediate product, combine with hydrogensulfate ions, HSO_4^-.

Dinitrobenzene

The nitration of benzene is carried out at about 50°C. Further substitution happens if the temperature is increased to 100°C, producing 1,3-dinitrobenzene.

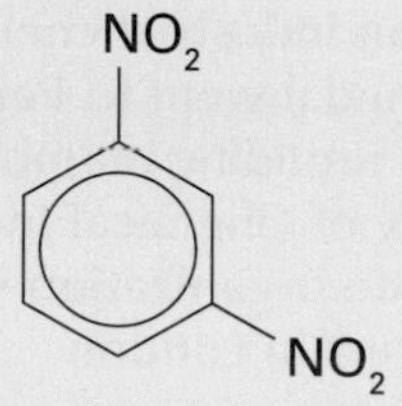

1,3-dinitrobenzene

Check your understanding

1 a What is an *electrophile*?
  b What is a *substitution reaction*?

2 a Name the electrophile involved in the nitration of benzene, and give its formula.
  b Write an equation to show how this electrophile is formed, and name the reagents needed.
  c Draw the mechanism for the nitration of benzene.

7.04 Alfred Nobel

OUTCOMES

already from A2 Level, you understand

- that electrophilic attack in arenes results in substitution
- the mechanism of nitration, including the generation of the nitronium ion

and after this spread you should understand

- that nitration is an important step in synthesis, e.g. the manufacture of explosives and formation of amines from which dyestuffs are manufactured

Mauveine – the first synthetic dye

The very first synthetic dye was made by William Perkin in 1856, when he was just eighteen years old. He was attempting to make quinine, which is used to treat malaria, from phenylamine. His experiment failed and he made a black solid instead. But it formed a purple solution when he rinsed out his flask with ethanol. This proved to be very effective at dyeing cloth.

Perkin patented his discovery to protect it. He started a business to make and sell the dye, called mauveine. The only dyes available previously were from natural sources such as plants. They easily washed out and faded in sunlight, so Perkin became very wealthy because of mauveine.

The Perkin Medal was established to commemorate the 50th anniversary of the discovery of mauveine. It is the highest honour in American industrial chemistry. The first medal went to Perkin, who was also a founding member of the Society of Chemical Industry, a worldwide organization with headquarters in London.

The nitration of benzene is an important step in the synthesis of many different substances. Explosive such as **TNT** contain nitro groups. Nitrobenzene can be reduced to form **phenylamine**, $C_6H_5NH_2$, an aromatic amine:

$$\underset{\text{nitrobenzene}}{C_6H_5NO_2} + 6[H] \rightarrow \underset{\text{phenylamine}}{C_6H_5NH_2} + 2H_2O$$

Synthetic dyes are made from aromatic amines such as phenylamine.

Ice dyes

Phenylamine can be converted to a reactive *diazonium salt* by reacting it with a mixture of sodium nitrate(III) and hydrochloric acid. The reaction mixture must be chilled in an ice bath to stop the diazonium salt decomposing.

$$C_6H_5NH_2 + NaNO_2 + 2HCl \rightarrow C_6H_5N^+{\equiv}N\ Cl^- + NaCl + 2H_2O$$

Phenylamine can be converted into benzenediazonium chloride.

The diazonium salt reacts with phenylamine and other aromatic compounds to produce coloured *azo dyes*. The reaction is called *coupling*, and the phenylamine is acting as a *coupling agent*. Different coupling agents produce dyes with different colours. For example, phenylamine produces a yellow azo dye, but benzene-1,3-diamine produces an orange azo dye.

$$H_2N{-}C_6H_4{-}H + Cl^-N{\equiv}N^+{-}C_6H_5 \rightarrow H_2N{-}C_6H_4{-}N{=}N{-}C_6H_5 + HCl$$

The coupling reaction between phenylamine and benzenediazonium chloride to produce (4-aminophenyl)azobenzene, a yellow compound.

Azo dyes are used to dye natural and synthetic fibres.

Trinitrotoluene, TNT

The aromatic nitro compound 2-methyl-1,3,5-trinitrobenzene is better known as TNT or trinitrotoluene. It was discovered in 1863 and was originally developed as a bright yellow dye. TNT is difficult to detonate so it is relatively safe to handle.

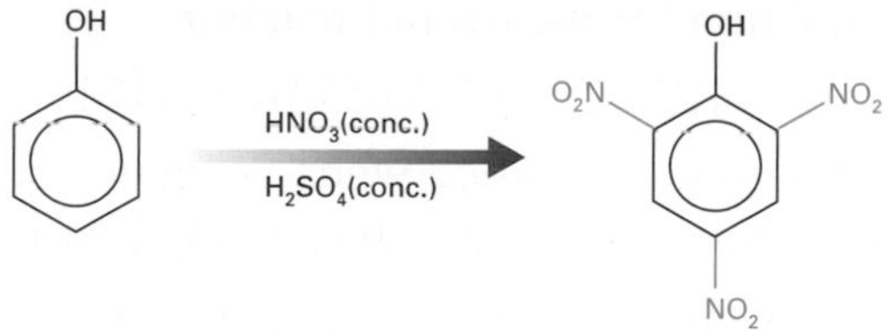

OH

HNO_3(conc.)
H_2SO_4(conc.)

OH
O_2N NO_2
NO_2

TNT is produced by the nitration of methylbenzene. The presence of the methyl group helps to stabilize the charged intermediates formed in the reaction and affects the positions of the nitro groups.

TNT is used to demolish old buildings.

Dynamite and the Nobel Prize

Nitroglycerine $C_3H_5(NO_3)_3$ was discovered in 1847 by Ascanio Sobrero, an Italian chemist. It is an oily liquid that is unstable and very difficult to handle without an explosion. Alfred Nobel's family ran a munitions business. After meeting Sobrero, Nobel decided to experiment with nitroglycerine to see if he could make it safer for demolition work.

Nobel experimented with nitroglycerine in his laboratory in Sweden. There were several accidental explosions, including one in 1864 that killed his younger brother. Nobel discovered that nitroglycerine could be absorbed by silica powder to make a safer explosive paste, which needed a detonator to make it explode. He patented his discovery, called **dynamite**, in 1867. Nobel went on to open factories and laboratories in more than twenty countries and became very wealthy.

Nobel's will contained a surprise when he died in 1896. He had bequeathed almost all his huge wealth to a prize fund for people who have made substantial contributions to science, literature, and world peace. The first Nobel Prizes were awarded in 1901 for Chemistry, Physics, Medicine, Literature, and Peace. A Prize in Economics was added later. Nobel Prizes are very prestigious. They include a gold medal and a grant of 10 million Swedish Krona.

Nobel Prize medals are made from gold and have a portrait of Alfred Nobel on the front.

Check your understanding

1. a. Write an equation to show the formation of phenylamine from nitrobenzene.
   b. Give two examples of how phenylamine can be used to produce dyes.
2. a. Write an equation to show how TNT is made from methylbenzene.
   b. Suggest why TNT is a more practical explosive than nitroglycerine.
3. Using the work of William Perkin and Alfred Nobel as examples, discuss how chemical discoveries can generate wealth and new industries.

7.05 Aromatic ketones from benzene

OUTCOMES

already from A2 Level, you

- know the reactions of water, alcohols, ammonia, and primary amines with acyl chlorides
- understand that electrophilic attack in arenes results in substitution
- understand the mechanism of nitration, including the generation of the nitronium ion
- understand that nitration is an important step in synthesis

and after this spread you should

- understand that Friedel–Crafts acylation reactions are important steps in synthesis
- understand the mechanism of acylation using $AlCl_3$ as a catalyst

Benzene undergoes electrophilic substitution reactions, such as nitration to produce nitrobenzene. It will also react with acyl chlorides in the presence of an aluminium chloride catalyst to form aromatic ketones. Here is the general equation for the overall reaction:

$$C_6H_6 + RCOCl \rightarrow C_6H_5COR + HCl$$

For example, phenylethanone is formed when benzene reacts with ethanoyl chloride in the presence of an aluminium chloride catalyst:

$$\underset{\text{benzene}}{C_6H_6} + \underset{\text{ethanoyl chloride}}{CH_3COCl} \rightarrow \underset{\text{phenylethanone}}{C_6H_5COCH_3} + HCl$$

phenylethanone

Reactions like this one are called **Friedel–Crafts acylation** reactions, after the two chemists who discovered them. They are electrophilic substitution reactions, like the nitration of benzene.

Friedel–Crafts acylation

The nitronium ion NO_2^+ is the electrophile in the nitration of benzene. The electrophile in a Friedel–Crafts acylation is called an **acylium ion**. The particular acylium ion generated depends upon the acyl chloride used. This then determines the aromatic ketone formed.

Generating the electrophile

The aluminium atom in aluminium chloride $AlCl_3$ has a **vacant orbital**. It can accept the pair of electrons from the C—Cl bond in the acyl chloride molecule.

The reaction between an acyl chloride and aluminium chloride produces the electrophile needed in a Friedel–Crafts acylation.

This step can also be written as an equation. For example, this is the equation for the generation of an electrophile from ethanoyl chloride:

$$CH_3COCl + AlCl_3 \rightarrow CH_3\overset{+}{C}O + [AlCl_4]^-$$

Note that the positive charge is placed above the carbon atom in the C=O bond.

The vacant orbital in aluminium chloride

The aluminium atom in aluminium chloride has a vacant orbital. This means that it can accept a pair of electrons, forming a co-ordinate bond. Aluminium chloride is acting here as a *Lewis acid*. You find out more about Lewis acids and bases in Unit 5.

A dot and cross diagram to illustrate the bonding in aluminium chloride.

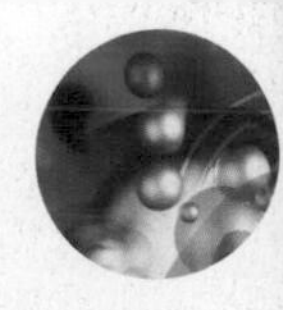

The reaction mechanism

Friedel–Crafts acylation has two main steps. These are very similar to the reaction mechanism for the nitration of benzene, but with a different electrophile.

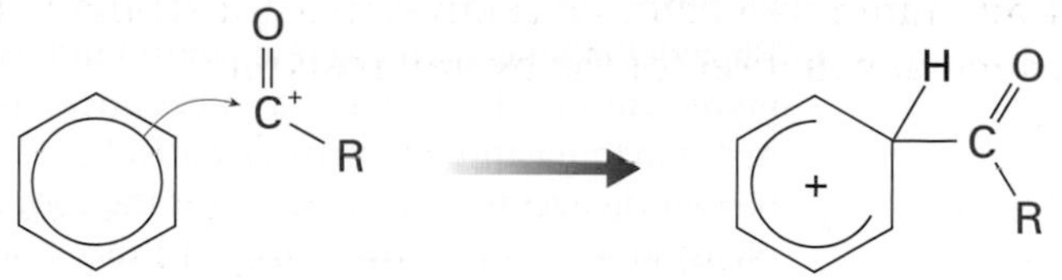

In the first step, the acylium ion is attracted to the delocalized electron cloud. The curly arrow is drawn from the circle and points towards the electrophile. A co-ordinate bond forms between one of the carbon atoms in the benzene ring and the positively charged carbon atom. An unstable positively charged intermediate is formed.

As with the reaction mechanism for the nitration of benzene, the partial circle should not extend any further than the two carbon atoms next to the one involved in the new bond. Again, a hydrogen atom is attached to each carbon atom in benzene, so one has not appeared from nowhere.

In the second step, the C—H bond breaks. Its pair of electrons restores the stable delocalized electron structure, and a hydrogen ion is eliminated.

Here is the reaction mechanism for the formation of phenylethanone. The generation of the electrophile from ethanoyl chloride has been left out for clarity.

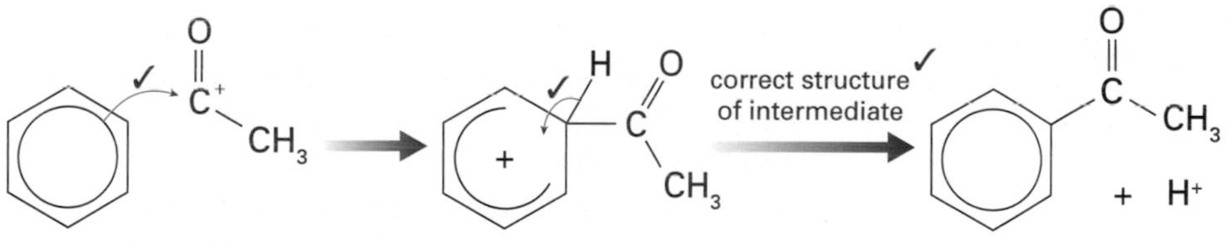

correct structure of intermediate ✓

The formation of phenylethanone from benzene and an acylium ion, with ticks to show the features an examiner is likely to look for in an examination.

Aluminium chloride as a catalyst

The tetrachloroaluminium ion $[AlCl_4]^-$ is formed when aluminium chloride reacts with the acyl chloride. It reacts with the hydrogen ion eliminated from the positively charged intermediate product. This regenerates aluminium chloride and forms hydrogen chloride.

$$[AlCl_4]^- + H^+ \rightarrow AlCl_3 + HCl$$

Check your understanding

1 Phenylpropan-1-one, $C_6H_5COCH_2CH_3$, can be produced from benzene and an acyl chloride in a Friedel–Crafts acylation reaction.

- **a** Name the acyl chloride needed and give its structural formula.
- **b** Write an equation to show how a reactive intermediate is formed when the acyl chloride reacts with aluminium chloride.
- **c** Draw the mechanism for the reaction between the reactive intermediate and benzene.
- **d** Name the reaction mechanism in part **c**.
- **e** Explain why aluminium chloride is able to form the reactive intermediate with the acyl chloride, and why it acts as a catalyst.

Questions: Chs 4–7

1 a **A**, **B**, and **C** have the molecular formula C_6H_{12}.

All three are branched-chain molecules and none is cyclic.

A can represent a pair of *E-Z* isomers.

B can represent another pair of *E-Z* isomers different from **A**.

C can represent a pair of optical isomers.

Identify compounds **A**, **B**, and **C**. [3]

b Pentan-2-one reacts with reagent **J** to form compound **K**, which exists as a racemic mixture. Dehydration of **K** forms **L**, C_6H_9N, which can represent a pair of *E-Z* isomers.

i State the meaning of the term *racemic mixture* and explain why such a mixture is formed in this reaction.

ii Identify reagent **J**, and draw a structural formula for each of **K** and **L**. [6]

[Total 9 marks]

2 a Consider the following pair of isomers.

X $H-\overset{\overset{O}{\|}}{C}-O-(CH_2)_3CH_3$

Y $HO-\overset{\overset{O}{\|}}{C}-(CH_2)_3CH_3$

i Name compound **X**.

ii Identify a reagent which could be used in a test-tube reaction to distinguish between **X** and **Y**. State what you would observe when you add your reagent, separately, to **X** and **Y**. [4]

b Look at the following pair of isomers.

M $H_3C-\overset{\overset{O}{\|}}{C}-O-(CH_2)_3CH_3$

N $H-\overset{\overset{O}{\|}}{C}-(CH_2)_4CH_3$

i Name compound **M**.

ii Name a reagent which could be used in a test-tube reaction to distinguish between **M** and **N**. State what you would see when this reagent is added first to **M** and then to **N**. [4]

c Draw the structure of a chain isomer of **N** which shows optical isomerism. [1]

[Total 9 marks]

3 a Draw out the mechanism for the reaction of CH_3CH_2CHO with HCN and name the product. [5]

b Draw out the mechanism for the reaction of CH_3OH with CH_3COCl and give the name of the organic product. [5]

c Phenylethanone is an organic compound used as an ingredient in perfumes and as a chemical intermediate in the manufacture of pharmaceuticals, resins, and flavouring agents. An equation for the formation of phenylethanone is shown below. In this reaction a reactive intermediate is formed from ethanoyl chloride. This intermediate then reacts with benzene.

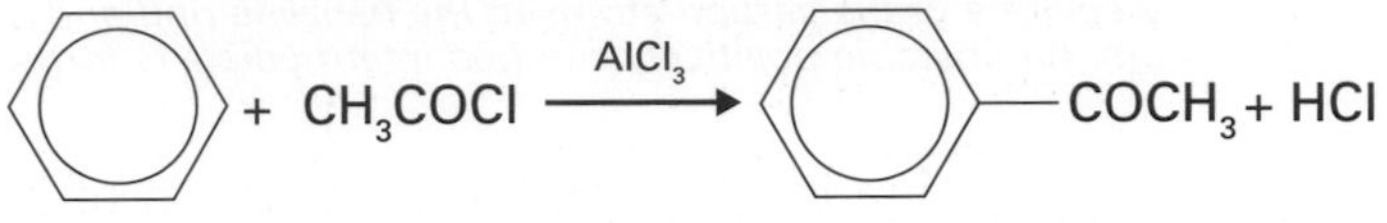

$$C_6H_6 + CH_3COCl \xrightarrow{AlCl_3} C_6H_5COCH_3 + HCl$$

i Give the formula of the reactive intermediate.

ii Draw out the mechanism for the reaction of this intermediate with benzene to form phenylethanone. [4]

[Total 14 marks]

4 a **i** Write a symbol equation for the formation of propanamine (propylamine) from propanenitrile.

ii Give the name or formula of a suitable reagent or a combination of reagent and catalyst for this reaction. [2]

b Name the type of reaction taking place between propanamine and an excess of bromoethane. Give the structures of the three organic products obtained from this reaction. [4]

c Draw out the mechanism to clearly show how molecules of propanamine and hydrogen bromide react together. [3]

d Write an equation for the formation of the compound $CH_3CONHCH_3$ from methanamine (methylamine) and a suitable reagent.

Name and draw out the mechanism for this reaction. [7]

[Total 16 marks]

5 Look at the following reaction scheme.

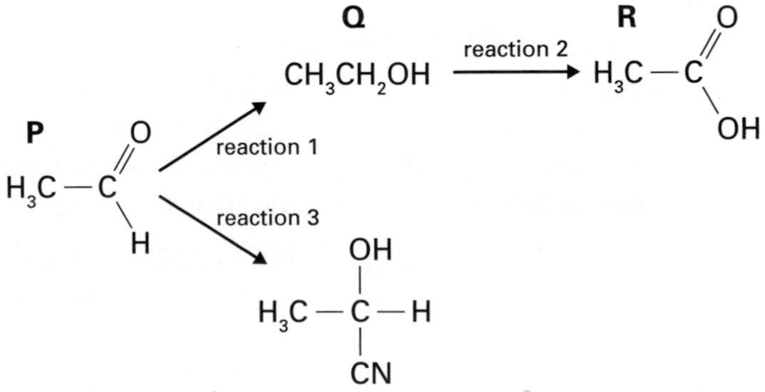

P H_3C-CHO (ethanal) —reaction 1→ **Q** CH_3CH_2OH —reaction 2→ **R** $H_3C-COOH$

P —reaction 3→ $H_3C-C(OH)(H)-CN$

a Name the reagent(s) used in reaction 1 and name the type of reaction involved. [2]

b **i** State the reagent(s) and conditions needed to carry out reaction 2.

ii Write an equation for this reaction using the symbol [O] to represent the oxidizing agent. [4]

c Give the reagent(s) and name the mechanism involved in reaction 3. [2]

d Reaction 3 produces a mixture of two stereoisomers.

i What form of stereoisomerism is shown by these two isomers?

ii How can separate samples of these isomers be distinguished? [3]

e **i** Draw the structure and give the name of the organic product formed when **Q** reacts with **R**.

ii Draw the structure of an isomer of **R** which forms methanol on hydrolysis. [3]

f Write an equation for the complete combustion of **Q**. [2]

[Total 16 marks]

6 *N*-phenylethanamide has pain-killing and fever-reducing properties but it can also have some very nasty side-effects. It has been replaced by less toxic drugs such as paracetamol. *N*-phenylethanamide can be prepared from benzene in three steps:

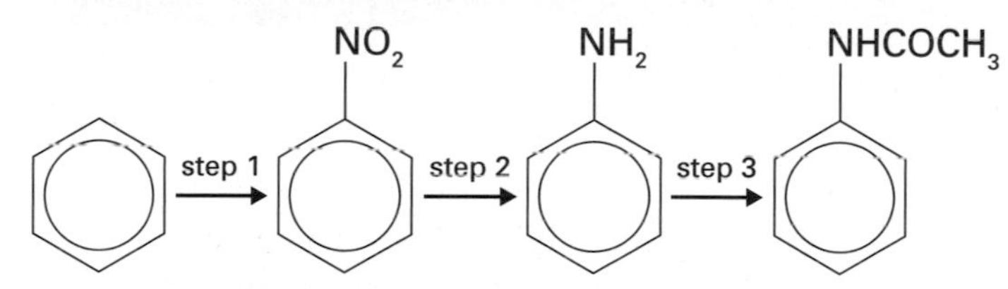

a State the reagents needed to carry out Step 1 and write an equation for the formation of the reactive inorganic species present.

Name and draw out the mechanism for the reaction between this species and benzene. [7]

b Name the type of reaction taking place in Step 2 and state a suitable reagent or combination of reagents that would be used in the reaction. [2]

c Write an equation for the reaction occurring in Step 3.

Name and draw out the mechanism for this reaction. [7]

[Total 16 marks]

7 Look at the following reaction scheme and then answer the questions below.

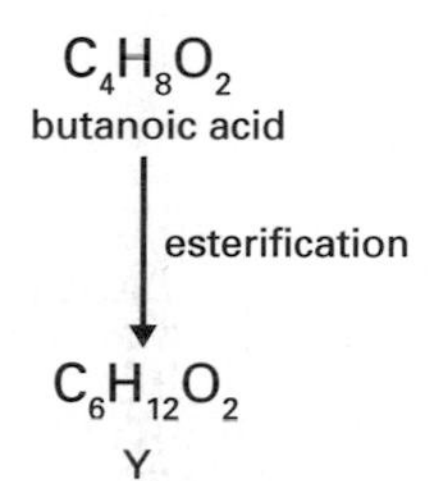

a Draw the displayed formula of butanoic acid. [1]

b Butanoic acid may be converted into compound **Y** by an esterification reaction.

i Give the reagent(s) and condition(s) required to make compound **Y** from butanoic acid. [3]

ii Give the name of compound **Y**. [1]

iii Write the equation for the esterification reaction. [2]

c Butyl ethanoate is an ester that is a structural isomer of compound **Y**. When butyl ethanoate is heated with aqueous sodium hydroxide, two products are formed.

i Explain what is meant by the term *structural isomerism*. [2]

ii Give the names and structures of the two products of this reaction. [4]

[Total 13 marks]

8 The molecular formulae of some compounds that can be prepared from propanoic acid are shown in the reaction scheme below.

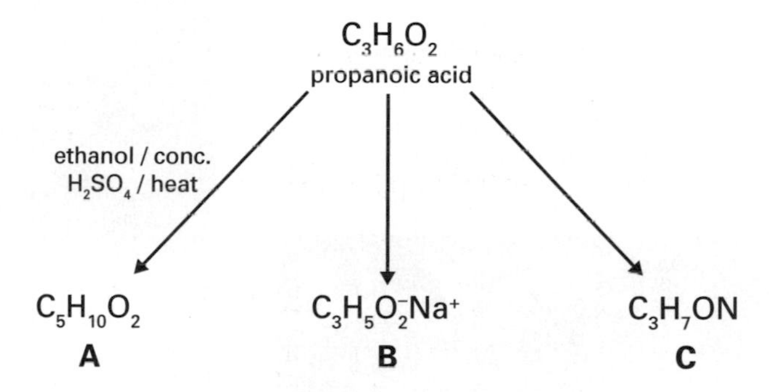

a **i** Give the name and displayed formula of **A**. [2]

ii Name of the type of reaction which occurs when **A** is formed from propanoic acid. [1]

b Propanoic acid can be obtained from **A**.

i Give the name of the reagent(s) and state the conditions needed. [2]

ii Write a balanced equation for the reaction [1]

c **i** Write the name or formula of the reagent and state the reaction conditions needed to convert propanoic acid into **B**. [2]

d **i** Give the name and displayed formula of the amide **C**. [2]

ii State the reagent(s) and reaction conditions that could be used for converting propanoic acid into **C**. [2]

iii Write a balanced equation for the reaction between **C** and aqueous hydrochloric acid. [1]

[Total 13 marks]

8.01 Naming amines

OUTCOMES

already from A2 Level, you

- know the reactions of primary amines with acyl chlorides
- understand the mechanism of nucleophilic addition–elimination reactions between primary amines and acyl chlorides

and after this spread you should be able to

- recognize the displayed formulae of primary, secondary, and tertiary amines
- apply IUPAC rules for nomenclature for amines
- recognize the displayed formulae of quaternary ammonium ions

Amines with a low relative formula mass smell rather like ammonia, but many have a fishy smell. Proteins break down to produce cadaverine, putrescine, and other foul-smelling amines.

Amines are organic compounds containing a nitrogen atom. They are derived from ammonia, NH_3, in which one or more of the hydrogen atoms has been replaced by an alkyl or aryl group. Amines are classified as primary, secondary, or tertiary amines according to how many hydrogen atoms have been substituted.

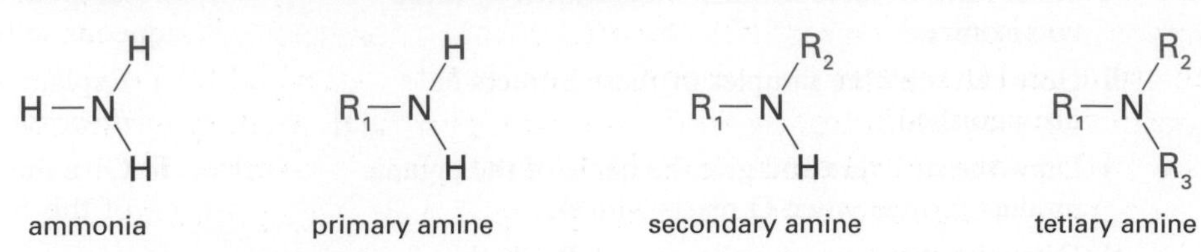

The number of alkyl or aryl groups attached to the nitrogen atom determines whether an amine is primary, secondary, or tertiary.

Molecular models of methanamine (a primary amine), N*-methylmethanamine (a secondary amine), and* N,N*-dimethylmethanamine (a tertiary amine).*

Primary amines

Primary amines have the general formula RNH_2. They contain the amino group $-NH_2$. They are named after the hydrocarbon group, with *amine* as the suffix. So CH_3NH_2 is methanamine and $CH_3CH_2NH_2$ is ethanamine. Note that the *e* from methane and ethane is left out.

methanamine

ethanamine

Displayed formulae of methanamine and ethanamine.

Position isomers of primary amines are possible. The longest carbon chain containing the amino group is identified and named, and the carbon atom to which the amino group is attached is numbered. For example, propanamine has two position isomers, propan-1-amine and propan-2-amine.

propan-1-amine

propan-2-amine

Displayed formulae of propan-1-amine and propan-2-amine.

Note that propan-2-amine is still a primary amine, even though the carbon atom to which the amino group is attached is directly attached to two other carbon atoms. This is because only one hydrogen atom on the parent ammonia molecule has been replaced. This is a different situation from the alcohols, where propan-2-ol is a secondary alcohol.

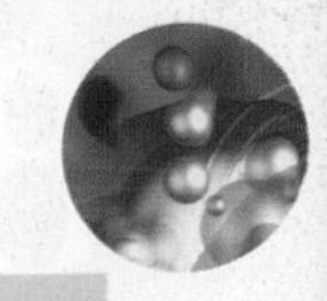

The prefix *amino* is used if the amino group is not the most important one when naming the compound. For example, $H_2NCH_2CH_2COOH$ is 3-aminopropanoic acid.

Secondary amines

Secondary amines have the general formula R_2NH. They are more complex to name than primary amines. The longest hydrocarbon chain attached to the nitrogen atom is identified. This is named in the same way as for a primary amine. The shorter hydrocarbon chain is named as usual, and the prefix *N*- is used to show that it is attached to the longer chain by the nitrogen atom. For example, the secondary amine containing two methyl groups is called *N*-methylmethanamine. The secondary amine containing a methyl group and an ethyl group is called *N*-methylethanamine.

```
                                       H
                                       |
                                     H—C—H
  H       H   H               H        |   H
  |       |   |               |        |   |
H—C — N — C — C—H           H—C — N — C — C—H
  |   |   |   |               |   |   |   |
  H   H   H   H               H   H   H   H
```

N-methylethanamine　　*N*-methylpropan-2-amine

Displayed formulae of N*-methylethanamine and* N*-methylpropan-2-amine.*

Tertiary amines

Tertiary amines have the general formula R_3N. They are named in a similar way to secondary amines. This time two *N*- prefixes are used. For example, the tertiary amine containing three methyl groups is called *N,N*-dimethylmethanamine. The tertiary amine containing two methyl groups and an ethyl group is called *N,N*-dimethylethanamine. *N*-ethyl-*N*-methylpropan-1-amine contains a methyl group, an ethyl group, and a propyl group. Note that the smaller hydrocarbon chains are named in alphabetical order.

```
                                              H
                                              |
                                            H—C—H
  H       H   H             H   H             |   H
  |       |   |             |   |             |   |
H—C — N — C — C—H         H—C — C — N — C — C—H
  |   |   |   |             |   |   |   |   |
  H   |   H   H             H   H   |   H   H
    H—C—H                         H—C—H
      |                             |
      H                             H
```

N,N-dimethylethanamine　　*N*-ethyl-*N*-methylpropan-2-amine

Displayed formulae of N,N*-dimethylethanamine and* N*-ethyl-*N*-methylpropan-2-amine.*

Check your understanding

1 Draw the displayed formulae of the following compounds.

  a butan-1-amine (1-aminobutane)

  b butan-2-amine (2-aminobutane)

  c 2-methylpropan-1-amine

  d *N*-ethyl-2-methylpropan-1-amine

  e *N*-ethyl-*N*-methylethanamine

2 Identify the primary, secondary, and tertiary amines in question **1**.

Other naming systems for amines

There are older ways to name amines. Methanamine is also called methylamine, and ethanamine is also called ethylamine. In this system position isomers are shown using the prefix *amino* instead of the suffix *amine*. For example, propan-1-amine is 1-aminopropane, and propan-2-amine is 2-aminopropane. The prefixes *di* and *tri* are used in some common names. For example, methanamine is methylamine, *N*-methylmethanamine is dimethylamine, and *N,N*-dimethylmethanamine is trimethylamine.

Science @ Work: Quaternary ammonium ions

Quaternary ammonium ions have the general formula R_4N^+. They are derived from the ammonium ion, NH_4^+, in which all of the hydrogen atoms have been replaced by an alkyl or aryl group. They are named in a similar way to tertiary amines, except three *N*-prefixes are needed, and the suffix is *aminium*.

```
        CH3
        |
 H3C —  N+ — CH3
        |
        CH3
```

The N,N,N*-trimethylmethanaminium ion, also called the tetramethylammonium ion.*

8.02 Making amines

OUTCOMES

already from AS Level, you understand

- that haloalkanes are susceptible to nucleophilic attack by CN^- and NH_3

already from A2 Level, you can

- recognize the displayed formulae of primary amines and apply IUPAC rules for their nomenclature

and after this spread you should know that

- primary aliphatic amines can be prepared from haloalkanes and by the reduction of nitriles
- aromatic amines are prepared by the reduction of nitro compounds

Primary aliphatic amines such as propan-1-amine can be prepared from haloalkanes, and by the reduction of nitriles. Aromatic amines are prepared by the reduction of nitro compounds, such as nitrobenzene (see Spread 7.03).

Primary aliphatic amines from haloalkanes

Ammonia can act as a nucleophile because its nitrogen atom has a lone pair of electrons. It reacts with primary haloalkanes to produce primary amines. For example 1-bromopropane reacts with excess ammonia to form propan-1-amine:

$$CH_3\,CH_2CH_2Br + 2NH_3 \rightarrow CH_3CH_2CH_2NH_2 + NH_4Br$$

Two ammonia molecules are needed to react with one haloalkane molecule. One acts as a nucleophile and the other acts as a base.

Step 1 Ammonia acting as a nucleophile:

$$CH_3CH_2CH_2Br + NH_3 \rightarrow CH_3CH_2CH_2\overset{+}{N}H_3 + Br^-$$

Step 2 The ammonia acting as a base:

$$CH_3CH_2CH_2\overset{+}{N}H_3 + NH_3 \rightarrow CH_3CH_2CH_2NH_2 + NH_4^+$$

Here the ammonia molecule accepts a hydrogen ion from the positively charged intermediate.

The ammonium ion from step 2 and the bromide ion from step 1 form ammonium bromide.

Primary aliphatic amines from nitriles

Amines can also be made by reducing nitriles. In general:

$$RCN + 4[H] \rightarrow RCH_2NH_2$$

Remember that [H] in the equation represents the reducing agent. This can be

- lithium tetrahydridoaluminate(III), $LiAlH_4$
- hydrogen gas in the presence of a nickel catalyst

For example, propan-1-amine is made by reducing propanenitrile:

$$CH_3CH_2CN + 4[H] \xrightarrow[\text{or } H_2/Ni]{LiAlH_4} CH_3CH_2CH_2NH_2$$

H H
| |
H—C—C—C≡N + 4[H]
| |
H H

↓

H H H   H
| | | /
H—C—C—C—N
| | | \
H H H   H

The formation of propan-1-amine from propanenitrile.

Haloalkanes or nitriles?

Reacting haloalkanes with ammonia is not an efficient way to make primary amines. The reaction also makes secondary amines, tertiary amines, and quaternary ammonium salts. Nitriles can provide a more efficient indirect route for making primary amines from haloalkanes. The nitriles themselves are made from haloalkanes. For example, bromoethane reacts with the cyanide ion in a nucleophilic substitution reaction to form propanenitrile:

$$CH_3CH_2Br + CN^- \rightarrow CH_3CH_2CN + Br^-$$

Propan-1-amine can then be formed by reducing the propanenitrile. Notice that the final primary amine contains one more carbon atom than the original haloalkane.

Aromatic amines from nitro compounds

Aromatic amines are made by reducing aromatic nitro compounds. In general:

$$RNO_2 + 6[H] \rightarrow RNH_2 + 2H_2O$$

Notice that water is also formed, and that you need 6[H] to balance the equation but only 4[H] to balance the equation for the reduction of a haloalkane. The reducing agent can be:

- granulated tin and concentrated hydrochloric acid
- hydrogen gas in the presence of a nickel catalyst

Scrap iron is used industrially instead of tin. Phenylamine, also called aniline, is an important chemical used in the manufacture of dyes. It is made by reducing nitrobenzene:

$$C_6H_5NO_2 + 6[H] \xrightarrow[\text{or conc. HCl/Fe}]{\text{conc. HCl/Sn}} C_6H_5NH_2 + 2H_2O$$

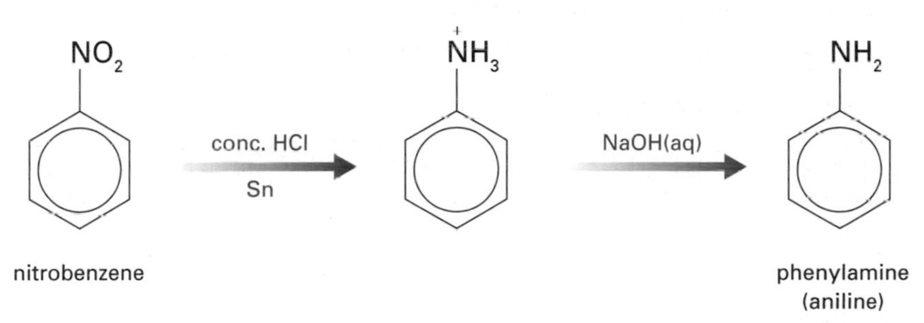

Phenylamine is made by reducing nitrobenzene. Phenylamine is weakly basic and its protonated form, $C_6H_5\overset{+}{N}H_3$, exists in the acidic reaction mixture. The amine is formed by adding a base such as sodium hydroxide.

Check your understanding

1 Describe two ways of producing butan-1-amine. Include the names of the reagents needed and suitable equations.

2 Describe how 4-amino-1-methylbenzene can be produced from 1-methyl-4-nitrobenzene. Include the names of the reagents needed and suitable equations.

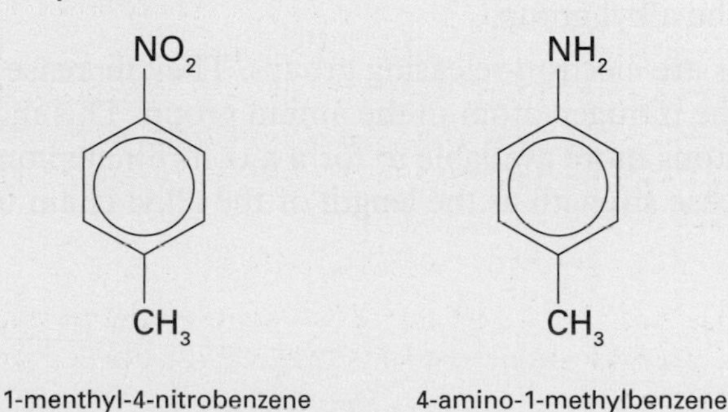

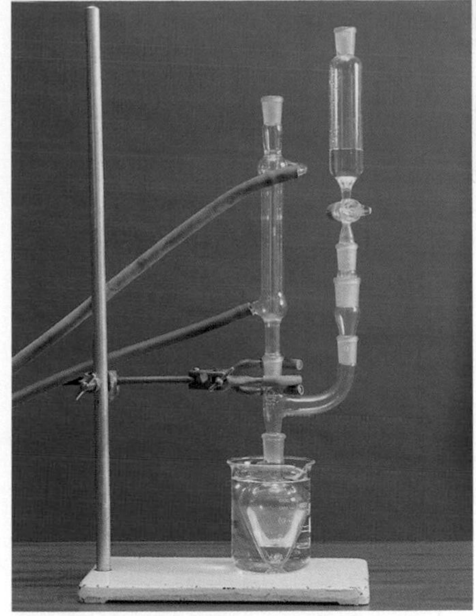

This apparatus is used to make phenylamine from nitrobenzene. Granulated tin and nitrobenzene are placed in the flask. The reflux condenser is fitted and concentrated hydrochloric acid is added slowly, cooling if needed. The condenser is removed and the flask heated in a boiling water bath for about an hour. The mixture is cooled and concentrated aqueous sodium hydroxide is added to produce the free amine.

Science @ Work

Aliphatic vs aromatic amines

You will have noticed that both aliphatic amines and aromatic amines are produced by reduction reactions. Aliphatic amines are also produced by reacting primary aliphatic haloalkanes with ammonia, but why are aromatic amines not produced by reacting haloarenes with ammonia?

Ammonia is a nucleophile. The lone pair of electrons on its nitrogen atom is repelled by the cloud of delocalized electrons in benzene. Delocalization also increases the strength of the carbon–halogen bond in haloarenes, and reduces its polarity. So ammonia reacts slowly with haloarenes such as bromobenzene, and the yield of aromatic amines is low.

8.03 Amines as bases

OUTCOMES

already from A2 Level, you know that

- an acid is a proton donor
- a base is a proton acceptor
- acid–base equilibria involve the transfer of protons

and after this spread you should

- be able to explain the difference in base strength between ammonia, primary aliphatic amines, and primary aromatic amines in terms of the availability of a lone pair on the nitrogen atom

Co-ordinate bonds

A covalent bond is a shared pair of electrons. Usually, the two atoms involved in a covalent bond each contribute one electron. But in a co-ordinate bond (also called a **dative covalent bond**) one of the two atoms contributes both electrons. Once formed, a co-ordinate bond behaves just the same as other covalent bonds.

For a co-ordinate bond to form, there must be

- a lone pair of electrons on one of the atoms; and
- a vacant orbital on the other atom.

There is a lone pair of electrons on the nitrogen atom in ammonia, and the proton has a vacant orbital. The co-ordinate bond they form is shown in the displayed formula as an arrow, pointing from the atom that contributes the bonding pair of electrons.

Ammonia, primary aliphatic amines, and primary aromatic amines are Brønsted–Lowry bases because they can accept protons. Their base strengths are different, and depend on the availability of the lone pair of electrons on their nitrogen atom.

Ammonia as a base

The ammonia molecule has a lone pair of electrons. It can form **co-ordinate bonds** with electron-deficient species such as the proton or hydrogen ion. When ammonia dissolves in water, it accepts protons from water molecules, forming ammonium ions and hydroxide ions:

$$NH_3(aq) + H_2O(l) \rightleftharpoons NH_4^+(aq) + OH^-(aq)$$

The solution is basic because of the presence of the hydroxide ions. The reaction does not go to completion, so ammonia is a weak base. For example, 1.0 mol dm^{-3} aqueous ammonia has a pH of 11.63 but aqueous sodium hydroxide at the same concentration has a pH of 14.00.

ammonia

ammonium ion

Dot and cross diagrams to show how ammonia accepts a proton to form the ammonium ion, and so acts as a Brønsted–Lowry base.

The displayed formula of the ammonium ion.

Amines as bases

The nitrogen atom in an amine molecule also has a lone pair of electrons. It can accept a proton, so amines act as Brønsted–Lowry bases. For example, aqueous methanamine accepts protons from water molecules, forming methanaminium ions and hydroxide ions:

$$CH_3NH_2(aq) + H_2O(l) \rightleftharpoons CH_3\overset{+}{N}H_3(aq) + OH^-(aq)$$

The solution is basic because of the presence of the hydroxide ions.

Primary aliphatic amines

Primary aliphatic amines are stronger bases than ammonia. For example, 1.0 mol dm^{-3} aqueous ammonia has a pH of 11.63 but aqueous methanamine at the same concentration has a pH of 12.32. This is because of the alkyl group.

Alkyl groups are electron-releasing groups. They increase the electron density on the nitrogen atom in the amino group. This makes its lone pair of electrons more available to form a co-ordinate bond. There is an increase in base strength as the length of the alkyl chain increases.

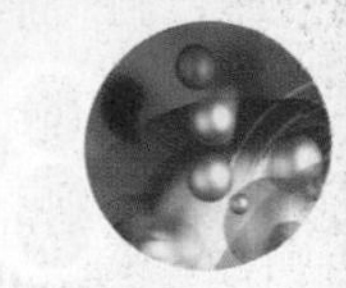

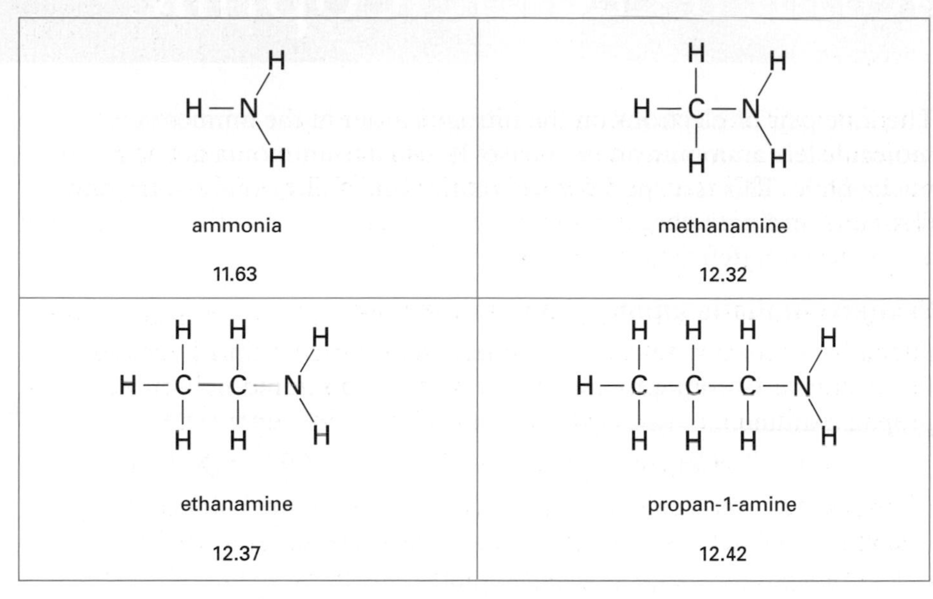

ammonia
11.63
methanamine
12.32
ethanamine
12.37
propan-1-amine
12.42

There is an increase in base strength from methanamine to propan-1-amine, due to the positive inductive effect of the alkyl groups. The numbers shown are the pH values of 1.0 mol dm^{-3} solutions at 25°C.

Primary aromatic amines

Primary aromatic amines are weaker bases than ammonia. For example, 1.0 mol dm^{-3} aqueous phenylamine has a pH of 9.32 but aqueous ammonia at the same concentration has a pH of 11.63. This is because of their benzene ring. The lone pair of electrons on the nitrogen atom is less available because it delocalizes with the delocalized electrons in the benzene ring.

Phenylamine (aniline) is a weaker base than ammonia because delocalization makes the lone pair of electrons on its nitrogen atom less available.

Check your understanding

1 a Write an equation for the reaction of ethanamine with water to produce an alkaline solution.

b Explain why ethanamine is a Brønsted–Lowry base.

c Explain why ethanamine is a stronger base than ammonia.

2 a Explain why phenylamine, $C_6H_5NH_2$, is a weaker base than ammonia.

b Suggest why cyclohexanamine, $C_6H_{11}NH_2$, is a stronger base than phenylamine (aniline).

Science @ Work: pK_a data and aliphatic amines

You learned in Spread 3.04 that pK_a values give a measure of the strength of a weak acid. The lower the pK_a value, the stronger the acid. The converse is true for bases. The higher the pK_a value, the stronger the base.

name	formula	pK_a
ammonia	NH_3	9.25
methanamine	CH_3NH_2	10.63
ethanamine	$CH_3CH_2NH_2$	10.73
propan-1-amine	$CH_3CH_2CH_2NH_2$	10.84

The pK_a values for ammonia and three primary aliphatic amines, arranged in order of increasing base strength.

Science @ Work: Phenylmethanamine

If the amino group is not directly attached to the benzene ring, the lone pair of electrons on the nitrogen atom does not delocalize with the electrons in the benzene ring. For example, the amino group in phenylmethanamine is attached via a CH_2 group. As a result, phenylmethanamine is a stronger base than ammonia. Its pK_a value is 9.33, whereas the pK_a of ammonia is 9.25.

phenylmethanamine

8.04 Ammonia as a nucleophile

OUTCOMES

already from AS level you understand

- the mechanism of nucleophilic substitution in primary haloalkanes

and after this spread you should understand

- the nucleophilic substitution reaction of excess ammonia with haloalkanes (including reaction mechanism) to form primary amines

Haloalkanes and reactivity

The reactivity of haloalkanes does not depend on the reactivity of the halogen. Instead it depends upon the strength of the carbon–halogen bond. The weaker it is, the more reactive the haloalkane. Fluoroalkanes are not readily attacked by nucleophiles because the carbon–fluorine bond is so strong.

bond	mean bond enthalpy ($kJ\ mol^{-1}$)
C—F	484
C—Cl	338
C—Br	276
C—I	238

The reactivity of the haloalkanes increases in the order fluoroalkanes < chloroalkanes < bromoalkanes < iodoalkanes.

The lone pair of electrons on the nitrogen atom in the ammonia molecule lets ammonia act as a base. It also lets ammonia act as a nucleophile. This is a species with a lone pair of electrons available to form a co-ordinate bond. It is attracted to regions of positive charge, such as the electron-deficient carbon atom in haloalkanes.

Primary aliphatic amines from haloalkanes

Haloalkanes react with excess ammonia to produce primary amines. For example, 1-bromopropane reacts with excess ammonia to form propan-1-amine:

$$CH_3CH_2CH_2Br + 2NH_3 \rightarrow CH_3CH_2CH_2NH_2 + NH_4Br$$

Two ammonia molecules are needed to react with one haloalkane molecule. One acts as a nucleophile and the other acts as a base.

Step 1 Ammonia acting as a nucleophile:

$$CH_3CH_2CH_2Br + NH_3 \rightarrow CH_3CH_2CH_2\overset{+}{N}H_3 + Br^-$$

Step 2 Ammonia acting as a base:

$$CH_3CH_2CH_2\overset{+}{N}H_3 + NH_3 \rightarrow CH_3CH_2CH_2NH_2 + NH_4^+$$

Here the ammonia molecule accepts a hydrogen ion from the positively charged intermediate.

The ammonium ion from step 2 and the bromide ion from step 1 form ammonium bromide.

The reaction mechanism for the formation of propan-1-amine (1-aminopropane) from 1-bromopropane and excess ammonia by nucleophilic substitution. The ticks show the features an examiner is likely to look for in an examination. You must show the correct intermediate to gain full marks.

Excess ammonia

Ammonia must be in excess to reduce the chance of further substitutions happening. The nitrogen atom in primary amines has a lone pair of electrons, just as ammonia does. This means that primary amines can also act as nucleophiles. Once formed in the reaction mixture, primary amines compete with ammonia to react with the haloalkane present. A secondary amine is formed as a result. For example, propan-1-amine reacts with 1-bromopropane to form *N*-propylpropan-1-amine:

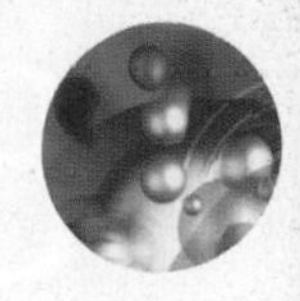

$$CH_3CH_2CH_2NH_2 + CH_3CH_2CH_2Br \rightarrow (CH_3CH_2CH_2)_2NH + HBr$$

propan-1-amine   1-bromopropane   *N*-propylpropan-1-amine

The nitrogen atom in secondary amines also has a lone pair of electrons. So secondary amines can act as nucleophiles, too. Further substitutions can happen, producing tertiary amines and finally quaternary ammonium salts.

Chilly ammonia

Ammonia is manufactured from nitrogen and hydrogen using the Haber Process:

$$N_2(g) + 3H_2(g) \xrightleftharpoons[]{\text{iron catalyst, 20 MPa / 450°C}} 2NH_3(g)$$

Over 100 million tonnes of ammonia is produced worldwide each year. Most of this is used in the manufacture of artificial fertilizers, such as ammonium nitrate and ammonium sulfate. Ammonia has some useful physical properties, too.

Although ammonia is a gas at room temperature, it is easily liquefied by reducing its temperature or putting it under pressure. Liquid ammonia itself may be used as fertilizer. Known as 'anhydrous ammonia', it is injected directly into the soil but away from the plants so that they are not damaged by its corrosive properties.

Anhydrous ammonia being applied as a fertilizer in the USA. The ammonia is stored under pressure in the large tank.

The ease with which ammonia is liquefied makes it useful as a refrigerant gas for commercial refrigerators and air conditioners. But it is corrosive and toxic, and its use declined when chlorofluorocarbons or CFCs became available. To help protect the Earth's ozone layer, the use of CFCs is now decreasing. As a result of this, and of advances in technology, ammonia is being widely used as a refrigerant again.

Ammonia is the refrigerant in air conditioning plants for many offices and large public buildings, such as Heathrow Airport's huge Terminal 5 building.

Check your understanding

1. a What is a nucleophile?
   b Explain why ammonia can act as a nucleophile.
2. a Write an equation to show the reaction between 2-bromopropane and excess ammonia to form propan-2-amine (2-aminopropane), $CH_3CH(NH_2)CH_3$.
   b Name the reaction mechanism involved.
   c Draw the reaction mechanism for the reaction between 2-bromopropane and excess ammonia.

8.05 Amines as nucleophiles

OUTCOMES

already from A2 Level, you understand

- the nucleophilic substitution reaction of excess ammonia with haloalkanes (including reaction mechanism) to form primary amines

and after this spread you should

- understand the nucleophilic substitution reactions (including mechanisms) of amines with haloalkanes to form secondary amines, tertiary amines, and quaternary ammonium salts
- know that quaternary ammonium salts can be used as cationic surfactants

Ammonia reacts with haloalkanes to produce primary amines. But further nucleophilic substitution can take place in the reaction mixture. Amines can act as nucleophiles because of the lone pair of electrons on their nitrogen atom. Their alkyl groups make this more available for bonding, so the amine produced in each successive substitution reaction is a better nucleophile than the amine from which it was formed. As a result, reactions between ammonia and haloalkanes produce a mixture of primary amines, secondary amines, tertiary amines, and quaternary ammonium salts.

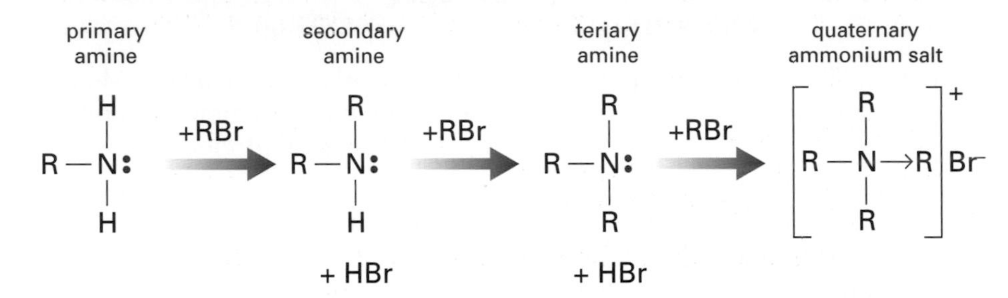

The nitrogen atom in primary amines, secondary amines, and tertiary amines has a lone pair of electrons. As a result, these compounds can act as nucleophiles in a series of reactions with haloalkanes.

Adjusting the reaction mixture

The production of primary amines is favoured if excess ammonia is used. This is because unreacted haloalkane molecules are more likely to react with the remaining ammonia, rather than with the primary amines produced.

The production of quaternary ammonium salts is favoured if excess haloalkane is used. This is because unreacted haloalkane molecules are more likely to react with the amines produced, rather than with ammonia.

Further substitution

Secondary amines from primary amines

Primary amines react with haloalkanes to produce secondary amines. For example ethanamine reacts with bromoethane to form *N*-ethylethanamine:

$$CH_3CH_2NH_2 + CH_3CH_2Br \rightarrow (CH_3CH_2)_2NH + HBr$$

The reaction mechanism for the formation of N*-ethylethanamine (diethylamine) from ethanamine (ethylamine) and bromoethane by nucleophilic substitution. The ticks show the features an examiner is likely to look for in an examination. You must show the correct intermediate to gain full marks.*

Tertiary amines from secondary amines

Secondary amines react with haloalkanes to produce tertiary amines. For example *N*-ethylethanamine reacts with bromoethane to form *N*,*N*-diethylethanamine:

$$(CH_3CH_2)_2NH + CH_3CH_2Br \rightarrow (CH_3CH_2)_3N + HBr$$

Check your understanding (1)

1 a Explain why methanamine CH_3NH_2 can act as a nucleophile.

b Explain why the *N,N,N*-trimethylmethanaminium ion (a quaternary ammonium ion) does not act as a nucleophile.

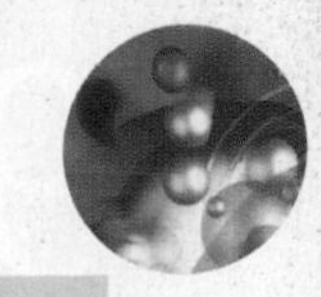

The reaction mechanism for the formation of N,N*-diethylethanamine (triethylamine) from* N*-ethylethanamine (diethylamine) and bromoethane by nucleophilic substitution.*

Quaternary ammonium salts from tertiary amines

Tertiary amines react with haloalkanes to produce quaternary ammonium salts. For example *N,N*-diethylethanamine reacts with bromoethane to form *N,N,N*-triethylethanaminium bromide:

$$(CH_3CH_2)_3N + CH_3CH_2Br \rightarrow (CH_3CH_2)_4N^+Br^-$$

The reaction mechanism for the formation of N,N,N*-triethylethanaminium bromide from* N,N*-diethylethanamine (triethylamine) and bromoethane by nucleophilic substitution.*

Uses of quaternary ammonium salts

Surface-active agents or **surfactants** reduce the surface tension of liquids. So the liquids spread further and wet materials more thoroughly. Quaternary ammonium ions act as **cationic surfactants**. Their positive charge lets them form bonds with water, and their alkyl groups let them form bonds with non-polar substances such as oils and fats.

Some quaternary ammonium salts have an antibacterial action. They dissolve bacterial cells and so kill the bacteria. For example, cetylpyridinium chloride is commonly used in disinfectants, mouthwashes, and toothpastes. Quaternary ammonium salts are also widely used in detergents and in the fabric dyeing industry, where they help the dye to pass into the fabric more efficiently.

Quaternary ammonium salts are used in fabric softeners, shampoos, and hair conditioners. They coat the fibres with positive charges, allowing the fibres to move past each other more easily.

Amines as bases

In the previous spread you found that ammonia has two roles in the reaction between excess ammonia and haloalkanes. It functions as a nucleophile and as a base. Amines can act in a similar way. For example, the reaction between ethanamine and bromoethane can be shown in two steps, rather than one.

Step 1 Amine acting as a nucleophile:

$$CH_3CH_2NH_2 + CH_3CH_2Br \rightarrow (CH_3CH_2)_2\overset{+}{N}H_2 + Br^-$$

Step 2 Amine acting as a base:

$$(CH_3CH_2)_2\overset{+}{N}H_2 + CH_3CH_2NH_2 \rightarrow (CH_3CH_2)_2NH + CH_3CH_2\overset{+}{N}H_3$$

The second reaction leads to the formation of a salt, in this case ethanaminium bromide. It is acceptable to leave this step out of your reaction mechanisms. Notice that the sequence of successive substitutions ends with quaternary ammonium salts. The nitrogen atom in the quaternary ammonium ion does not have a lone pair of electrons, and there are no more hydrogen ions to remove and be accepted by a base.

Check your understanding (2)

2 Draw the reaction mechanism for the reaction between *N,N*-dimethylmethanamine $(CH_3)_3N$ and chloromethane to form *N,N,N*-trimethylmethanaminium chloride $(CH_3)_4N^+Cl^-$.

3 State three uses of quaternary ammonium salts.

9.01 Amino acids

OUTCOMES

already from A2 Level, you know that

- carboxylic acids are weak acids
- primary aliphatic amines are bases

and after this spread you should

- understand that amino acids have both acidic and basic properties, including the formation of zwitterions

Amino acids are organic compounds that contain two functional groups, an amino group —NH_2 and a carboxyl group —COOH. They are the building blocks for proteins.

Amino acid structure

There are about twenty naturally occurring amino acids. Here is their general structure:

$H_2N-CH(R)-C(=O)-O-H$

Different amino acids have different R groups. The amino acids that form proteins are called α-amino (alpha-amino) acids, because the R group is attached to the alpha carbon atom (the one directly attached to the functional groups). The structures of some amino acids are shown below.

glycine: $H_2N-CH(H)-COOH$

alanine: $H_2N-CH(CH_3)-COOH$

serine: $H_2N-CH(CH_2OH)-COOH$

threonine: $H_2N-CH(CH(OH)CH_3)-COOH$

phenylalanine: $H_2N-CH(CH_2C_6H_5)-COOH$

glutamic acid: $H_2N-CH(CH_2CH_2COOH)-COOH$

histidine: $H_2N-CH(CH_2\text{-imidazolyl})-COOH$ (ring with NH and N)

cysteine: $H_2N-CH(CH_2SH)-COOH$

Some of the amino acids found in proteins.

GABA

Proteins only contain α-amino acids but other amino acids are also important to living organisms. One of these is 4-aminobutanoic acid or GABA, named after its older non-systematic name of gamma-aminobutyric acid. It is an important neurotransmitter, a molecule involved in passing nerve impulses from one nerve cell to another. Certain drugs to treat epilepsy increase the brain's production of GABA, so reducing brain activity.

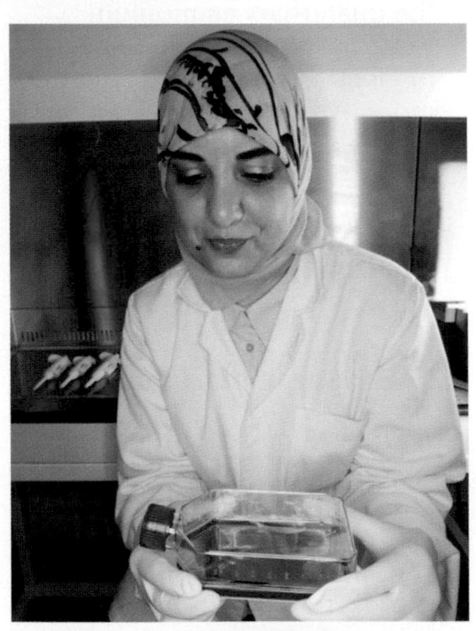

The cells growing in this flask are used in research into Alzheimer's disease. They have receptors for GABA, and the cells' responses to different drugs can be tested.

Zwitterions

The amino group —NH_2 is basic, and the carboxyl group —COOH is acidic. So amino acids are amphoteric – they can behave both as bases and as acids. For example, glycine (aminoethanoic acid) can both accept protons and donate them:

Acting as a base:

$$NH_2CH_2COOH(aq) + H^+(aq) \rightarrow \overset{+}{N}H_3CH_2COOH(aq)$$

Acting as an acid:

$$NH_2CH_2COOH(aq) + OH^-(aq) \rightarrow NH_2CH_2COO^-(aq) + H_2O(l)$$

Amino acids can also form salts with both acids and bases:

With an acid:

$$NH_2CH_2COOH(aq) + HCl(aq) \rightarrow [\overset{+}{N}H_3CH_2COOH]Cl^-(aq) + H_2O(l)$$

With a base:

$$NH_2CH_2COOH(aq) + NaOH(aq) \rightarrow NH_2CH_2COO^-Na^+(aq) + H_2O(l)$$

Amino acids also have another interesting property. In solution and in the solid state they exist as **zwitterions** (from the German word *zwitter*, pronounced 'tsvitter'). Both functional groups are charged in the zwitterion. The amino group is positively charged and the carboxyl group is negatively charged. The diagram shows how changing the pH affects the zwitterion.

$$H_3\overset{+}{N}-CH(R)-COOH \xleftarrow[\text{decreasing pH}]{+H^+(aq)} H_3\overset{+}{N}-CH(R)-COO^- \xrightarrow[\text{increasing pH}]{+OH^-(aq)} H_2N-CH(R)-COO^- + H_2O$$

zwitterion

The effect on the zwitterion of changing the pH. Remember to replace the R group with the formula for the appropriate group if you are asked a question about a particular amino acid.

Isoelectric points

The pH at which an amino acid is neutral overall is called its **isoelectric point**. You might expect this to be pH 7 but it is not. The side groups themselves may be acidic or basic. Different amino acids have different isoelectric points, varying from pH 2.85 for aspartic acid to pH 10.76 for arginine.

Melting points of amino acids

Amino acids dissolve readily in water and they are white crystalline solids at room temperature. These properties are a result of the existence of zwitterions. For example, glycine, propanoic acid CH_3CH_2COOH, and butan-1-amine $CH_3(CH_2)_3NH_2$ have similar relative formula masses (75, 74, and 73 respectively). Propanoic acid and butan-1-amine are liquids at room temperature but glycine is solid. The ionic bonds between the zwitterions in glycine are stronger than the hydrogen bonds between molecules of propanoic acid or butan-1-amine.

Optical isomerism

Except for glycine, amino acids have optical isomers or enantiomers. Their alpha carbon atom is asymmetrical. It has four different groups attached to it, so it acts as a chiral centre. Naturally occurring amino acids are L isomers (laevorotatory isomers, see Spread 4.03).

H, H_3C, COOH, NH_2 — H, HOOC, CH_3, H_2N

The two enantiomers of alanine (2-aminopropanoic acid) $NH_2CH(CH_3)COOH$.

Check your understanding

1 a Write the structural formulae for glycine and alanine, and state their systematic names.
  b Explain why alanine has enantiomers but glycine does not.

2 a Draw the species present in solid glycine.
  b Draw the species formed by aqueous glycine at **i** pH 1, and **ii** pH 14.

3 Explain why alanine has a much higher melting point than 2-methylpropanoic acid $(CH_3)_2CHCOOH$, even though they have similar relative formula masses.

9.02 Proteins from amino acids

OUTCOMES

already from A2 Level, you understand

- that amino acids have both acidic and basic properties, including the formation of zwitterions

and after this spread you should understand

- that condensation polymers may be formed by reactions between amino acids
- that proteins are sequences of amino acids joined by peptide links
- the importance of hydrogen bonding in proteins (detailed structures not required)

A computer model of trypsin, an enzyme that breaks down proteins. The blue areas represent alanine, lysine, and arginine residues. The red areas represent other amino acid residues.

A planar peptide link

It is easiest to draw the bond angles in the peptide link at 90°. But in reality the bond angles are approximately 120°. The peptide link itself is planar because of delocalization involving the carbon, oxygen, and nitrogen atoms.

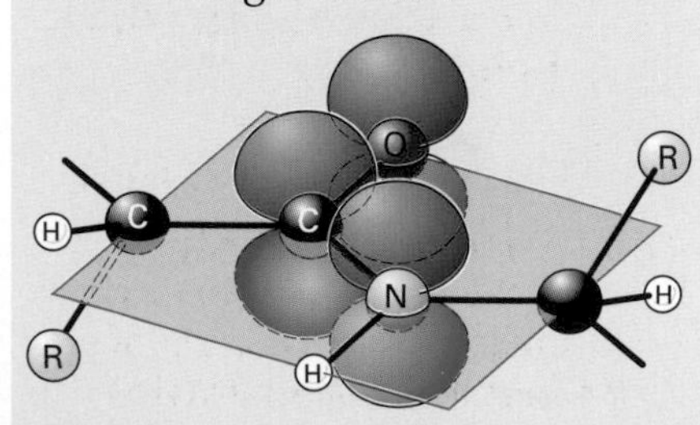

The peptide link is planar.

Two amino acids can react together to form a larger molecule called a **dipeptide**. Water is also formed in the reaction, so it is a **condensation reaction**. The two amino acid **residues** in the dipeptide are joined by a **peptide link**. When many amino acids join together by peptide links, a **condensation polymer** is formed. These are called **polypeptides** or proteins, depending on how many amino acid residues there are. Polypeptides generally contain less than twenty amino acid residues and proteins contain more than this.

Proteins are important biological molecules. The antibodies involved in the body's immune response are proteins, and so are some hormones. **Enzymes** are proteins that catalyse reactions in the cell. Transport proteins control the movement of substances into and out of the cell, and structural proteins are found in tissues such as muscle and skin.

Dipeptides and the peptide link

The amino group of one amino acid can react with the carboxyl group of another amino acid, producing a dipeptide. For example, two glycine molecules can react together:

$$NH_2CH_2COOH + NH_2CH_2COOH \rightarrow NH_2CH_2CONHCH_2COOH + H_2O$$

The sequence —CONH— represents the peptide link. Notice that one water molecule is produced for each peptide link formed. Also note that the dipeptide has an amino group and a carboxyl group. This means it can react with more amino acid molecules to form longer **polymer** chains.

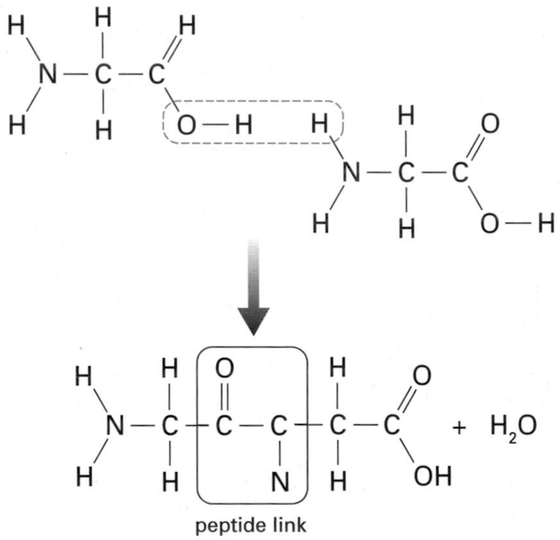

peptide link

The reaction of two glycine molecules to form a dipeptide and water.

Two different dipeptides are possible when two different amino acid molecules react together.

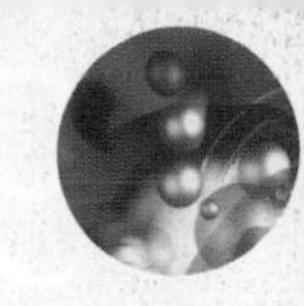

glycine-alanine

alanine-glycine

Glycine and alanine react together to produce two different dipeptides.

Proteins

The **primary structure** of a protein is the particular order of amino acid residues in its chain. This chain is folded into a **secondary structure**, which involves features such as the α-helix and β-pleated sheet. Hydrogen bonds maintain the secondary structure. They occur between the N—H group of one peptide link and the C═O group of another peptide link.

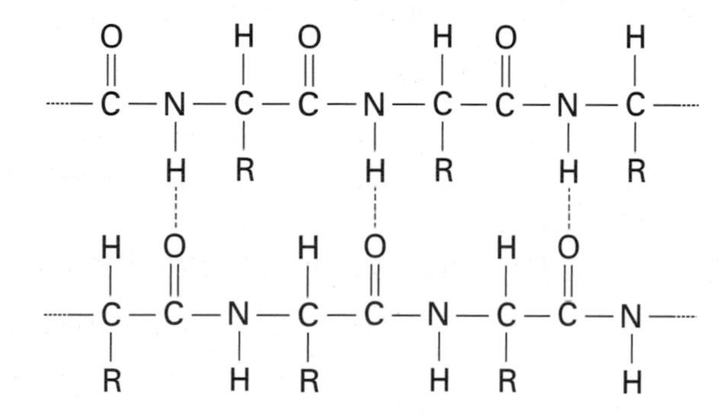

Hydrogen bonds can form between amino acid residues in the protein chain.

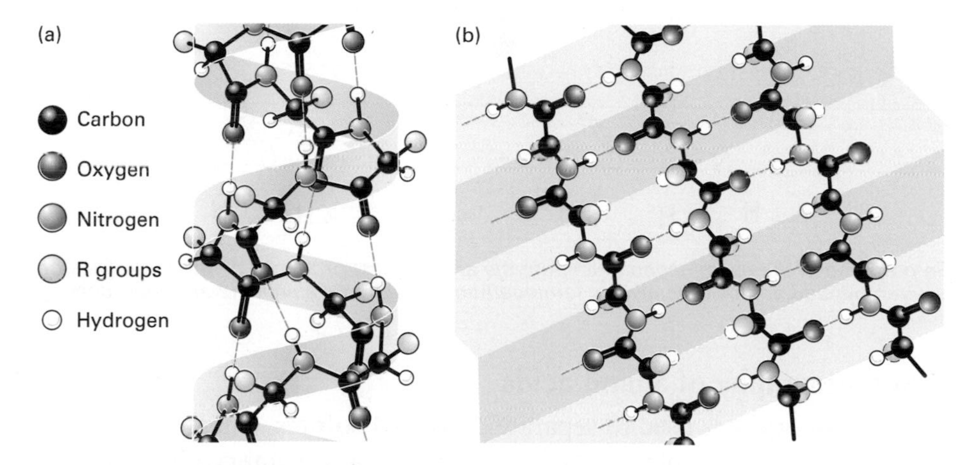

The secondary structure contains features such as: (a) the α-helix, and (b) the β-pleated sheet. The hydrogen bonds are shown by dashed lines.

Distant parts of the protein chain may be brought together to form the **tertiary structure**. This is also maintained by hydrogen bonds. Covalent bonds between two cysteine amino acid residues, called disulfide bridges, are important too. Some complex proteins may be made up of separate protein chains. These are brought together to form the **quaternary structure**.

When a protein is heated or exposed to extremes of pH, its hydrogen bonds are broken. This disrupts its structure and prevents the protein carrying out its function properly. For example, most enzymes are denatured above about 45°C and then cannot catalyse reactions efficiently.

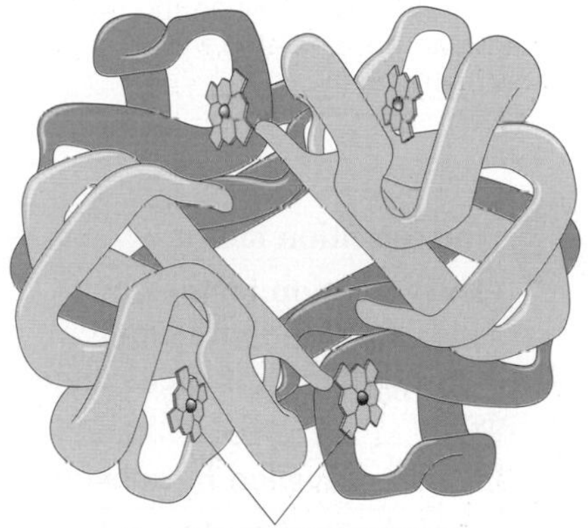

Haemoglobin comprises four protein chains. Oxygen binds to the four haem groups.

Check your understanding

1 Glycine NH_2CH_2COOH and serine $NH_2CH(CH_2OH)COOH$ can form two dipeptides.
  - a Show the displayed formula of serine.
  - b State the systematic name for serine.
  - c Draw the two dipeptides formed and identify the peptide links.
  - d State the type of polymerization involved.

2 Proteins are often folded into the shape of a helix or a sheet. Name the type of interaction that maintains these structures.

9.03 Amino acids from proteins

OUTCOMES

already from A2 Level, you understand

- that condensation polymers may be formed by reactions between amino acids
- that proteins are sequences of amino acids joined by peptide links

and after this spread you should

- understand that hydrolysis of the peptide link produces the constituent amino acids
- know that mixtures of amino acids can be separated by chromatography

Proteins are broken down to amino acids by enzymes in the body's digestive system. They can also be broken down in the laboratory, and the individual amino acids separated by chromatography. Analysis of the results of such experiments provides information about the structure of the original protein.

Hydrolysis of the peptide link

Proteins can be **hydrolysed** (split by their reaction with water) by heating them in 6 mol dm^{-3} hydrochloric acid for 24 hours. The products of hydrolysis are the amino acids that formed the protein. These are the protein's **constituent amino acids**. For example, here is the equation for the hydrolysis of a dipeptide formed by glycine and alanine:

$$NH_2CH_2CONHCH(CH_3)COOH + H_2O \rightarrow$$
$$NH_2CH_2COOH + NH_2CH(CH_3)COOH$$

6 mol dm^{-3} HCl, heat, 24 hours

Be prepared in the examination to identify the amino acids produced by the hydrolysis of a given peptide, in this case glycine (aminoethanoic acid) and alanine (2-aminopropanoic acid).

Chromatography of amino acids

Chromatography is used to separate dissolved **solutes**. There are several types of chromatography. In each one, there is a **stationary phase** that does not move, and a **mobile phase** that moves through the stationary phase, carrying the dissolved solutes with it. Different solutes in a mixture are attracted to the two phases to different extents, so they become separated as the mobile phase moves.

Paper chromatography

In paper chromatography, the stationary phase consists of water molecules attached by hydrogen bonds to the cellulose fibres of a sheet of paper. The mobile phase is a liquid such as water, an organic solvent such as ethanol, or a mixture of solvents. Paper chromatography is often used to separate coloured substances such as food dyes or plant pigments.

The retention factor

The retention factor or R_f value is used to identify an individual substance. It is calculated like this:

$$R_f = \frac{\text{distance travelled by substance}}{\text{distance travelled by solvent}}$$

Different substances have different R_f values in different solvents. The values vary from 0, where the substance remains at the starting point, to 1 where it travels with the solvent front. Chemists run known substances or 'standards' next to the test mixture. They can also refer to published tables of R_f values to help them identify each substance in the mixture.

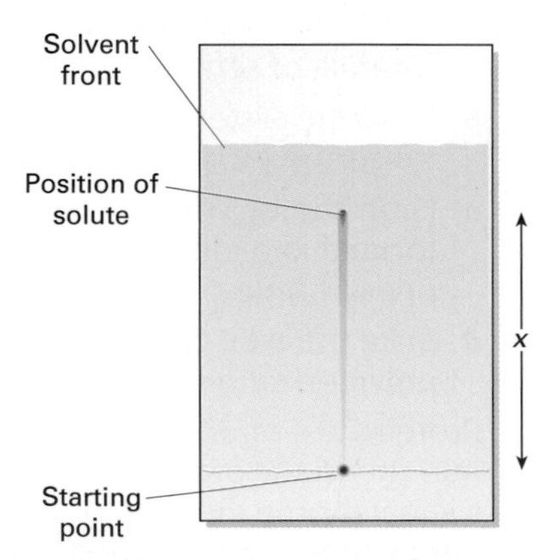

The R_f value is calculated in this example by dividing distance x by distance y.

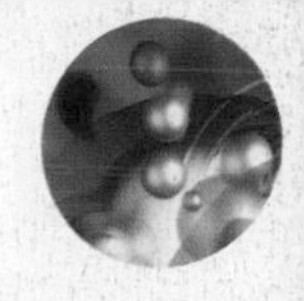

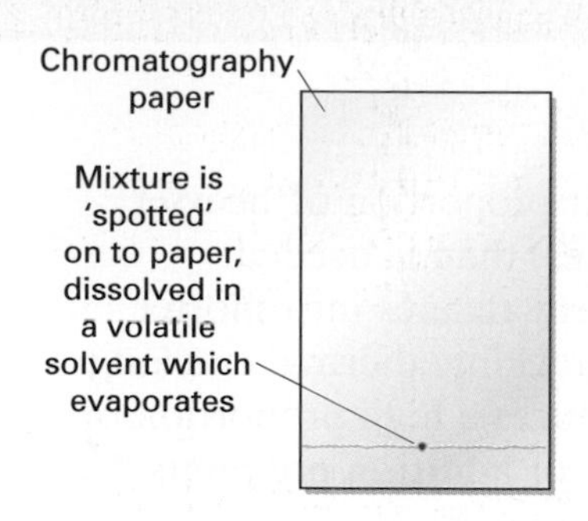

The mixture to be separated is added as a small spot near the bottom of the paper. A glass capillary tube is often used to do this.

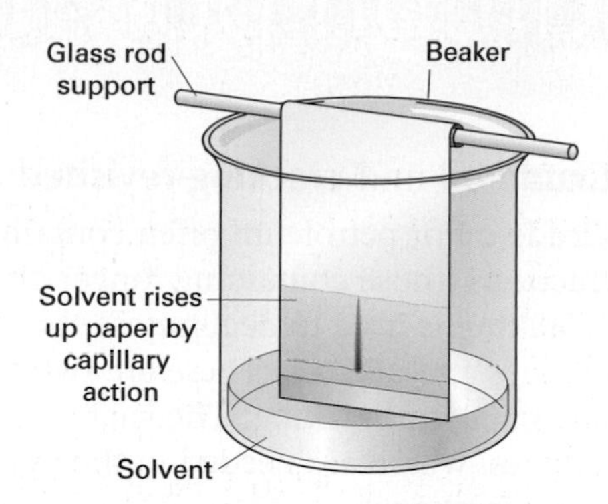

The solvent rises through the paper and carries the different components of the mixture with it. When the solvent front is near the top of the paper, the paper is taken out and dried.

Thin-layer chromatography

In **thin-layer chromatography (TLC)** the stationary phase is a coating of powdered silica, SiO_2, or alumina, Al_2O_3, on a sheet of glass or plastic. Individual spots can be scraped off the completed chromatogram for further analysis.

The resolution can be improved by two-dimensional (2D) chromatography. The chromatogram is completed with one solvent mixture, dried, and turned through 90°. It is then completed a second time but with a different mixture of solvents. A much better separation of individual substances is achieved this way.

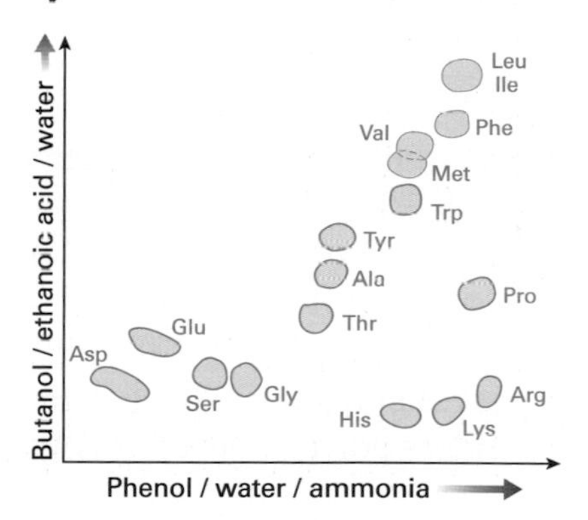

Amino acids can be separated by 2D TLC. They cannot be seen on the chromatogram without using a developing chemical such as ninhydrin solution. This is lightly sprayed over the plate, which is then heated to 110°C for 10 minutes in an oven. The amino acids appear as purple spots.

Science @ Work

Sanger and insulin

Frederick Sanger won the 1958 Nobel Prize in Chemistry for his work on the structure of proteins, especially that of insulin. The English biochemist developed a way to chemically tag the amino end of a peptide chain. He hydrolysed insulin using trypsin or acid, and used various types of chromatography to analyse the peptide fragments and amino acids released. Eventually he was able to piece together the structure of insulin, which contains 51 amino acids in two polypeptide chains, joined by disulfide bridges.

People with Type 1 diabetes, and some with Type 2 diabetes, need injections of insulin to keep their blood glucose levels under control.

Check your understanding

1 a Explain how a protein can be hydrolysed in the laboratory.

b Write an equation to show the hydrolysis of the dipeptide $NH_2CH(CH_2OH)CONHCH(CH_3)COOH$.

c Name the two constituent amino acids of the dipeptide in part **b**.

2 Outline how amino acids can be separated.

10.01 Addition polymers

OUTCOMES

already from AS Level, you

- know how addition polymers are formed from alkenes
- can recognize the repeating unit in a poly(alkene)
- know some typical uses of poly(ethene) and poly(propene)

and after this spread you should

- be able to draw the repeating unit of addition polymers from monomer structures and vice versa

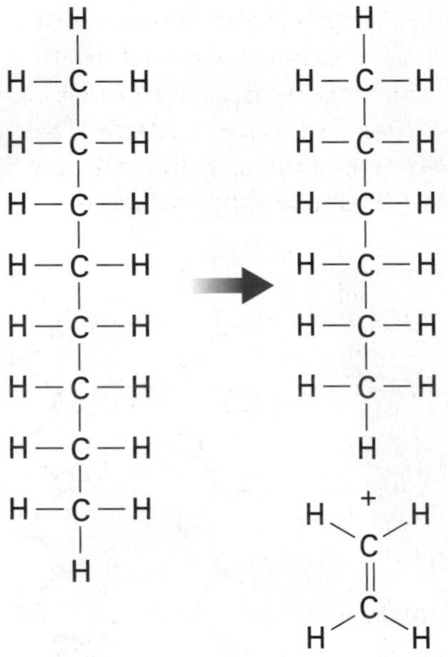

An example of cracking in which octane C_8H_{18} is decomposed to form hexane C_6H_{14} and ethene C_2H_4.

Crude oil and cracking revisited

Crude oil or petroleum often contains a higher proportion of heavier fractions (those containing longer chain alkanes) than is needed. **Cracking** is used to decompose the longer chain alkanes into shorter alkanes and alkenes. There are two types of cracking, thermal cracking and catalytic cracking. Thermal cracking produces a high proportion of alkenes, which are needed in the manufacture of **addition polymers** such as poly(ethene).

	thermal cracking	catalytic cracking
temperature	450°C to 900°C	450°C
pressure	high	moderate
catalyst	none	zeolites (minerals with microscopic pores)
products	high proportion of straight-chain alkanes and alkenes, useful as raw materials for the chemical industry	high proportion of branched alkanes and alkenes, and cyclic and aromatic hydrocarbons, useful for motor fuels

Thermal cracking and catalytic cracking of petroleum fraction favour the formation of different products.

Addition polymers revisited

Unsaturated hydrocarbons such as alkenes can join together to form addition polymers. The polymer itself is the only product of addition polymerization. This is different from condensation polymerization, where a smaller molecule such as water is also produced. Polypeptides and proteins are condensation polymers, formed from amino acids.

Ethene is the **monomer** that makes poly(ethene) or polythene, and propene is the monomer that makes poly(propene) or polypropylene.

Repeating units

The idea of the **repeating unit** makes it easier to represent the structure of polymers. The diagram shows the formation of a section of poly(propene) from three propene monomers. The same structure is repeated three times in the section of poly(propene). This structure is the repeating unit.

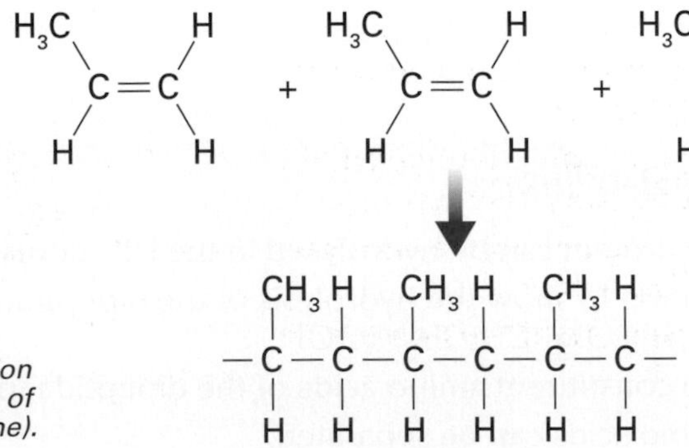

The formation of a section of poly(propene).

The repeating unit from a monomer

Follow these steps to work out the corresponding repeating unit from a given monomer. Draw the same structure, but change it like this:

- change the double bond to a single bond;
- draw the bond angles at 90°; and
- draw a trailing bond to the left and right of the carbon atoms that were involved in the double bond.

The repeating unit of poly(propene) from the displayed formula of propene. Note that repeating units are sometimes shown inside brackets. You can leave these out, but you must show the trailing bonds to the left and right. Only draw one repeating unit unless you are asked to show more.

The monomer from the repeating unit

Follow these steps to work out the corresponding monomer from a given repeating unit. Draw the same structure, but change it like this:

- draw a double bond between the two carbon atoms with the trailing bonds;
- draw bond angles at 120° where appropriate; and
- do not draw the trailing bonds.

repeating unit → monomer

Drawing the monomer from the repeating unit of polystyrene.

Check your understanding

1. State the difference between the number of products formed in addition polymerization and condensation polymerization.
2. The diagram shows the repeating unit of a polymer.
   - a Draw the monomer used to form the polymer.
   - b Name the monomer.
   - c What feature of the monomer allows it to form addition polymers?
3. The diagram shows a monomer used in the manufacture of a certain polymer.
   - a Draw the repeating unit of the polymer formed by the monomer.
   - b Name the monomer and the polymer it forms.

Ziegler–Natta catalysts

The German chemist Karl Ziegler and the Italian chemist Giulio Natta were awarded the Nobel Prize in Chemistry in 1963. They discovered that certain catalysts containing titanium could affect the way in which the polymer chain forms during addition polymerization. These Ziegler–Natta catalysts can produce 'isotactic' polymers where all the side groups are on the same side, or 'syndiotactic' polymers where the side groups alternate regularly from one side of the chain to the other. Without such catalysts, 'atactic' polymer chains form, with the side groups in a random arrangement. The strength of the intermolecular forces between polymer chains varies between the three types of polymer. For example, isotactic polystyrene has a higher melting point than atactic polystyrene.

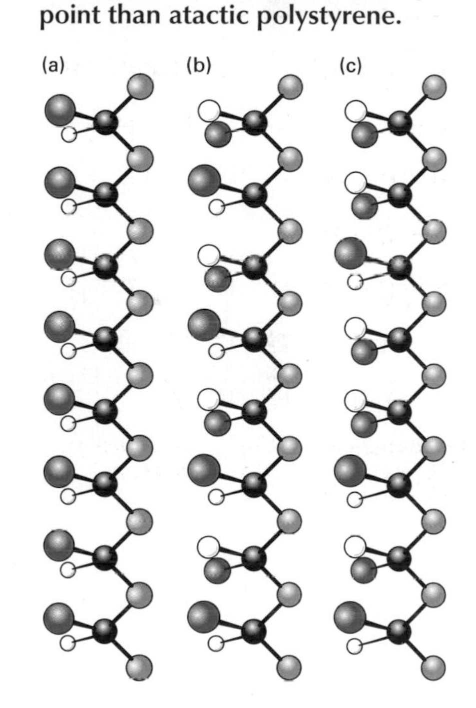

Polymer geometry: (a) isotactic polymer, (b) syndiotactic, and (c) atactic polymer chains.

10.02 Polyesters

OUTCOMES

already from A2 Level, you

- can draw the repeating unit of addition polymers from monomer structures and vice versa
- understand that condensation polymers may be formed by reactions between amino acids
- know that carboxylic acids and alcohols react, in the presence of a strong acid catalyst, to give esters

and after this spread you should

- understand that condensation polymers may be formed by reactions between dicarboxylic acids and diols
- know the linkage of the repeating units of polyesters (e.g. Terylene)

Addition polymerization involves just one functional group, the carbon–carbon double bond. But condensation polymerization involves two different functional groups. For example, polypeptides and proteins are condensation polymers formed from amino acids, which have an amino group and a carboxyl group. **Polyesters** are condensation polymers formed from a **dicarboxylic acid** and a **diol**.

Dicarboxylic acids and diols

These organic compounds contain two identical functional groups. Dicarboxylic acids have two carboxyl groups and diols have two hydroxyl groups.

Naming dicarboxylic acids

Dicarboxylic acids are named from the carbon chain that includes the two carboxyl groups, with the suffix *dioic acid* added. For example, HOOCCOOH is called ethanedioic acid. Notice that the number of carbon atoms used to name the main chain includes the two carboxyl groups, and that the letter e is included in the name. There is no need to identify the carboxyl groups with a number, unless the carboxyl groups are attached to a benzene ring.

ethanedioic acid

benzene-1,4-dicarboxylic acid

Two dicarboxylic acids

Naming diols

Ethane-1,2-diol

Diols are named from the carbon chain that includes the two hydroxyl groups, with the suffix *diol* added. For example, $HOCH_2CH_2OH$ is called ethane-1,2-diol. The position of the hydroxyl groups must be identified using numbers unless there is only one carbon atom, as in methanediol.

Making polyesters

Carboxylic acids react with alcohols to produce esters and water. For example, ethanoic acid reacts with methanol to produce methyl ethanoate and water:

$$CH_3COOH(aq) + CH_3OH(aq) \rightarrow CH_3COOCH_3(aq) + H_2O(l)$$

Dicarboxylic acids and diols also react together. For example, benzene-1,4-dicarboxylic acid reacts with ethane-1,2-diol. When this happens, one end of the ester has a carboxyl group and the other end has a hydroxyl group. This means that further esterification reactions can happen, producing a long polyester molecule.

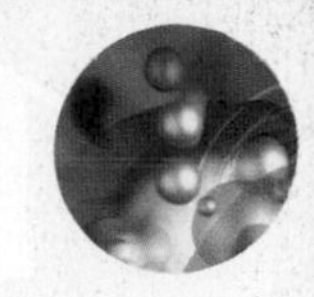

ester link

$+ H_2O$

The formation of an ester from benzene-1,4-dicarboxylic acid and ethane-1,2-diol.

Repeating units

It is more difficult to draw the repeating unit of a polyester. You need to take care over where to place the trailing bonds. You should draw a trailing bond from

- a C=O group on the dicarboxylic acid side
- an oxygen atom on the diol side

from dicarboxylic acid | from diol | from dicarboxylic acid | from diol

The structure of polyesters follows this general pattern. Notice that the direction of the ester links alternates along the chain. Remember that the carboxyl group loses a hydroxyl group when the ester link forms, and the hydroxyl group in the diol loses a hydrogen atom.

The polyester formed from benzene-1,4-dicarboxylic acid and ethane-1, 2-diol has very many uses and names, including PET and Terylene™.

PET and more

The non-systematic name for benzene-1,4-dicarboxylic acid is terephthalic acid, and the non-systematic name for ethane-1,2-diol is ethylene glycol. PET (polyethylene terephthalate) is the polyester made from these two compounds. It is used to make plastic bottles for fizzy drinks. It can be drawn into filaments to make fibres such as Terylene™, which is used in clothing. It can be made into thin films such as Melinex®, which is used in food packaging.

Mylar® is a tough, waterproof polyester used to make the sails on windsurfers.

The repeating unit for the polyester formed from benzene-1,4-dicarboxylic acid and ethane-1,2-diol.

Check your understanding

1. Draw the repeating unit for the polyester formed from butanedioic acid and propane-1,3-diol.
2. The diagram shows the repeating unit of a polyester.

```
  O   H   H   O       H  CH3 H
  ||  |   |   ||      |   |   |
—C — C — C — C — O — C — C — C — O—
      |   |           |   |   |
     CH3 CH3          H   H   H
```

   a. Draw the structures of the two compounds which react together to form this polyester, and give their names.
   b. Name the type of polymerization involved.
3. Diacyl chlorides may be used instead of dicarboxylic acids because they are more reactive. Name the inorganic product of the reaction between a diacyl chloride and a diol.

10.03 Polyamides

OUTCOMES

already from A2 Level, you

- understand that condensation polymers may be formed by reactions between dicarboxylic acids and diols
- know the linkage of the repeating units of polyesters (e.g. Terylene™)
- understand that proteins are sequences of amino acids joined by peptide links

and after this spread you should

- understand that condensation polymers may be formed by reactions between dicarboxylic acids and diamines
- know the linkage of the repeating units of polyamides (e.g. nylon 6,6 and Kevlar®)

Naming diamines

Diamines are named from the carbon chain that includes the two amino groups, with the suffix *diamine* added. For example, $H_2NCH_2CH_2NH_2$ is called ethane-1,2-diamine. The position of the amine groups must be identified using numbers unless there is only one carbon atom, as in methanediamine.

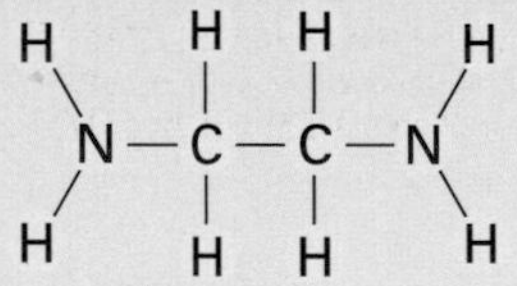

Ethane-1,2-diamine

Polypeptides and proteins are natural examples of **polyamides**. The carboxyl group of one amino acid reacts with the amino group of another amino acid. A peptide link forms with the elimination of a water molecule. The peptide link is identical to the **amide link** formed in polyamides. The difference between the two types of condensation polymer lies in their monomers.

Amino acids are **difunctional compounds**: they each have a carboxyl group and an amino group. But synthetic polyamides form in the reaction between a dicarboxylic acid and a **diamine**, molecules that each have two identical functional groups. Nylon and Kevlar® are two very common polyamides.

Making polyamides

Carboxylic acids react with amines to produce N-substituted amides and water. For example, ethanoic acid reacts with methanamine to produce *N*-methylethanamide and water:

$$CH_3COOH(aq) + CH_3NH_2(aq) \rightarrow CH_3CONHCH_3(aq) + H_2O(l)$$

Dicarboxylic acids and diamines also react together. For example, hexane-1,6-dioic acid reacts with hexane-1,6-diamine. When this happens, one end of the N-substituted amide has a carboxyl group and the other end has an amino group. This means that further reactions can happen, producing a long polyamide molecule.

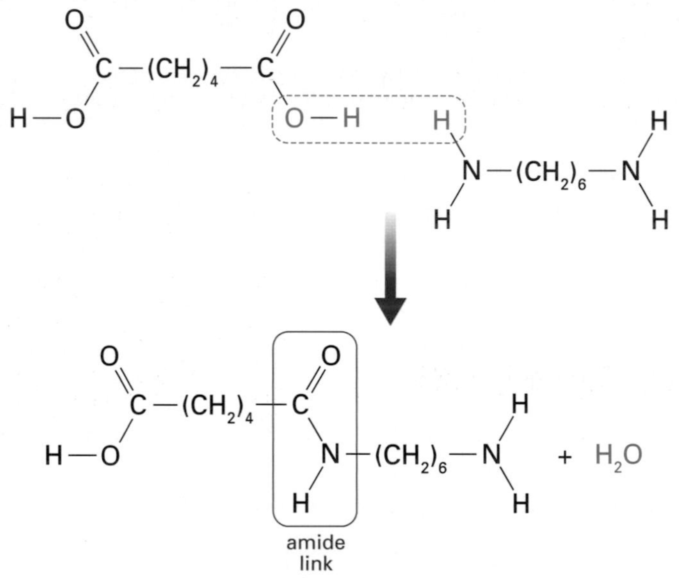

amide link

The formation of an amide from hexane-1,6-dioic acid and hexane-1,6-diamine. Remember that the amide link is identical to the peptide link.

Repeating units

When drawing the repeating unit of a polyamide, you should draw a trailing bond from

- a C=O group on the dicarboxylic acid side
- an N—H group on the diamine side

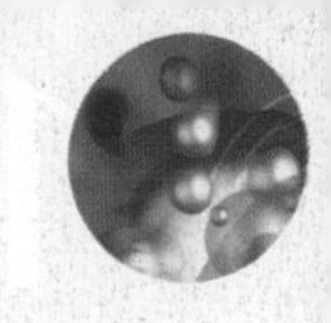

The general diagram shows why this is the case. Remember that the carboxyl group loses a hydroxyl group when the amide link forms, and the amino group in the diamine loses a hydrogen atom.

$$-\overset{\overset{O}{\|}}{C}-\boxed{\text{from dicarboxylic acid}}-\overset{\overset{O}{\|}}{C}-\underset{\underset{H}{|}}{N}-\boxed{\text{from diamine}}-\underset{\underset{H}{|}}{N}-\overset{\overset{O}{\|}}{C}-\boxed{\text{from dicarboxylic acid}}-\overset{\overset{O}{\|}}{C}-\underset{\underset{H}{|}}{N}-\boxed{\text{from diamine}}-\underset{\underset{H}{|}}{N}-$$

The structure of polyamides follows this general pattern. Notice that the direction of the amide links alternates along the chain.

The polyamide formed from the reaction of hexane-1,6-dioic acid with hexane-1,6-diamine is called **nylon 6,6**.

$$-\overset{\overset{O}{\|}}{C}-(CH_2)_4-\overset{\overset{O}{\|}}{C}-\underset{\underset{H}{|}}{N}-(CH_2)_6-\underset{\underset{H}{|}}{N}-$$

The repeating unit for nylon 6,6. Take care to show the correct number of carbon atoms when you draw the diagram like this.

Kevlar

Kevlar® is a an **aramid**, a polyamide in which the amide links are directly attached to benzene rings. Aramids are very strong, and have high melting points and low densities. Kevlar® is used to reinforce tyres and aircraft parts, and in brake linings and cables. It is used in the body armour worn by military personnel and police officers.

Body armour like this bullet-proof and stab-proof vest contains layers of Kevlar®, titanium, and plastic reinforced with fibreglass.

Science @ Work

Naming nylon

Nylon was discovered in 1935 by Wallace Carothers, a chemist working for the DuPont company in the US. He had made and evaluated over one hundred different polyamides before settling on nylon 6,6. Its potential as an artificial fibre to replace silk was realized, and women's stockings made from nylon went on sale in 1940. The name 'nylon' comes from changing 'norun' to something more catchy.

The numbers in the name indicate how many carbon atoms are present in the original monomers. The first number refers to the number of carbon atoms in the diamine, and the second one to the number of carbon atoms in the dicarboxylic acid. So nylon 6,10 is made from hexane-1,6-diamine and decanedioic acid.

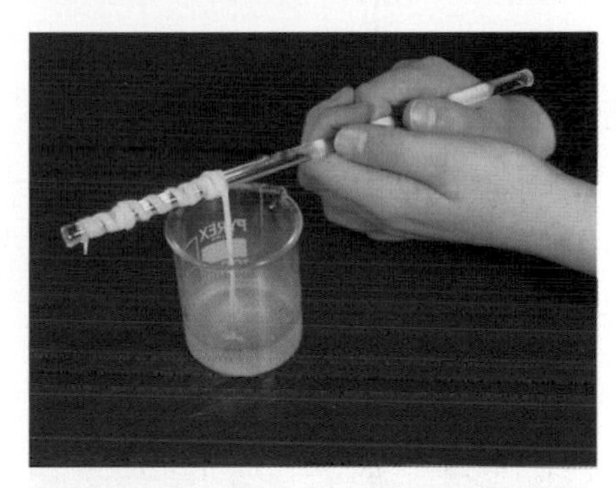

Nylon 6,6 is made in the laboratory using hexanedioyl chloride rather than hexanedioic acid, because acyl chlorides are more reactive. Hexanedioyl chloride dissolved in cyclohexane is carefully poured onto aqueous hexane-1,6-diamine. Nylon 6,6 forms at the interface. It can be drawn out and wound around a glass rod.

Check your understanding

1. Draw the repeating unit for the polyamide formed from decanedioic acid $HOOC(CH_2)_8COOH$ and hexane-1,6-diamine $H_2N(CH_6)_2NH_2$.
2. The diagram shows the repeating unit of Kevlar®.

$$-\overset{\overset{O}{\|}}{C}-C_6H_4-\overset{\overset{O}{\|}}{C}-\underset{\underset{H}{|}}{N}-C_6H_4-\underset{\underset{H}{|}}{N}-$$

   a. Draw the structures of the two compounds which react together to form this polyamide, and give their names.
   b. Name the type of polymerization involved.
   c. What type of polyamide is Kevlar®?

10.04 Biodegradable polymers

OUTCOMES

already from AS Level, you

- recognize that poly(alkenes) are unreactive, like alkanes

already from A2 Level, you

- know that esters can be hydrolysed
- understand that hydrolysis of the peptide link in proteins produces the constituent amino acids

and after this spread you should understand

- that poly(alkenes) are chemically inert and therefore non-biodegradable
- that polyesters and polyamides can be broken down by hydrolysis and are, therefore, biodegradable (mechanisms not required)

Poly(ethene) bags do not degrade quickly in the environment.

Biodegradable substances can be broken down by micro-organisms such as bacteria and fungi. But many substances are not biodegradable. They may be chemically inert, or decay organisms may not have the enzymes needed to break them down.

Poly(alkenes)

Poly(alkenes) such as poly(ethene) and poly(propene) are addition polymers. Their monomers are unsaturated compounds, but the polymers themselves are saturated and are far less reactive. For example, ethene readily reacts with bromine to form 1,2-dibromoethane, but poly(ethene) does not react with bromine. The most significant reaction of poly(alkenes) is combustion. They burn in air to produce carbon dioxide and water vapour.

These polymers persist in the environment because they are chemically inert. Poly(ethene) shopping bags, for example, take between 10 and 20 years to break down in the environment after being thrown away. Thicker plastic items such as plastic bottles are expected to take hundreds of years to break down.

Polyesters and polyamides

Esters and amides can be hydrolysed by acids, alkalis, or certain enzymes. Here are two examples.

Methyl ethanoate is hydrolysed to ethanoic acid and methanol:

$$CH_3COOCH_3(aq) + H_2O(l) \rightarrow CH_3COOH(aq) + CH_3OH(aq)$$

N-methylethanamide is hydrolysed to ethanoic acid and methanamine:

$$CH_3CONHCH_3(aq) + H_2O(l) \rightarrow CH_3COOH(aq) + CH_3NH_2(aq)$$

Polyesters and polyamides can be hydrolysed, too, and this means that they are biodegradable. Polyesters react slowly with acids, but they are readily attacked by strong alkalis such as aqueous sodium hydroxide. The sodium salt of a dicarboxylic acid and a diol are released as a result.

from dicarboxylic acid
from diol
from dicarboxylic acid
from diol

Ester links can be hydrolysed, which breaks down the polyester chain.

Polyamides react slowly with alkalis, but they are readily attacked by strong acids. For example, sulfuric acid or nitric acid spilled on nylon clothing will cause a hole to appear. A dicarboxylic acid and a diamine are released as a result.

from dicarboxylic acid — from diamine

from dicarboxylic acid — from diamine

Amide links can be hydrolysed, which breaks down the polyamide chain.

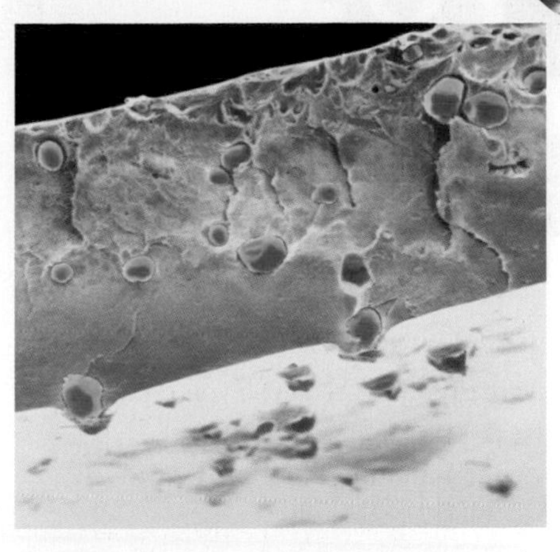

Electron micrograph of a section through a sheet of a biodegradable plastic. The starch granules (coloured orange here) are embedded in the plastic.

Degradable polymers

Although poly(ethene) and poly(propene) are not biodegradable, chemists have developed ways to modify them so they do break down more quickly in the environment. For example, starch can be incorporated into the poly(ethene) used for carrier bags. Micro-organisms can hydrolyse starch, which also absorbs water and swells, so a bag made of this material breaks up once it is in the soil. Small pieces of poly(ethene) are left behind but these do not cause the litter nuisance of a plastic bag.

Degradable polymers can be made by including an additive that promotes breakdown by oxidation. Poly(ethene) and poly(propene) films containing these additives are 'oxo-biodegradable'. They can be programmed to start breaking down by including a controlled amount of an antioxidant. Once this is used up, the degradation starts. It can be delayed for several months to several years, depending upon the intended use of the plastic. Degradation then continues as long as the polymer is in contact with oxygen.

An oxo-degradable carrier bag.

Check your understanding

1 a What does the term 'biodegradable' mean?

b Explain why poly(ethene) is not biodegradable but polyesters and polyamides are.

2 The diagram shows the repeating unit of a condensation polymer.

$$-C(=O)-CH_2-C(=O)-N(H)-(CH_2)_4-N(H)-$$

Draw the structures of the two compounds released by the acid hydrolysis of this polymer, and give their names.

3 Describe two ways in which poly(ethene) can be modified at the time of manufacture so that it is degradable.

10.05 Polymer problems

OUTCOMES

already from AS Level, you know

- that poly(propene) is recycled

already from A2 Level, you understand

- that poly(alkenes) are chemically inert and therefore non-biodegradable
- that polyesters and polyamides can be broken down by hydrolysis and are, therefore, biodegradable

and after this spread you should appreciate

- the advantages and disadvantages of different methods of disposal of polymers
- the advantages and disadvantages of recycling polymers

The UK produces about 2 million tonnes of plastic waste each year. There are three main ways to deal with waste polymers. These are burial in a **landfill** site, incineration, and recycling. Each one has advantages and disadvantages.

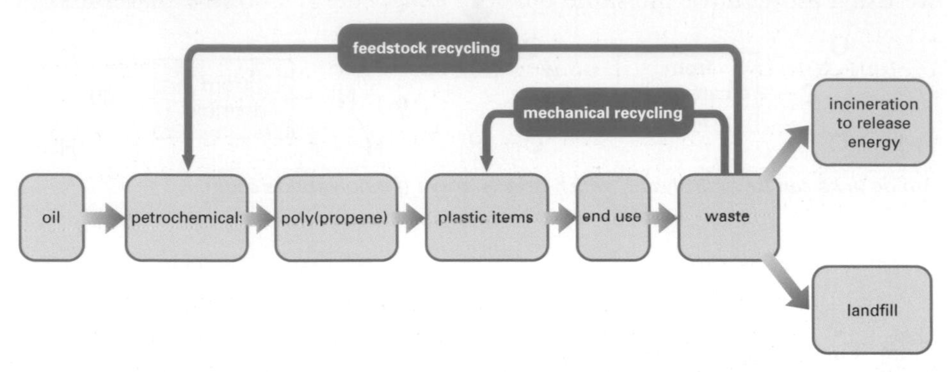

A summary of the fates of poly(propene).

It is becoming increasingly difficult to find new landfill sites.

Landfill disposal

A landfill site is where waste is buried, either on the surface or in a large hole in the ground. Around 80% of polymer waste is disposed of this way. The waste is delivered, spread out, and then buried. Polyesters and polyamides are biodegradable, but poly(alkenes) such as poly(ethene) and poly(propene) are not biodegradable. They form around two-thirds of household waste and are expected to take many years to degrade in the ground.

New landfill sites are increasingly difficult to find. Local people may complain about the potential for nuisance from a proposed site. Vermin such as gulls are attracted to landfill sites. The wind can carry away foul smells and waste materials, and there is a lot of heavy traffic from lorries delivering the waste.

Rising fuel costs are making the transport of waste to landfill sites more expensive. European Union directives have set targets for reducing the amount of waste disposed of in landfill sites. Waste must be treated and sorted before disposal, further increasing costs. The amount of metals, and biodegradable materials such as waste food and paper, going to landfills must be reduced. As a result, sites that were once expected to fill up quickly now stay open longer, which prolongs the nuisance for local people.

Incineration

Around 6% of UK polymer waste is disposed of by **incineration**, which involves burning the waste at high temperatures in a furnace. The combustion of polymers releases useful amounts of thermal energy. This is used for heating and for generating electricity.

Incinerating waste polymers instead of burying them in landfill sites reduces the need to find new sites. But the combustion products include carbon dioxide, which adds to the **greenhouse effect** and so contributes to **global warming**. They may also include toxic gases. These must be removed from the waste gases, adding to the cost of incineration.

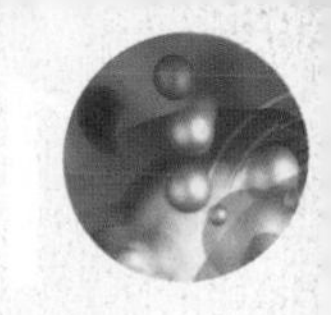

Recycling

There are several different ways to recycle waste polymers. These include simply re-using the plastic item, mechanical recycling, and feedstock recycling.

Re-use

Re-using an item for the same purpose saves energy and raw materials, and reduces waste. There is no need to synthesize the polymer, manufacture the item, and deliver it to customers. But the polymer items must be sorted out from other waste and cleaned. This needs the co-operation of individuals when they are throwing their waste away, or it must be done at the recycling centre. A lot of sorting is done by hand, which is expensive.

Mechanical recycling

Polymers can be recycled to make new materials. After sorting and cleaning, the polymer items are shredded into flakes, then processed to form granules. These are melted down and moulded into new items such as trays, garden furniture, or artificial fibres, depending on the polymer. For example, drinks bottles made from PET are recycled to make artificial fibres for clothing. But there may be limited markets for the recycled objects, and for safety reasons they cannot be used to contain food.

Used drinks bottles made from the polyester PET are shredded to form chips. These are used to make polyester fibre for clothing. Poly(ethene) chips can be recycled to make new bottles.

Feedstock recycling

Feedstock recycling involves decomposing the polymer, rather like cracking heavy oil fractions. The polymer breaks down in the absence of air at around 500°C. This produces a fraction similar to naphtha, which is used as a feedstock for the chemical industry.

Check your understanding

1 a Outline the advantages and disadvantages of dealing with waste polymers by landfill and by recycling.

b To what extent are the costs of fuel and raw materials likely to influence how waste polymers are dealt with?

Science @ Work

Tyres

Half a million tonnes of tyres are scrapped in the UK each year. They are no longer allowed to be dumped in landfill sites and must be recycled instead. Old tyres can be chopped into granules that are used in the manufacture of carpet underlay. Cement manufacturers use old tyres as the fuel for their kilns.

The soft surface of this children's playground is made from granulated tyres.

11.01 Aliphatic reaction pathways

OUTCOMES

already from AS and A2 Level, you understand

- that aliphatic compounds may be chemically converted into different organic compounds

and after this spread you should have

- a summary of aliphatic reactions required by the Specification

The carbon atoms in aliphatic compounds may be joined together in chains and rings, with or without branches. The rings do not contain delocalized electrons. The two diagrams shown here summarize most of the aliphatic reaction pathways you need. They are identical except that the one on the left shows the types of compounds and the type of reactions involved, and the one on the right shows typical reaction conditions and formulae, starting with propene near the top.

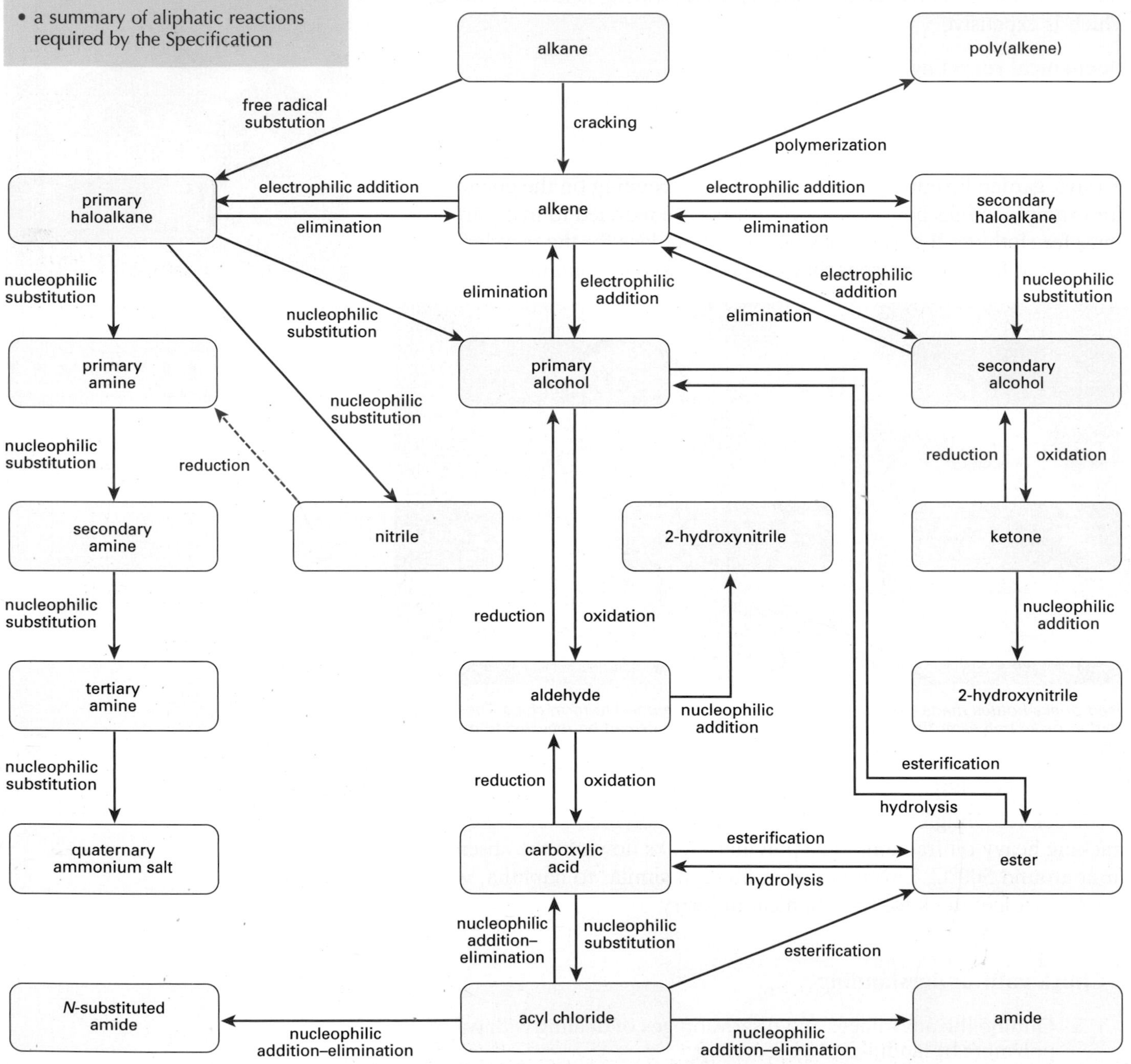

Aliphatic reaction pathways to show types of compounds and reactions. Note that the dashed purple arrow indicates that the number of carbon atoms has increased by one.

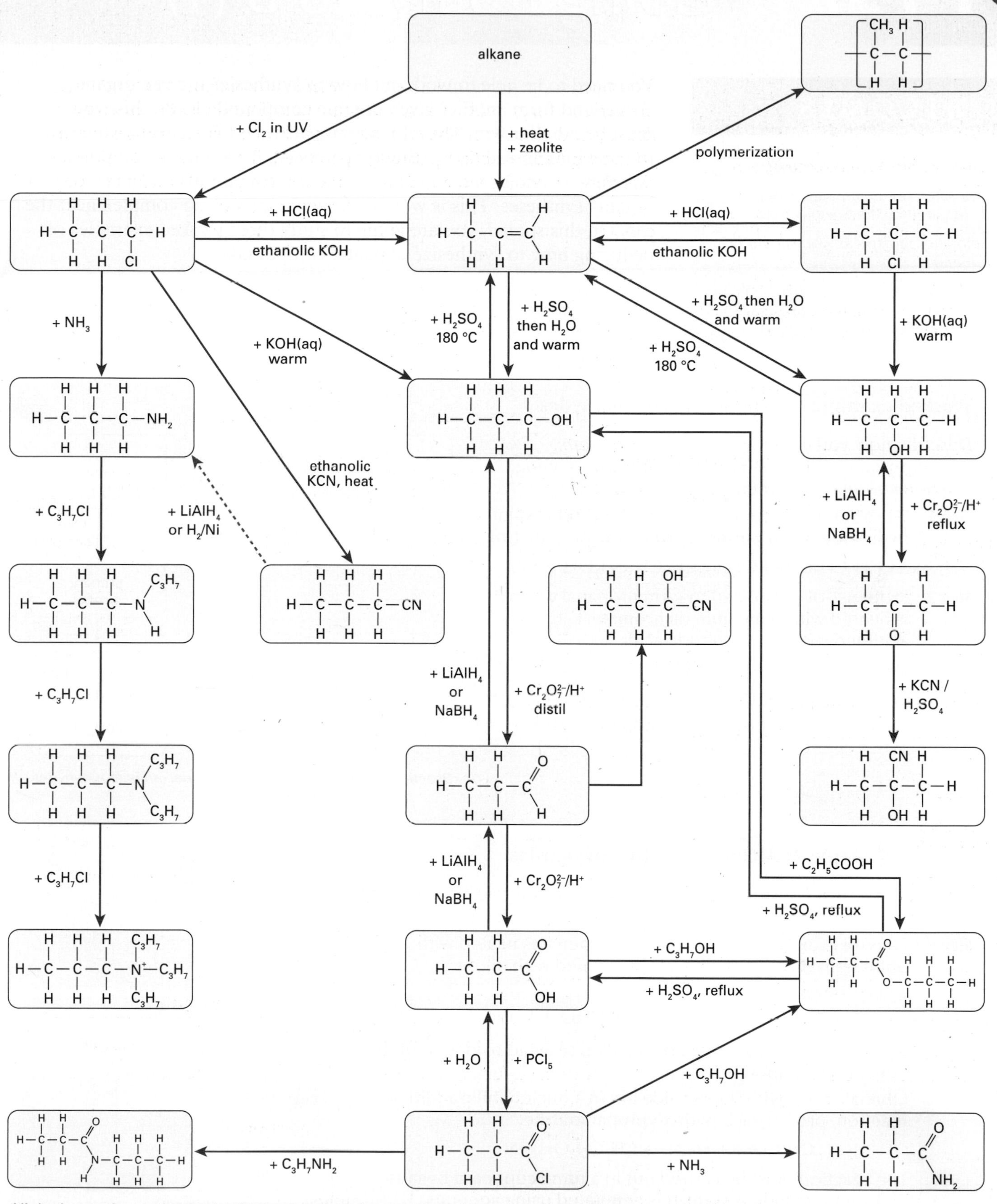

Aliphatic reaction pathways to show examples of conversions from one compound to another. Note that the dashed purple arrow indicates that the number of carbon atoms has increased by one, so the compound at the end of it is a primary amine but it should have an extra —CH_2— group.

11.02 Going from A to B (1)

OUTCOMES

already from A2 Level, you have

- a summary of aliphatic reactions required by the Specification

and after this spread you should be able to

- deduce how to synthesize aliphatic organic compounds using reactions in the Specification

You need to be able to work out how to synthesize a given organic compound from another given organic compound. To do this, you must be familiar with several reaction pathways. The two flowcharts on the previous spread show reaction pathways for aliphatic compounds. Alkenes, alcohols, and haloalkanes are important intermediates for many organic syntheses. This is why the flow charts look so complex near the top. On this spread, you are going to study three worked examples of deducing how to synthesize aliphatic compounds.

Worked example 1: Haloalkanes to aldehydes

Describe how you could produce propanal from 1-bromopropane. Include equations and any essential reaction conditions needed.

Step 1 1-bromopropane reacts with aqueous potassium hydroxide when warmed, producing propan-1-ol:

$$CH_3CH_2CH_2Br + KOH \rightarrow CH_3CH_2CH_2OH + KBr$$

Step 2 Propan-1-ol is oxidized to form propanal when it is heated with potassium dichromate, $K_2Cr_2O_7$, acidified with dilute sulfuric acid:

$$CH_3CH_2CH_2OH + [O] \rightarrow CH_3CH_2CHO + H_2O$$

The reaction mixture must be distilled to avoid oxidation of the propanal to propanoic acid.

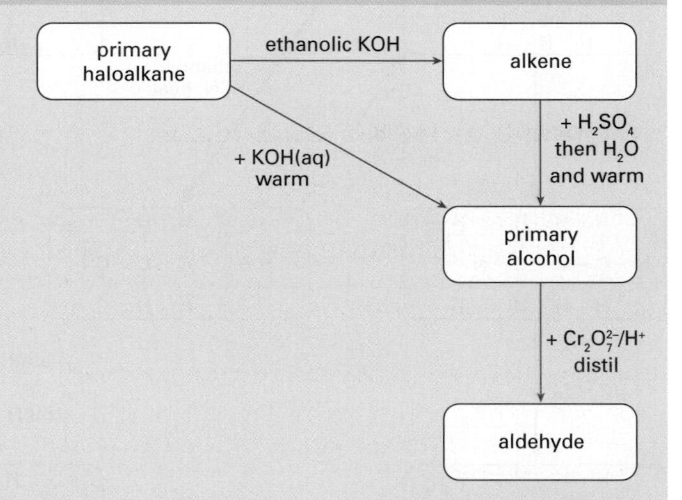

Reaction pathways from primary haloalkanes to aldehydes.

Worked example 2: Alcohols to 2-hydroxynitriles

Describe how you could produce 2-hydroxypropanenitrile from ethanol. Include equations and any essential reaction conditions needed.

Step 1 Ethanol is oxidized to form ethanal when it is heated with potassium dichromate, $K_2Cr_2O_7$, acidified with dilute sulfuric acid:

$$CH_3CH_2OH + [O] \rightarrow CH_3CHO + H_2O$$

The reaction mixture must be distilled to avoid oxidation of the ethanal to ethanoic acid.

Step 2 Ethanal reacts with the cyanide ion in a nucleophilic addition reaction, producing 2-hydroxypropanenitrile:

$$CH_3CHO + HCN \rightarrow CH_3CH(OH)CN$$

The reaction must be carried out in a fume cupboard because hydrogen cyanide is toxic. It is generated using aqueous potassium cyanide with the addition of sulfuric acid.

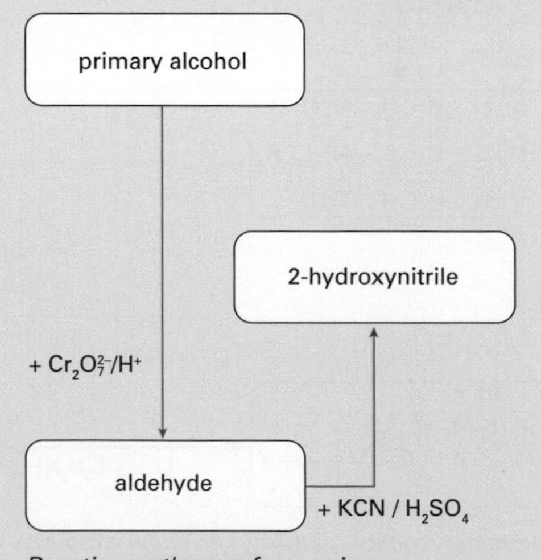

Reaction pathways from primary alcohols to 2-hydroxynitriles.

Worked example 3: Haloalkanes to primary amines

Describe how you could produce propanamine, $CH_3CH_2CH_2NH_2$ from bromoethane. Include equations and any essential reaction conditions needed.

Step 1 Bromoethane reacts with the cyanide ion in a nucleophilic substitution reaction, producing propanenitrile:

$$CH_3CH_2Br + HCN \rightarrow CH_3CH_2CN + HBr$$

As in example 2, the reaction must be carried out in a fume cupboard. Hydrogen cyanide is generated using aqueous potassium cyanide with the addition of sulfuric acid.

Step 2 Propanenitrile is reduced to form propanamine:

$$CH_3CH_2CN + 4[H] \rightarrow CH_3CH_2CH_2NH_2$$

The reducing agent is $LiAlH_4$ in anhydrous conditions, or hydrogen in the presence of a nickel catalyst.

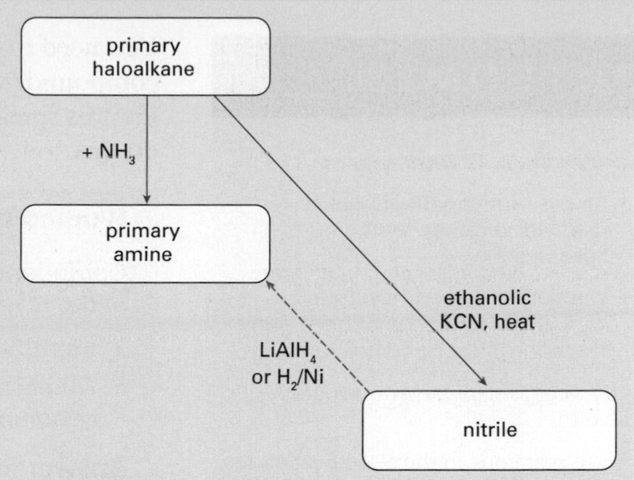

Reaction pathways from primary haloalkanes to primary amines. Remember that the pathway including the dashed line increases the chain length by one carbon atom.

Comments on the worked examples

Example 1: Notice that it would be possible to make propanal by converting 1-bromopropane into propene first, then hydrating this to propan-1-ol. But there is a problem in doing this, as propan-2-ol would be the major product, not propan-1-ol. This is because the secondary carbocation intermediate is more stable. Instead of making propanal, we would make propanone, a ketone.

Example 2: The product, 2-hydroxypropanenitrile, has an asymmetric carbon atom. This means that it has optical isomers or enantiomers.

Notice that ethanol has two carbon atoms but 2-hydroxypropanenitrile has three carbon atoms. The number of carbon atoms in organic syntheses is increased by nucleophilic addition of cyanide ions to aldehydes or ketones.

2-Hydroxypropanenitrile has a chiral centre.

Example 3: As in example 2, the product has one more carbon atom than the starting reactant. So the reaction pathway includes a step to form a nitrile. Note that primary amines can be produced directly from primary haloalkanes by nucleophilic substitution involving ammonia. But the reaction between bromoethane and ammonia would produce ethanamine, not propanamine.

Check your understanding

1 Outline how you could achieve the following conversions in one step:

a ethanoic acid from ethanal
b chloromethane from methane
c butanone from butan-2-ol
d propanamide from propanoyl chloride

2 Outline how you could achieve the following conversions in two steps:

a propan-2-ol from propene
b ethanal from bromoethane
c 2-hydroxy-2-methylpropanenitrile from propan-2-ol

11.03 Going from A to B (2)

OUTCOMES

already from A2 Level, you

- have a summary of aliphatic reactions required by the Specification
- can deduce how to synthesize aliphatic organic compounds using reactions in the Specification

and after this spread you should be able to

- use reactions in the Specification to deduce the intermediate aliphatic organic compounds in a given synthesis

You need to be able to work out the identity of any intermediate aliphatic compounds when you are given the starting and finishing compounds. On this spread, you are going to study two worked examples of how to do this.

Worked example 1

*Compound **X** reacts with compound **Y** to form compound **Z**, as shown in the reaction pathway shown below.*

*a Identify compounds **X** and **Y**.*

b For each step, name the type of reaction, and state the reagents and conditions needed.

H_3C-CHO —Step 1→ X

$H_3C-CH_2-CO-CH_3$ —Step 2→ Y

X + Y —Step 3→ Z: $H_3C-CO-O-CH(CH_2CH_3)(CH_3)$

Answer for part a

Compound **X** is ethanoic acid, CH_3COOH.

Compound **Z** is an ester. The part on the left has come from an unbranched carboxylic acid with two carbon atoms. This must be ethanoic acid, because compound **X** is formed from ethanal, and aldehydes can be oxidized to form carboxylic acids.

Compound **Y** is butan-2-ol, $CH_3CH_2CH(OH)CH_3$.

The right hand part of compound **Z** must have come from a secondary alcohol, as the oxygen atom is not joined to one of the two end carbon atoms. The secondary alcohol is unbranched and has four carbon atoms, so it must be butan-2-ol.

Answer for part b

Step 1 oxidation reaction
Reflux with aqueous potassium dichromate(VI), acidified with dilute sulfuric acid.

Step 2 reduction reaction
dry $LiAlH_4$

Step 3 esterification reaction
Warm in the presence of hydrochloric acid as a catalyst.

Learning the reactions in detail

This part of Unit 4 involves a lot of synoptic knowledge and understanding. You must also revise the work you did in Unit 2 to tackle such questions confidently. To help you do this, the flow chart shows the spread references from *AS Chemistry for AQA* and this book for each reaction pathway.

Check your understanding

1 Butane-1,4-diamine can be synthesized from ethene in three steps, as shown. Step 1 happens in the presence of UV light.

a Name compound **X** and give its structure.

b Name compound **Y** and give its structure.

c For each step, state the reagents and conditions needed.

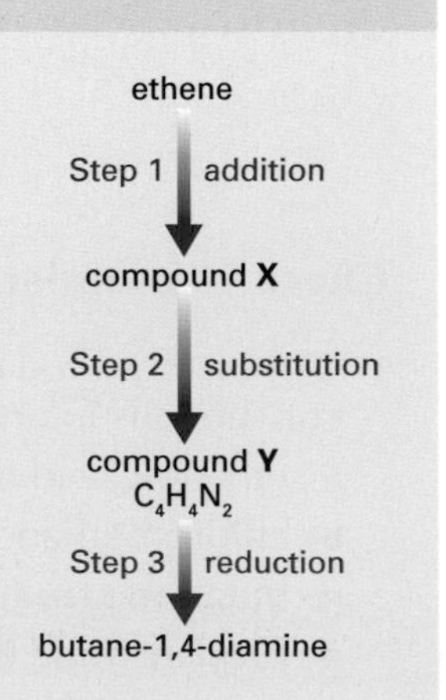

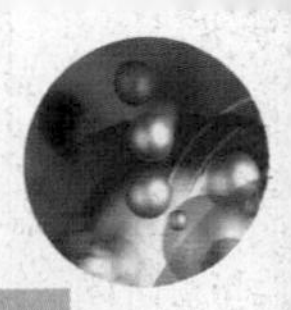

Comments on worked examples

Example 1: There are often combinations of different reagents and conditions that could do the same conversion. For example, in step 2 you could also use aqueous $NaBH_4$. In step 3, you could use sulfuric acid instead of hydrochloric acid, and you could heat the mixture under reflux.

Example 2: In this question you need to identify an alkene that can be formed from substance **A**, and in turn will go on to form substance **B**. Instead of being told that substance **X** was an alkene, you might be told something about it instead. For example, that it can change bromine water from brown to colourless.

Worked example 2

*Compound **A** is converted into compound **B** in two steps, via an alkene **X**, as shown in the reaction pathway shown below.*

A $H_3C{-}CH_2{-}CH(CH_3){-}CH_2Br$ $\xrightarrow{\text{step 1}}$ **X** $\xrightarrow{\text{step 2}}$ **B** $H_3C{-}CH_2{-}CBr(CH_3){-}CH_3$

a *Identify alkene **X** and give its structure.*

b *For each step, name the type of reaction, and state the reagents and conditions needed.*

Answer for part a

Compound **X** is 2-methylbut-1-ene.

Compounds **A** and **B** are branched with a methyl group, so **X** should be branched, too. Haloalkanes can undergo elimination reactions to form alkenes, and alkenes can undergo electrophilic addition reactions with hydrogen bromide to form bromoalkanes.

$$H_3C{-}CH_2{-}C(CH_3){=}CH_2$$

Answer for part b

Step 1 elimination reaction
Warm with ethanolic potassium hydroxide.

Step 2 electrophilic addition reaction
Add hydrogen bromide.

Aliphatic pathways (in these books)

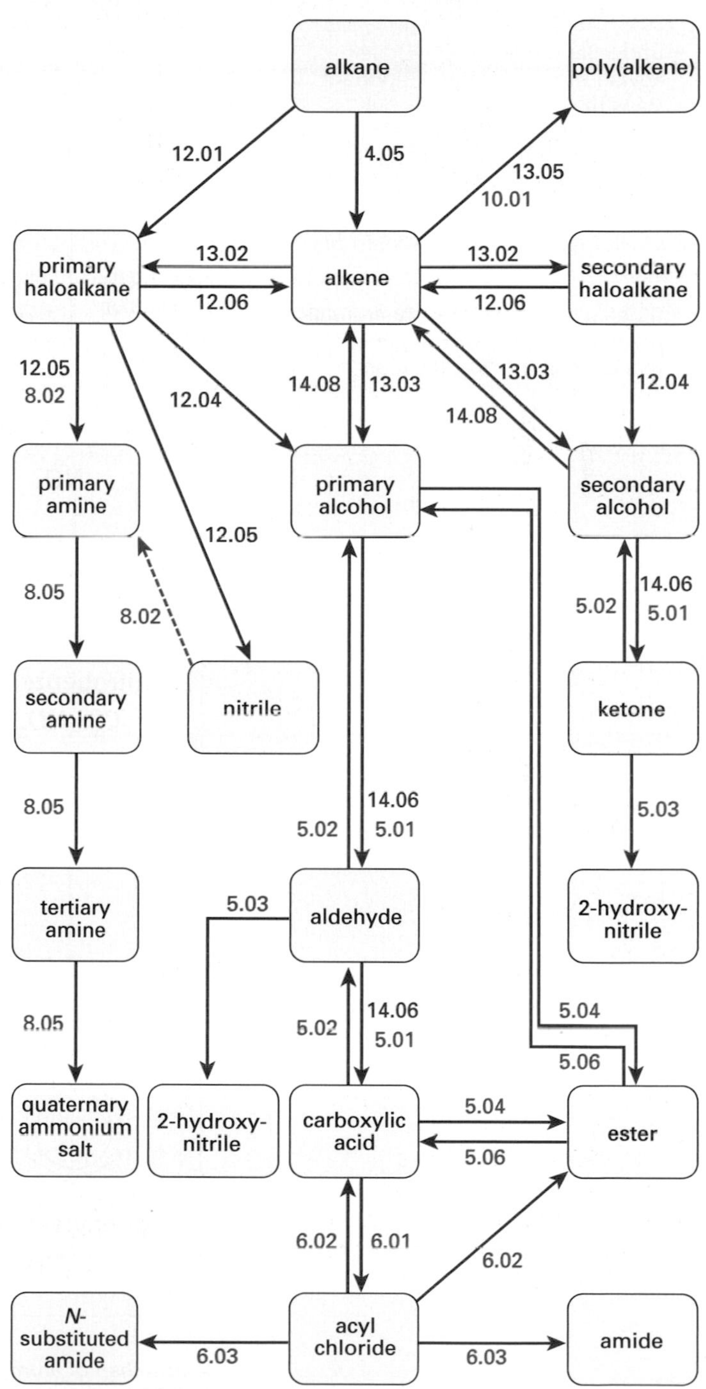

Aliphatic reaction pathways labelled to show the spreads where each one is covered in detail. The red numbers refer to the AS book and the blue numbers refer to this book.

11.04 Aromatic reaction pathways

OUTCOMES

already from A2 Level, you

- understand that nitration is an important step in synthesis, for example in the manufacture of explosives and in the formation of amines from which dyestuffs are manufactured
- understand that Friedel–Crafts acylation reactions are important steps in synthesis
- know that aromatic amines are prepared by the reduction of nitro compounds

and after this spread you should be able to

- deduce how to synthesize aromatic organic compounds using the reactions in the Specification

On this spread, you are going to study two simple worked examples of deducing how to synthesize aromatic compounds.

Carbon atoms in aromatic compounds may be joined together in rings that contain delocalized electrons. In your A Level studies, this means compounds based on benzene, C_6H_6. The diagram shown here summarizes the aromatic reaction pathways you need.

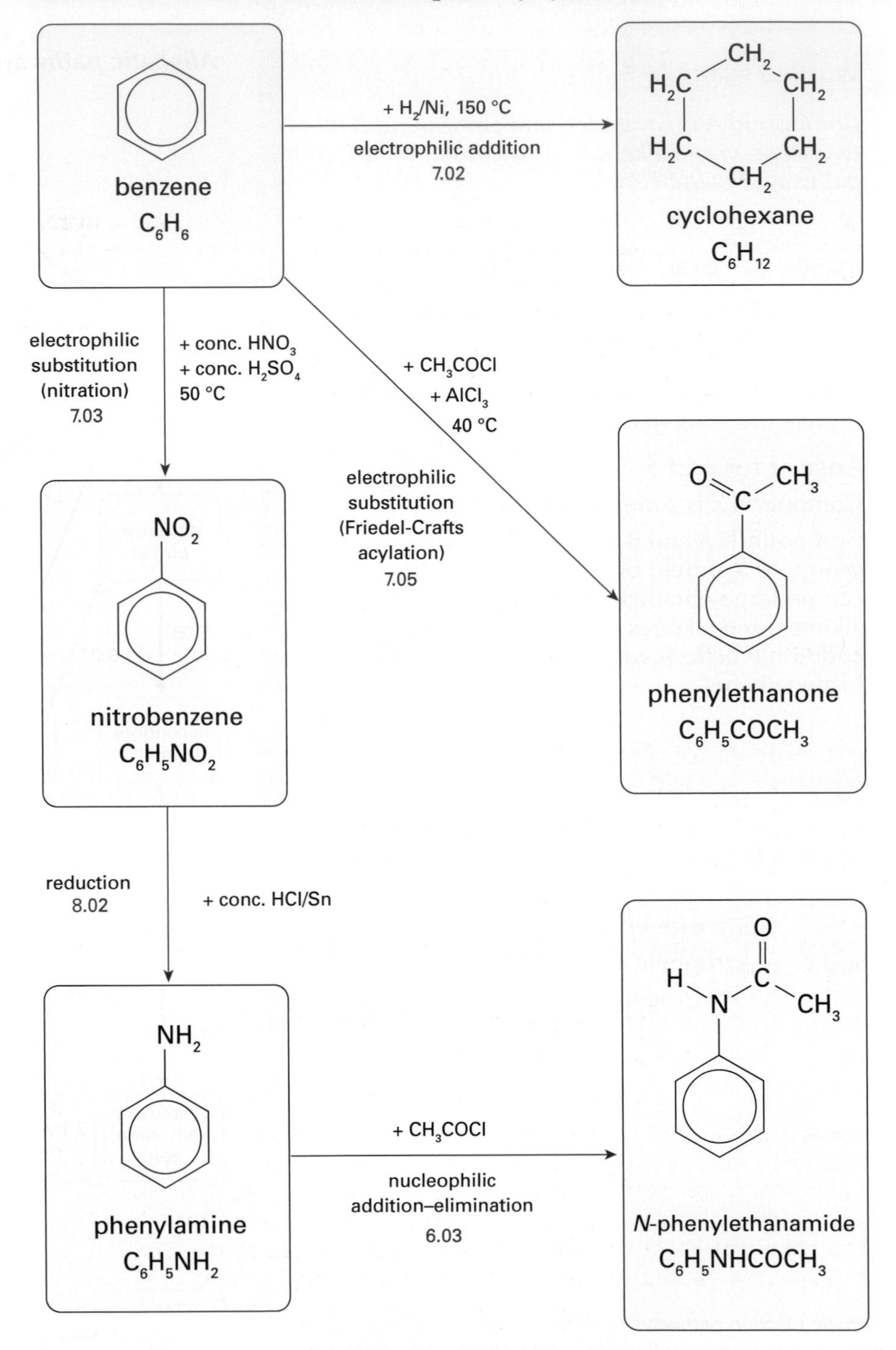

A flowchart of aromatic reaction pathways. Each reaction is labelled to show the spread in this book where it is covered in detail.

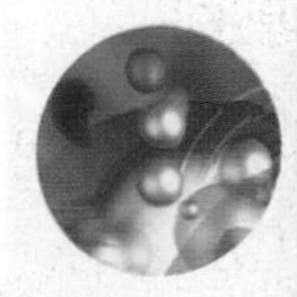

Worked example 1: Benzene to aromatic amine

Describe how you could produce phenylamine, $C_6H_5NH_2$, from benzene. Include equations and any essential reaction conditions needed.

Step 1 Benzene reacts with a mixture of concentrated nitric acid and concentrated sulfuric acid at around 40°C, producing nitrobenzene:

$$C_6H_6 + NO_2^+ \rightarrow C_6H_5NO_2 + H^+$$

Step 2 Nitrobenzene is reduced by reaction with concentrated hydrochloric acid and tin, producing phenylamine:

$$C_6H_5NO_2 + 6[H] \rightarrow C_6H_5NH_2 + 2H_2O$$

Worked example 2: Benzene to aromatic ketone

Describe how you could produce phenylpropanone $C_6H_5COCH_2CH_3$ from benzene. Include equations and any essential reaction conditions needed.

Step 1 Benzene reacts with propanoyl chloride in the presence of aluminium chloride at around 40°C, producing phenylpropanone:

$$C_6H_6 + CH_3CH_2COCl \rightarrow C_6H_5COCH_2CH_3 + HCl$$

O=C(CH$_2$CH$_3$)–C$_6$H$_5$ (structure)

phenylpropanone

Comments on worked examples

Example 1: In Step 1, you might also be asked to explain how the nitronium ion NO_2^+ is formed. It is formed in two steps:

Step 1 Sulfuric acid protonates nitric acid:

$$H_2SO_4 + HNO_3 \rightarrow H_2NO_3^+ + HSO_4^-$$

Step 2 The protonated nitric acid breaks down to form the nitronium ion:

$$H_2NO_3^+ + H_2SO_4 \rightarrow NO_2^+ + H_3O^+ + HSO_4^-$$

Here is the overall equation:

$$HNO_3 + 2H_2SO_4 \rightarrow NO_2^+ + H_3O^+ + 2HSO_4^-$$

Example 2: This is an example of a Friedel–Crafts acylation. You might also be asked to explain the role of aluminium chloride as a catalyst in the reaction. The tetrachloroaluminium ion, $[AlCl_4]^-$, is formed when aluminium chloride reacts with propanoyl chloride. It reacts with the hydrogen ion eliminated from the intermediate carbocation. This regenerates aluminium chloride and forms hydrogen chloride:

$$[AlCl_4]^- + H^+ \rightarrow AlCl_3 + HCl$$

Check your understanding

1 Work out the steps needed for each of the following conversions. For each step, name the type of reaction, and state the reagents and conditions needed.

C$_6$H$_5$–CH$_3$ (structure)

methylbenzene

a methylbenzene to 1-(2-methylphenyl)ethanone

(structure: benzene ring with COCH$_3$ and adjacent CH$_3$)

1-(2-methylphenyl) ethanone

b methylbenzene to 2-methylphenylamine

(structure: benzene ring with NH$_2$ and adjacent CH$_3$)

2-methylphenylamine

11.05 Going from A to B (3)

OUTCOMES

already from A2 Level, you can

- deduce how to synthesize aromatic organic compounds using the reactions in the Specification
- deduce how to synthesize aliphatic organic compounds using the reactions in the Specification

and after this spread you should be able to

- deduce how to synthesize aromatic organic compounds with side groups using the reactions in the Specification

Organic syntheses can involve making changes to a group attached to the benzene ring. The reduction of nitrobenzene to phenylamine is an example of this type of change. But other changes are possible, in particular from phenylamine and organic ketones. On this spread, you are going to study two worked examples of deducing how to synthesize aromatic compounds. Both of these involve some knowledge of aliphatic reaction pathways, too.

Worked example 1: Benzene to aromatic secondary alcohol

Describe how you could produce phenylethanol, $C_6H_5CH(OH)CH_3$, from benzene. Include equations and any essential reaction conditions needed.

Step 1 Benzene reacts with ethanoyl chloride in the presence of aluminium chloride at around 40°C, producing phenylethanone:

$$C_6H_6 + CH_3COCl \rightarrow C_6H_5COCH_3 + HCl$$

Step 2 Phenylethanone is reduced to phenylethanol using dry $LiAlH_4$ or aqueous $NaBH_4$:

$$C_6H_5COCH_3 + 2[H] \rightarrow C_6H_5CH(OH)CH_3$$

phenylethanone

phenylethanol

Comments on worked examples

Example 1: Although the compounds in the synthesis are aromatic, the reaction in step 2 involves the aliphatic side group, not the benzene ring. Take care with your choice of reducing agent. Concentrated hydrochloric acid and tin are used to reduce nitrobenzene to phenylamine, but not to reduce aldehydes and ketones.

Example 2: Ethanoic anhydride could be used instead in Step 1. Ethanoyl chloride is more reactive but ethanoic anhydride is safer. It is cheaper, less susceptible to hydrolysis, and less corrosive. It may be the reagent of choice in industrial applications, such as the manufacture of aspirin (see Spread 6.04).

The reaction at step 3 is not in the Specification. It is called the *Hofmann degradation*, and involves heating with a mixture of aqueous sodium hydroxide and bromine. Notice that the product has one less carbon atom.

Take care with your choice of reducing agent in Step 4. Dry $LiAlH_4$ or aqueous $NaBH_4$ are used to reduce aldehydes and ketones, but not aromatic nitro compounds.

Worked example 2: Phenylamine to benzene-1, 4-diamine

Benzene-1,4-diamine can be synthesized from phenylamine in four steps.

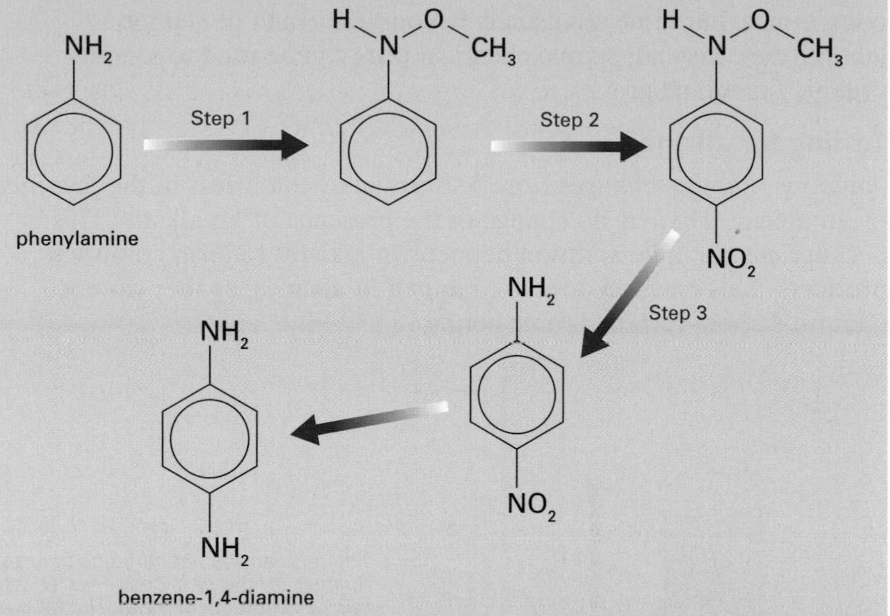

phenylamine

Step 1

Step 2

Step 3

benzene-1,4-diamine

Reaction pathway from phenylamine to benzene-1,4-diamine.

Identify the type of reaction and the reagents needed for each of the steps 1, 2, and 4.

Step 1 nucleophilic addition–elimination reaction
ethanoyl chloride

Step 2 electrophilic substitution or nitration
concentrated nitric acid and concentrated sulfuric acid

Step 4 reduction
concentrated hydrochloric acid and tin, or hydrogen and nickel

Check your understanding

1 Phenylethene (also called ethenylbenzene or styrene) is the monomer used in the manufacture of polystyrene. It can be synthesized from benzene in three steps.

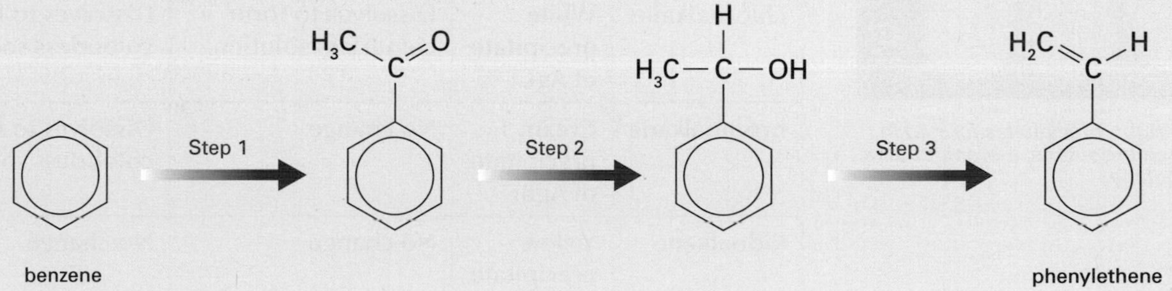

benzene

Step 1

Step 2

Step 3

phenylethene

For each step, name the type of reaction, and state the reagents and conditions needed.

11.06 Identifying functional groups

OUTCOMES

already from AS Level, you

- know that bromine can be used to test for unsaturation
- understand that haloalkanes are susceptible to nucleophilic attack by OH^- ions
- understand why acidified silver nitrate solution is used as a reagent to identify and distinguish between F^-, Cl^-, Br^-, and I^-
- can use Fehling's solution or Tollens' reagent to distinguish between aldehydes and ketones
- understand that primary alcohols can be oxidized to aldehydes and carboxylic acids, that secondary alcohols can be oxidized to ketones, and that tertiary alcohols are not easily oxidized

and after this spread you should be able to

- identify organic functional groups using the reactions in the Specification

From left to right, precipitates formed in tests with 1-chlorobutane, 1-bromobutane, and 1-iodobutane.

Instrumental methods of analysis can be used to identify organic compounds, but these are usually beyond the reach of school laboratories. Instead, simple chemical tests can be used to identify organic functional groups.

Testing for alkenes

Aqueous bromine changes from red-brown to colourless in the presence of an alkene. There is no change in the presence of an alkane. This is because electrophilic addition happens in alkenes to form colourless products. This reaction does not happen in alkanes, as they do not contain carbon–carbon double bonds.

The tube on the left contains benzene floating on top of bromine water, which has not been decolorized. The tube on the right contains cyclohexene floating on top of bromine water, which has been decolorized.

Testing for haloalkanes

Haloalkanes can undergo nucleophilic substitution. For example, they form alcohols when they are warmed with aqueous sodium hydroxide:

$$RCH_2{-}Hal + OH^- \rightarrow RCH_2{-}OH + Hal^-$$

The halide ion released, Hal^-, depends upon the identity of the haloalkane. The halide ion itself can be identified using aqueous silver nitrate, which forms different coloured precipitates.

Haloalkanes are warmed with aqueous sodium hydroxide, which is then neutralized using aqueous nitric acid. Aqueous silver nitrate is added, followed by aqueous ammonia as a confirmatory test. The table summarizes the results of these tests.

haloalkane	colour of precipitate	after adding dilute aqueous ammonia	after adding concentrated aqueous ammonia
chloroalkane	White precipitate of AgCl	Dissolves to form colourless solution	Dissolves to form colourless solution
bromoalkane	Cream precipitate of AgBr	No change	Dissolves to form colourless solution
iodoalkane	Yellow precipitate of AgI	No change	No change

Fluoroalkanes do not react readily with aqueous sodium hydroxide, and fluoride ions do not form precipitates with aqueous silver nitrate.

Testing for carboxylic acids

Carboxylic acids are weak acids with a pH of about 3, so they turn universal indicator red or orange. They react with carbonates to produce bubbles of carbon dioxide gas. If necessary, the identity of the gas can be confirmed using limewater. This turns cloudy white when carbon dioxide is bubbled through it.

Reaction with sodium

Carboxylic acids react with sodium to produce bubbles of hydrogen gas. But water also reacts with sodium, so the presence of a carboxylic acid can only be inferred if the rate of bubbling is much higher than it is with water.

Testing for aldehydes and ketones

Aldehydes and ketones both contain the carbonyl group >C=O, but they react differently with acidified potassium dichromate(VI), Fehling's solution, and Tollens' reagent.

Potassium dichromate(VI)

Aldehydes are readily oxidized to form carboxylic acids but ketones are not. When aldehydes are heated with aqueous $K_2Cr_2O_7$, acidified with dilute sulfuric acid, the colour of the mixture changes from orange to green. There is no change with ketones.

Fehling's solution

When aldehydes are heated with Fehling's solution, the colour of the mixture changes from blue to brown-green, and finally a brick-red precipitate is formed. There is no change with ketones.

Tollens' reagent

When aldehydes are heated with Tollens' reagent, a silver mirror forms on the inside of the test tube. There is no change with ketones.

Fehling's solution stays blue when it is heated with a ketone, but produces a red precipitate with aldehydes. The exact colour achieved depends on the amount of aldehyde present.

Testing for primary, secondary, and tertiary alcohols

Alcohols can be oxidized by heating them with acidified potassium dichromate(VI):

- primary alcohols produce aldehydes
- secondary alcohols produce ketones
- tertiary alcohols are resistant to oxidation

The three types of alcohol can be identified by attempting to oxidize them, then testing the oxidation products using Fehling's solution or Tollens' reagent. The table summarizes how this works.

alcohol	observation when heated with acidified $K_2Cr_2O_7$	observations from tests on oxidation product	
		Fehling's solution	Tollens' reagent
primary	Changes from orange to green	Changes from blue solution to red precipitate	Silver mirror forms
secondary	Changes from orange to green	Stays blue	No change
tertiary	Stays orange	n/a	n/a

Check your understanding

1 Outline how you would distinguish between:

a ethane and ethene

b chloropropane and iodopropane

c ethanol and ethanoic acid

d propanal and propanone

e hexan-1-ol and hexan-2-ol

12.01 Mass spectrometry

OUTCOMES

already from AS Level, you

- understand the principles of a simple mass spectrometer
- can interpret simple mass spectra of elements and calculate relative atomic mass from isotopic abundance
- know that mass spectrometry can be used to identify elements, and to determine the relative molecular mass and molecular formula of compounds

and after this spread you should

- understand that the fragmentation of a molecular ion $M^{+\bullet} \rightarrow X^+ + Y^\bullet$ gives rise to a characteristic relative abundance spectrum that may give information about the structure of the molecule
- know that the more stable X^+ species give higher peaks, limited to carbocation and acylium (RCO^+) ions

The **mass spectrometer** analyses gaseous samples of elements and compounds. It provides information about the relative atomic mass of elements and the relative molecular mass of compounds. The molecular formula of a compound can be determined using high-resolution mass spectrometry. Mass spectrometry can also give information about the structure of a molecule.

Fragmentation

After injection, a gaseous sample passes through four main stages in the mass spectrometer. These are ionization, acceleration, deflection, and detection.

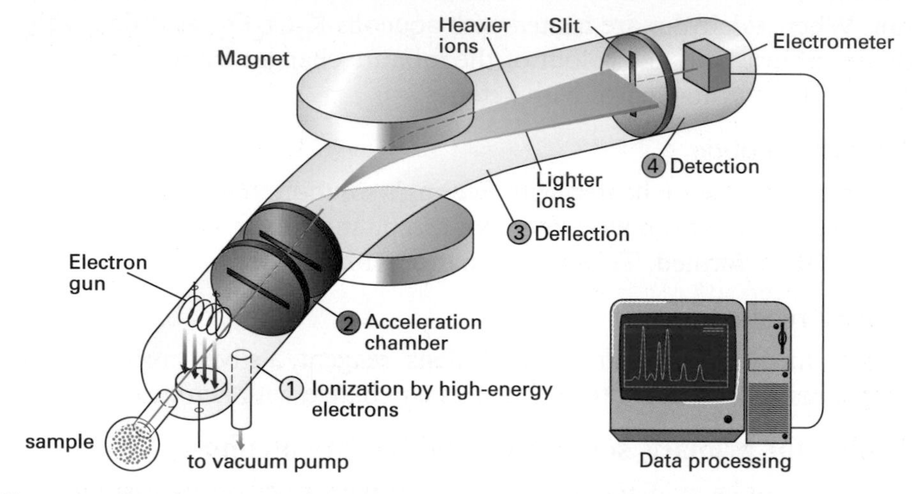

The main parts of a mass spectrometer.

A high-energy electron from an electron gun knocks an electron off an atom or molecule in the sample to produce a **cation**, M^+. This is accelerated in an electric field, deflected by a variable magnetic field, and detected electrically.

The base peak

The greater the abundance of a particular cation, the higher the peak it causes in the **mass spectrum**. The **base peak** in the mass spectrum is the highest peak, caused by the most abundant ion.

The molecular ion peak

The cation produced when the complete molecule loses an electron is called the **molecular ion**. It produces the **molecular ion peak** in the mass spectrum, which is usually the one furthest to the right. The position of a particular peak depends on its **mass to charge ratio**, shown as m/z. For a cation with a single positive charge, the m/z ratio is the same as the relative molecular mass of the ion. This is labelled M in the mass spectrum.

Molecular fragments

When an electron is knocked off a sample molecule, the cation formed has an unpaired electron. This is represented by a dot, giving the symbol $M^{+\bullet}$. So the cation is also a **free radical**. Here is the general equation for the ionization stage:

$$M(g) + e^- \rightarrow M^{+\bullet}(g) + 2e^-$$

The molecular ion can then break apart to form fragments:

$$M^{+\bullet} \rightarrow X^{+} + Y^{\bullet}$$

Only the cation X^+ will be accelerated, deflected, and detected. The uncharged free radical $Y^\bullet$ will not. Different molecules give different characteristic **fragmentation patterns** in their mass spectra, depending on the relative stability and abundance of the fragments they produce.

Fragmentation patterns

The more stable cations including acylium ions RCO^+ and carbocations, give the highest peaks in mass spectra. Tertiary carbocations R_3C^+ are more stable than secondary carbocations R_2CH^+, which are in turn more stable than primary carbocations RCH_2^+.

Butane isomers

Butane and methylpropane are isomers of C_4H_{10}. They produce different fragmentation patterns.

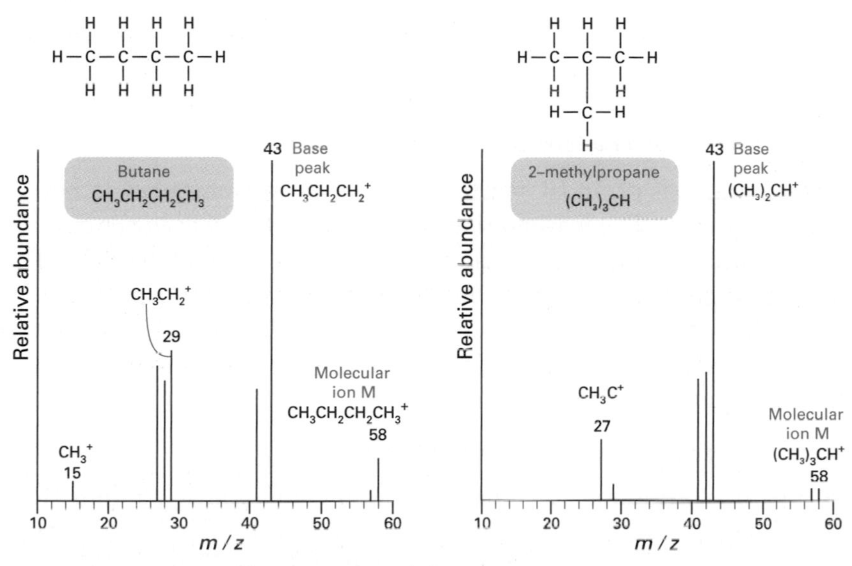

Fragmentation patterns of butane and methylpropane.

Both isomers have a molecular ion peak at $m/z = 58$. The base peaks of butane and methylpropane represent the loss of a methyl group ($58 - 15 = 43$). There is a peak in the butane spectrum at $m/z = 29$, representing $CH_3CH_2^+$. But this cannot form in the methylpropane spectrum. Instead there is a peak at $m/z = 27$, representing CH_3C^+.

Check your understanding

1 a Explain why the fragmentation $CH_4^{+\bullet} \rightarrow CH_3^+ + H^\bullet$ gives rise to a mass spectrum peak at $m/z = 15$ but not at $m/z = 1$.

b Suggest why it is possible to detect a small peak one unit greater than the peak from the molecular ion.

c Write an equation for the fragmentation of $CH_3CH_2COOH^{+\bullet}$ that gives a peak at $m/z = 57$.

2 Assuming that only one bond breaks at a time when ethanol, CH_3CH_2OH, fragments, six peaks in its mass spectrum are possible with an m/z ratio between 15 and 46. Work out the identity of each fragment and explain how it could form.

M+1 and *M*+2 peaks

Most carbon atoms are ^{12}C atoms but 1.07% are ^{13}C instead. The presence of this isotope in organic compounds produces a small peak one unit to the right of the molecular ion peak, the *M*+1 peak.

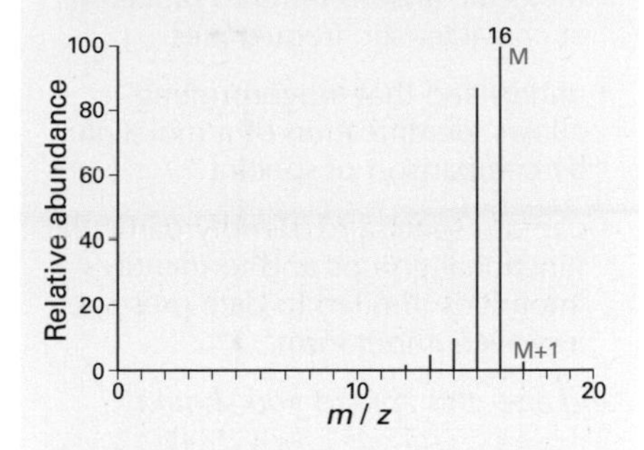

The mass spectrum of methane CH_4 showing the molecular ion peak at m/z = 16 and the smaller M+1 peak at m/z = 17.

Organic compounds containing a chlorine or bromine atom have two molecular ion peaks, at *M* and *M*+2. This is because of the common isotopes ^{35}Cl and ^{37}Cl, and ^{79}Br and ^{81}Br.

Some common fragments

m/z ratio	fragment
77	$C_6H_5^+$ (from benzene ring)
57	$CH_3CH_2CO^+$
43	CH_3CO^+ or $(CH_3)_2CH^+$
31	CH_3O^+ or CH_2OH^+
29	CHO^+ or $CH_3CH_2^+$
15	CH_3^+

The m/z ratios of some fragments commonly met in fragmentation patterns. Note that a decrease of 14 is due to the loss of a CH_2 group.

Infrared spectroscopy

OUTCOMES

already from AS Level, you

- understand that certain groups in a molecule absorb infrared radiation at characteristic frequencies
- understand that 'fingerprinting' allows identification of a molecule by comparison of spectra
- can use spectra to identify particular functional groups and to identify impurities, limited to data presented in wavenumber form

and after this spread you should

- be able to use infrared spectra to identify functional groups in the A Level Chemistry specification

The **infrared spectrometer** uses infrared (IR) radiation to provide information about the type of bonds present in a molecule. Different bonds absorb different frequencies of IR radiation. The percentage transmission through the sample is measured by the machine, so an IR spectrum contains troughs rather than peaks. A reference containing the solvent alone is used so that the spectrum only shows the troughs caused by the sample.

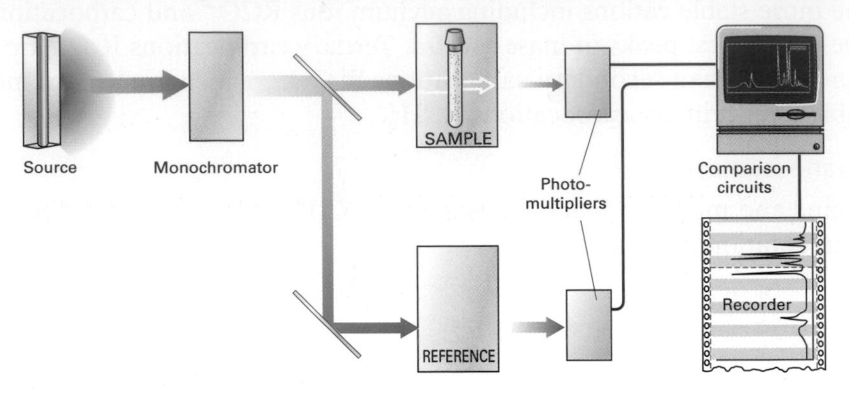

An outline of how the infrared spectrometer works.

The horizontal axis of an IR spectrum shows the **wavenumber** rather than the frequency. The wavenumber is the number of waves per cm. The wavenumber (and so the frequency) decreases from left to right. The **fingerprint region** is the part of the spectrum between 1500 cm^{-1} and 400 cm^{-1}. Its pattern of troughs is unique to each compound. An unknown compound is identified when its fingerprint region exactly matches the fingerprint region of a known compound.

Functional groups from wavenumbers

Most organic compounds contain C—H and C—C bonds, so their IR spectra show troughs at 2850–3300 cm^{-1} and 750–1100 cm^{-1}. Alkenes also have C=C bonds, producing troughs at 1620–1680 cm^{-1}. Other bonds can be identified using **correlation charts** like the one in the box. Note that the functional groups corresponding to these groups cannot always be identified from the IR spectrum, unless their fingerprint region matches a known compound.

compound	bond					
	C=O	C—O	O—H	C—N	N—H	N—O
aldehyde	✓					
ketone	✓					
acyl chloride	✓					
acid anhydride	✓	✓				
ester	✓	✓				
carboxylic acid	✓	✓	✓			
alcohol		✓	✓			
amine				✓	✓	
amide	✓			✓	✓	
aromatic nitro compound				✓		✓

Bonds founds in some common organic compounds, other than C—H, C—C, and C=C bonds.

Wavenumbers

You are not expected to recall the values of any wavenumbers, apart from the range of the fingerprint region. You will be provided with appropriate data in the examination, similar to the data here.

bond	wavenumber (cm^{-1})
C—H	2850–3300
C—C	750–1100
C=C	1620–1680
C=O	1680–1750
C—O	1000–1300
O—H in alcohols	3230–3550
O—H in acids	2500–3000
C—N	1180–1360
N—H	3320–3560
C≡N	2210–2260
N—O	1330–1550

Some common bonds and their wavenumbers in infrared spectra.

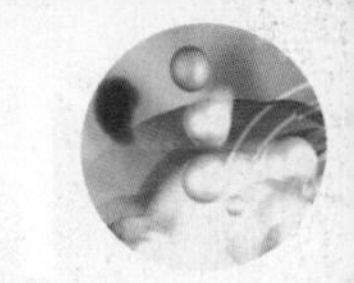

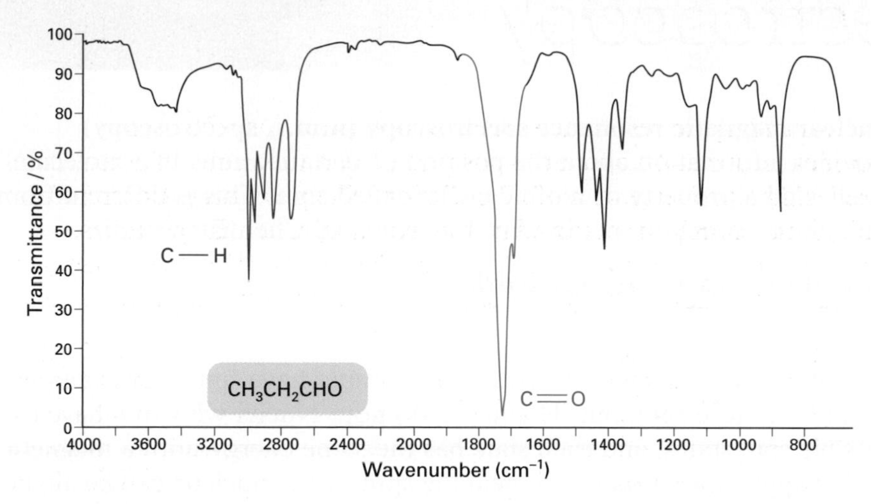

The IR spectrum of propan-2-ol has a large trough between 3230 and 3550 cm^{-1}, indicating a hydroxyl group.

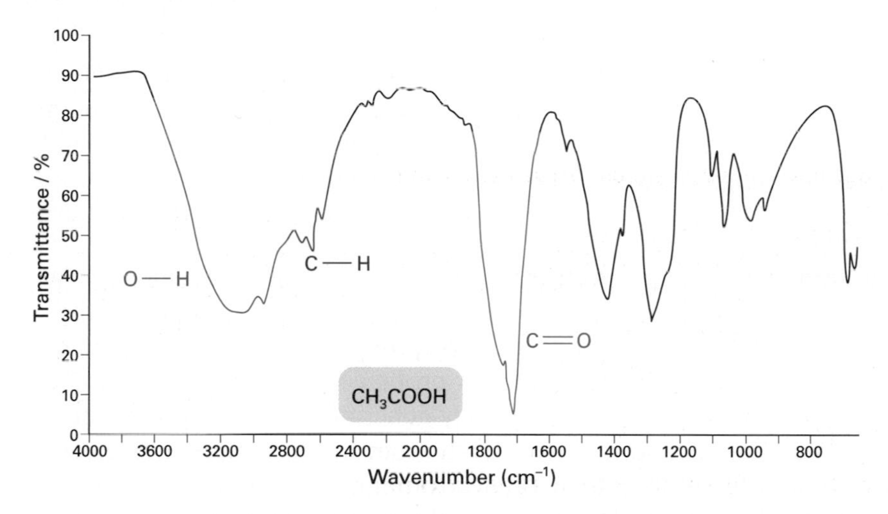

The IR spectrum of propanal has a large trough between 1680 and 1750 cm^{-1} indicating a carbonyl group, but none that would indicate the presence of a hydroxyl group, as in ethanoic acid.

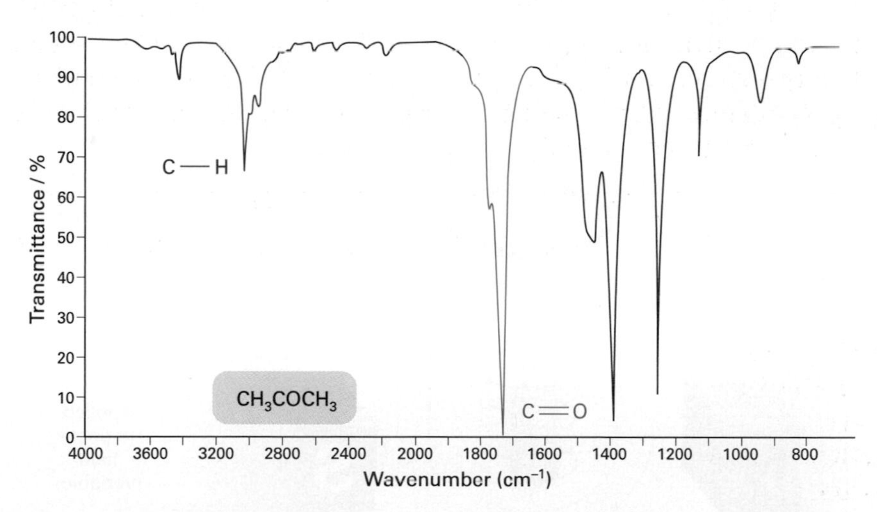

The IR spectrum of ethanoic acid has a large trough between 2500 and 3000 cm^{-1}, indicating a hydroxyl group. The presence of a trough between 1680 and 1750 cm^{-1} indicates a carbonyl group. The two pieces of evidence suggest that a carboxyl group COOH is present.

Check your understanding

1 a State the bonds present in an ester link, —COO—.

b Predict the wavenumber range of absorption by the bonds in an ester link.

c Suggest how an ester could be distinguished from a carboxylic acid using infrared spectroscopy.

2 a The infrared spectrum of substance Z shows a strong absorption between 2210 and 2260 cm^{-1}. Name the functional group most likely to cause this.

b Substance Z is reduced to an amine Y, and then analysed. Describe the differences between the IR spectra of substances Z and Y.

12.03 Nuclear magnetic resonance spectroscopy

OUTCOMES

already from AS Level, you

- can recall the meaning of mass number (*A*) and atomic (proton) number (*Z*)

and after this spread you should

- understand that nuclear magnetic resonance gives information about the position of ^{13}C or ^{1}H atoms in a molecule
- know the use of the δ scale for recording chemical shift
- understand that chemical shift depends on the molecular environment

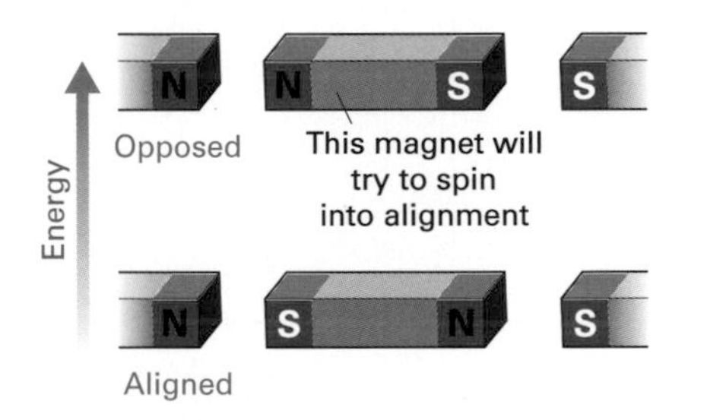

A bar magnet opposed to a magnetic field is at a higher energy than one aligned with it.

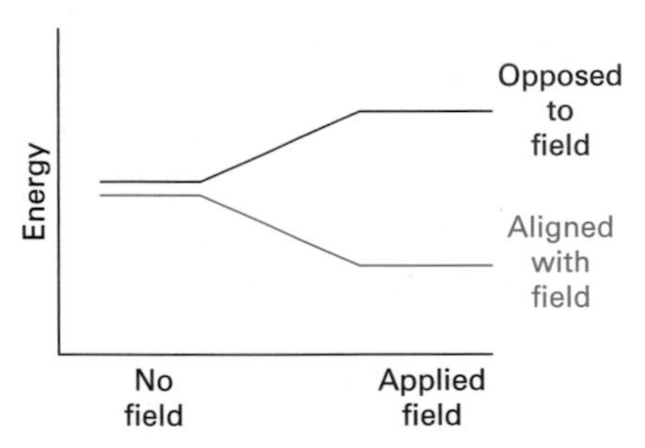

The spin states of a nucleus with an odd mass number in the absence and presence of a magnetic field.

Nuclear magnetic resonance spectroscopy (n.m.r. spectroscopy) provides information about the position of certain atoms in a molecule. It relies on a property of atomic nuclei called spin. This is different from the electron spin you met in Unit 1 of your AS Chemistry studies.

How n.m.r. spectroscopy works

Nuclear spin

The nuclei of atoms with an odd mass number have spin. For example, ^{1}H and ^{13}C have spin, but ^{2}H and ^{12}C do not. Nuclei with spin have two possible spin states, and each state has the same energy until a magnetic field is applied. In a magnetic field the spin of the nucleus can be aligned with the field or opposed to it. A nucleus in the spin state opposed to the field is at a higher energy than a nucleus aligned with the field. The difference in energy depends upon the strength of the magnetic field.

Nuclei aligned with the magnetic field can change their spin to the opposed state if they absorb the correct amount of energy. The energy needed is very small and it needs electromagnetic radiation in the radio-frequency range, between 20 MHz and 100 MHz.

Making an n.m.r. spectrum

The sample for n.m.r. spectroscopy must be a liquid or dissolved in a solvent. It is put into a cylindrical sample tube, and lowered into the **n.m.r. spectrometer**. Once there, it is spun on its axis between the poles of a very powerful electromagnet. A probe coil surrounds the sample, and is connected to a radio-frequency generator and a receiver. The applied magnetic field is varied. Each atom in the sample molecules absorbs a different frequency of electromagnetic radiation, called its **resonance frequency**. This produces the **n.m.r. spectrum**.

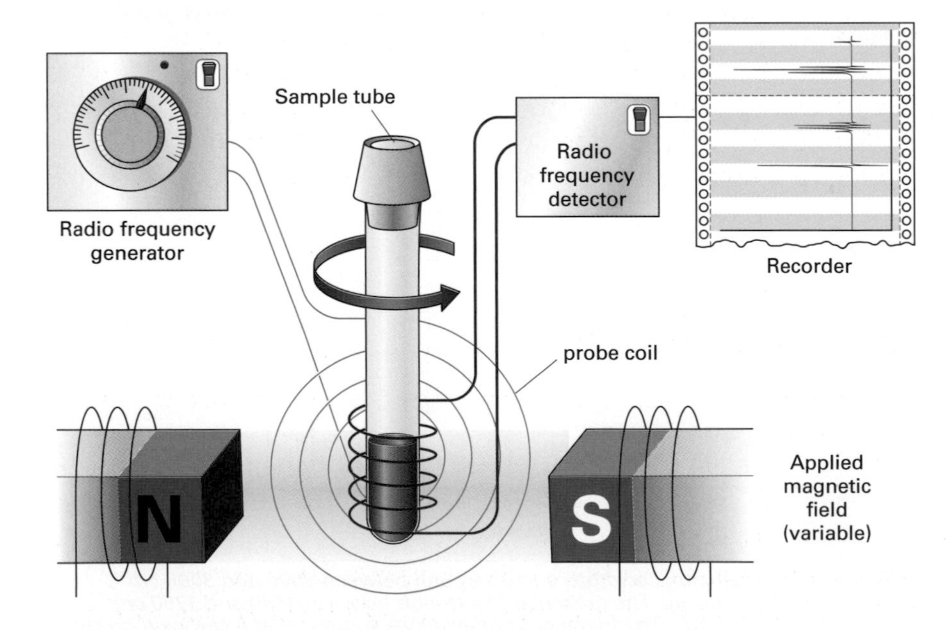

The main parts of an n.m.r. spectrometer.

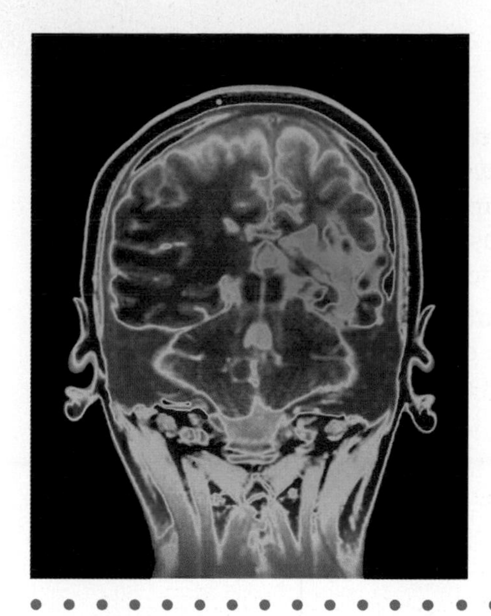

MRI scanners

Magnetic resonance imaging (MRI) uses a similar technology to n.m.r. spectroscopy. It is used in medicine to make detailed images of the different tissues inside the body. MRI scanners often detect ^{31}P, because phosphorus is a common component of the lipids (fats) in cell membranes and the nucleic acids in DNA. Whole-body scanners can make images of the entire body if needed.

Computer software analyses the data from an MRI scanner to produce false-coloured images that doctors can use.

A chemist adds a sample to an n.m.r. spectrometer. This may also be done automatically so that the machine can analyse many samples one after the other. The powerful electromagnets are cooled with liquid nitrogen and liquid helium.

Chemical shift

It is difficult to measure the precise resonance frequencies of the ^{1}H or ^{13}C atoms in an organic compound. Instead a **reference compound** is added to the sample. The differences or shifts between the resonance frequencies of the ^{1}H or ^{13}C atoms in the sample and the reference compound are measured.

The size of each shift depends on the spectrometer frequency, as well as the position of the ^{1}H or ^{13}C atoms in sample molecules. To allow results from different spectrometers to be compared, the **chemical shift**, δ, is calculated:

$$\delta = \frac{\text{shift (Hz)}}{\text{spectrometer frequency (MHz)}}$$

The units of δ are parts per million (ppm) because 1 MHz is one million Hz. Different ^{1}H or ^{13}C atoms in a molecule have different chemical shifts depending on their chemical environment. For example, δ is about 0.9 ppm for a proton (^{1}H atom) in an alkane, but about 7.3 ppm for a proton in benzene.

Check your understanding

1 a Explain why ^{1}H and ^{13}C atoms produce n.m.r. spectra but ^{2}H and ^{12}C atoms do not.

b What happens when a nucleus with spin aligned with a magnetic field absorbs energy at its resonance frequency?

c What is a proton n.m.r. spectrum?

2 a State the equation for calculating chemical shift, δ.

b ^{13}C atoms at a certain position in an organic compound have a shift of 900 Hz from the ^{13}C atoms in the reference compound. If the spectrometer frequency is 50 Mz, calculate the chemical shift and state the units.

12.04 Low-resolution n.m.r. spectroscopy

OUTCOMES

already from A2 Level, you

- understand that nuclear magnetic resonance gives information about the position of ^{13}C or ^{1}H atoms in a molecule
- know the use of the δ scale for recording chemical shift
- understand that chemical shift depends on the molecular environment

after this spread you should understand

- why tetramethylsilane (TMS) is used as a standard
- that ^{1}H n.m.r. spectra are obtained using samples dissolved in proton-free solvents (e.g. deuterated solvents and CCl_4)
- that ^{13}C n.m.r. gives a simpler spectrum than ^{1}H n.m.r.
- how integrated spectra indicate the relative numbers of ^{1}H atoms in different environments

The chemical shift produced by an atom in an n.m.r. spectrum depends on its **chemical environment**. This involves any atoms covalently bonded to it, and neighbouring atoms or groups of atoms. Atoms in the same chemical environment are described as **equivalent atoms** and those in different chemical environments are described as **non-equivalent atoms**. In cyclohexane, for example, all the carbon atoms are in the same chemical environment. The ^{13}C n.m.r. spectrum of cyclohexane has a single peak at $\delta = 27$ ppm. The chemical shift of the reference standard is always 0 ppm.

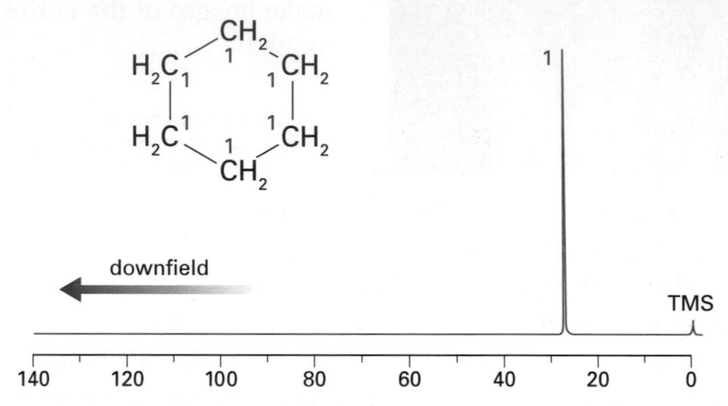

The ^{13}C n.m.r. spectrum of cyclohexane. The peak produced by its carbon atoms is shifted downfield compared to the reference standard.

Shielding and chemical shift

The resonance frequency of a particular ^{1}H or ^{13}C atom in a compound is affected by the local electron density. Electrons oppose the applied magnetic field, reducing its effect on the resonating nucleus and causing it to be **shielded**. Atoms that are covalently bonded to electronegative elements such as oxygen become **deshielded**. Their chemical shift is increased as a result.

The ^{13}C n.m.r. spectrum of cyclohexene shows this effect (above). Like cyclohexane, it has six carbon atoms. But the presence of the carbon–carbon double bond produces three different chemical environments not just one. The carbon atoms joined by the double bond are deshielded, and produce a peak with a greater chemical shift than the other carbon atoms.

TMS – the reference standard

CH_3
$H_3C—Si—CH_3$
CH_3

Tetramethylsilane, TMS. The chemical shift of TMS is 0 ppm by definition, so its peak is often left out of n.m.r. spectra.

The reference compound used in n.m.r. spectroscopy is **tetramethylsilane (TMS)**, $Si(CH_3)_4$. It is chosen for several reasons, including:

- Its ^{1}H and ^{13}C nuclei are highly shielded because the electronegativity of silicon is low. So their peaks in n.m.r. spectra are upfield of the peaks from most other ^{1}H and ^{13}C nuclei.
- All 12 hydrogen atoms, and all four carbon atoms, are equivalent. So TMS produces strong, sharp peaks in spectra.
- TMS is cheap and non-toxic and it does not react with the sample.

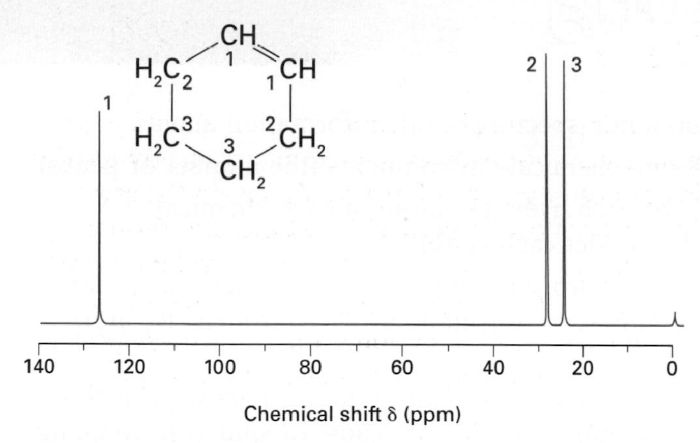

The ^{13}C n.m.r. spectrum of cyclohexene.

^1H n.m.r spectra

In general, ^{13}C n.m.r. spectra are simpler than ^1H n.m.r. spectra (also called **proton n.m.r. spectra**). For example, the carbon atoms in ethanol have two chemical environments, but its hydrogen atoms have three chemical environments.

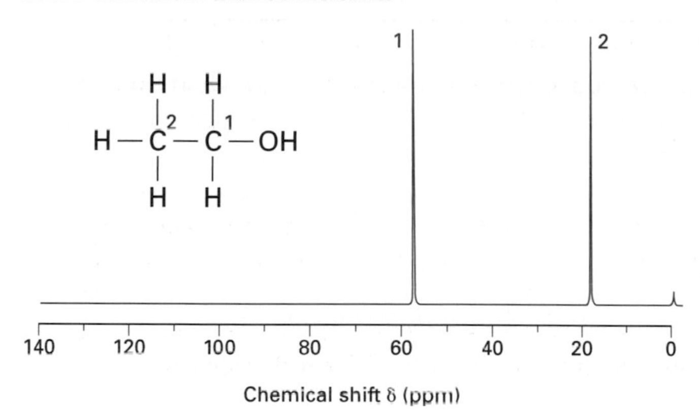

The ^{13}C n.m.r. spectrum of ethanol.

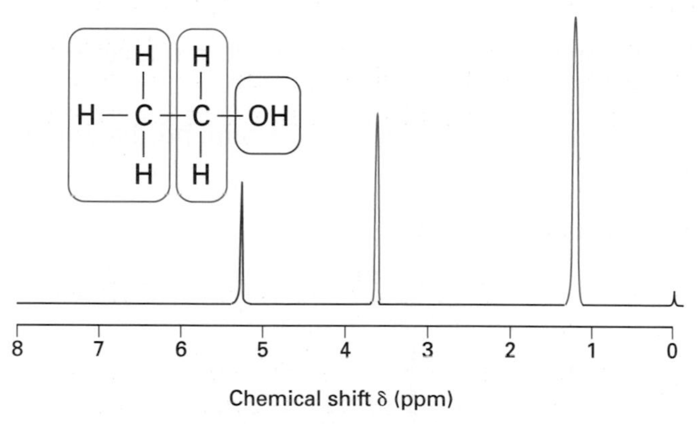

The ^1H n.m.r. spectrum of ethanol.

The area under each peak in an n.m.r. spectrum is proportional to the number of atoms in the same chemical environment. **Integrated spectra** include a stepped line. The height of this **integral** is proportional to the area under the peak. This helps you to work out the relative number of atoms in each chemical environment in the molecule. Note that that you usually need other data about the compound such as its relative formula mass, to work out the actual number of atoms.

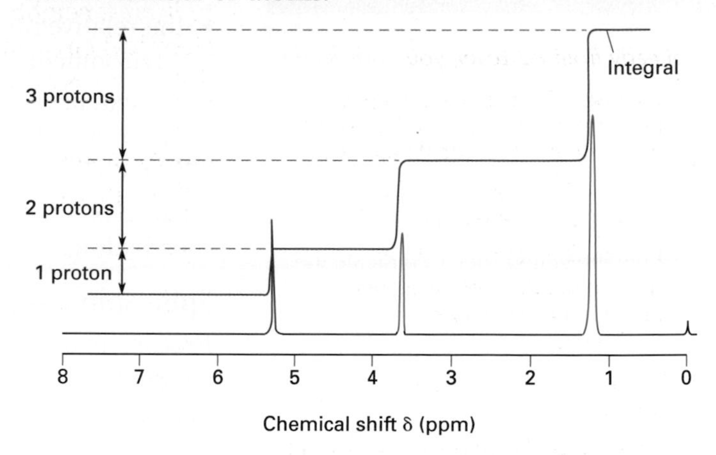

The integrated ^1H n.m.r. spectrum of ethanol.

Solvents for ^1H n.m.r.

The solvents used in ^1H n.m.r. spectroscopy must be proton-free or they will produce their own peaks in the spectrum. Tetrachloromethane, CCl_4, does not contain any hydrogen atoms, so it is proton-free. Deuterium, ^2H, is a hydrogen isotope that does not produce peaks in ^1H n.m.r. spectra. So **deuterated solvents** such as $CDCl_3$ ($CHCl_3$ containing ^2H not ^1H) can also be used.

Check your understanding

1 How many different chemical environments are there for the carbon atoms in the following compounds?
  a benzene, C_6H_6
  b propene, $CH_3CH{=}CH_2$
  c phenylamine, $C_6H_5NH_2$

2 How many different chemical environments are there for the hydrogen atoms in the following compounds?
  a ethane, CH_3CH_3
  b propan-1-ol, $CH_3CH_2CH_2OH$
  c propan-2-ol, $CH_3CH(OH)CH_3$

3 a Explain why TMS is used as the reference compound in n.m.r. spectroscopy.
  b Explain why CCl_4 and $CDCl_3$ are used as solvents in proton n.m.r. spectroscopy.
  c What information does the integral in an integrated n.m.r. spectrum provide?

12.05 Peak splitting

OUTCOMES

already from A2 Level, you understand

- that nuclear magnetic resonance gives information about the position of ^1H atoms in a molecule
- that chemical shift depends on the molecular environment
- how integrated spectra indicate the relative numbers of ^1H atoms in different environments

after this spread you should

- be able to use the $n + 1$ rule to deduce the spin–spin splitting patterns of adjacent, non-equivalent protons (limited to doublet, triplet, and quartet formation in simple aliphatic compounds)

Low-resolution proton n.m.r. spectra provide information about

- the number of different chemical environments (the number of peaks)
- the relative number of each hydrogen atom in each chemical environment (the area under each peak)
- the extent to which the hydrogen atoms in each chemical environment are deshielded, for example by a neighbouring electronegative atom (the chemical shift)

High-resolution proton n.m.r. gives fine detail about the relative position of each hydrogen atom in a compound. This is because of **spin–spin coupling**.

Spin–spin coupling and peak splitting

Protons in different neighbouring chemical environments are said to be coupled. A neighbouring proton with its spin aligned with the magnetic field will have a slightly deshielding effect, and one with its spin opposed to the field will have a slightly shielding effect. The effect of this spin–spin coupling is to split a single peak into a cluster of peaks in high-resolution proton n.m.r. spectra.

Working out the number of peaks

Spin–spin coupling, and so peak splitting, only happens when there are adjacent non-equivalent hydrogen atoms. It does not happen where two identical chemical environments are next to each other.

This is how you work out the splitting pattern for a peak corresponding to a chemical environment:

- Look at the non-equivalent group next to it.
- Count the number of hydrogen atoms in this group.
- The peak is split into this number plus one (the ***n* + 1 rule**).

This means that

- One adjacent hydrogen atom, such as in a CH group, produces two peaks called a **doublet**.
- Two adjacent hydrogen atoms, such as in a CH_2 group, produces three peaks called a **triplet**.
- Three adjacent hydrogen atoms, such as in a CH_3 group, produces four peaks called a **quartet**.

The ratios of peak heights

The ratio of peak heights in peak splitting can be worked out using Pascal's triangle.

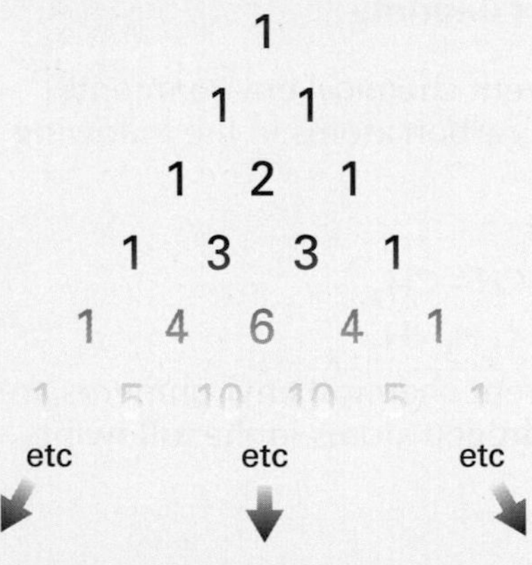

In Pascal's triangle, each number is the sum of the two numbers above it immediately to the left and right.

The peak area ratio of the two peaks in a doublet is 1:1. The ratio of the three peaks in a triplet is 1:2:1, and the ratio of the four peaks in a doublet is 1:3:3:1.

The proton n.m.r spectrum of ethanol

Ethanol, CH_3CH_2OH, has three different chemical environments for its protons. These give three peaks of ratio 3:2:1 at low-resolution,

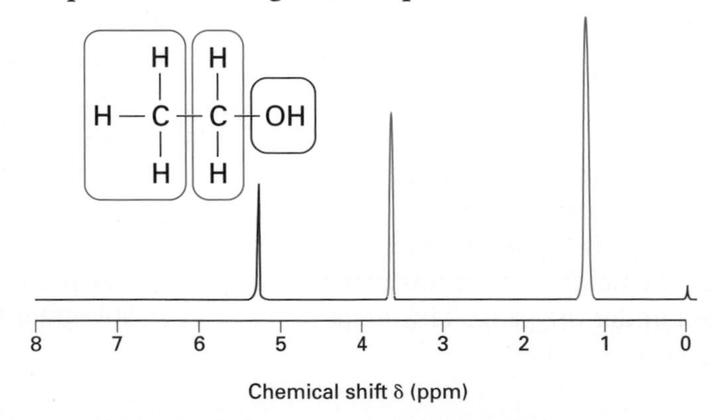

The low-resolution proton n.m.r. spectrum of ethanol.

corresponding to the $-CH_3$ group, the $-CH_2$ group, and the $-OH$ group respectively. The hydrogen atom in the $-OH$ group is deshielded by the oxygen atom, shifting its resonance peak downfield.

At high resolution, the hydrogen atom in the $-OH$ group is not adjacent to other hydrogen atoms. It is separated from them by the oxygen atom, so its peak stays as a **singlet**.

The two hydrogen atoms in the $-CH_2$ group are adjacent to the three non-equivalent hydrogen atoms in the $-CH_3$ group. Their peak is split into $3 + 1 = 4$, a quartet with the peak area ratio 1:3:3:1.

The three hydrogen atoms in the $-CH_3$ group are adjacent to the two non-equivalent hydrogen atoms in the $-CH_2-$ group. Their peak is split into $2 + 1 = 3$, a triplet with the ratio 1:2:1.

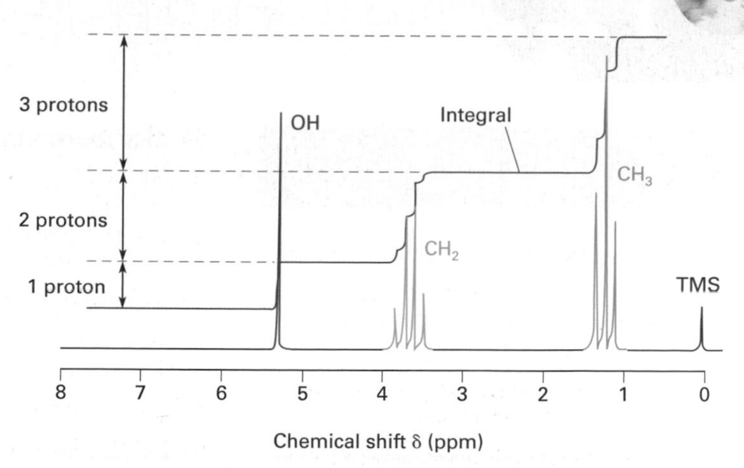

The high-resolution proton n.m.r. spectrum of ethanol.

Another example

The low-resolution proton n.m.r. spectrum of 1,1,2-trichloroethane has two peaks in the ratio 1:2. This shows that there are two chemical environments, with one containing twice as many hydrogen atoms as the other. The hydrogen atom in the $CHCl_2$ group is more deshielded by chlorine than the two hydrogen atoms in the CH_2Cl group.

The high-resolution proton n.m.r. spectrum of 1,1,2-trichloroethane shows peak splitting. The peak corresponding to the hydrogen atom in the $CHCl_2$ group is split into a triplet, showing that the atom is adjacent to a group containing two hydrogen atoms. The peak corresponding to the hydrogen atoms in the CH_2Cl group is split into a doublet, showing that those atoms are adjacent to a group containing one hydrogen atom.

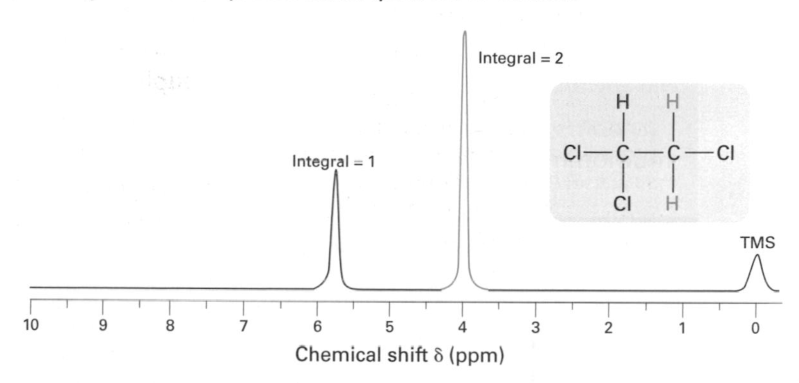

The low-resolution proton n.m.r. spectrum of 1,1,2-trichloroethane.

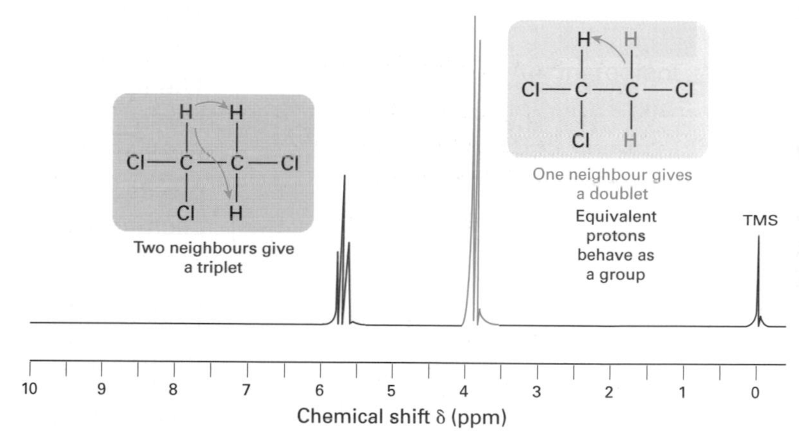

The high-resolution proton n.m.r. spectrum of 1,1,2-trichloroethane.

Check your understanding

1 a Outline why peak splitting happens.

b Use the $n + 1$ rule to explain the splitting seen in the high-resolution proton n.m.r spectrum of iodoethane, CH_3CH_2I.

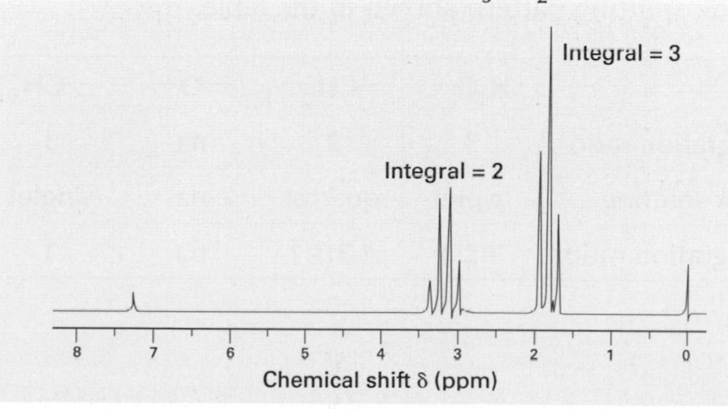

12.06 High-resolution n.m.r.

OUTCOMES

already from A2 Level, you

- understand that chemical shift depends on the molecular environment
- understand how integrated spectra indicate the relative numbers of ^{1}H atoms in different environments
- can use the n + 1 rule to deduce the spin–spin splitting patterns of adjacent, non-equivalent protons

and after this spread you should

- be able to predict the peaks in proton n.m.r. spectra from the structure of a molecule
- be able to interpret data from high-resolution proton n.m.r. spectra

Peaks from structures

Low-resolution proton n.m.r. spectra

Draw the displayed formula first, unless the compound has a simple structure. Then circle all the groups that contain hydrogen atoms. Check to see if any of the groups are in the same chemical environment. If they are, draw a line between the circles involved.

Consider butane, C_4H_{10}. It has two identical chemical environments, containing CH_3 or CH_2 groups. Its low-resolution proton n.m.r. spectrum has two peaks with an integration (peak area) ratio of 6:4. Remember that this shows the *relative numbers* of hydrogen atoms in each chemical environment, and not necessarily the *actual number*. You can simplify 6:4 to 3:2 or just leave it as 6:4.

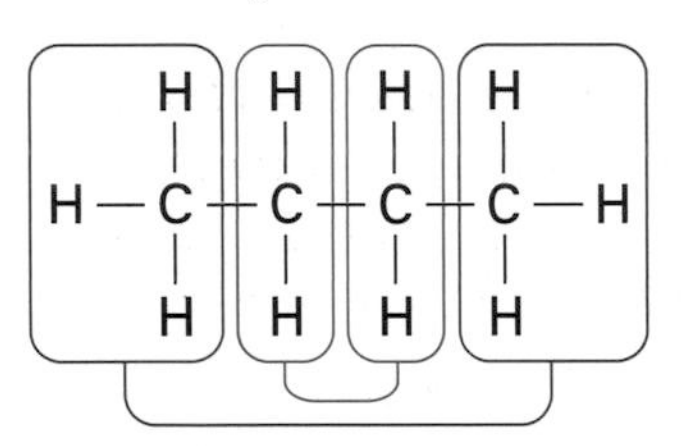

Butane and its chemical environments.

1-methoxypropane and its chemical environments.

Consider 1-methoxypropane $CH_3CH_2CH_2OCH_3$. It has just one more atom than butane but its low-resolution proton n.m.r. spectrum has four peaks with an integration ratio of 3:2:2:3. This is because the presence of the oxygen atom has produced four different chemical environments.

Diethyl ether $CH_3CH_2OCH_2CH_3$ is an isomer of 1-methoxypropane. Its low-resolution proton n.m.r. spectrum has two peaks with an integration ratio of 6:4. Notice that this is the same ratio as for butane. But the presence of the oxygen atom will deshield the protons, increasing the chemical shift of the peaks compared to those of butane.

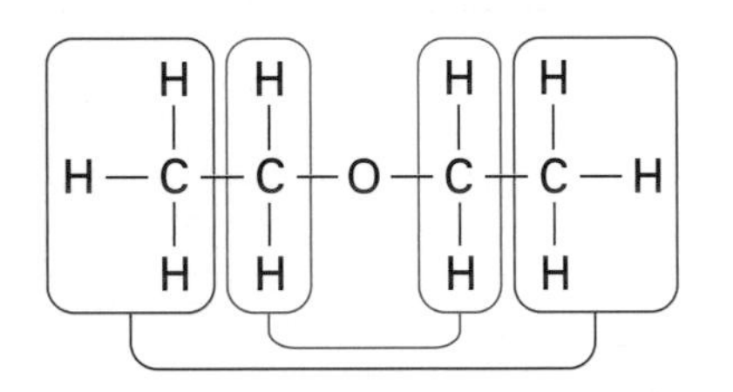

Diethyl ether and its chemical environments.

Methoxyethane and its chemical environments.

High-resolution proton n.m.r. spectra

Peak splitting patterns become complex when a group containing hydrogen atoms lies between two adjacent non-equivalent groups containing hydrogen atoms. So oxygen atoms are often used in examination questions to avoid this complication.

Consider methoxyethane, $CH_3CH_2OCH_3$. Its low-resolution spectrum contains three peaks with an integration ratio of 3:2:3. Its high-resolution spectrum has the peak splitting pattern shown in the table.

group	H_3C—	—CH_2—	—O—	—CH_3
low-resolution integration ratio	3	2	na	3
high resolution peak splitting	triplet	quartet	na	singlet
high resolution integration ratio	1:2:1	1:3:3:1	na	1

Consider 1-methoxyethanol $CH_3CH(OH)OCH_3$. Its low-resolution spectrum contains four peaks with an integration ratio of 3:1:1:3. Its high-resolution spectrum has the peak splitting pattern shown in the table.

group	$H_3C—$	—OH	—CH	—O—	$—CH_3$
low-resolution integration ratio	3	1	1	na	3
high resolution peak splitting	doublet	singlet	quartet	na	singlet
high resolution integration ratio	1:1	1	1:3:3:1	na	1

Structures from peaks

High-resolution proton n.m.r. spectra

You should be able to work out the structure of a compound if you are given its molecular formula, and a table of peak splitting with integration ratios. The chemical shift data may be given, too, to use with the table of chemical shifts given on the examination Data Sheet. These data can help to pin-point the group responsible for a particular peak.

Worked example

Consider a substance Z that has the molecular formula $C_6H_{12}O_3$. It contains an $RCOCH_3$ group and an ester functional group. The peak splitting pattern in its high-resolution proton n.m.r. spectrum is shown in the table. Work out its structure.

low-resolution integration ratio	2	3	2	2	3
high resolution peak splitting	quartet	singlet	triplet	triplet	triplet
chemical shift, δ (ppm)	4.1	3.4	2.7	2.5	1.2

The integration ratios add up to 12, the same as the number of hydrogen atoms in the molecular formula. So each integration ratio must be the same as the number of hydrogen atoms in that corresponding group.

The quartet represents a CH_2 group adjacent to a CH_3 group. The triplet at $\delta = 1.2$ ppm represents a group with three hydrogen atoms, adjacent to a CH_2 group. So there must be a CH_3CH_2 group.

The two triplets, each with two hydrogen atoms, must represent two adjacent non-equivalent CH_2 groups. So there must be a CH_2CH_2 group. Looking at the chemical shift data, singlet at $\delta = 3.4$ ppm must represent an $ROCH_3$. Putting all the pieces together gives this structure.

```
    H       H  H  O        H  H
    |       |  |  ‖        |  |
H — C — O — C — C — C — O — C — C — H
    |       |  |           |  |
    H       H  H           H  H
```

The deduced structure of substance Z.

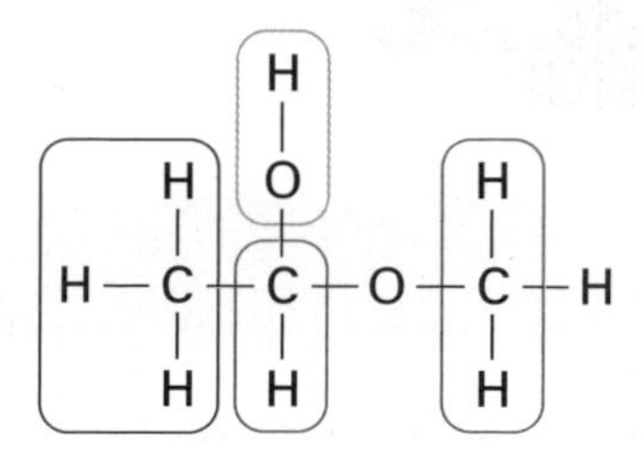

1-methoxyethanol and its chemical environments.

Proton n.m.r. chemical shift data

type of proton	δ (ppm)
RCH_3	0.7–1.1
R_2CH_2	1.2–1.4
R_3CH	1.4–1.6
$RCOCH_3$	2.1–2.6
$ROCH_3$	3.1–3.9
$RCOOCH_3$	3.7–4.1
ROH	0.5–5.0

Check your understanding

1 For each of the following compounds, work out the number of peaks in the low-resolution proton n.m.r. spectrum and their integration ratios.
  a propane
  b 2,2-dimethylpropane
  c propanoic acid

2 For each of the following compounds, work out the low-resolution integration ratios, and high-resolution peak splitting patterns.
  a propanoic acid
  b ethyl ethanoate, $CH_3COOCH_2CH_3$
  c 1-hydroxypentan-3-one, $HOCH_2CH_2COCH_2CH_3$

12.07 Column chromatography

OUTCOMES

already from A2 Level, you know

- that mixtures of amino acids can be separated by chromatography

and after this spread you should know

- that separation by column chromatography depends on the balance between solubility in the moving phase and retention in the stationary phase
- that gas–liquid chromatography can be used to separate mixtures of volatile liquids

Thin-layer chromatography or TLC is discussed in Spread 9.03. It can be used to separate dissolved solutes such as a mixture of amino acids. Powdered silica or alumina on the glass or plastic sheet is the stationary phase, and the solvent is the mobile phase. The stationary phase does not move but the mobile phase does. It carries the dissolved solutes with it.

The stationary phase can be packed into a glass or metal tube, called a column, allowing **column chromatography**. One of the advantages of column chromatography is that the separated solutes can be recovered, if needed, for further analysis using other techniques.

In column chromatography the different components in a mixture are **eluted** by the solvent (leave the bottom of the column) at different times. The **retention time** of a solute is the time taken for it to leave the column after adding the mixture. The retention time for a solute depends on

- its attraction to the stationary phase; and
- its **solubility** in the mobile phase.

The retention time is low when the attraction to the stationary phase is weaker than the attraction to the solvent in the mobile phase. The retention time is high when the attraction to the stationary phase is stronger than the attraction to the solvent in the mobile phase.

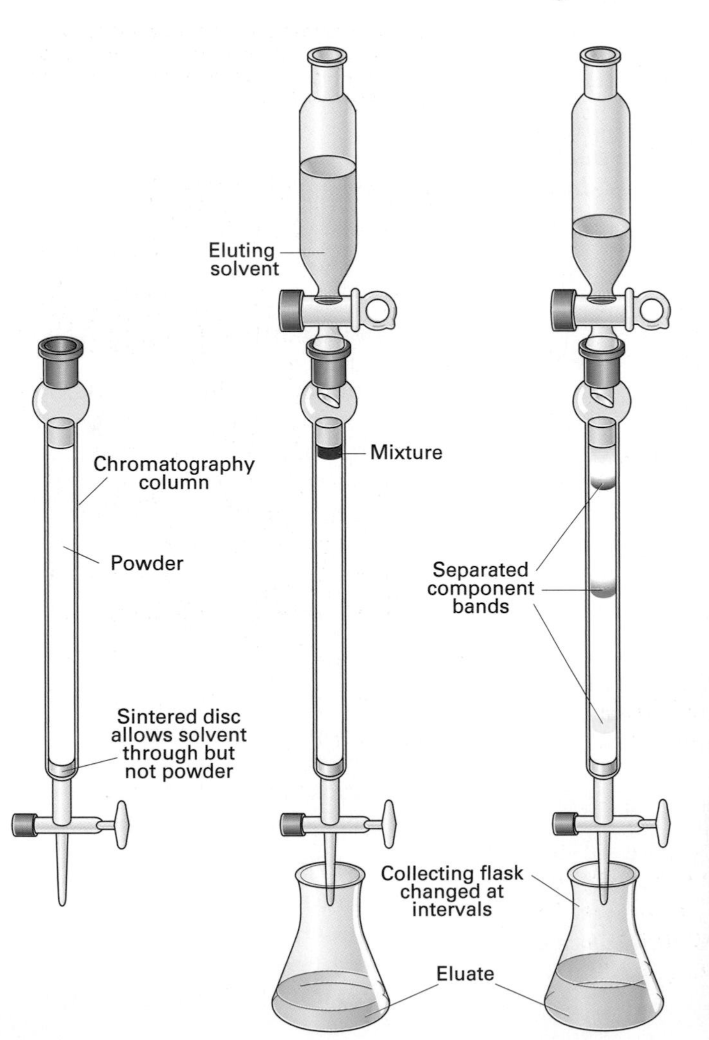

Simple column chromatography

The stationary phase is a powder packed into a glass column about 1–2 cm in diameter. Solvent is run into the column under gravity to wet the stationary phase. The sample mixture is then applied to the top and allowed to run into the stationary phase. More solvent is run through to elute the different components. Different components are eluted one after the other and collected in flasks or test tubes. This process can be automated to collect a series of identical volumes of **eluate** for analysis.

A chemist examining a test tube containing a liquid obtained using column chromatography.

Gas–liquid chromatography

Gas-liquid chromatography in action. It can be used to provide forensic evidence such as the amount of alcohol in a driver's breath or the identity of an unknown sample.

Gas–liquid chromatography or **GLC** is used to separate mixture of volatile liquids. The column is a long glass or stainless steel tube, coiled so that it fits into the oven of the GLC machine. The stationary phase is a liquid with a high boiling point, such as a waxy polymer, coated onto a powder. It can be packed into the column or coated onto its inside walls. The mobile phase is an unreactive gas such as nitrogen, called the **carrier gas**.

The sample vaporizes when it is injected into the hot column, and is pushed through the column by the carrier gas. Different components emerge from the column at different times depending on the boiling points of the components compared to the temperature of the column, and their solubility in the liquid of the stationary phase. As in HPLC the individual components are identified by their retention times.

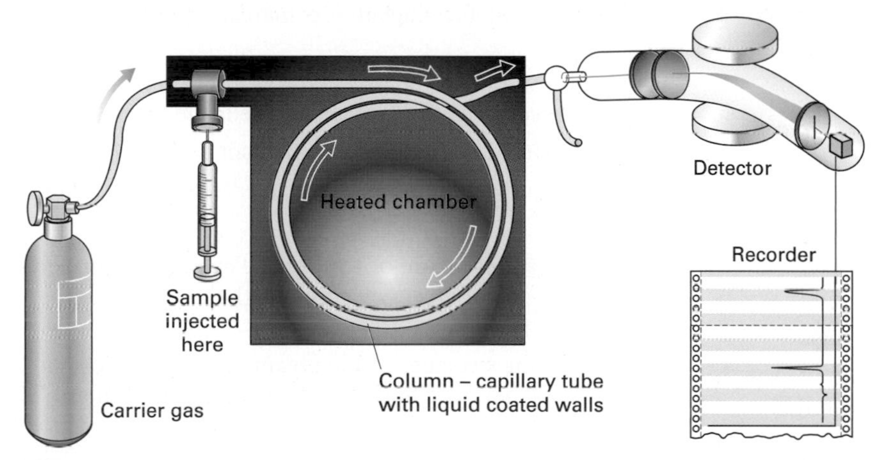

The different components separated by GLC may be detected by a simple flame ionization detector. *But using a mass spectrometer allows for very detailed analysis.*

Check your understanding

1 a What is the *retention time* of a substance?

b A mixture of two substances X and Y is separated by column chromatography. X is eluted first. What does this tell you about the balance between its solubility in the mobile phase and its retention in the stationary phase?

2 a What sort of mixtures can be separated by gas–liquid chromatography, GLC?

b In GLC, quick separations can be achieved at high temperatures but the components are poorly separated. At low temperatures the components are separated well but it takes a long time. Suggest why the machine might be programmed to gradually increase the temperature during a run.

High-performance liquid chromatography

High-performance liquid chromatography or HPLC allows faster and more precise separations. Instead of relying on gravity to elute the various components of a mixture, the mobile phase is forced through the stationary phase under pressure. The eluate is analysed continuously as it passes through a detector that measures the absorption of ultraviolet light. Under the same conditions, identical substances have identical retention times.

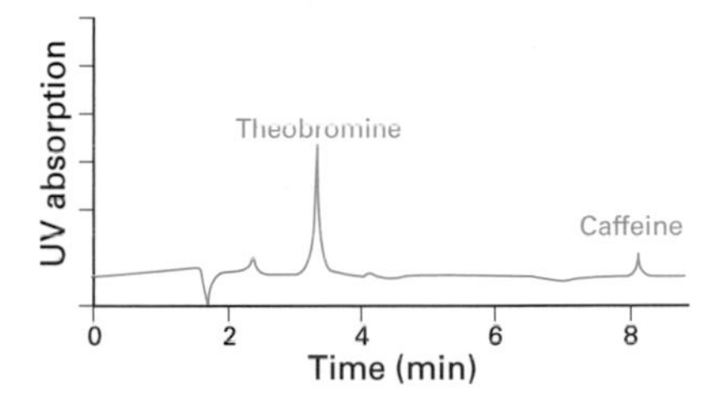

An HPLC chromatogram of an extract from drinking chocolate. The area under each peak is proportional to the amount of each substance.

HPLC in action.

Questions: Chs 8–12

1 a Explain why propan-1-amine (1-aminopropane) is a Brønsted–Lowry base. [2]

b Why is phenylamine a weaker base than propan-1-amine? [2]

c Propan–1-amine can be prepared from the reaction between bromopropane and ammonia.

i Name the type of reaction taking place.

ii Give the structures of three other organic substitution products which can be obtained from the reaction between bromopropane and ammonia. [4]

d Write an equation for the conversion of propanitrile into propan–1-amine, and give one reason why this method of synthesis is better than the one in part **c**. [2]

[Total 10 marks]

2 Look at the following reaction scheme and then use it to answer the questions below.

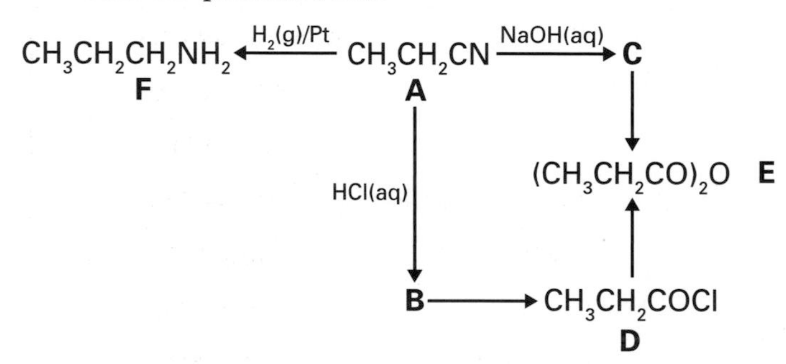

a The reaction of compound **C** with **D** produces compound **E**.

i Draw the structure of **E**. [1]

ii State the compound type to which **E** belongs. [1]

b Compound **F** may be prepared by the reaction of **A** with hydrogen gas in the presence of a platinum catalyst.

i Give the name of compound **F**. [1]

ii Name the type of reaction involved in the conversion of **A** into **F**. [1]

c Compounds **D** and **F** react together. Write an equation for the reaction between them. [2]

[Total 6 marks]

3 a Give the name(s) or formula(e) of a suitable reagent or combination of reagents for the formation of propan-1-amine (1-aminopropane) from propanenitrile. Name the type of reaction involved and write an equation for the conversion. [3]

b Give the structure of the final substitution product obtained when propan-1-amine reacts with an excess of bromopropane. Name the type of compound formed and suggest a use for this type of product. [3]

c Name and outline a mechanism for the reaction between propan-1-amine and ethanoyl chloride. [5]

d Write an equation for the reaction between propan-1-amine and ethanoic anhydride. [1]

[Total 12 marks]

4 The following reaction scheme shows the formation of two amines, **G** and **H**, from ethylbenzene.

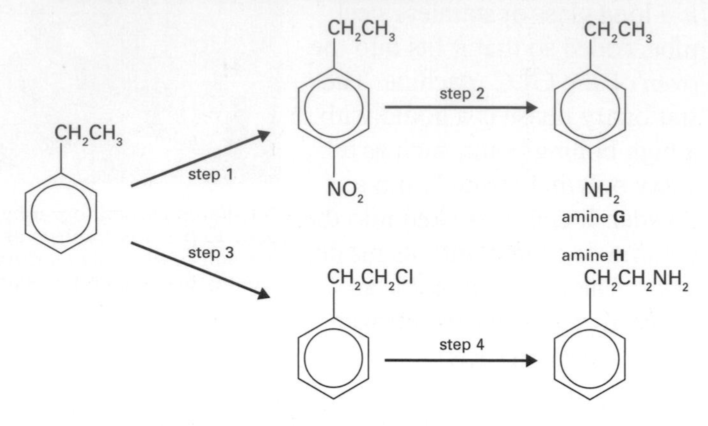

a i State the reagents needed to carry out step 1. Write an equation to show the inorganic species formed when these reagents react.

ii Name and outline a mechanism for the reaction between this inorganic species and ethylbenzene. [7]

b Give a suitable reagent or combination of reagents for step 2. [1]

c i Give the reagent for step 4 and state a condition to ensure that the primary amine is the major product.

ii Name and outline a mechanism for step 4. [7]

d Explain why amine **G** is a weaker base than ammonia. [2]

e Draw the structure of the organic compound formed when a large excess of bromomethane reacts with amine **H**. [1]

f Draw the structure of the organic compound formed when ethanoyl chloride reacts with amine **H** in an addition–elimination reaction. [1]

[Total 19 marks]

5 a The repeating units of two polymers, **J** and **K**, are shown below.

J $-CH(CH_3)-CH(CH_3)-$

K $-O-CH(CH_3)-CH(CH_3)-O-C(=O)-CH(CH_3)-CH(CH_3)-C(=O)-$

i Draw the structure of the monomer used to form polymer **J**. and name the type of polymerization involved.

ii Draw the structures of two compounds which react together to form polymer **K**. Name these two compounds and name the type of polymerization involved.

iii Give the name or formula of a compound which, in aqueous solution, will break down polymer **K** but not polymer **J**. [8]

b Draw the structures of the two dipeptides which can form when one of the amino acids shown below reacts with the other. [2]

H
$H_2N—C—COOH$
CH_2CH_3

amino acid 1

H
$H_2N—C—COOH$
CH_2CH_2OH

amino acid 2

[Total 10 marks]

6 The structures of two amino acids and are shown below.

CH_3
$H_2N—C—COOH$
H

L

H
$H_2N—C—COOH$
H

M

a Give the systematic name for amino acid **L**. [1]

b Give the structural formula of the species formed by amino acid **M** at pH 14. [1]

c Give the structural formula of the organic compound formed when amino acid **M** reacts with methanol in the presence of a small amount of concentrated sulfuric acid. [1]

d A sample of amino acid **L** is dissolved in water.

i Draw the structure of the main amino acid **L** species present in this aqueous solution and give the name of this type of species.

ii Draw the structure of the species formed when an excess of hydrochloric acid is added to the solution of amino acid **L**. [3]

e Molecules of amino acid **L** may be reacted together to form a polypeptide. Give the repeating unit of this polypeptide and name the type of polymerization involved in its formation. [2]

[Total 8 marks]

7 **a** The ester formed when ethanol reacts with ethanoyl chloride was analysed by high-resolution proton n.m.r. spectroscopy.

i Give the structural formula of the ester.

ii State the number of different types of proton present in this ester.

iii Describe the splitting pattern of the ethyl group in the high resolution n.m.r. spectrum of this ester. [4]

b A substance with the formula $CH_3CH_2OCH_2CH_2OH$ was analysed by high-resolution proton n.m.r. spectroscopy.

i How many different types of proton are present in this product?

ii State the two splitting patterns that will be seen in the n.m.r. spectrum of this compound. Any OH protons are seen as singlets. [5]

[Total 9 marks]

8 **a** Compound **X** has the molecular formula $C_4H_8O_2$. Its infrared absorption spectrum is shown below.

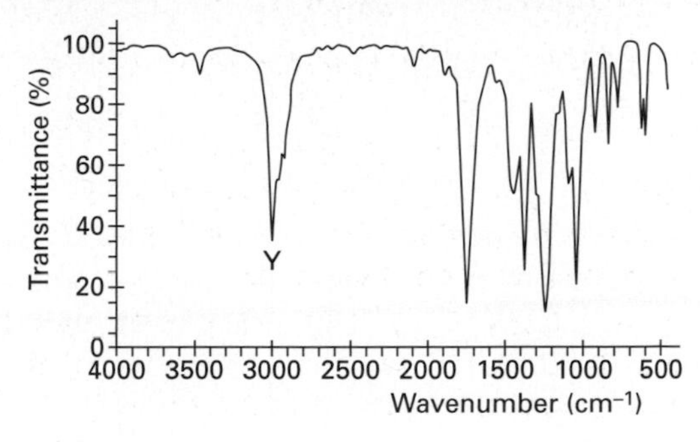

Use data from this table of infra-red absorption to help you answer the questions.

bond	wavenumber (cm^{-1})
C—H	2850–3300
C—C	750–1100
C=C	1620–1680
C=O	1680–1750
C—O	1000–1300
O—H (alcohols)	3230–3550
O—H (acids)	2500–3000

i Name the bond responsible for the absorption marked **Y**.

ii Draw the structures of two carboxylic acids having the molecular formula $C_4H_8O_2$ and explain why compound **X** cannot be either of these. [5]

b The fingerprint regions of the infrared spectra of compound **X** and of three other compounds are shown below labelled **I**, **II**, **III**, and **IV**.

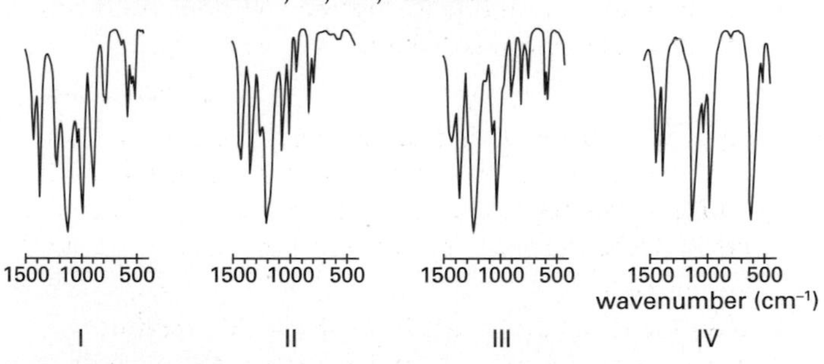

Which one of these fingerprint regions is that of compound **X**? [1]

c The low-resolution proton n.m.r. spectrum of compound **X** contains three peaks with the integration ratio of 3:2:3. Draw two possible structures for compound **X**. [2]

[Total 8 marks]

Unit 5: *Energetics, Redox, and Inorganic Chemistry* further develops the concepts of physical chemistry introduced at AS Level. You find out to make and use energy cycles for inorganic compounds, and how to explain spontaneous endothermic reactions. You also discover how electrochemistry can be used to predict the direction of redox reactions, and how it is used to design batteries and fuel cells.

Your study of inorganic chemistry is extended to include a survey of the elements in period 3 and the reactions of their oxides. The chemistry of transition metals is diverse and fascinating. You discover how transition metals can act as catalysts, why they form coloured compounds, and how they form complex ions. You will also study the reactions of iron, cobalt, copper, and chromium compounds in solution, including their hydrolysis and substitution reactions.

Energetics, redox, and inorganic chemistry

Silver crystals growing from a piece of copper wire.

13.01 Enthalpy cycles

OUTCOMES

already from AS Level, you

- know that reactions can be endothermic or exothermic
- understand that enthalpy change (ΔH) is the heat energy change measured under conditions of constant pressure
- know that standard enthalpy changes refer to standard conditions, i.e. 100 kPa and a stated temperature (e.g. ΔH_{298})
- can recall the definition of standard enthalpy of formation ($\Delta H_f^\ominus$)
- know Hess's Law and can use it to perform simple calculations

and after this spread you should

- have familiarized yourself again with the key concepts of enthalpy changes and Hess's Law from *AS Level Chemistry Unit 2*
- be able to define and apply the term enthalpy of formation

Standard states

The standard state of a substance is its state under standard conditions. Sometimes a temperature other than 298 K is used. For example, if a temperature of 300 K were used, the symbol for the standard enthalpy change would be $\Delta H^\ominus_{300}$. Remember that you subtract 273 to convert from K to °C. So 298 K is 25°C.

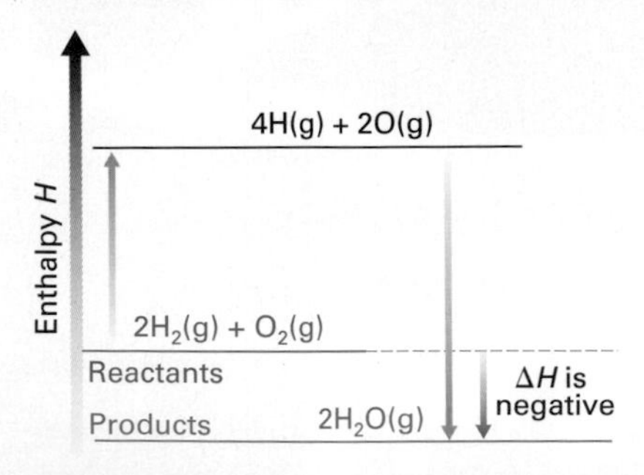

The energy level diagram for the formation of water.

Standard enthalpy change, $\Delta H^\ominus$

Enthalpy change, ΔH, is defined as the heat energy change measured under conditions of constant pressure. The value of the enthalpy change for a particular reaction depends on the pressure, the temperature, and the amount of substance used. So enthalpy changes are quoted for **standard conditions** of temperature and pressure. **Standard enthalpy changes** have the symbol $\Delta H^\ominus$ and are measured in units of kJ mol^{-1}. The symbol $\ominus$ shows the substances are in their **standard states** and the conditions used are:

- a pressure of 100 kPa
- a temperature of 298 K

and the enthalpy change is measured per mole of the specified substance.

Exothermic and endothermic reactions

Reactions can be **exothermic** or **endothermic**, depending on whether there is an overall transfer of energy to the surroundings or from them.

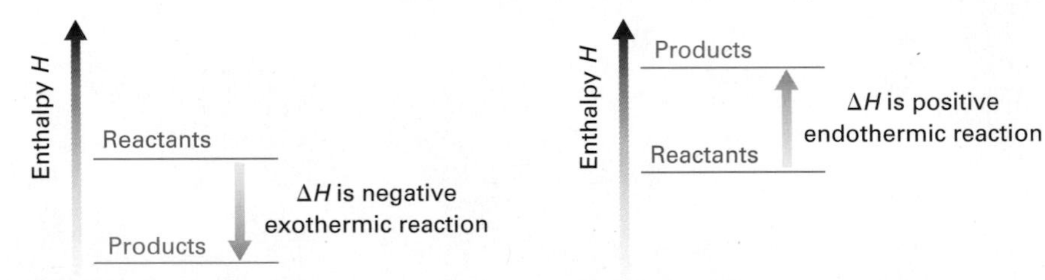

In an exothermic reaction, more energy is released when new bonds form in the products than is needed to break existing bonds in the reactants.

In an endothermic reaction, more energy is needed to break existing bonds in the reactants than is released when new bonds form in the products.

Standard enthalpy of formation, $\Delta H_f^\ominus$

The standard enthalpy of formation, $\Delta H_f^\ominus$ is the enthalpy change when *one mole* of a compound is formed from *its elements* in their *standard states* under standard conditions. By definition, this is equal to zero for an element in its standard state.

For example, the standard enthalpy of formation for water is:

$$H_2(g) + \tfrac{1}{2}O_2(g) \rightarrow H_2O(l) \qquad \Delta H_f^\ominus = -285.8 \text{ kJ mol}^{-1}$$

The negative sign for $\Delta H_f^\ominus$ shows that the reaction is exothermic.

Hess's Law

Hess's Law states that if a reaction can happen by more than one route, the overall enthalpy change is the same whichever route is taken. This lets you calculate the enthalpy change for reactions that are difficult to achieve in practice.

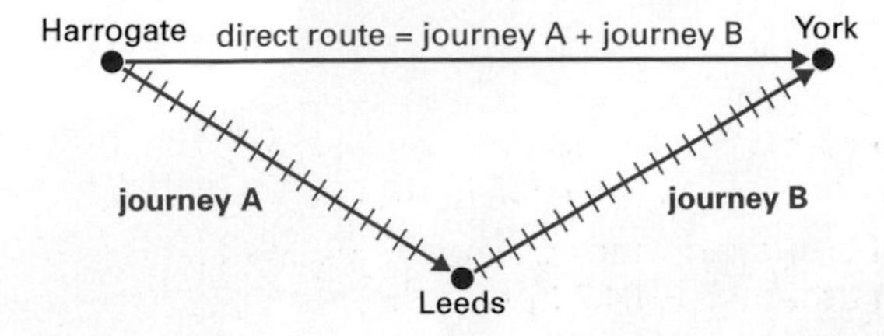

If you want to travel from Harrogate to York by train you can take the direct route. Or you can take the indirect route via Leeds. Whichever way you do it, the end result is the same.

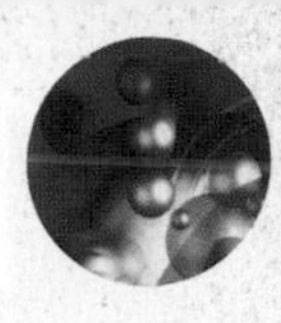

The standard enthalpy of formation of methane is the enthalpy of the reaction

$$C(s) + 2H_2(g) \rightarrow CH_4(g)$$

if it could be carried out under standard conditions.

It can be calculated using an **enthalpy cycle diagram** and the data in the table.

quantity	reaction	$\Delta H^\ominus$ (kJ mol^{-1})
$\Delta H_f^\ominus(CO_2(g))$	$C(s) + O_2(g) \rightarrow CO_2(g)$	−393.5
$\Delta H_f^\ominus(H_2O(l))$	$H_2(g) + \frac{1}{2}O_2(g) \rightarrow H_2O(l)$	−285.8
$\Delta H_c^\ominus(CH_4(g))$	$CH_4(g) + 2O_2(g) \rightarrow CO_2(g) + 2H_2O(l)$	−890.3

$\Delta H_c^\ominus$ represents the standard enthalpy of combustion, the enthalpy change when one mole of a substance is completely burned in oxygen under standard conditions.

When you draw an enthalpy cycle diagram, it is easiest to put the reaction for the enthalpy change you are trying to calculate across the top.

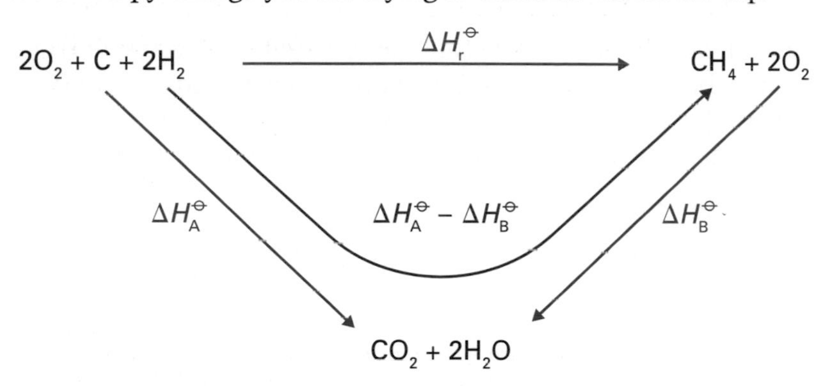

The enthalpy cycle for the formation of methane.

From the enthalpy cycle diagram, $\Delta H_f^\ominus = \Delta H_A^\ominus - \Delta H_B^\ominus$ (remember you reverse the sign if you go against the arrow).

$\Delta H_A^\ominus = \Delta H_f^\ominus(CO_2(g)) + 2\Delta H_f^\ominus(H_2O(l))$

$\Delta H_A^\ominus = -393.5 + 2(-285.8) = -393.5 - 571.6 = \mathbf{-965.1\ kJ\ mol^{-1}}$

$\Delta H_B^\ominus = \Delta H_c^\ominus(CH_4(g)) = \mathbf{-890.3\ kJ\ mol^{-1}}$

So $\Delta H_f^\ominus = -965.1 - (-890.3) = \mathbf{-74.8\ kJ\ mol^{-1}}$
(an exothermic reaction)

Check your understanding

1 a Define the terms *enthalpy change, standard enthalpy change,* and *standard enthalpy of formation.*

b Define Hess's Law.

2 Use the data in the table to calculate

a the standard enthalpy of formation for ethene, C_2H_4, if the enthalpy of combustion of ethene is −1410.8 kJ mol^{-1}.

b the standard enthalpy of combustion for ethane, C_2H_6, if the enthalpy of formation of ethane is −84.7 kJ mol^{-1}.

13.02 Mean bond enthalpies

OUTCOMES

already from AS Level, you can

- use mean bond enthalpies to calculate a value of ΔH for simple reactions

and after this spread you should be able to

- define and apply the term bond dissociation enthalpy
- use mean bond enthalpies to calculate an approximate value of ΔH for reactions
- explain why values from mean bond enthalpy calculations differ from those determined from enthalpy cycles

When a reaction happens, heat energy is absorbed from the surroundings to break bonds in the reactants, and heat energy is transferred to the surroundings when new bonds are formed in the products. The enthalpy change for the reaction is the difference between the energy absorbed and the energy released.

Standard bond dissociation enthalpy, $\Delta H^{\ominus}_{diss}$

The **standard bond dissociation enthalpy**, $\Delta H^{\ominus}_{diss}$, is the enthalpy change when one mole of bonds of the same type in gaseous molecules is broken under standard conditions, producing gaseous fragments:

$$X—Y(g) \rightarrow X(g) + Y(g)$$

For example, the standard bond dissociation enthalpy of the H—Cl bond is the enthalpy change when one mole of gaseous HCl molecules forms gaseous H and Cl atoms:

$$H—Cl(g) \rightarrow H(g) + Cl(g) \qquad \Delta H^{\ominus}_{diss} = +432 \text{ kJ mol}^{-1}$$

Bond dissociation enthalpy values are always positive, because energy must be absorbed to break bonds. The enthalpy change for making the same bond is the same amount but with the opposite sign. So the enthalpy change when one mole of H—Cl bonds forms is -432 kJ mol^{-1}.

Radicals

The gaseous fragments from bond dissociation are **radicals.** They form because the covalent bond breaks by **homolytic fission.** Each fragment receives one of the electrons from the bonding pair. The general equation for bond dissociation is written like this:

$$X—Y(g) \rightarrow \cdot X(g) + \cdot Y(g)$$

Mean bond enthalpies

Some bonds only occur in one substance. For example, H—H bonds only occur in hydrogen molecules, H_2, and Cl—Cl bonds only occur in chlorine molecules, Cl_2. But most bonds can occur in more than one substance. For example, the C—H bond can occur in almost every organic compound. The strength of the C—H bond varies, depending on its chemical environment. It will differ from one compound to the next, and even within the same compound if it occurs at different positions. The idea of mean bond enthalpies gets around this problem.

The **mean bond enthalpy** is the enthalpy change when one mole of a specified type of bond is broken, averaged over different compounds. Mean bond enthalpies can be used to calculate an *approximate* value of ΔH for reactions.

bond	mean bond enthalpy (kJ mol^{-1})
H—H	436
O—H	463
O=O	498
C—H	413
C—C	348
C=C	612
C=O	743

Mean bond enthalpy values vary slightly from one source to another, depending on the range of compounds included.

Science @ Work

Chipping away at methane

The bond dissociation enthalpies for the four C—H bonds in methane are not identical:

$CH_4(g) \rightarrow CH_3(g) + H(g)$ $\quad \Delta H = 428$ kJ mol^{-1}

$CH_3(g) \rightarrow CH_2(g) + H(g)$ $\quad \Delta H = 470$ kJ mol^{-1}

$CH_2(g) \rightarrow CH(g) + H(g)$ $\quad \Delta H = 416$ kJ mol^{-1}

$CH(g) \rightarrow C(g) + H(g)$ $\quad \Delta H = 338$ kJ mol^{-1}

The mean of these four values is 413 kJ mol^{-1}.

Mean bond enthalpies and ΔH

The approximate value of ΔH for a reaction is calculated like this:

- Add the mean bond enthalpies together for the bonds in the reactants.
- Add the mean bond enthalpies together for the bonds in the products.
- Then subtract the second answer from the first one.

Worked example

Calculate the enthalpy change for the combustion of methane using mean bond enthalpy values.

$$CH_4(g) + 2O_2(g) \rightarrow CO_2(g) + 2H_2O(g)$$

Step 1 Add together the mean bond enthalpies for all the bonds in the reactants.

$$[4 \times C\text{—}H] + [2 \times O\text{=}O]$$
$$(4 \times 413) + (2 \times 498) = 2648 \text{ kJ mol}^{-1}$$

Step 2 Add together the mean bond enthalpies for all the bonds in the products.

$$[2 \times C\text{=}O] + [4 \times O\text{—}H]$$
$$(2 \times 743) + (4 \times 463) = 3338 \text{ kJ mol}^{-1}$$

Step 3 ΔH = energy in − energy out

$\Delta H = 2648 - 3338 = -690 \text{ kJ mol}^{-1}$ (an exothermic reaction)

Calculations like these give an approximate value for the enthalpy change. They are not as accurate as ones using enthalpy cycles. This is because mean bond enthalpies are average values from a range of compounds, and may not be the exact values for the substances in the reaction.

The combustion of methane can be very dangerous. This building was wrecked by a gas explosion.

Check your understanding

1. a Define the term *bond dissociation enthalpy*.
   b Explain why values from mean bond enthalpy calculations are different from those calculated from enthalpy cycles.
2. Use the mean bond enthalpy values in the table to calculate the enthalpy change for the
   a combustion of hydrogen: $H_2(g) + \frac{1}{2}O_2(g) \rightarrow H_2O(g)$
   b hydrogenation of ethene: $C_2H_4(g) + H_2(g) \rightarrow C_2H_6(g)$
   c cracking of propane: $C_3H_8(g) \rightarrow C_2H_4(g) + CH_4(g)$

13.03 Dissolving

OUTCOMES

already from AS level you can

- define and apply the term enthalpy of formation

and after this spread you should

- be able to define and apply the terms lattice enthalpy (defined as either lattice dissociation or lattice formation), enthalpy of hydration, and enthalpy of solution

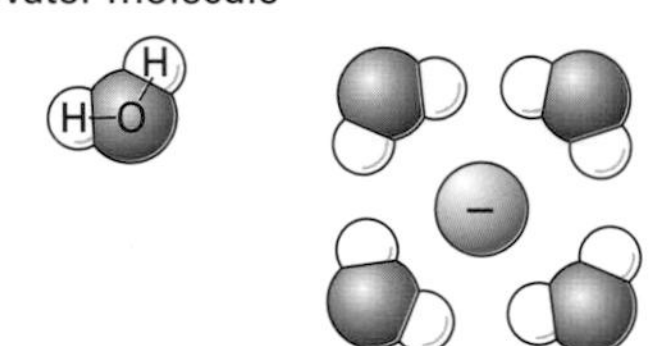

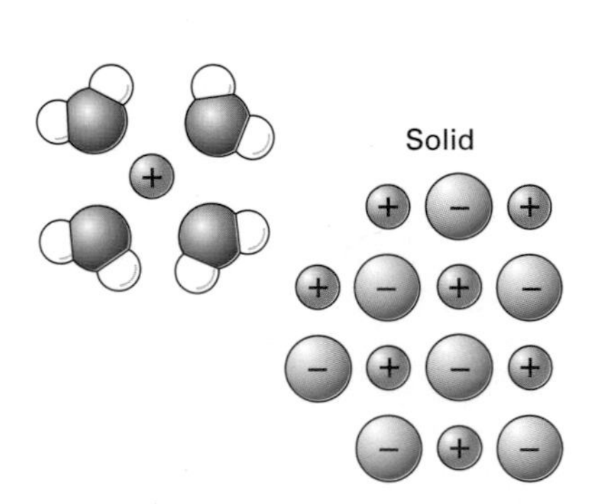

Dissolving needs two processes: the break-up of the crystal lattice followed by the hydration of the ions.

compound	$\Delta H_L^\ominus$ (kJ mol^{-1})
NaF	918
NaCl	780
NaBr	742
KCl	711
RbCl	685
CsCl	661
MgO	3791

Two processes happen when an ionic solid such as sodium chloride dissolves in water. The ions in the crystal **lattice** are separated from each other, and the separate ions become surrounded by water molecules. The first process is endothermic and the second one is exothermic. The overall enthalpy change when an ionic solid dissolves is the difference between the enthalpy changes for these two processes. So dissolving can be an exothermic process or an endothermic one.

Lattice enthalpy, $\Delta H_L^\ominus$

There are two conflicting definitions for lattice enthalpy. It is important that you realize this and take care when answering questions involving lattice enthalpy, so always look at the defining equation.

Lattice dissociation enthalpy

The **lattice dissociation enthalpy**, $\Delta H_L^\ominus$, is the enthalpy change when one mole of an ionic solid is separated into its gaseous ions. For example:

$$NaCl(s) \rightarrow Na^+(g) + Cl^-(g) \qquad \Delta H_L^\ominus = +787 \text{ kJ mol}^{-1}$$

Energy must be absorbed to overcome the strong **ionic bonds** in the ionic solid. So lattice dissociation is an endothermic process, and lattice dissociation enthalpies have positive values. We will use lattice dissociation enthalpies in this book.

Lattice formation enthalpy

The **lattice formation enthalpy** is the enthalpy change when one mole of an ionic solid is formed from its gaseous ions. For example:

$$Na^+(g) + Cl^-(g) \rightarrow NaCl(s) \qquad \Delta H_L^\ominus = -787 \text{ kJ mol}^{-1}$$

Energy is released when ionic bonds form. So lattice formation is an exothermic process, and lattice formation enthalpies have negative values. Notice that the enthalpy change is the same amount as the lattice dissociation enthalpy but with the opposite sign.

Factors affecting $\Delta H_L^\ominus$

The table (left) shows some values for lattice dissociation enthalpies. Two main factors influence the size of the lattice dissociation enthalpy: the distance between the ions in the crystal, and the charges on the ions.

Distance between the ions

The halide ions increase in size in the order $F^- < Cl^- < Br^-$. Notice that the lattice dissociation enthalpies for the sodium halides NaF, NaCl, and NaBr decrease in the same order. The larger the distance between the oppositely charged ions in a crystal lattice, the weaker the force of attraction between them.

The charges on the ions

The greater the charges on the ions in a crystal lattice the greater the force of attraction between them. Sodium fluoride and magnesium oxide have similar structures, but the lattice dissociation enthalpy of magnesium oxide is around four times larger. This is because the product of the charges in $Mg^{2+}O^{2-}$ is four times larger than the product of charges in Na^+F^-.

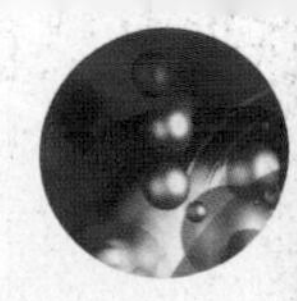

Enthalpy of hydration, $\Delta H^{\ominus}_{hyd}$

The enthalpy of hydration $\Delta H^{\ominus}_{hyd}$ is the enthalpy change when one mole of separated gaseous ions is dissolved completely in water to form one mole of aqueous ions. For example:

$$Na^+(g) + (aq) \rightarrow Na^+(aq) \qquad \Delta H^{\ominus}_{hyd} = -406\ kJ\ mol^{-1}$$

$$Cl^-(g) + (aq) \rightarrow Cl^-(aq) \qquad \Delta H^{\ominus}_{hyd} = -377\ kJ\ mol^{-1}$$

Energy is released when bonds form between the ions and water molecules. So hydration is an exothermic process, and enthalpies of hydration have negative values.

Enthalpy of solution, $\Delta H^{\ominus}_{soln}$

The enthalpy of solution, $\Delta H^{\ominus}_{soln}$, is the enthalpy change when one mole of an ionic substance is dissolved in a volume of water large enough to ensure that the ions are separated and do not interact with each other. For example:

$$NaCl(s) + (aq) \rightarrow Na^+(aq) + Cl^-(aq) \qquad \Delta H^{\ominus}_{soln} = +4\ kJ\ mol^{-1}$$

An enthalpy of solution can be positive or negative, depending on the values for lattice dissociation enthalpy and enthalpies of hydration.

Science @ Work

Determining $\Delta H^{\ominus}_{hyd}$

The enthalpy of hydration of a single ion cannot be determined directly because the ion will always be accompanied by an oppositely charged ion. Instead, the enthalpies of hydration for pairs of ions are determined. For example, this is how you could determine the values of $\Delta H^{\ominus}_{hyd}$ for Na^+ and Cl^-.

The enthalpy change for the hydration of HCl can be measured:

$$H^+Cl^-(g) + (aq) \rightarrow H^+(aq) + Cl^-(aq) \qquad \Delta H^{\ominus} = -1467\ kJ\ mol^{-1}$$

The enthalpy of hydration of the hydrogen ion has an accepted value of $-1090\ kJ\ mol^{-1}$. So for the chloride ion, $\Delta H^{\ominus}_{hyd} = -1467 - (-1090) = -377\ kJ\ mol^{-1}$.

The enthalpy change for the hydration of NaCl can be measured, too:

$$Na^+Cl^-(g) + (aq) \rightarrow Na^+(aq) + Cl^-(aq) \qquad \Delta H^{\ominus} = -783\ kJ\ mol^{-1}$$

So for the sodium ion, $\Delta H^{\ominus}_{hyd} = -783 - (-377) = -406\ kJ\ mol^{-1}$.

Check your understanding

1 Define the terms *lattice dissociation enthalpy, lattice formation enthalpy, enthalpy of hydration*, and *enthalpy of solution*.

2 Look at the table of lattice dissociation enthalpies. Describe and explain the trend in the values for the group 1 chlorides NaCl, KCl, RbCl, and CsCl.

13.04 Enthalpy of solution

OUTCOMES

already from AS Level you know

- Hess's Law and can use it to perform simple calculations

already from A2 Level you can

- define and apply the terms lattice enthalpy (defined as either lattice dissociation or lattice formation), enthalpy of hydration, and enthalpy of solution

and after this spread you should be able to

- calculate enthalpies of solution for ionic compounds from lattice enthalpies and enthalpies of hydration

When an ionic substance dissolves in water, its enthalpy of solution depends on the difference between its lattice enthalpy and the enthalpies of hydration of its ions. The enthalpy of solution can be calculated with the help of an enthalpy cycle diagram.

Calculating the enthalpy of solution, $\Delta H^{\ominus}_{soln}$

Remember that the enthalpy of solution, $\Delta H^{\ominus}_{soln}$, is the enthalpy change when one mole of an ionic substance is dissolved in a volume of water large enough to ensure that the ions are separated and do not interact with each other. Here is the equation for sodium chloride dissolving in water:

$$NaCl(s) + (aq) \rightarrow Na^{+}(aq) + Cl^{-}(aq)$$

Two processes can be identified when sodium chloride dissolves.

1 Breaking the bonds in the sodium chloride crystal lattice to produce gaseous ions:

$$NaCl(s) \rightarrow Na^{+}(g) + Cl^{-}(g)$$

The enthalpy change that accompanies this process is the lattice dissociation enthalpy, $\Delta H^{\ominus}_{L}$.

2 The separated gaseous ions become surrounded by water molecules:

$$Na^{+}(g) + (aq) \rightarrow Na^{+}(aq)$$

$$Cl^{-}(g) + (aq) \rightarrow Cl^{-}(aq)$$

The enthalpy changes that accompany this process are the enthalpies of hydration, $\Delta H^{\ominus}_{hyd}$, for sodium ions and chloride ions.

These equations can be combined to produce an enthalpy cycle diagram.

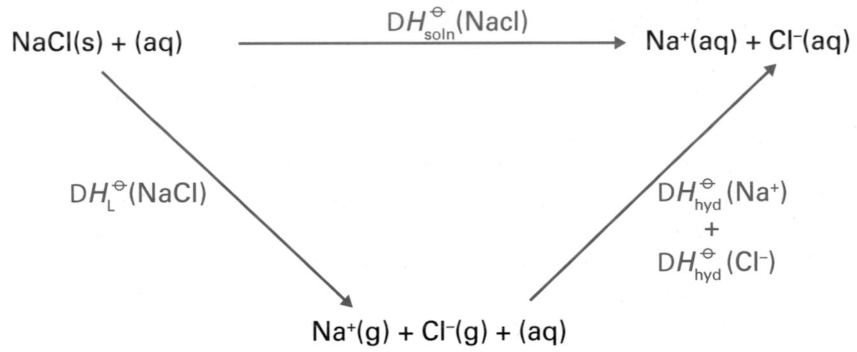

NaCl(s) + (aq) —— $DH^{\ominus}_{soln}$(Nacl) ——→ Na^{+}(aq) + Cl^{-}(aq)

$DH^{\ominus}_{L}$(NaCl)

$DH^{\ominus}_{hyd}$ (Na^{+}) + $DH^{\ominus}_{hyd}$ (Cl^{-})

Na^{+}(g) + Cl^{-}(g) + (aq)

An enthalpy cycle diagram for the dissolving of sodium chloride in water.

quantity	kJ mol^{-1}
$\Delta H^{\ominus}_{L}$(NaCl)	+787
$\Delta H^{\ominus}_{hyd}$ (Na^{+})	−406
$\Delta H^{\ominus}_{hyd}$ (Cl^{-})	−377

From the enthalpy cycle diagram:

$$\Delta H^{\ominus}_{soln}(NaCl) = \Delta H^{\ominus}_{L}(NaCl) + \Delta H^{\ominus}_{hyd}(Na^{+}) + \Delta H^{\ominus}_{hyd}(Cl^{-})$$

The enthalpy of solution can be calculated using the values in the table.

$$\Delta H^{\ominus}_{sol}(NaCl) = +787 + (-406) + (-377)\ \text{kJ mol}^{-1}$$
$$= +787 - 406 - 377\ \text{kJ mol}^{-1}$$
$$= \mathbf{+4\ kJ\ mol^{-1}}$$

Dissolving sodium chloride in water is an endothermic process. The solution will become colder as the sodium chloride dissolves. Energy is transferred from the surroundings if the process is carried out at constant temperature.

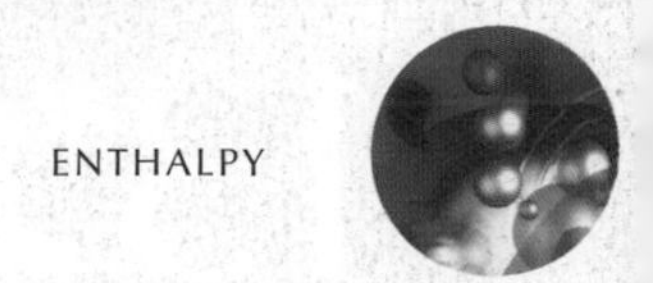

Enthalpy level diagrams for dissolving

Enthalpy level diagrams are another way to represent the processes involved in dissolving. These are charts in which exothermic processes are shown with downwards pointing arrows, and endothermic processes are shown with upwards pointing arrows. They may be shown drawn to scale, where the length of each arrow is proportional to the enthalpy change it represents.

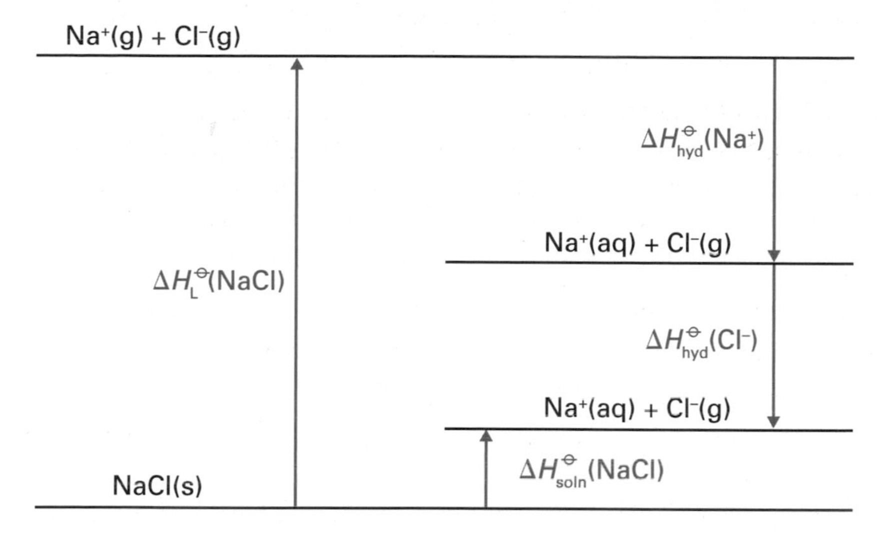

An enthalpy level diagram for the dissolving of sodium chloride in water.

You should be prepared to interpret data shown in either way. This includes naming the enthalpy change represented by each arrow, and writing the equation that defines it. You may also be asked to calculate any of the quantities if you are given information about the other two. For example, the lattice dissociation enthalpy can be calculated from the enthalpy of solution and the enthalpies of hydration.

Instant cold

Cold packs are used to treat sports injuries. Instant cold packs use endothermic reactions such as the dissolving of ammonium nitrate to achieve a low temperature quickly. Water and ammonium nitrate are held in two separate compartments in the pack. The pack is activated by breaking one of the compartments so that the contents mix together. The amounts of water and ammonium nitrate are calculated to achieve the maximum cooling effect, and the packs can stay cold for around 20 minutes.

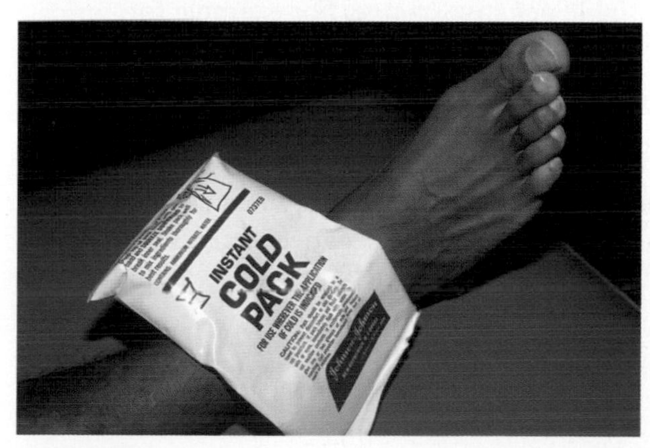

Instant cold packs rely on a highly endothermic dissolving process to bring relief from injuries such as ankle sprains.

Check your understanding

1 a Draw an enthalpy cycle for dissolving of ammonium nitrate in water.

b Use the data in the table to calculate the enthalpy of solution of ammonium nitrate.

quantity	$kJ\ mol^{-1}$
$\Delta H^{\ominus}_{L}(NH_4NO_3)$	+647
$\Delta H^{\ominus}_{hyd}(NH_4^+)$	−307
$\Delta H^{\ominus}_{hyd}(NO_3^-)$	−314

c Suggest why ammonium nitrate is a better ingredient of instant cold packs than sodium chloride.

d Suggest why the solubility of ammonium nitrate increases as the temperature increases.

2 Calcium chloride, $CaCl_2$, is used in some instant heat packs.

a Use the data in the table to calculate the lattice dissociation enthalpy of calcium chloride.

quantity	$kJ\ mol^{-1}$
$\Delta H^{\ominus}_{soln}(CaCl_2)$	−82
$\Delta H^{\ominus}_{hyd}(Ca^{2+})$	−1560
$\Delta H^{\ominus}_{hyd}(Cl^-)$	−377

b Explain why calcium chloride is used in heat packs.

13.05 Born–Haber cycles (1)

OUTCOMES

already from AS Level you know

- the meaning of the term ionization energy
- Hess's Law and can use it to perform simple calculations

already from A2 Level you can

- define and apply the terms lattice enthalpy, enthalpy of formation, and bond dissociation enthalpy

and after this spread you should be able to

- define and apply the terms ionization enthalpy, electron affinity, and enthalpy of atomization of an element and of a compound

Born and Haber

The Born–Haber cycle is named after Max Born and Fritz Haber, who developed the idea and published their work in 1919. Born worked in Germany in the early part of his career but left before the Second World War. He became a British citizen in 1939 and went on to share the 1954 Nobel Prize in Physics for his research in quantum mechanics. Haber was awarded the 1918 Nobel Prize in Chemistry for developing the Haber Process, the synthesis of ammonia from nitrogen and hydrogen.

You have used enthalpy cycles to calculate enthalpy changes involved in dissolving ionic compounds, and to calculate enthalpy changes involved in the formation of simple covalent compounds. A **Born–Haber cycle** is an enthalpy level diagram that lets you calculate enthalpy changes involved in the formation of ionic compounds, rather than covalent compounds. There are three further enthalpy changes to understand before you can use Born–Haber cycles.

Ionization enthalpy, $\Delta H_i^\ominus$

Ionization enthalpy, $\Delta H_i^\ominus$, is the enthalpy change when cations are formed. It is the enthalpy change when an electron is removed from a gaseous atom, ion, or molecule to form a gaseous cation. Ionization enthalpies have positive values.

First ionization enthalpy

The **first ionization enthalpy** is the enthalpy change when the highest energy electrons are removed from a mole of gaseous atoms or molecules to form a mole of gaseous ions, each with a single positive charge. For example for sodium atoms:

$$Na(g) \rightarrow Na^+(g) + e^- \qquad \Delta H_i^\ominus = +496\ kJ\ mol^{-1}$$

Second ionization enthalpy

The **second ionization enthalpy** is the enthalpy change when the highest energy electrons are removed from a mole of gaseous ions with single positive charges to form a mole of gaseous ions, each with two positive charges. For example for sodium ions:

$$Na^+(g) \rightarrow Na^{2+}(g) + e^- \qquad \Delta H_i^\ominus = +4563\ kJ\ mol^{-1}$$

Electron affinity, $\Delta H_{ea}^\ominus$

Electron affinity, $\Delta H_{ea}^\ominus$, is the enthalpy change when anions are formed. It is the enthalpy change when an electron is gained by a gaseous atom, ion, or molecule to form a gaseous anion. Electron affinities can have positive or negative values.

First electron affinity

The **first electron affinity** is the enthalpy change when electrons are gained by a mole of gaseous atoms or molecules to form a mole of gaseous ions, each with a single negative charge. For example for chlorine atoms:

$$Cl(g) + e^- \rightarrow Cl^-(g) \qquad \Delta H_{ea}^\ominus = -349\ kJ\ mol^{-1}$$

Second electron affinity

The **second electron affinity** is the enthalpy change when electrons are gained by a mole of gaseous ions with single negative charges to form a mole of gaseous ions, each with two negative charges. For example for O^- ions:

$$O^-(g) + e^- \rightarrow O^{2-}(g) \qquad \Delta H_{ea}^\ominus = +798\ kJ\ mol^{-1}$$

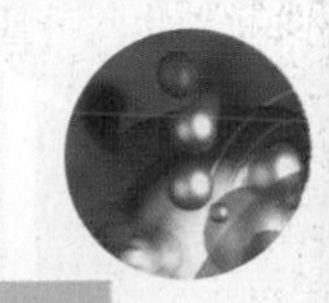

Enthalpy of atomization, $\Delta H^{\ominus}_{at}$

The **enthalpy of atomization**, $\Delta H^{\ominus}_{at}$, is the enthalpy change when one mole of gaseous atoms is formed from an element or compound. Enthalpies of atomization have positive values because they represent processes in which energy must be absorbed to break bonds.

Enthalpies of atomization can appear in several different ways. The key to success here is to identify the equation that describes the process. For example, enthalpies of atomization can also appear as bond dissociation enthalpies. Consider atomizing chlorine molecules, Cl_2. Here are the equations that describe the enthalpy of atomization of chlorine and the bond dissociation enthalpy of chlorine:

$$\tfrac{1}{2}\,Cl_2(g) \rightarrow Cl(g) \qquad \Delta H^{\ominus}_{at} = +122\ \text{kJ mol}^{-1}$$

$$Cl_2(g) \rightarrow 2Cl(g) \qquad \Delta H^{\ominus}_{diss} = +244\ \text{kJ mol}^{-1}$$

Notice that the enthalpy of atomization of chlorine is half its bond dissociation enthalpy. You could be supplied data describing either process, so make sure you know the definitions for the various enthalpy changes and the equations that go with them. For example, the enthalpy of atomization of a metal might be shown as an enthalpy of sublimation. **Sublimation** is the process where a solid turns directly into a gas without passing through the liquid state.

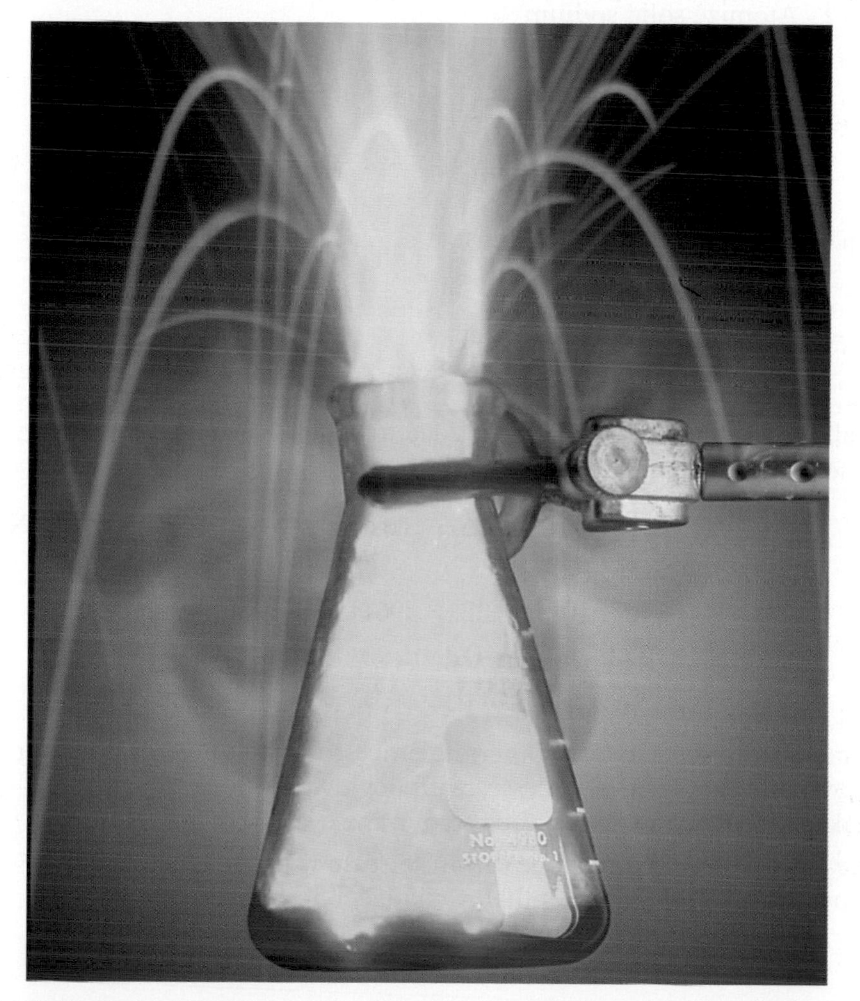

Born–Haber cycles describe the enthalpy changes that happen when ionic compounds form, such as the formation of sodium bromide from the reaction of sodium with bromine.

Bond dissociation enthalpies

Remember that the standard bond dissociation enthalpy, $\Delta H^{\ominus}_{diss}$, is the enthalpy change when one mole of bonds of the same type in gaseous molecules is broken under standard conditions, producing gaseous fragments:

$$X{-}Y(g) \rightarrow X(g) + Y(g)$$

Check your understanding

1 With the help of suitable equations, define the following:
   - a the first ionization enthalpy of magnesium
   - b the first electron affinity of sulfur
   - c the second electron affinity of sulfur

2 a Write an equation to describe the enthalpy of atomization of oxygen.
   - b The bond dissociation enthalpy of oxygen is +498 kJ mol^{-1}. Calculate the enthalpy of atomization of oxygen, and explain your answer.

13.06 Born–Haber cycles (2)

OUTCOMES

already from AS Level you know

- Hess's Law and can use it to perform simple calculations

already from A2 Level you can

- define and apply the terms lattice enthalpy, enthalpy of formation, bond dissociation enthalpy, ionization enthalpy, electron affinity, and enthalpy of atomization

and after this spread you should be able to

- construct Born–Haber cycles to calculate lattice enthalpies from experimental data

Other enthalpy changes

Other enthalpy changes may be given for some of the steps.

Step	alternative description(s)
2	enthalpy of sublimation of sodium
3	enthalpy of atomization of chlorine

Always write the defining equations to be sure about the data you are given. Remember that you might have to find the lattice formation enthalpy.

You have already seen an enthalpy level diagram for dissolving an ionic compound in Spread 13.04. These are charts in which exothermic processes are shown with downwards pointing arrows, and endothermic processes are shown with upwards pointing arrows. A Born–Haber cycle is an enthalpy level diagram that lets you calculate enthalpy changes involved in the formation of ionic compounds. You must be able to draw Born–Haber cycles and use them to calculate lattice enthalpies.

Drawing a Born–Haber cycle

We will look at the Born–Haber cycle for sodium chloride. The lattice dissociation enthalpy of sodium chloride is represented by this process:

$$NaCl(s) \rightarrow Na^+(g) + Cl^-(g) \qquad \Delta H^\ominus_L = +787\ kJ\ mol^{-1}$$

The description

In the Born–Haber cycle there are several steps in the indirect route for dissociated the sodium chloride crystal lattice.

Step 1 The reverse of the enthalpy of formation

$$NaCl(s) \rightarrow Na(s) + \tfrac{1}{2}Cl_2(g) \qquad -\Delta H^\ominus_f = +410\ kJ\ mol^{-1}$$

Step 2 Atomize solid sodium

$$Na(s) \rightarrow Na(g) \qquad \Delta H^\ominus_{at} = +108\ kJ\ mol^{-1}$$

Step 3 Atomize gaseous chlorine

$$\tfrac{1}{2}Cl_2(g) \rightarrow Cl(g) \qquad \tfrac{1}{2}\Delta H^\ominus_{diss} = +122\ kJ\ mol^{-1}$$

Step 4 Form gaseous sodium ions

$$Na(g) \rightarrow Na^+(g) + e^- \qquad \Delta H^\ominus_i = +496\ kJ\ mol^{-1}$$

Step 5 Form gaseous chloride ions

$$Cl(g) + e^- \rightarrow Cl^-(g) \qquad \Delta H^\ominus_{ea} = -349\ kJ\ mol^{-1}$$

If all these enthalpy changes are added together, they equal the lattice dissociation enthalpy of sodium chloride:

$$\begin{aligned}\Delta H^\ominus_L(NaCl) &= -\Delta H^\ominus_f(NaCl) + \Delta H^\ominus_{at}(Na) + \tfrac{1}{2}\Delta H^\ominus_{diss}(Cl_2) + \Delta H^\ominus_i(Na) + \Delta H^\ominus_{ea}(Cl)\\ &= +410 + 108 + 122 + 496 + (-349)\ kJ\ mol^{-1}\\ &= \mathbf{+787\ kJ\ mol^{-1}}\end{aligned}$$

The enthalpy cycle diagram

Remember to represent endothermic processes with upwards pointing arrows and exothermic processes with downwards pointing arrows. Make sure you can identify each step in the cycle.

The Born–Haber cycle can be drawn to scale with the length of each arrow proportional to the enthalpy change, but it does not have to be.

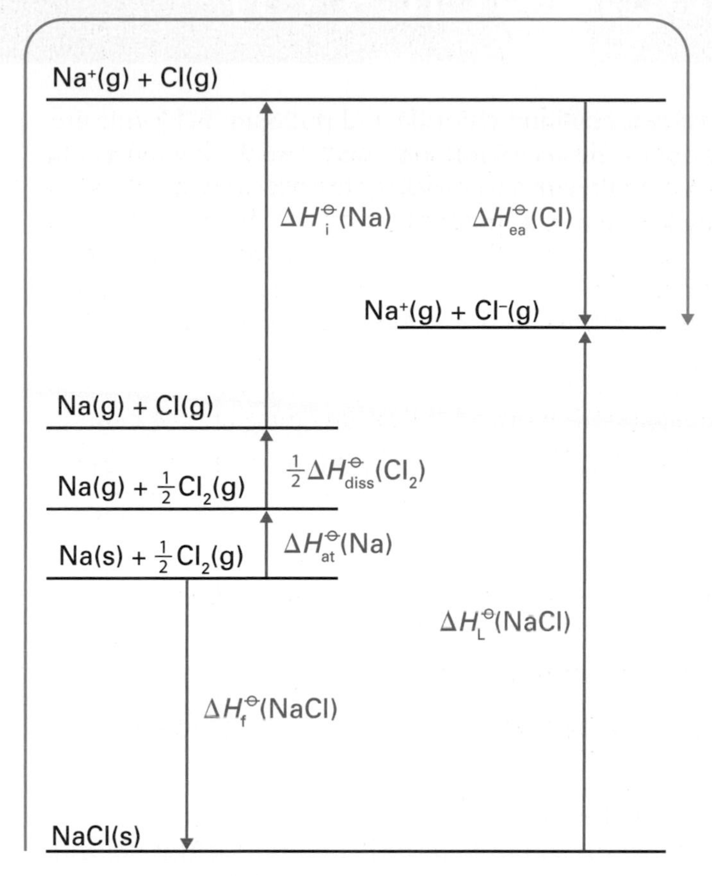

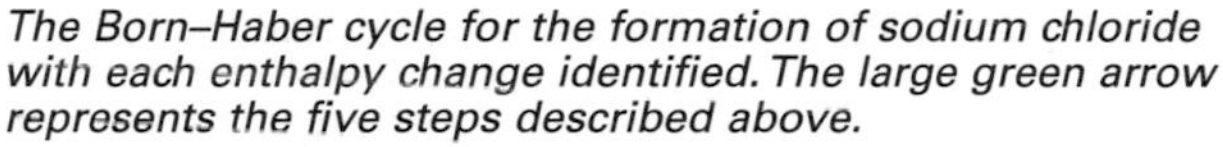

The Born–Haber cycle for the formation of sodium chloride with each enthalpy change identified. The large green arrow represents the five steps described above.

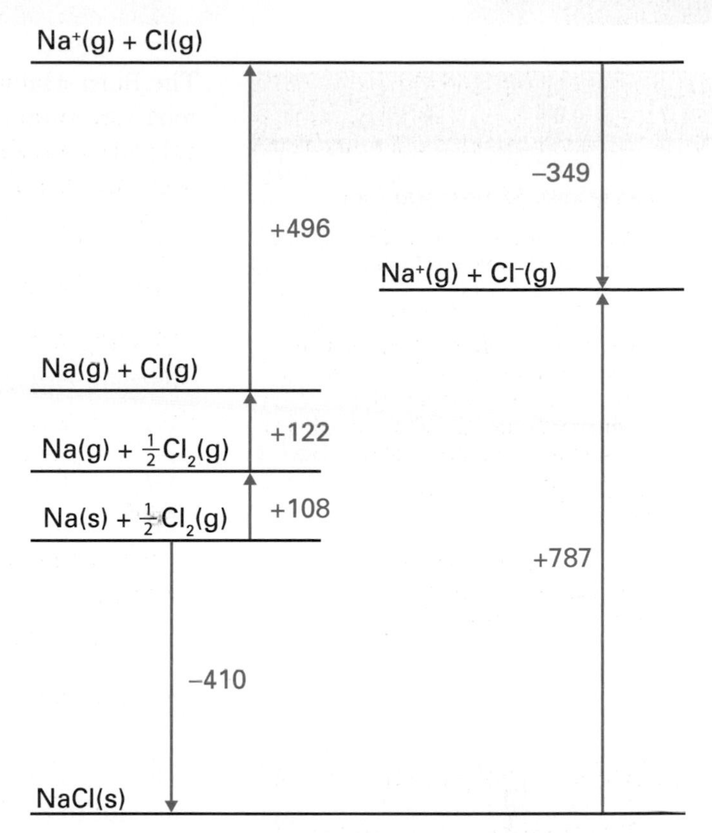

The Born–Haber cycle for the formation of sodium chloride with the value in kJ mol– 1 for each enthalpy change. You have to go in the opposite direction to the arrow in the enthalpy of formation step, so in the calculation the sign is reversed to give +410 kJ mol– 1 rather than –410 kJ mol– 1.

Check your understanding

1 Use the information in the table to answer the questions below.

standard enthalpy changes		$kJ\ mol^{-1}$
$\Delta H_{sub}^\ominus$	Enthalpy of sublimation of potassium	+89
$\Delta H_{diss}^\ominus$	Bond dissociation enthalpy of chlorine	+244
$\Delta H_i^\ominus$	First ionization enthalpy of potassium	+419
$\Delta H_{ea}^\ominus$	First electron affinity of chlorine	−349
$\Delta H_f^\ominus$	Enthalpy of formation of potassium chloride	−437

a Draw the Born–Haber cycle for the formation of potassium chloride, KCl. Label each step with the appropriate symbol.

b Use your cycle to calculate the lattice dissociation enthalpy, $\Delta H_L^\ominus$, of potassium chloride.

c What is the value of the lattice formation enthalpy of potassium chloride?

13.07 Born–Haber cycles (3)

OUTCOMES

already from A2 Level you can

- construct simple Born–Haber cycles to calculate lattice enthalpies from experimental data

and after this spread you should be able to

- construct more complex Born–Haber cycles to calculate lattice enthalpies from experimental data

The Born–Haber cycles for sodium chloride and potassium chloride are relatively simple to draw. The cycles for compounds such as magnesium chloride and sodium oxide are more complex. They have more steps because ions with two charges are involved.

Magnesium chloride

The table shows the data needed to draw the Born–Haber cycle for magnesium chloride $MgCl_2$.

standard enthalpy changes		$kJ\ mol^{-1}$
$\Delta H^{\ominus}_{f}$	Enthalpy of formation of magnesium chloride	−641
$\Delta H^{\ominus}_{at}$	Enthalpy of atomization of magnesium	+148
$\Delta H^{\ominus}_{diss}$	Bond dissociation enthalpy of chlorine	+244
$\Delta H^{\ominus}_{i}$	First ionization enthalpy of magnesium	+738
$\Delta H^{\ominus}_{i}$	Second ionization enthalpy of magnesium	+1451
$\Delta H^{\ominus}_{ea}$	First electron affinity of chlorine	−349

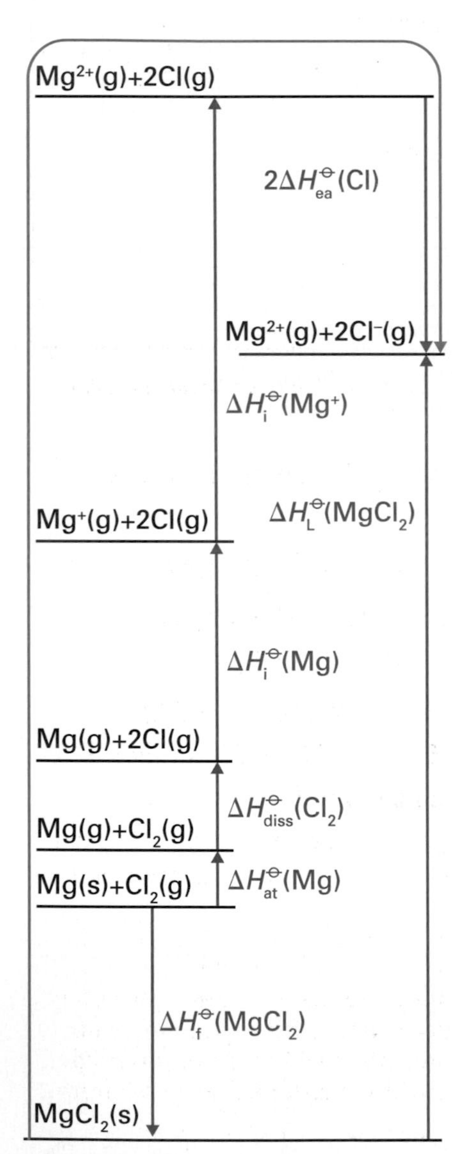

The Born–Haber cycle for the formation of magnesium chloride with each enthalpy change identified.

Notice that there is an extra ionization enthalpy step compared to the data in Spread 13.06 for sodium chloride and potassium chloride. This is because magnesium atoms ionize to form Mg^{2+} ions:

$Mg(g) \rightarrow Mg^{+}(g) + e^{-}$ first ionization enthalpy

$Mg^{+}(g) \rightarrow Mg^{2+}(g) + e^{-}$ second ionization enthalpy

Here is the calculation for the lattice dissociation enthalpy:

$$\Delta H^{\ominus}_{L}(MgCl_2) = -\Delta H^{\ominus}_{f}(MgCl_2) + \Delta H^{\ominus}_{at}(Mg) + \Delta H^{\ominus}_{diss}(Cl_2) + \Delta H^{\ominus}_{i}(Mg) + \Delta H^{\ominus}_{i}(Mg^{+}) + 2\Delta H^{\ominus}_{ea}(Cl)$$

$$= -(-641) + 148 + 244 + 738 + 1451 + 2(-349)\ kJ\ mol^{-1}$$

$$= \mathbf{+2524\ kJ\ mol^{-1}}$$

Check your understanding

1 Use the information on this spread to answer the questions below.

a Draw the Born–Haber cycle for the formation of magnesium oxide, MgO. Label each step with the appropriate symbol.

b Calculate the lattice formation enthalpy of MgO, if the enthalpy of formation is $-602\ kJ\ mol^{-1}$.

2 The third ionization enthalpy of magnesium is $+7733\ kJ\ mol^{-1}$ and the calculated lattice dissociation enthalpy of $MgCl_3$ is $+5440\ kJ\ mol^{-1}$. Confirm that the enthalpy of formation of $MgCl_3$ would be $+3949\ kJ\ mol^{-1}$.

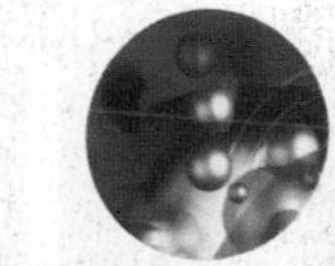

Why is it $MgCl_2$, and not MgCl or $MgCl_3$?

Born–Haber cycles can be used to calculate other enthalpy changes, not just lattice dissociation enthalpies. For example, the enthalpy of formation can be calculated if the other enthalpy changes are known. This is useful for calculating the enthalpies of formation of hypothetical compounds such as MgCl. The table shows the data needed.

standard enthalpy changes		$kJ\ mol^{-1}$
$\Delta H^{\ominus}_{at}$	Enthalpy of atomization of magnesium	+148
$\Delta H^{\ominus}_{diss}$	Bond dissociation enthalpy of chlorine	+244
$\Delta H^{\ominus}_{i}$	First ionization enthalpy of magnesium	+738
$\Delta H^{\ominus}_{ea}$	First electron affinity of chlorine	−349
$\Delta H^{\ominus}_{L}$	Lattice dissociation enthalpy of MgCl	+753

Here is the calculation for the enthalpy of formation of MgCl:

$$\Delta H^{\ominus}_{f}(MgCl) = \Delta H^{\ominus}_{at}(Mg) + \tfrac{1}{2}\Delta H^{\ominus}_{diss}(Cl_2) + \Delta H^{\ominus}_{i}(Mg) + \Delta H^{\ominus}_{ea}(Cl) - \Delta H^{\ominus}_{L}(MgCl)$$

$$= 148 + \tfrac{1}{2}(244) + 738 + (-349) - 753\ kJ\ mol^{-1}$$

$$= \mathbf{-94\ kJ\ mol^{-1}}$$

The formation of MgCl is just exothermic, whereas the formation of $MgCl_2$ is much more exothermic. Similarly, the calculated enthalpy of formation of $MgCl_3$ is +3949 $kJ\ mol^{-1}$, which is very endothermic. So $MgCl_2$ is formed in preference to MgCl or $MgCl_3$ because this transfers the most energy to the surroundings.

Sodium oxide

The table shows the data needed to draw the Born–Haber cycle for sodium oxide, Na_2O.

standard enthalpy changes		$kJ\ mol^{-1}$
$\Delta H^{\ominus}_{f}$	Enthalpy of formation of sodium oxide	−414
$\Delta H^{\ominus}_{at}$	Enthalpy of atomization of sodium	+108
$\Delta H^{\ominus}_{diss}$	Bond dissociation enthalpy of oxygen	+498
$\Delta H^{\ominus}_{i}$	First ionization enthalpy of sodium	+496
$\Delta H^{\ominus}_{ea}$	First electron affinity of oxygen	−141
$\Delta H^{\ominus}_{ea}$	Second electron affinity of oxygen	+798

Notice that there is an extra electron affinity step. This is because oxygen atoms ionize to form O^{2-} ions:

$O(g) + e^- \rightarrow O^-(g)$ first electron affinity

$O^-(g) + e^- \rightarrow O^{2-}(g)$ second electron affinity

Notice too that the second electron affinity of oxygen has a positive value. It represents an endothermic process, so it is shown in the Born–Haber cycle by an upwards pointing arrow.

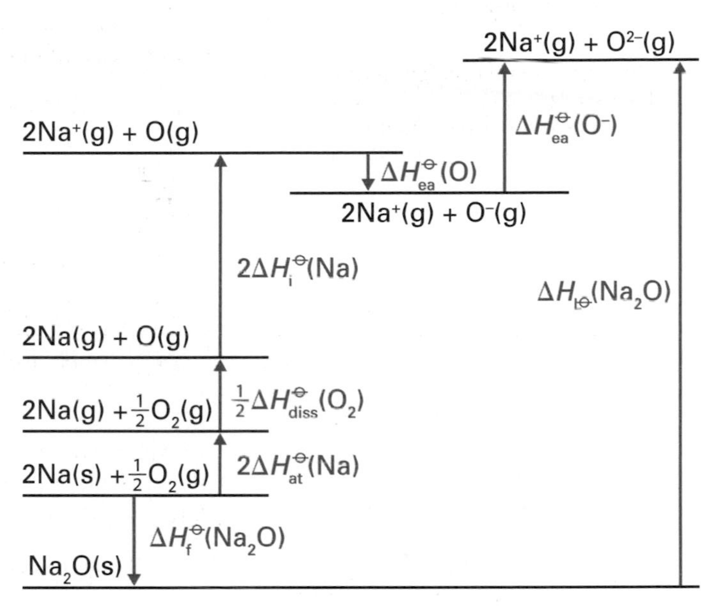

The Born–Haber cycle for the formation of sodium oxide with each enthalpy change identified.

Here is the calculation for the lattice dissociation enthalpy:

$$\Delta H^{\ominus}_{L}(Na_2O) = -\Delta H^{\ominus}_{f}(Na_2O) + 2\Delta H^{\ominus}_{at}(Na) + \tfrac{1}{2}\Delta H^{\ominus}_{diss}(O_2) + 2\Delta H^{\ominus}_{i}(Na) + \Delta H^{\ominus}_{ea}(O) + \Delta H^{\ominus}_{ea}(O^-)$$

$$= -(-414) + 2(108) + \tfrac{1}{2}(498) + 2(496) + (-141) + 798\ kJ\ mol^{-1}$$

$$= \mathbf{+2528\ kJ\ mol^{-1}}$$

It is very easy to forget to multiply or divide by two where necessary in these calculations, so take care.

Sneaky bromine

Take care in a question involving bromine, which is a liquid in its standard state. Unless you are given the enthalpy of atomization of bromine, you might need two steps to produce gaseous bromine atoms:

- vaporization of bromine: $Br_2(l) \rightarrow Br_2(g)$
- bond dissociation enthalpy: $Br_2(g) \rightarrow 2Br(g)$

You will be given the data you need.

13.08 Covalent character in ionic compounds

OUTCOMES

already from A2 Level you can

- construct Born–Haber cycles to calculate lattice enthalpies from experimental data

and after this spread you should be able to

- compare lattice enthalpies from Born–Haber cycles with those from calculations based on a perfect ionic model to provide evidence for covalent character in ionic compounds

Born–Haber cycles are used to calculate lattice enthalpies from experimental data. These data include enthalpies of formation, enthalpies of atomization, bond dissociation enthalpies, ionization enthalpies, and electron affinities. It is also possible to calculate lattice enthalpies using a model of the way oppositely charged ions interact with each other. Some interesting differences arise when you compare lattice enthalpies calculated using both these methods.

Scientific models

Scientists use models to attempt to explain observations. Models form the basis for experimental work and are used to make predictions that can be tested. Progress is made in science when validated evidence is found that supports a new model. In turn, scientists can be more confident that they understand the factors involved if the experimental data confirm their predictions.

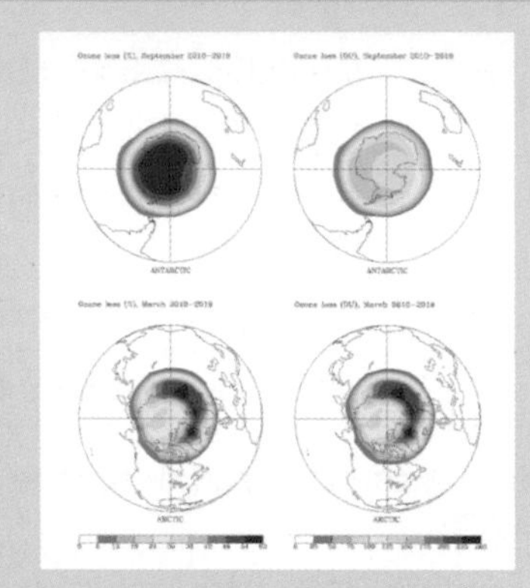

A computer model was used to make these predictions of the damage to the ozone layer over the Antarctic and Arctic between the years 2010 and 2019. The maps show percentage losses on the left and absolute losses on the right. Damage to the ozone layer is predicted to peak during this period and the ozone layer should then start to recover. If the predictions turn out to be true, it will show that scientists have a good understanding of the processes involved.

An ionic model

A lattice dissociation enthalpy is essentially the enthalpy change when ions that are packed together in an ionic lattice become widely separated. The factors involved include:

- the charge on an electron
- the number of charges on the ions
- the distance from the centre of one ion to the next

When developing a model of ionic bonding it is easiest to assume that the ions are spheres and pack together regularly. The table shows some lattice enthalpies, calculated using a model and calculated using Born–Haber cycles and experimental data.

compound	lattice dissociation enthalpy (kJ mol^{-1})		% difference
	from experiments	from model	
NaCl	787	769	2.3
NaBr	747	732	2.0
NaI	704	682	3.2
AgCl	915	864	5.9
AgBr	904	830	8.9
AgI	889	808	10.0

The sodium halides NaCl, NaBr, and NaI

For the sodium halides, there is a good agreement between lattice dissociation enthalpies calculated from experimental data and those calculated from the model. This is evidence in support of the model. It suggests that the structure of the sodium halides closely resembles the model for perfect ionic bonding. Electron density maps of sodium chloride confirm that its sodium ions and chloride ions are almost spheres. They are discrete charged particles with very little electron density between them.

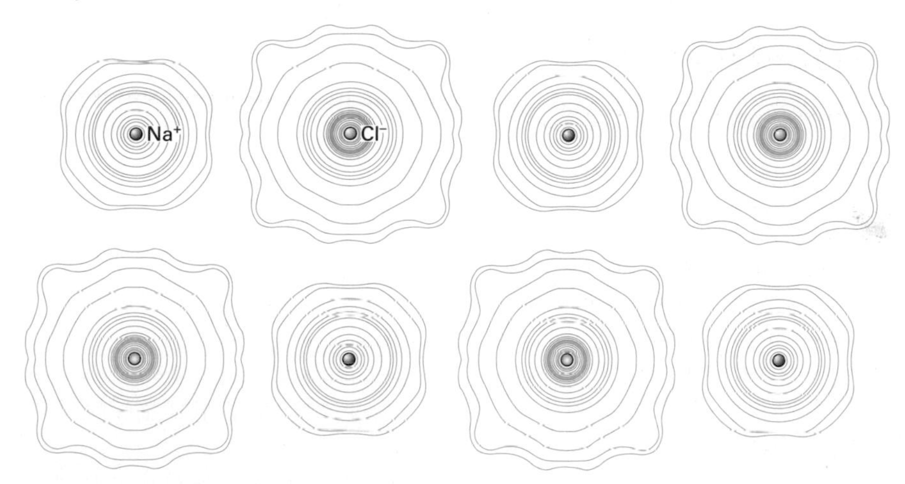

An electron density map for sodium chloride, made by studying the diffraction of X-rays by the crystal.

The silver halides AgCl, AgBr, and AgI

There is less agreement between the experimental values and those from the model for the silver halides. The differences suggest that the model for perfect ionic bonding is not enough to fully explain the bonding in the silver halides. It is likely that the ions are not discrete spheres and that some electron density is concentrated between the oppositely charged ions. As a result, the silver halides show a degree of covalent character in their bonding.

Fajans' rules

The electron cloud around a negative ion can be distorted and withdrawn towards a positive ion. The negative ion becomes polarized by the positive ion and the ionic bond can show a degree of covalent character. How much is described by rules developed by Kasimir Fajans.

- Negative ions are most easily polarised if they are large and have a high charge.
- Positive ions have the most polarizing power if they are small and have a high charge (they have a high charge density).

Notice that the largest discrepancy between experimental and calculated values for lattice enthalpies happens with the iodides in the table. Iodide ions are larger and more easily polarized than chloride ions.

Check your understanding

1 a Outline the factors involved in the model of perfect ionic bonding.

b Explain why the difference between the lattice enthalpy value calculated using the model and the value calculated using experimental data is larger for silver iodide than it is for silver chloride.

2 The table shows lattice dissociation enthalpies for potassium chloride and zinc sulfide.

compound	lattice dissociation enthalpy (kJ mol^{-1})	
	from experiments	from model
KCl	718	709
ZnS	3615	3417

a For each compound, calculate the percentage differences between the lattice dissociation enthalpy calculated from experimental data and that calculated from the model.

b Suggest an explanation for the differences noted in part **a**.

14.01 Entropy

OUTCOMES

already from AS Level, you can

- explain the energy changes associated with changes of state

and after this spread you should understand

- the concept of increasing disorder (entropy change ΔS), illustrated by physical change, e.g. melting and evaporation

Is your room like this? If so, it has suffered from increasing entropy since its last tidy up.

Units for entropy

Enthalpy and enthalpy change are measured in units of kJ mol^{-1}, but entropy and entropy change are measured in units of J K^{-1} mol^{-1}. Take care not to muddle them up. Note that the units for enthalpy involve kilojoules but the units for enthalpy involve joules. Also note that entropy involves absolute temperature, measured in kelvin, K.

How tidy is your room at home? If it is very tidy with everything neatly arranged on shelves and in drawers, it will have a lower **entropy** than if everything is really messy. Entropy is a measure of disorder in a system. The more disordered something is, the higher its entropy. When substances change state there is a change in entropy. There is an increase in entropy when a substance melts or boils, and a decrease in entropy when it freezes or condenses.

Entropy and state

The arrangement and movement of particles in the different states of matter are different from each other. As a result, the degree of disorder or entropy is different in the different states.

Solids

The particles in solids are close together. They are held together in regular arrangements by strong forces, and are only able to vibrate about fixed positions. Solids tend to have low entropies because they are ordered. There are only limited ways in which the particles in a solid could be arranged differently.

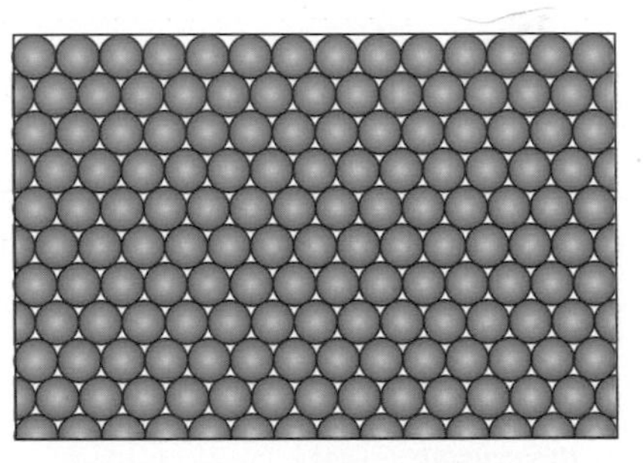

The arrangement of particles in a solid.

Liquids

The particles in liquids are close together but there are fewer forces attracting them together than in a solid. The particles can move randomly around each other. Liquids tend to have higher entropies than solids because they are disordered. There are many ways in which the particles in a liquid could be arranged differently.

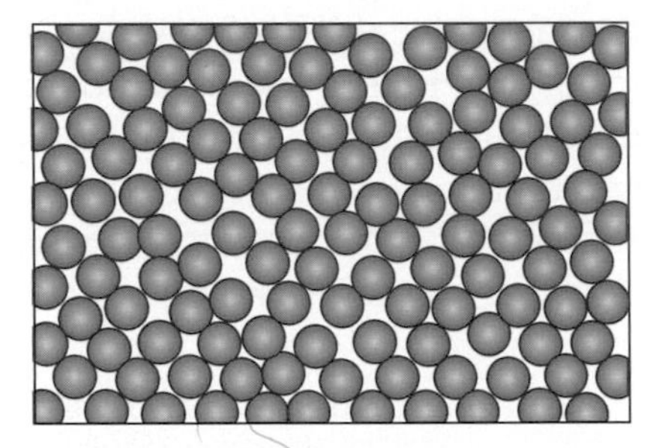

The arrangement of particles in a liquid.

Gases

The particles in gases are far apart and they have no permanent forces attracting them together. The particles can move randomly in any direction. Gases tend to have the highest entropies because they are highly disordered. There are very many ways in which the particles in a gas could be arranged differently.

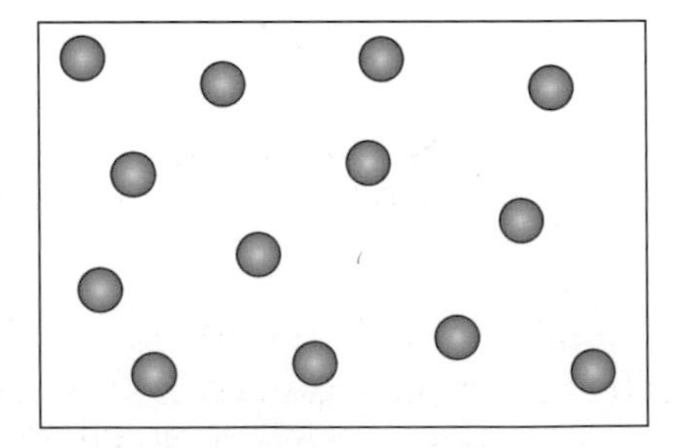

The arrangement of particles in a gas.

Entropy and change of state

Melting and evaporation

The entropy of a solid increases when it melts because some of the bonds between its particles are broken, allowing particles to move around each other. The entropy of a liquid increases when it evaporates or boils (a substance evaporates as fast as it can at its boiling point). This is because all the bonds between particles are broken, letting the particles escape from the liquid as a gas.

The table shows the standard entropies S of water as a solid, liquid, and gas.

Substance	Standard entropy, S ($J\ K^{-1}\ mol^{-1}$)
$H_2O(s)$	62.1
$H_2O(l)$	69.9
$H_2O(g)$	188.7

Notice that there is a larger increase in entropy when water boils than when ice melts. The **entropy change**, ΔS, when ice melts is $+7.8\ J\ K^{-1}\ mol^{-1}$ but ΔS when water boils is $+118.8\ J\ K^{-1}\ mol^{-1}$.

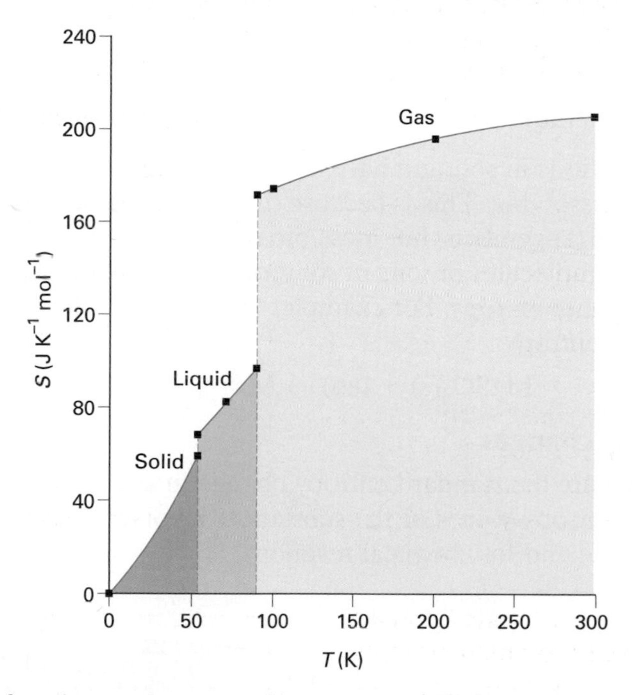

The entropy of a substance increases with temperature. This graph shows the change in entropy for oxygen as it is warmed up. Notice the large jump in entropy when liquid oxygen boils.

Condensing and solidifying

The entropy of a gas decreases when it condenses. Bonds form between its particles, reducing the motion of the particles and decreasing its disorder. The entropy of a liquid decreases when it solidifies. Bonds form between its particles, restricting their motion and producing an ordered arrangement.

Check your understanding

1 a What is entropy?
  b Explain why steam has a higher entropy than ice.
  c Suggest why ΔS when water boils is much greater than ΔS when ice melts.

14.02 Entropy in chemical changes

OUTCOMES

already from A2 Level, you understand

- the concept of increasing disorder (entropy change ΔS) illustrated by physical change, e.g. melting and evaporation

and after this spread you should

- understand the concept of increasing disorder (entropy change ΔS) illustrated by chemical change, e.g. dissolution, evolution of carbon dioxide from hydrogencarbonates with acid
- be able to calculate entropy changes from absolute entropy values

Complex ions

$[Cu(H_2O)_6]^{2+}$ and $[Cu(EDTA)]^{2-}$ are examples of complex ions. You will find out about complex ions in Spread 18.1.

The detonation of an explosive is accompanied by a very large increase in entropy.

Physical changes such as melting and evaporating are accompanied by entropy changes. Chemical changes are accompanied by entropy changes, too. Entropy can increase or decrease, depending on the nature of the reactants and products involved in the reaction.

Qualitative changes

You have already seen that, in general, gases have the highest entropies and solids have the lowest entropies. There is likely to be an increase in entropy in reactions in which

- the reactants are solids or liquids and the products include a gas
- there are more particles of products than reactants and they are in the same state

For example, these reactions are accompanied by an increase in entropy:

$$N_2O_4(g) \rightarrow 2NO_2(g)$$

$$CaCO_3(s) \rightarrow CaO(s) + CO_2(g)$$

$$C_2H_6(g) + 3\tfrac{1}{2}O_2(g) \rightarrow 2CO_2(g) + 3H_2O(g)$$

$$CH_3COCl(l) + H_2O(l) \rightarrow CH_3COOH(l) + HCl(g)$$

$$NaHCO_3(s) + HCl(aq) \rightarrow NaCl(aq) + H_2O(l) + CO_2(g)$$

$$[Cu(H_2O)_6]^{2+}(aq) + EDTA^{4-}(aq) \rightarrow [Cu(EDTA)]^{2-}(aq) + 6H_2O(l)$$

Molecules and ions in solution have higher entropies than the original solids before dissolving. This is because the molecules or ions in solids have ordered arrangements, but these breaks down as the solid dissolves. The separated molecules or ions in solution are highly disordered, so they have a higher entropy. For example, this change is accompanied by an increase in entropy:

$$MgSO_4(s) + (aq) \rightarrow Mg^{2+}(aq) + SO_4^{2-}(aq)$$

Quantitative changes

You can calculate the standard entropy change in a process if you know the standard entropy values of the substances involved. This works for changes of state and for chemical reactions.

element	entropy, $S^\ominus$ ($J\,K^{-1}\,mol^{-1}$)	compound	entropy, $S^\ominus$ ($J\,K^{-1}\,mol^{-1}$)
$H_2(g)$	65.3	$CuO(s)$	42.6
$O_2(g)$	102.4	$H_2O(g)$	188.7
$Fe(s)$	27.2	$CO_2(g)$	213.6
$Cu(s)$	33.2	$C_2H_4(g)$	219.5
C(graphite)	5.7	$C_2H_6(g)$	229.5
C(diamond)	2.4	$C_9H_{20}(l)$	393.7

The standard entropy values for some elements and compounds.

Substances comprising elements with large relative atomic masses tend to have higher entropies than those with small relative atomic masses. For example, hydrogen and oxygen both exist as diatomic molecules but oxygen has a higher entropy. Substances with ordered structures tend to have lower entropies than those where more arrangements are possible. For example, graphite and diamond have lower entropies than iron or copper.

Here is the equation for calculating standard entropy change $\Delta S^{\ominus}$:

$$\Delta S^{\ominus} = \Sigma S^{\ominus}(\text{products}) - \Sigma S^{\ominus}(\text{reactants})$$

You add the standard entropy values of all the products, then subtract the standard entropy values of the all reactants. Remember to include the signs + or − in your answer.

Sigma Σ

The Greek letter Σ (sigma) represents 'the sum of'. So '$\Sigma S^{\ominus}$ (products)' means 'the sum of $S^{\ominus}$ of the products'.

Worked example 1

Calculate the standard entropy change for the reaction between ammonia and hydrogen chloride.

$$NH_3(g) + HCl(g) \rightarrow NH_4Cl(s)$$

substance	$NH_3(g)$	$HCl(g)$	$NH_4Cl(s)$
entropy, $S^{\ominus}$ (J K^{-1} mol^{-1})	192.3	186.8	94.6

$\Sigma S^{\ominus}(\text{products}) = 94.6\ \text{J K}^{-1}\ \text{mol}^{-1}$

$\Sigma S^{\ominus}(\text{reactants}) = 192.3 + 186.8 = 379.1\ \text{J K}^{-1}\ \text{mol}^{-1}$

$\Delta S^{\ominus} = \Sigma S^{\ominus}(\text{products}) - \Sigma S^{\ominus}(\text{reactants})$

$= 94.6 - 379.1 = -284.5\ \text{J K}^{-1}\ \text{mol}^{-1}$

Worked example 2

Calculate the standard entropy change for the combustion of methane.

$$CH_4(g) + 2O_2(g) \rightarrow CO_2(g) + 2H_2O(g)$$

substance	$CH_4(g)$	$O_2(g)$	$CO_2(g)$	$H_2O(g)$
entropy, $S^{\ominus}$ (J K^{-1} mol^{-1})	186.2	102.4	213.6	188.7

$\Sigma S^{\ominus}(\text{products}) = (1 \times 213.6) + (2 \times 188.7) = 591.0\ \text{J K}^{-1}\ \text{mol}^{-1}$

$\Sigma S^{\ominus}(\text{reactants}) = (1 \times 186.2) + (2 \times 102.4) = 391.0\ \text{J K}^{-1}\ \text{mol}^{-1}$

$\Delta S^{\ominus} = \Sigma S^{\ominus}(\text{products}) - \Sigma S^{\ominus}(\text{reactants})$

$= 591.0 - 391.0 = +200.0\ \text{J K}^{-1}\ \text{mol}^{-1}$

Check your understanding

1 For each of the following reactions, predict whether the change in entropy will be positive or negative.

- **a** $2H_2(g) + O_2(g) \rightarrow 2H_2O(g)$
- **b** $C_6H_{12}O_6(aq) \rightarrow 2C_2H_5OH(l) + 2CO_2(g)$
- **c** $C_2H_5OH(g) \rightarrow C_2H_4(g) + H_2O(g)$
- **d** the polymerization of ethene

2 Use the data on this spread to calculate the standard entropy changes accompanying the following reactions.

- **a** $C_2H_4(g) + H_2(g) \rightarrow C_2H_6(g)$
- **b** $Cu(s) + \frac{1}{2}O_2(g) \rightarrow CuO(s)$
- **c** $CH_4(g) + 2H_2O(g) \rightarrow CO_2(g) + 4H_2(g)$
- **d** $C_9H_{20}(l) + 14O_2(g) \rightarrow 9CO_2(g) + 10H_2O(g)$
- **e** the conversion of graphite into diamond

14.03 Spontaneous reactions

OUTCOMES

already from A2 Level, you

- understand the concept of increasing disorder (entropy change ΔS) illustrated by physical and chemical changes
- can calculate entropy changes from absolute entropy values

and after this spread you should understand

- that ΔH, whilst important, is not sufficient to explain spontaneous change (e.g. spontaneous endothermic reactions)
- that the concept of entropy change, ΔS, accounts for this deficiency
- that the balance between entropy and enthalpy determines the feasibility of a reaction

Rusting is a spontaneous reaction.

A **spontaneous change** happens in one direction only and needs an input of energy to reverse it. For example, a sugar lump will dissolve and spread out in a cup of tea spontaneously. Sweet tea is not going to spontaneously separate itself into a sugar lump and non-sweet tea. Similarly, if you forget to tie the end of a party balloon after blowing it up, the air inside escapes and the balloon flies around the room with a rude noise. Air will not spontaneously inflate the balloon, so a lot of effort is needed to blow it up again.

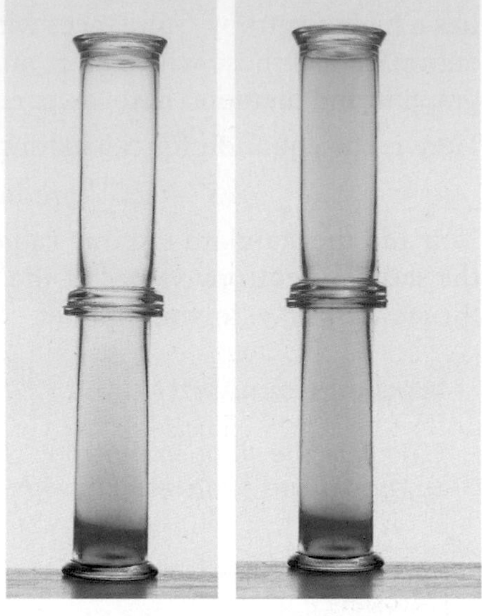

A gas spontaneously diffuses to fill its container uniformly.

Chemical reactions

Rusting is a spontaneous reaction.
Iron and steel react with oxygen in the presence of water, and the reaction continues until the metal has completely oxidized. It would take a large input of energy to reverse the process to produce the metal again.

The reaction between an acid and an alkali is spontaneous. For example, dilute hydrochloric acid and dilute aqueous sodium hydroxide react readily when they are mixed, producing sodium chloride and water:

$$HCl(aq) + NaOH(aq) \rightarrow NaCl(aq) + H_2O(l) \quad \Delta H^{\ominus} = -57.9 \text{ kJ mol}^{-1}$$

The reverse reaction is not easily achieved. Neutralization is exothermic, as is rusting. It would be tempting to conclude that spontaneous reactions must be exothermic, but some spontaneous reactions are also endothermic. For example, ammonium nitrate dissolves in water spontaneously:

$$NH_4NO_3(s) + (aq) \rightarrow NH_4^+(aq) + NO_3^-(aq) \quad \Delta H^{\ominus} = +25.8 \text{ kJ mol}^{-1}$$

So enthalpy changes are important in describing chemical changes, but they are not enough on their own to explain spontaneous changes such as spontaneous endothermic reactions. You may have noticed that the spontaneous changes described here involve energy or matter becoming spread out. Entropy change, ΔS, is a measure of disorder. It is the concept that is needed to complete the explanation of spontaneous change.

Feasibility, enthalpy, and entropy

A **feasible reaction** is a reaction that is possible when the entropy change and enthalpy change are taken into account. Here is the equation that links entropy change and enthalpy change:

$$\Delta S = \frac{\Delta H}{T}$$

This can be rewritten as

$$\Delta H = T\Delta S.$$

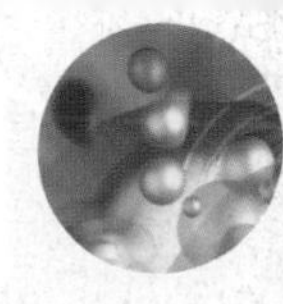

The size of ΔH compared to $T\Delta S$ determines whether a reaction is feasible or not. The table summarizes the four possibilities.

		ΔS	
		negative	**positive**
ΔH	**negative**	feasible if $\Delta H > T\Delta S$	always feasible
	positive	never feasible	feasible if $\Delta H < T\Delta S$

Consider the combustion of sugar. The standard enthalpy change of reaction is negative, and the standard entropy change is positive. So the combustion of sugar is a feasible reaction. Note that a feasible reaction is not spontaneous if its activation energy is large. This is why sugar lumps do not burst into flames at room temperature.

Worked example

Calcium carbonate decomposes to form calcium oxide and carbon dioxide:

$$CaCO_3(s) \rightarrow CaO(s) + CO_2(g) \qquad \Delta H^{\ominus} = +178\ \text{kJ mol}^{-1}$$
$$\Delta S^{\ominus} = +165\ \text{J K}^{-1}\ \text{mol}^{-1}$$

Could it decompose under standard conditions?

From the table, when both ΔH and ΔS are positive, the reaction will be feasible if $\Delta H < T\Delta S$. Remember to divide ΔS by 1000 to take into account that its units are J K^{-1} mol^{-1} but the units of ΔH are kJ mol^{-1}.

$$T\Delta S = 298 \times \frac{165}{1000} = +49.17\ \text{kJ mol}^{-1}$$

So $\Delta H > T\Delta S$ and the reaction is not feasible (or spontaneous) under standard conditions.

A traditional lime kiln in which limestone (calcium carbonate) is decomposed to form quicklime (calcium oxide).

Standard conditions

Remember that standard conditions include a temperature of 298 K (25°C) and a pressure of 100 kPa.

Check your understanding

1 Explain why a spontaneous reaction must be feasible, but a feasible reaction need not be spontaneous.

2 Under what conditions of ΔH and ΔS will a reaction
  a never be feasible?
  b always be feasible?

3 Use the supplied values for $\Delta H^{\ominus}$ and $\Delta S^{\ominus}$ to determine whether the following reactions will be feasible under standard conditions.
  a The synthesis of ammonia from nitrogen and hydrogen.
  $\Delta H^{\ominus} = -46.1$ kJ mol^{-1}
  $\Delta S^{\ominus} = +92.9$ J K^{-1} mol^{-1}
  b The decomposition of sodium hydrogencarbonate.
  $\Delta H^{\ominus} = +65.1$ kJ mol^{-1}
  $\Delta S^{\ominus} = +167.4$ J K^{-1} mol^{-1}
  c The conversion of graphite into diamond.
  $\Delta H^{\ominus} = +1.9$ kJ mol^{-1}
  $\Delta S^{\ominus} = -3.3$ J K^{-1} mol^{-1}

14.04 Free energy and feasibility

OUTCOMES

already from A2 Level, you

- understand the concept of increasing disorder (entropy change ΔS) illustrated by physical and chemical changes
- can calculate entropy changes from absolute entropy values
- understand that the balance between entropy and enthalpy determines the feasibility of a reaction

and after this spread you should

- understand that the balance between entropy and enthalpy is given by the relationship $\Delta G = \Delta H - T\Delta S$ (derivation not required)
- be able to use this relationship to determine the temperature at which a reaction is feasible
- be able to use this equation to determine how ΔG varies with temperature

Why is it ΔG?

Willard Gibbs was the American scientist who developed the idea of free energy change during the 1870s. It is also called Gibbs energy.

Willard Gibbs (1839–1903).

In the previous spread you were shown a relationship between entropy change and enthalpy change, $\Delta H = T\Delta S$, and used it to decide if reactions are feasible under standard conditions. Where a reaction is not feasible under standard conditions, it would be useful to determine if there might be a temperature at which it does become feasible. This is where the idea of **free energy change** helps.

Free energy change, ΔG

Free energy change links ΔH, T, and ΔS together. Here is the expression for free energy change, ΔG:

$$\Delta G = \Delta H - T\Delta S$$

The units of ΔG are kJ mol^{-1}. You do not need to be able to show how to arrive at this expression, but you do need to know how to use it.

ΔG decreases in any feasible change, so for a reaction to be feasible, $\Delta G \leq 0$. A reaction that is not feasible is one where $\Delta G > 0$. The temperature at which a reaction just becomes feasible is the one which gives $\Delta G = 0$. This means the expression for ΔG can be usefully rearranged like this:

$$\Delta G = \Delta H - T\Delta S = 0$$

$$\text{so} \quad \Delta H = T\Delta S$$

and $T = \dfrac{\Delta H}{\Delta S}$ where T is the absolute temperature in kelvin.

Worked example 1

Calcium carbonate decomposes to form calcium oxide and carbon dioxide:

$$CaCO_3(s) \rightarrow CaO(s) + CO_2(g) \qquad \Delta H^{\ominus} = +178 \text{ kJ mol}^{-1}$$
$$\Delta S^{\ominus} = +165 \text{ J K}^{-1} \text{ mol}^{-1}$$

What is the minimum temperature at which the reaction becomes feasible?

$$\Delta S^{\ominus} = +165 \text{ J K}^{-1} \text{ mol}^{-1} = +0.165 \text{ kJ K}^{-1} \text{ mol}^{-1}$$

$$T = \frac{\Delta H}{\Delta S} = \frac{178}{0.165} = 1079 \text{ K}$$

This temperature is 806°C, which is why calcium carbonate must be heated strongly before it will decompose.

Worked example 2

Sulfur dioxide reacts with oxygen to form sulfur trioxide:

$$2SO_2(g) \rightarrow O_2(g) \rightarrow 2SO_3(g) \qquad \Delta H^{\ominus} = -198 \text{ kJ mol}^{-1}$$
$$\Delta S^{\ominus} = -186 \text{ J K}^{-1} \text{ mol}^{-1}$$

a *Calculate $\Delta G^{\ominus}$.*

$\Delta S^{\ominus} = -186 \text{ J K}^{-1} \text{ mol}^{-1} = -0.186 \text{ kJ K}^{-1} \text{ mol}^{-1}$

$\Delta G^{\ominus} = \Delta H^{\ominus} - T\Delta S^{\ominus}$

$\Delta G^{\ominus} = -198 - 298(-0.186) = -198 + 55 = -143 \text{ kJ mol}^{-1}$

b *Explain why the reaction is feasible under standard conditions.*

$\Delta G^{\ominus}$ is negative.

c *At what temperature does the reaction stop being feasible?*

$$T = \frac{\Delta H}{\Delta S} = \frac{-198}{-0.186} = 1065 \text{ K}$$

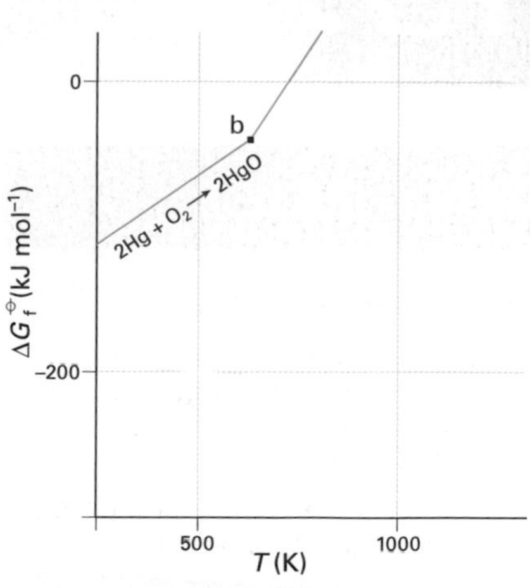

The variation of ΔG with temperature for the formation of mercury(II) oxide. The gradient becomes steeper at b, the boiling point of mercury. This is because the mercury(II) oxide is then being formed from two gaseous elements.

ΔG of elements

$\Delta G^{\ominus}_f$, the standard Gibbs free energy change of formation for an element, is defined as zero. Remember that $\Delta H^{\ominus}_f$, the standard enthalpy of formation for an element, is also zero.

ΔG and temperature

Here is the expression for free energy change again:

$$\Delta G = \Delta H - T\Delta S.$$

It can be rewritten like this:

$$\Delta G = -\Delta S T + \Delta H$$

This follows the general formula for a straight-line graph, $y = mx + c$. A graph of ΔG on the vertical axis against T on the horizontal axis gives a straight line with a gradient of $-\Delta S$.

The graph shows how ΔG varies with temperature for the formation of mercury(II) oxide from its elements:

$$2Hg(l) + O_2(g) \rightarrow 2HgO(s)$$

The gradient is positive, so ΔS is negative. ΔG becomes less negative as the temperature increases. It becomes zero above about 750 K, when the mercury(II) oxide spontaneously decomposes instead.

Mercury(II) oxide decomposes above about 750 K to form mercury and oxygen.

Check your understanding

1 Write the expression that links free energy change, enthalpy change, and entropy change.

2 **a** What is a feasible reaction in terms of free energy change ΔG?

b Iron is extracted from iron ore by reducing it using carbon:

$$2Fe_2O_3(s) + 3C(s) \rightarrow 4Fe(s) + 3CO_2(g)$$

$\Delta H^{\ominus} = -356.3 \text{ kJ mol}^{-1}$

$\Delta S^{\ominus} = -558.1 \text{ J K}^{-1} \text{ mol}^{-1}$

Calculate the temperature at which this reaction becomes feasible.

3 One of the stages in the extraction of titanium involves this reaction:

$$TiO_2(s) + C(s) + 2Cl_2(g) \rightarrow TiCl_4(g) + CO_2(g)$$

$\Delta H^{\ominus} = -257.8 \text{ kJ mol}^{-1}$

$\Delta S^{\ominus} = +245.0 \text{ J K}^{-1} \text{ mol}^{-1}$

Explain why this reaction is always feasible.

15.01 Sodium and magnesium

OUTCOMES

already from A2 Level, you

- can describe the trends in atomic radius, first ionization energy, melting points, and boiling points of the elements Na–Ar
- understand the reasons for the trends in these properties
- know the reactions of the elements magnesium and chlorine with water

and after this spread you should be able to

- describe trends in the reactions of the period 3 elements with water, limited to sodium and magnesium

Period 3 elements

Period 3 contains eight elements, from sodium to argon. Sodium and magnesium are in the s block of the periodic table, and the other six elements are in the p block.

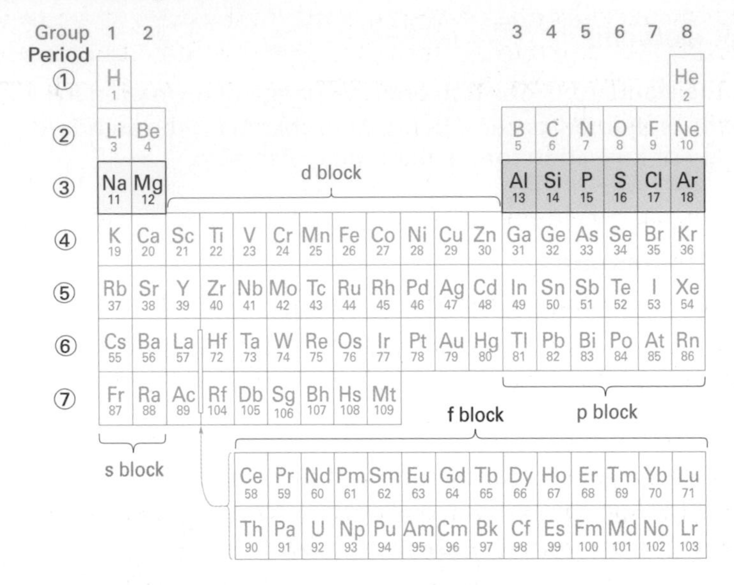

The position of period 3 in the periodic table.

The structure and bonding of the elements change across the period. This affects their physical and chemical properties. The elements are solids, except for chlorine and argon, which are gases at room temperature. Sodium, magnesium, and aluminium are metals. Silicon is a **metalloid** and the remaining elements are non-metals.

element	Na	Mg	Al	Si	P_4	S_8	Cl_2	Ar
structure		metallic		giant covalent		simple molecular		monatomic
bonds broken		metallic		covalent		van der Waals' forces		

The type of structure and bonding changes across period 3.

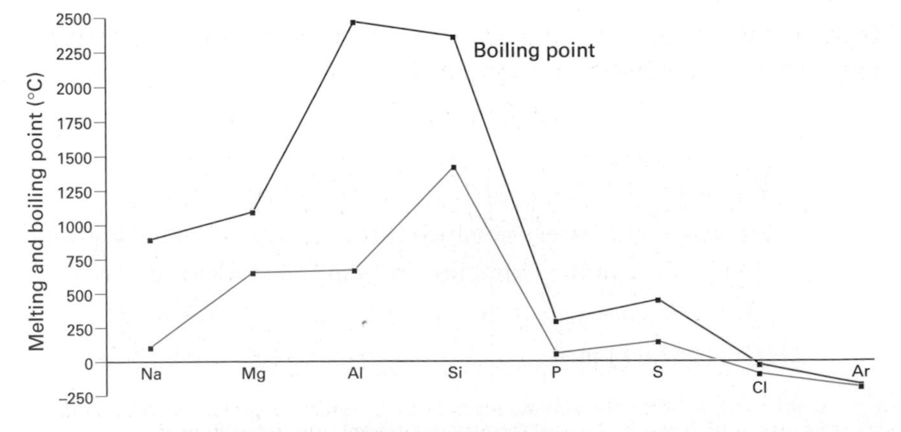

Trends in the melting and boiling points across period 3.

First ionization enthalpy

The first ionization enthalpy generally increases across the period.

Atomic radius

The atomic radius decreases across the period.

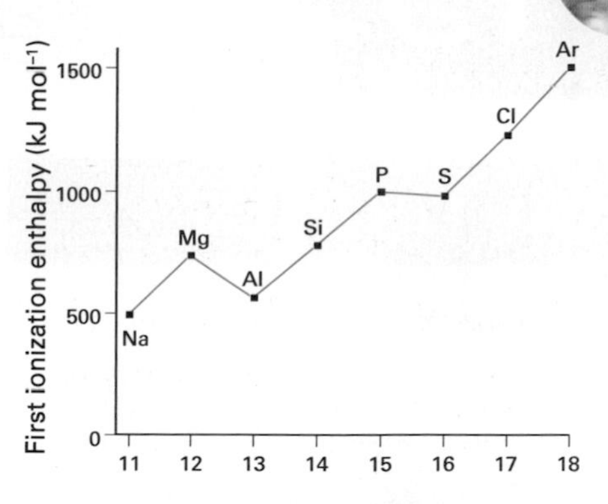

First ionization enthalpy across period 3.

Reactions of period 3 elements with water

Aluminium and silicon are both protected from reacting with water by a thin layer of their oxides. Only sodium, magnesium, and chlorine in period 3 react with water. Sodium reacts more vigorously with water than magnesium does.

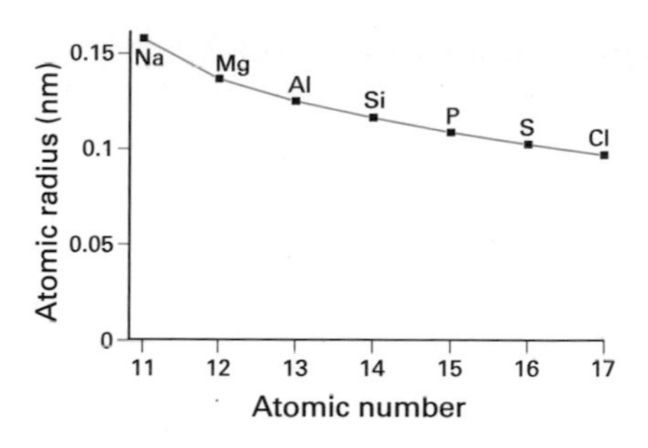

Atomic radius across period 3.

Chlorine

Chlorine reacts with water to form a mixture of hydrochloric acid and chloric(I) acid:

$$Cl_2(aq) + H_2O(l) \rightleftharpoons HCl(aq) + HClO(aq)$$

The mixture is often called chlorine water. Chloric(I) acid is a disinfectant and drinking water is treated with chlorine to make it safe to drink.

Sodium

Sodium is less dense than water and floats. It reacts vigorously with water to produce sodium hydroxide and hydrogen:

$$2Na(s) + 2H_2O(l) \rightarrow 2NaOH(aq) + H_2(g)$$

The heat from the exothermic reaction melts the sodium, which forms a ball of molten metal. The hydrogen released in the reaction pushes the sodium around on the surface of the water. A trail of white sodium hydroxide is left behind. The sodium may ignite with an orange flame if it is prevented from moving around. The pH of the resulting solution of sodium hydroxide is typically 12 – 14.

Sodium reacts vigorously with water.

Magnesium

Magnesium is denser than water and sinks. It reacts very slowly with water to produce magnesium hydroxide and hydrogen:

$$Mg(s) + 2H_2O(l) \rightarrow Mg(OH)_2(aq) + H_2(g)$$

It may take several days to collect enough hydrogen to test. Magnesium hydroxide is sparingly soluble and the pH of the resulting solution is typically 8 – 10.

Magnesium does react vigorously with steam, producing magnesium oxide rather than magnesium hydroxide:

$$Mg(s) + H_2O(g) \rightarrow MgO(s) + H_2(g)$$

Magnesium reacts vigorously with steam to produce white magnesium oxide and hydrogen gas, which can be led out through a tube and ignited with care.

Check your understanding

1 a Write an equation to represent the reaction between sodium and water.

b State the approximate pH of the solution produced.

2 a Under which conditions does magnesium react vigorously with water?

b Write an equation to represent the reaction in part a.

15.02 Reactions of period 3 elements with oxygen

OUTCOMES

already from A2 Level, you can

- describe trends in the reactions of the period 3 elements with water, limited to sodium and magnesium

and after this spread you should be able to

- describe the trends in the reactions of the elements Na, Mg, Al, Si, P, and S with oxygen, limited to the formation of Na_2O, MgO, Al_2O_3, SiO_2, P_4O_{10}, and SO_2

Chlorine and argon do not react with oxygen. The remaining elements in period 3 do react with oxygen, and you need to be able to describe their reactions.

Reactions with oxygen

Sodium

Sodium burns vigorously in air or oxygen with a yellow flame. White sodium oxide is produced in the reaction:

$$4Na(s) + O_2(g) \rightarrow 2Na_2O(s)$$

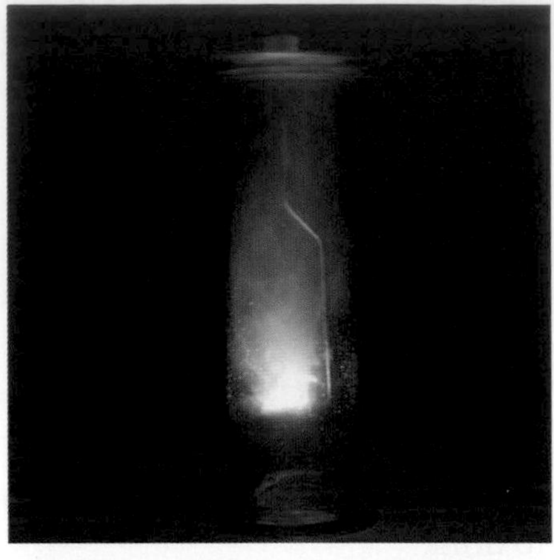

Sodium reacts vigorously with oxygen to produce white sodium oxide.

Magnesium

Magnesium burns very vigorously in air or oxygen with a brilliant white flame. White magnesium oxide is produced in the reaction:

$$2Mg(s) + O_2(g) \rightarrow 2MgO(s)$$

Magnesium reacts very vigorously with oxygen to produce white magnesium oxide.

Not just oxides

The reaction between magnesium and oxygen in the air is so exothermic that magnesium also reacts with nitrogen. This produces magnesium nitride, a greenish yellow powder:

$$3Mg(s) + N_2(g) \rightarrow Mg_3N_2(s)$$

Aluminium

Aluminium reacts with oxygen to form a thin layer of aluminium oxide. Unlike the rust on the surface of iron, this layer is tightly bound to the surface of the metal and protects it from further attack by oxygen. So pieces of aluminium and aluminium foil do not burn in air. But powdered aluminium burns vigorously in oxygen to produce white aluminium oxide:

$$4Al(s) + 3O_2(g) \rightarrow 2Al_2O_3(s)$$

Aluminium powder reacts vigorously with oxygen.

Anodizing

The thickness of the layer of aluminium oxide on the surface of aluminium can be increased by electrolysis. This process is called anodizing. Anodized aluminium is tougher and more corrosion resistant than untreated aluminium. It is commonly used for window frames, cookware, and sports equipment. The anodized layer takes up dyes, producing an attractive and permanently coloured surface.

Phosphorus

White phosphorus (see box) ignites spontaneously in air. It reacts very vigorously with oxygen, burning with a bright light and producing white phosphorus(V) oxide:

$$P_4(s) + 5O_2(g) \rightarrow P_4O_{10}(s)$$

Sulfur

Sulfur easily melts when heated. It burns readily in air with a blue flame, producing choking white fumes of sulfur(IV) oxide:

$$S(s) + O_2(g) \rightarrow SO_2(g)$$

Sulfur(IV) oxide or sulfur dioxide can trigger asthma attacks in some people, so the reaction should be carried out with plenty of ventilation, for example in an efficient fume cupboard.

Sulfur burns in oxygen with a blue flame, producing choking white fumes of sulfur dioxide.

Silicon

Like aluminium, silicon is protected by a thin layer of its oxide. Silicon resists attack by oxygen at temperatures up to around 900 °C, when it reacts to form silicon(IV) oxide:

$$Si(s) + O_2(g) \rightarrow SiO_2(s)$$

Silicon(IV) oxide is also called silica or silicon dioxide, but it has a giant molecular structure rather than a simple molecular structure.

Incendiary phosphorus

Phosphorus exists as different forms with different structures, called allotropes. Red phosphorus is used in matches, but white phosphorus is very hazardous. It is toxic, and it ignites spontaneously in air. This property has been exploited in the manufacture of incendiary bombs, which are intended to start fires.

Bomb disposal experts with protective clothing defusing a phosphorus shell. The bomb is designed to spread phosphorus over a wide area, where it will spontaneously ignite, injuring and killing people in its way.

Formulae of the oxides

There is a trend in the formulae of the oxides formed by these period 3 elements, summarized in the table.

element	Na	Mg	Al	Si	P	S
formula of oxide	Na_2O	MgO	Al_2O_3	SiO_2	P_4O_{10}	SO_2
ratio of period 3 element to oxygen	1:0.5	1:1	1:1.5	1:2	1:2.5	1:2

If you follow the trend from left to right, you would expect sulfur to form an oxide in the ratio 1:3. Sulfur(VI) oxide, SO_3, does form in the presence of a vanadium(V) oxide catalyst. But normally sulfur forms SO_2 and does not use its highest oxidation state.

Check your understanding

1 a Write equations for the reactions of period 3 elements to form Na_2O, MgO, Al_2O_3, SiO_2, P_4O_{10}, or SO_2.

b Outline the trend in reactivity of the period 3 elements towards oxygen.

15.03 Period 3 oxides: physical properties

OUTCOMES

already from A2 Level, you can

- describe the trends in the reactions of the elements Na, Mg, Al, Si, P and S with oxygen

and after this spread you should be able to

- explain the link between the physical properties of the highest oxides of the elements sodium to sulfur in terms of their structure and bonding

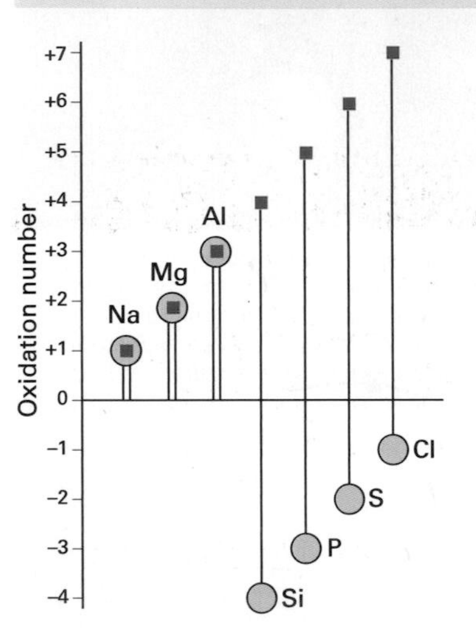

The oxidation number for the elements sodium to chlorine.

The trend of the formulae of the period 3 oxides was discussed on the previous spread. The reason for this trend lies in the highest **oxidation states** of the period 3 elements. These increase from +1 for sodium in group 1, to +7 for chlorine in group 7.

This means that the highest oxides of the elements sodium to sulfur are Na_2O, MgO, Al_2O_3, SiO_2, P_4O_{10}, and SO_3 (not SO_2).

Melting points

The period 3 oxides under study are all solids, apart from sulfur(VI) oxide, SO_3, which is a liquid at room temperature. The melting points of the period 3 oxides increase from Na_2O to MgO, then decrease steadily.

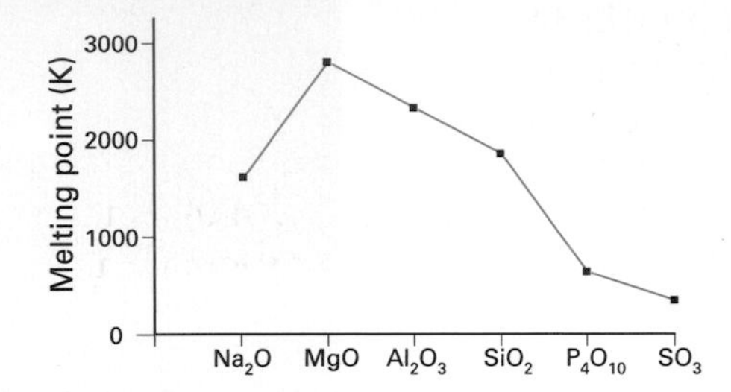

The trend in melting points of the period 3 oxides.

Sodium, magnesium, and aluminium oxides

The three metal oxides have the highest melting points, and exist as giant ionic lattices. Very many strong ionic bonds must be overcome for these oxides to melt. Magnesium oxide has a higher melting point than sodium oxide because its metal ions are smaller and more highly charged. So the electrostatic forces of attraction between oxide ions and magnesium ions are stronger than those between oxide ions and sodium ions.

Aluminium ions are even smaller and more highly charged than magnesium ions. They polarize oxide ions to a greater extent, so aluminium oxide has a degree of covalent character. As a result, its melting point is lower than that of magnesium oxide.

Silica

Silicon(IV) oxide or silica has a **giant covalent structure**. It has very many strong covalent bonds that must be broken to melt it, so silica has a high melting point. The formula SiO_2 is its **empirical formula**, showing the simplest whole number ratio of its elements.

The very high melting point of magnesium oxide makes it useful as a heat-resistant refractory lining, for example in blast furnaces.

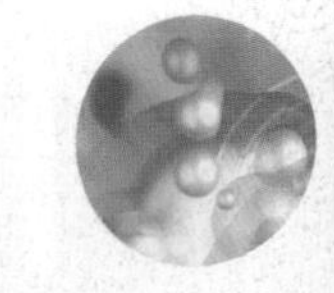

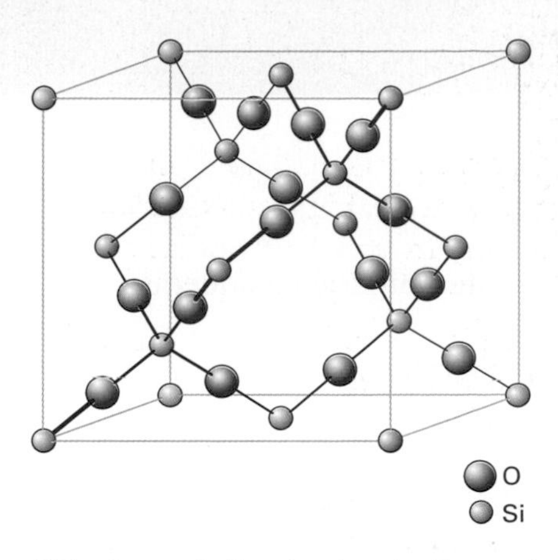

Silica has a similar structure to diamond in which each silicon atom is attached to two oxygen atoms.

Quartz is a form of silica. It is piezoelectric, which means that it produces a potential difference when a stress is applied. This property allows it to be used in the oscillators in watches.

Phosphorus(V) oxide

Phosphorus(V) oxide, P_4O_{10}, exists as **simple covalent molecules**. These have a large relative formula mass (M_r = 284.0) so the **van der Waals' forces** are strong. As a result, it is solid at room temperature.

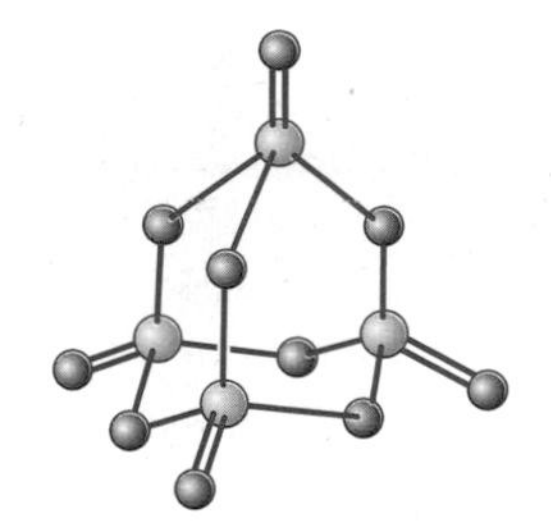

The structure of a P_4O_{10} molecule.

Sulfur(VI) oxide

Sulfur(VI) oxide, SO_3, exists as rings of three molecules (M_r = 240.3). The van der Waals' forces between these *trimers* are weaker than the forces between phosphorus(V) oxide molecules, so sulfur(VI) oxide is a liquid at room temperature.

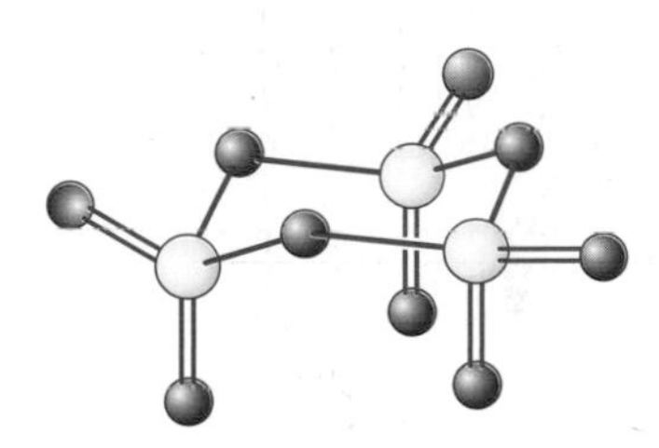

The structure of a sulfur trioxide trimer.

Electrical conductivity

The ionic compounds

Ionic compounds conduct electricity when molten, or dissolved in water, because their ions are free to move. Sodium oxide, magnesium oxide, and aluminium oxide all conduct electricity when molten. Sodium oxide dissolves readily in water to produce aqueous sodium hydroxide, which can conduct electricity. But magnesium oxide is only sparingly soluble, and aluminium oxide is insoluble, so they do not conduct electricity when added to water.

The covalent compounds

In general, covalent compounds do not conduct electricity because they do not have mobile charge carriers, such as free electrons or ions. Silica is a covalent compound and insoluble in water, so it does not conduct electricity. Phosphorus(V) oxide and sulfur(VI) oxide do not conduct electricity when they are molten, but they both react with water to produce acidic solutions. The aqueous ions in these solutions conduct electricity.

Sulfur(IV) oxide

Sulfur(IV) oxide, SO_2, exists as simple molecules. The van der Waals' forces between sulfur(IV) oxide molecules are weaker than those between sulfur(VI) oxide trimers, so it is a gas at room temperature. Sulfur(IV) oxide boils at −10°C whereas sulfur(VI) oxide melts at 17°C and boils at 45°C.

Check your understanding

1 a Sulfur forms an oxide in which it is not in its highest oxidation state. Give the oxidation state of sulfur in this oxide.

b Explain why phosphorus(V) oxide is a solid at room temperature but sulfur(VI) oxide is a liquid.

2 a Explain why magnesium oxide has a higher melting point than aluminium oxide.

b Suggest how you could use electrolysis to distinguish between aluminium oxide and silica.

15.04 Period 3 oxides: reactions with water

OUTCOMES

already from A2 Level, you can

- describe the trends in the reactions of the elements sodium to sulfur with oxygen
- explain the link between the physical properties of the highest oxides of the elements sodium to sulfur in terms of their structure and bonding

and after this spread you should

- be able to describe the reactions of the oxides of the elements sodium to sulfur with water, limited to Na_2O, MgO, Al_2O_3, SiO_2, P_4O_{10}, SO_2, and SO_3
- know the change in pH of the resulting solutions across period 3
- be able to write equations for the reactions which occur between these oxides and given simple acids and bases
- be able to explain the trends in these properties in terms of the type of bonding present

Oxide ions and hydrolysis

The oxide ions in sodium oxide and magnesium oxide are hydrated when the compounds are added to water. They then react with water in a hydrolysis reaction to produce hydroxide ions:

$$O^{2-}(aq) + H_2O(l) \rightarrow 2OH^-(aq)$$

Since magnesium oxide is sparingly soluble in water, there is a limited amount of hydrolysis. The pH of the resulting solution is lower than that formed from sodium oxide.

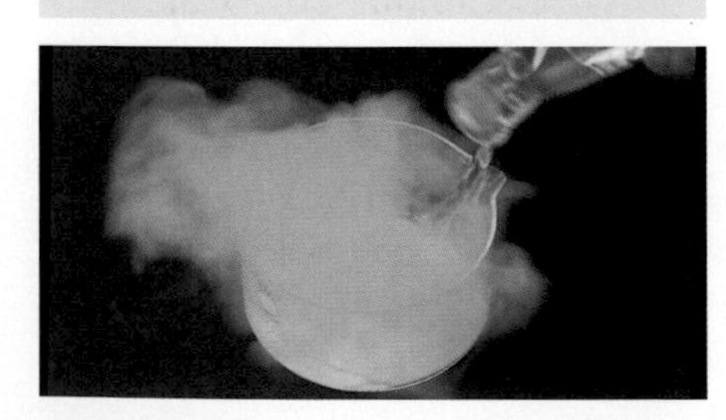

Phosphorus(V) oxide reacts vigorously with water to form phosphoric(V) acid.

Reactions with water

Sodium oxide

Sodium oxide reacts exothermically with water, producing aqueous sodium hydroxide:

$$Na_2O(s) + H_2O(l) \rightarrow 2NaOH(aq)$$

The pH of the resulting solution is typically 14.

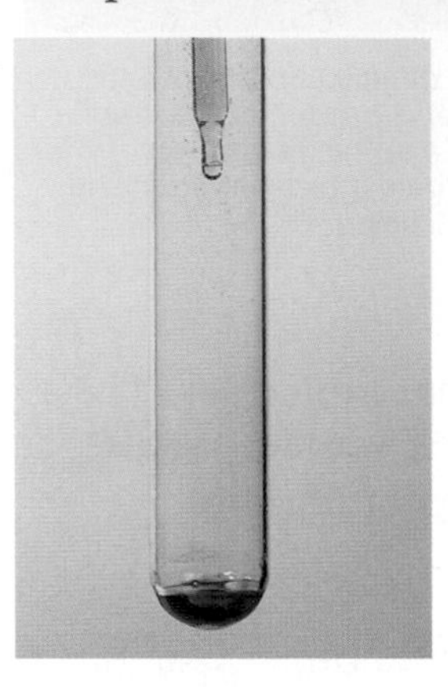

Sodium oxide reacts with water to produce an alkaline solution that turns universal indicator blue.

Magnesium oxide

Magnesium oxide is only slightly soluble in water:

$$MgO(s) + H_2O(l) \rightarrow Mg(OH)_2(aq)$$

The pH of the resulting solution is typically 10.

Aluminium oxide and silica

Aluminium oxide and silica are insoluble in water, and there are no reactions.

Phosphorus(V) oxide

Phosphorus(V) oxide reacts vigorously with water in a very exothermic reaction. In excess water, phosphoric(V) acid is formed:

$$P_4O_{10}(s) + 6H_2O(l) \rightarrow 4H_3PO_4(aq)$$

The pH of the resulting solution is typically 0.

Sulfur oxides

Sulfur(IV) oxide dissolves readily in water and reacts with it to form sulfuric(IV) acid, also called sulfurous acid:

$$SO_2(g) + H_2O(l) \rightleftharpoons H_2SO_3(aq)$$

The pH of the resulting solution is typically 1. Sulfuric(IV) acid decomposes to form sulfur(IV) oxide if attempts are made to dehydrate it, so it cannot be isolated in its anhydrous form.

Sulfur(VI) oxide reacts vigorously with water form sulfuric(VI) acid:

$$SO_3(g) + H_2O(l) \rightarrow H_2SO_4(aq)$$

The pH of the resulting solution is typically 0. Unlike sulfuric(IV) acid, sulfuric(VI) acid can be dehydrated to form an oily liquid called *oleum* or *fuming sulfuric acid*.

Reactions with acids and bases

Sodium oxide and magnesium oxide

Sodium oxide and magnesium oxide are bases. They react with acids to form salts and water. Here are three examples:

$$Na_2O(s) + 2HCl(aq) \rightarrow 2NaCl(aq) + H_2O(l)$$

$$Na_2O(s) + H_2SO_4(aq) \rightarrow Na_2SO_4(aq) + H_2O(l)$$

$$3MgO(s) + 2H_3PO_4(aq) \rightarrow Mg_3(PO_4)_2(aq) + 3H_2O(l)$$

Aluminium oxide

Aluminium oxide is amphoteric because it acts as a base and as an acid. For example:

$$Al_2O_3(s) + 3H_2SO_4(aq) \rightarrow Al_2(SO_4)_3(s) + 3H_2O(l)$$

$$Al_2O_3(s) + 2NaOH(aq) + 3H_2O(l) \rightarrow 2NaAl(OH)_4(aq)$$

Silica

Although silica is insoluble in water, it acts as an acid and reacts with bases. For example:

$$SiO_2(s) + 2NaOH(aq) \rightarrow Na_2SiO_3(aq) + H_2O(l)$$

This reaction can cause glass stoppers to get stuck in reagent bottles containing aqueous sodium hydroxide.

Phosphorus(V) oxide

Phosphorus(V) oxide reacts with bases to form salts. For example:

$$P_4O_{10}(s) + 12NaOH(aq) \rightarrow 4Na_3PO_4(aq) + 6H_2O(l)$$

Sulfur(IV) oxide

Sulfur(IV) oxide reacts with bases to form salts. For example:

$$SO_2(g) + CaCO_3(s) \rightarrow CaSO_3(s) + CO_2(g)$$

Trends and bonding

The table summarizes the acid–base properties of the period 3 oxides and the bonding present.

oxide	Na_2O	MgO	Al_2O_3	SiO_2	P_4O_{10}	SO_2	SO_3
typical pH in water	14	10	7	7	0	3	0
structure	giant				molecular		
bonding	ionic		ionic with covalent character	covalent			
nature of oxide	basic		amphoteric	acidic			

Science @ Work

Flue-gas desulfurization

Fossil fuels such as coal often naturally contain sulfur compounds. These oxidize when the fuel is burned, releasing sulfur(IV) oxide gas, a major cause of acid rain. Flue gases from power stations are desulfurized or scrubbed using powdered wet limestone (calcium carbonate):

$$SO_2(g) + CaCO_3(s) + 2H_2O(l) + \tfrac{1}{2}O_2(g) \rightarrow CaSO_4{\cdot}2H_2O + CO_2(g)$$

Hydrated calcium sulfate $CaSO_4{\cdot}2H_2O$ is gypsum, used in plaster and plasterboard.

Check your understanding

1. State the relationship between the type of bonding in period 3 oxides and the pH of the solutions formed when they react with water.
2. a Explain why aluminium oxide is said to be amphoteric.
   b Explain why silica behaves as an acid even though it is insoluble in water.
3. a Write equations to show the reactions of nitric acid, HNO_3, with sodium oxide, magnesium oxide, and aluminium oxide.
   b Write equations to show the reactions of aqueous potassium hydroxide with silica and phosphorus(V) oxide.

16.01 Oxidation states

OUTCOMES

already from AS Level, you

- know that oxidation is the loss of electrons and reduction is the gain of electrons
- can apply the rules for assigning oxidation states to work out the oxidation state of an element in a compound
- can write half-equations and combine them to give an overall redox equation

and after this spread you should have

- familiarized yourself again with the key concepts of oxidation states and redox equations from AS Level Chemistry

Oxidation states

The oxidation state of an element is the number of electrons that must be gained to make a neutral atom. For example, the oxidation state of sodium in Na^+ is +1 because one electron would need to be gained to make a neutral atom:

$$Na^+ + e^- \rightarrow Na$$

The oxidation state of chlorine in Cl^- is −1 because one electron would need to be lost to make a neutral atom:

$$Cl^- - e^- \rightarrow Cl \text{ ... which is better written as ... } Cl^- \rightarrow Cl + e^-$$

Atoms and simple ions

For a simple ion formed from an atom, the oxidation state is equal to the charge on the ion, as seen above. So the oxidation state of magnesium in the Mg^{2+} ion is +2, and the oxidation state of nitrogen in the N^{3-} ion is −3. The oxidation state for the atoms in an uncombined element is always zero. For example, it is zero for sodium in Na, for chlorine in Cl_2, and for sulfur in S_8.

Compounds and polyatomic ions

In a compound, the sum of all the oxidation states is zero. In a polyatomic ion, such as NH_4^+ or SO_4^{2-}, the sum of all the oxidation states is the same as the charge on the ion. You can work out the oxidation states of the different elements in a compound or polyatomic ion because some elements have fixed oxidation states, as seen in the table.

element	oxidation state in compounds and ions	exceptions
H	+1	oxidation state is −1 in metal hydrides, e.g. NaH
Li, Na, K	+1	
Mg, Ca, Ba	+2	
Al	+3	
F	−1	
Cl	−1	oxidation state is not −1 when Cl is combined with F or O
O	−2	oxidation state is not −2 when O is combined with F oxidation state is −1 in peroxides such as H_2O_2

Here are the four steps you need to work out the oxidation state of an element in a compound or polyatomic ion.

Step 1 Write down the formula.

Step 2 For the known oxidation states (as in the table), write each oxidation state above the symbol for each element.

Step 3 For the known oxidation states, write the total oxidation state below the symbol.

Step 4 Subtract the total known oxidation states from zero if the problem is about a compound, or from the charge if the problem is about a polyatomic ion. Divide your answer by the number of atoms of the element in the formula.

Worked example 1

What is the oxidation state of iron in Fe_2O_3?

Step 1 Fe_2O_3

Step 2 $Fe_2\overset{-2}{O_3}$

Step 3 $Fe_2\underset{-6}{\overset{-2}{O_3}}$

Step 4 Total oxidation state of Fe = 0 −(−6) = +6

Oxidation state of each Fe = +6 ÷ 2 = +3

Worked example 2

What is the oxidation state of nitrogen in NO_3^-?

Step 1 NO_3^-

Step 2 $N\overset{-2}{O_3^-}$

Step 3 $N\underset{-6}{\overset{-2}{O_3^-}}$

Step 4 Total oxidation state of N = −1 −(−6) = +5

Half-equations

Half-equations show the loss or gain of electrons by one substance. For example, this is the full equation for the reaction between magnesium and chlorine:

$$Mg(s) + Cl_2(g) \rightarrow MgCl_2(s)$$

These are the two half-equations:

$Mg \rightarrow Mg^{2+} + 2e^-$ ... an oxidation reaction

$Cl_2 + 2e^- \rightarrow 2Cl^-$ ... a reduction reaction

They can be recombined to give the full equation again. If the number of electrons in each half-equation is different, you multiply each equation by a factor needed to get the same number of electrons in each one. Notice that one half-equation is an **oxidation** reaction and the other is a **reduction** reaction. The overall reaction is a **redox** reaction.

Magnesium is oxidized by chlorine in a vigorous exothermic reaction.

OIL RIG

Oxidation is the loss of electrons, and reduction is the gain of electrons. The mnemonic 'OIL RIG' helps you to remember this (Oxidation Is Loss, Reduction Is Gain).

Check your understanding

1 In terms of electrons, what is: **a** oxidation and **b** reduction?

2 Work out the oxidation states of the underlined element in:

  a $\underline{S}O_2$, $\underline{P}Cl_3$, $Na_2\underline{S}_2O_3$.

  b $\underline{S}^{2-}$, $\underline{Cl}O_3^-$, $\underline{P}Cl_4^+$.

3 **a** Zinc reacts with copper chloride:

  $Zn + CuCl_2 \rightarrow ZnCl_2 + Cu$

  Write the two half-equations, and identify which species is oxidized and which is reduced.

  b Combine these two half-equations:

  $Al \rightarrow Al^{3+} + 3e^-$ ... and ... $Cl_2 + 2e^- \rightarrow 2Cl^-$

Worked example

Combine these two half-equations:

$Al^{3+} + 3e^- \rightarrow Al$ ... and ... $2O^{2-} \rightarrow O_2 + 4e^-$

$4Al^{3+} + 12e^- \rightarrow 4Al$ (multiply by 4 to get $12e^-$)

$6O^{2-} \rightarrow 3O_2 + 12e^-$ (multiply by 3 to get $12e^-$)

The overall redox equation is: $4Al^{3+} + 6O^{2-} \rightarrow 4Al + 3O_2$

16.02 Half-equations and the d block elements

OUTCOMES

already from AS Level, you

- know that oxidation is the loss of electrons and reduction is the gain of electrons
- can work out the oxidation state of an element in a compound
- can write half-equations and combine them
- understand oxidation and reduction reactions of s and p block elements

and after this spread you should be able to

- apply the electron transfer model of redox, including oxidation states and half equations, to d block elements

The d block elements

The **d block** is in the central part of the periodic table. It lies between the **s block** to the left and the **p block** to the right.

d block

Mg 12											Al 13
Ca 20	Sc 21	Ti 22 [Ar]$3d^2 4s^2$	V 23 [Ar]$3d^3 4s^2$	Cr 24 [Ar]$3d^5 4s^1$	Mn 25 [Ar]$3d^5 4s^2$	Fe 26 [Ar]$3d^6 4s^2$	Co 27 [Ar]$3d^7 4s^2$	Ni 28 [Ar]$3d^8 4s^2$	Cu 29 [Ar]$3d^{10} 4s^1$	Zn 30	Ga 31
Sr 38	Y 39	Zr 40	Nb 41	Mo 42	Tc 43	Ru 44	Rh 45	Pd 46	Ag 47	Cd 48	In 49
Ba 56	La 57	Hf 72	Ta 73	W 74	Re 75	Os 76	Ir 77	Pt 78	Au 79	Hg 80	Tl 81
Ra 88	Ac 89	Rf 104	Db 105	Sg 106	Bh 107	Hs 108	Mt 109				

f block

Ce	Pr	Nd	Pm	Sm	Eu	Gd	Tb	Dy

The position of the d block in the periodic table.

The atoms of d block elements contain electrons in the d **sub-level**. The **transition metals** are d block elements that can form at least one stable ion with a partially filled d sub-level. So only the elements titanium to copper in the first row of the d block are transition metals. Scandium and zinc are d block elements but they are not transition metals.

The transition metals have several characteristics, including having variable oxidation states. You can find out more about these characteristics in Chapters 17 and 18. In this spread, you are going to find out how to work out the oxidation states of d block elements in compounds and ions, and how to combine half-equations involving their reactions.

Oxidation states

Manganese, for example, has five common oxidation states, +2, +3, +4, +6, and +7. The table summarizes these oxidation states and the species they are found in. (See Worked examples 1 + 2 opposite page)

species	oxidation state of Mn
Mn^{2+}	+2
$Mn(OH)_3$	+3
MnO_2	+4
MnO_4^{2-}	+6
MnO_4^-	+7

The oxidation states of manganese in various species. Check that you can confirm each oxidation state (remember that the oxidation state of oxygen in most compounds is −2).

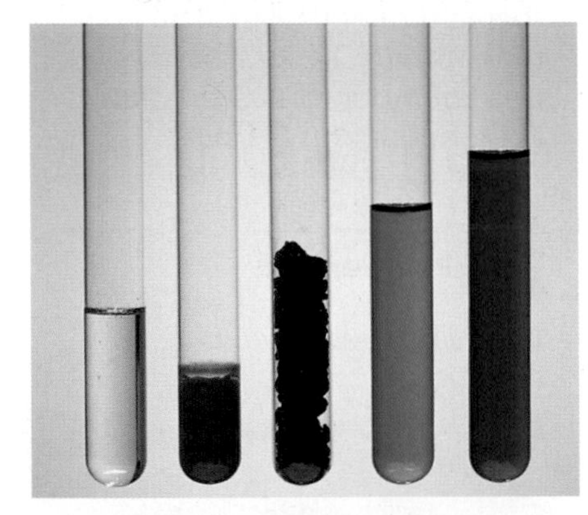

Compounds showing the five common oxidation states of manganese. From left to right, $Mn^{2+}(aq)$, $Mn(OH)_3(s)$, $MnO_2(s)$, $K_2MnO_4(aq)$, and $KMnO_4(aq)$.

Half-equations

Half-equations involving d block elements can be trickier than others, as they may involve hydrogen ions H^+. This happens when the d block element is combined with oxygen in a polyatomic ion, such as the manganate(VII) ion MnO_4^-.

Manganate(VII) can be reduced to manganese(II). Here is how you can work out the corresponding half-equation.

Step 1	Write out the two ions	$MnO_4^- \rightarrow Mn^{2+}$
Step 2	Add enough water molecules to the right hand side to use the oxygen atoms	$MnO_4^- \rightarrow Mn^{2+} + 4H_2O$
Step 3	Add enough hydrogen ions to the left hand side to account for the hydrogen atoms on the right	$MnO_4^- + 8H^+ \rightarrow Mn^{2+} + 4H_2O$
Step 4	Add enough electrons to make the total charges on each side the same	$MnO_4^- + 8H^+ + 5e^- \rightarrow Mn^{2+} + 4H_2O$

When you combine two half-equations to make a full redox equation, you usually have to multiply each half-equation by the factors needed to give equal numbers of electrons. (See example 3 below)

Worked example 3

Combine these two half-equations:

$MnO_4^- + 8H^+ + 5e^- \rightarrow Mn^{2+} + 4H_2O$

$Fe^{2+} \rightarrow Fe^{3+} + e^-$

$MnO_4^- + 8H^+ + 5e^- \rightarrow Mn^{2+} + 4H_2O$

$5Fe^{2+} \rightarrow 5Fe^{3+} + 5e^-$ (multiply by 5 to get $5e^-$)

Add together and cancel the electrons:

$MnO_4^- + 8H^+ + \cancel{5e^-} + 5Fe^{2+} \rightarrow Mn^{2+} + 4H_2O + 5Fe^{3+} + \cancel{5e^-}$

The overall redox equation is:

$$MnO_4^- + 8H^+ + 5Fe^{2+} \rightarrow Mn^{2+} + 4H_2O + 5Fe^{3+}$$

Check your understanding

1 What do the terms *d block element* and *transition metal* mean?

2 Work out the oxidation states of chromium in:

a Cr^{2+}

b CrO_4^{2-}

3 a Dichromate(VI) ions, $Cr_2O_7^{2-}$, can be reduced to chromium(III) ions, Cr^{3+}. Work out the half-equation for this change.

b Sulfur(IV) oxide, SO_2, can be oxidized to sulfate(VI):

$$SO_2 + 2H_2O \rightarrow SO_4^{2-} + 4H^+ + 2e^-$$

Combine this half-equation with your half-equation from part **a** to produce the overall redox reaction. (Hint: you will be able to cancel some H^+ and H_2O).

Worked example 1

Work out the oxidation state of vanadium in VO_2^+.

Step 1 VO_2^+

Step 2 $V\overset{-2}{O}{}_2^+$

Step 3 $V\overset{-2}{\underset{-4}{O}}{}_2^+$

Step 4 Total oxidation state of V = 1 −(−4) = +5

Worked example 2

What is the oxidation state of chromium in $Cr_2O_7^{2-}$.

Step 1 $Cr_2O_7^{2-}$

Step 2 $Cr_2\overset{-2}{O}{}_7^{2-}$

Step 3 $Cr_2\overset{-2}{\underset{-14}{O}}{}_7^{2-}$

Step 4 Total oxidation state of Cr = −2 −(−14) = +12

Oxidation state of each Cr = +12 ÷ 2 = +6

Naming ions

An ion only containing the metal itself is named after the metal and the oxidation state. For example, Mn^{2+} is manganese(II) and Fe^{3+} is iron(III). If the ion contains oxygen, it is named after the metal with the ending 'ate', and the oxidation state. For example, MnO_4^- is manganate(VII), VO_2^+ is vanadate(V), and $Cr_2O_7^{2-}$ is dichromate(VI).

16.03 Electrochemical cells

OUTCOMES

already from A2 Level, you

- know that oxidation is the loss of electrons and reduction is the gain of electrons
- can write half-equations and combine them

and after this spread you should know

- the IUPAC convention for writing half-equations for electrode reactions
- and be able to use the conventional representation of cells

The reaction between zinc and aqueous copper(II) sulfate. The blue colour gradually fades as copper is deposited and colourless aqueous zinc sulfate forms.

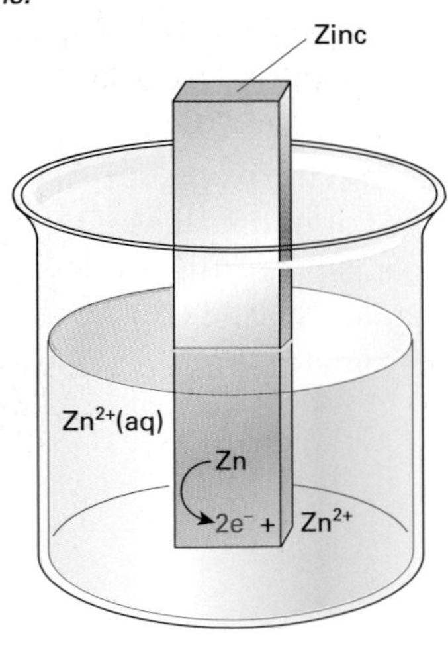

A simple zinc half-cell.

If a strip of zinc metal is dipped into aqueous copper(II) sulfate, a redox reaction happens:

$$Zn(s) + CuSO_4(aq) \rightarrow ZnSO_4(aq) + Cu(s)$$

The zinc becomes coated with copper and slowly disappears as aqueous zinc sulfate forms. The equation can be separated into two half-equations:

$$Zn(s) \rightarrow Zn^{2+}(aq) + 2e^- \quad \text{... an oxidation reaction}$$

$$Cu^{2+}(aq) + 2e^- \rightarrow Cu(s) \quad \text{... a reduction reaction}$$

The sulfate(VI) ions SO_4^{2-} are **spectator ions** and do not take part in the reaction. The zinc atoms act as a reducing agent because they donate electrons to the copper(II) ions and are oxidized. In turn, the copper(II) ions act as an oxidizing agent because they accept electrons from the zinc atoms and are reduced.

Nothing will happen if you attempt to carry out the reverse reaction, because copper is not a sufficiently powerful reducing agent to reduce zinc ions to zinc. The idea of **electrochemical cells** lets you measure the relative reducing ability of different species, and also lets you predict the direction of spontaneous change.

Half-cells

A typical **half-cell** comprises a piece of metal dipped in an aqueous solution of its ions. In the zinc and copper example above, there would be two half-cells. One of these could be a strip of zinc dipped into aqueous zinc sulfate as a source of aqueous zinc ions. A **dynamic equilibrium** forms between zinc atoms and zinc ions:

$$Zn^{2+}(aq) + 2e^- \rightleftharpoons Zn(s)$$

The zinc metal strip acts as an **electrode** because electrons can enter or leave through it. Note that the IUPAC convention for electrode reactions like this one is to write the half-equation as a reduction reaction, showing electrons gained by a species.

The other half-cell could be a strip of copper dipped into aqueous copper sulfate. A dynamic equilibrium forms between copper atoms and copper ions:

$$Cu^{2+}(aq) + 2e^- \rightleftharpoons Cu(s)$$

If the position of equilibrium in one of the half-cells lies further to the left than that in the other half-cell, the electrodes will have a **potential difference**. This can be measured if the half-cells are connected to make an electrochemical cell.

An electrochemical cell

The two electrodes in an electrochemical cell are connected by wires to a high-resistance voltmeter. This measures the potential difference or p.d. between them and shows the direction in which the current flows. The electric circuit is completed using a **salt bridge**, which lets ions flow between the two solutions. The diagram shows the electrochemical cell formed by a zinc half-cell and a copper half-cell.

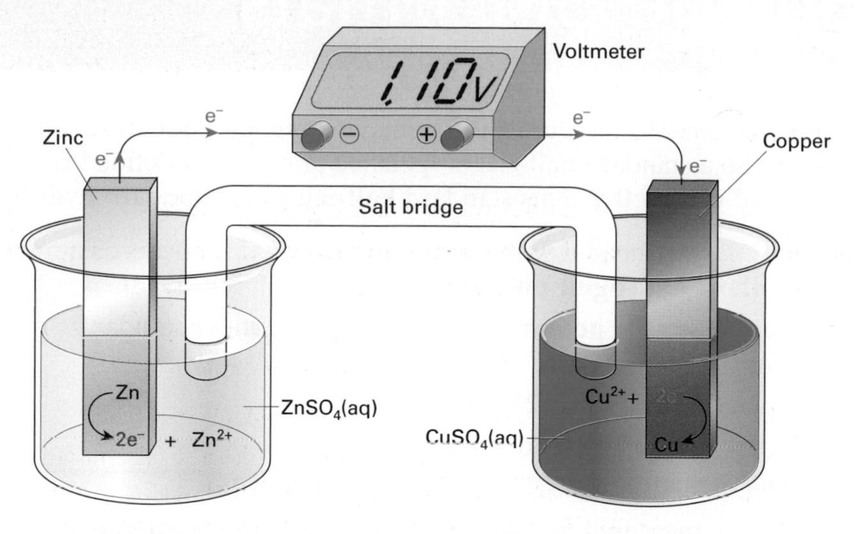

In this electrochemical cell, electrons flow in the external circuit from left to right, giving a potential difference of + 1.10 V.

The zinc electrode loses electrons, so the position of equilibrium lies to the left:

$$Zn^{2+}(aq) + 2e^- \rightleftharpoons Zn(s)$$

The copper electrode gains electrons, so the position of equilibrium lies to the right:

$$Cu^{2+}(aq) + 2e^- \rightleftharpoons Cu(s)$$

Overall, the concentration of zinc ions increases and copper is deposited, just as in the simple redox reaction described at the start. But note that the copper is deposited on the copper electrode and not on the zinc.

Representing cells

There is a simple way to represent cells that saves you from drawing them out. Here is the standard cell notation or **cell diagram** for the zinc/copper cell:

$$Zn(s) \mid Zn^{2+}(aq) \parallel Cu^{2+}(aq) \mid Cu(s)$$

A single vertical line separates two *phases*, such as a solid and an aqueous solution. A double vertical line represents the salt bridge. The two aqueous solutions are written next to the salt bridge symbol, and the two solid electrodes at each end. If a half-cell contains a mixture of aqueous ions, a comma is used to separate them. For example:

$$Fe(s) \mid Fe^{2+}(aq), Fe^{3+}(aq)$$

Electromotive force, e.m.f.

A high-resistance voltmeter lets very little current flow. This ensures that the maximum potential difference that the cell can develop is measured. This is called its **e.m.f.**

The salt bridge

A salt bridge is often just a strip of filter paper soaked in saturated aqueous potassium chloride or aqueous potassium nitrate. This ensures that ions can move without precipitates forming.

Check your understanding

1 Write half-equations for the electrode reactions that happen when:
- **a** iron is dipped into aqueous iron(III) nitrate
- **b** nickel is dipped into nickel chloride

2 a Explain the functions in an electrochemical cell of the *salt bridge* and the *high-resistance voltmeter*.
- **b** Describe the electrochemical cell represented by:

$$Pt(s) \mid Fe^{2+}(aq), Fe^{3+}(aq) \parallel Ag^{+}(aq) \mid Ag(s)$$

16.04 Electrode potentials

OUTCOMES

already from A2 Level, you know

- the IUPAC convention for writing half-equations for electrode reactions
- and can use the conventional representation of cells

and after this spread you should

- understand how cells are used to measure electrode potentials by reference to the standard hydrogen electrode
- know the importance of the conditions when measuring the electrode potential, *E* (Nernst equation not needed)
- know that standard electrode potential, $E^\ominus$, refers to conditions of 298 K, 100 kPa, and 1.00 mol dm^{-3} solution of ions

A standard hydrogen electrode.

It is only possible to measure a difference in potential, not an absolute potential. So a standard half-cell is specified and given a defined standard potential of exactly 0 V. This standard half-cell is the **standard hydrogen electrode**.

The standard hydrogen electrode

A **standard electrode potential**, $E^\ominus$, is measured under standard conditions of

- a temperature of 298 K (25°C)
- a pressure of 100 kPa (100,000 Pa)
- solutions of ions at a concentration of 1.00 mol dm^{-3}

$E^\ominus$ values are measured for half-cells connected to a standard hydrogen electrode. This comprises a platinum electrode dipped into acid, where the concentration of hydrogen ions is 1.00 mol dm^{-3}. Hydrogen gas at 100 kPa is bubbled through, and the temperature is maintained at 298 K. This is the half-equation for the electrode reaction that takes place.

$$2H^+(aq) + 2e^- \rightleftharpoons H_2(g) \qquad E^\ominus = 0\ V\ \text{(defined)}$$

The corresponding cell diagram is:

$$Pt(s)\ |\ H_2(g)\ |\ H^+(aq)$$

The calomel electrode

The hydrogen electrode is tricky to use, so the *standard calomel electrode* may be used instead. This has an $E^\ominus$ value of +0.27 V. The half-equation for its electrode reaction is:

$$Hg_2Cl_2(s) + 2e^- \rightleftharpoons 2Hg(l) + 2Cl^-(aq)$$

'Calomel' is the old name for mercury(I) chloride, Hg_2Cl_2.

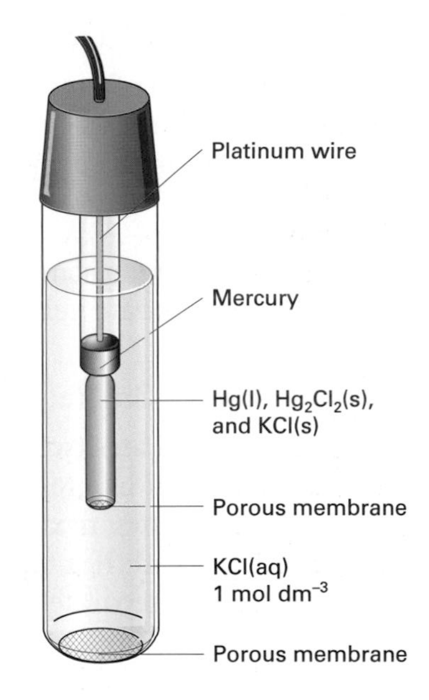

The calomel electrode, a secondary standard electrode that is easier to use than the standard hydrogen electrode.

Measuring an $E^\ominus$ value

The e.m.f. of an electrochemical cell measured under standard conditions has the symbol $E^\ominus_{cell}$. It is calculated by subtracting the $E^\ominus$ value for the left hand half-cell from the $E^\ominus$ value for the right hand half-cell:

$$E^\ominus_{cell} = E^\ominus_R - E^\ominus_L$$

Since the $E^\ominus$ value for the standard hydrogen electrode is zero by definition, it makes good sense to measure $E^\ominus$ values by putting the standard hydrogen electrode on the left hand side. This way, the $E^\ominus$ value you want to find is equal to the $E^\ominus_{cell}$ value measured by the high-resistance voltmeter. The diagram shows how the $E^\ominus$ value for a zinc half-cell could be found.

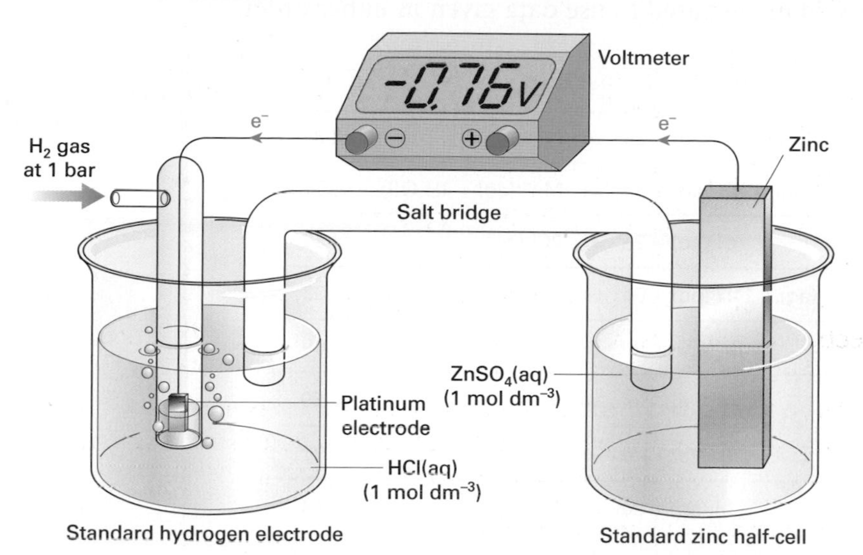

Measuring the standard electrode potential $E^\ominus$ for the electrode reaction $Zn^{2+}(aq) + 2e^- \rightleftharpoons Zn(s)$. The reading on the high-resistance voltmeter shows that $E^\ominus = -0.76$ V.

Check your understanding

1 State the standard conditions needed to measure $E^\ominus$ values.

2 a Describe the standard hydrogen electrode. State its $E^\ominus$ value and include the cell diagram.

b Explain why the calomel electrode might be used instead.

3 Describe how you could use the standard hydrogen electrode to measure the $E^\ominus$ value for this electrode reaction:

$Ni^{2+}(aq) + 2e^- \rightleftharpoons Ni(s)$

The $E^\ominus$ value for a zinc half-cell is -0.76 V. The negative sign shows that the potential on the zinc electrode is more negative than the potential on the standard hydrogen electrode. The $E^\ominus$ value for a copper half-cell can be measured in the same way. It is $+0.34$ V, which shows that its potential is more positive than that of the standard hydrogen electrode, and even more positive than that of the zinc half-cell.

The zinc and copper cell — the return

You are more likely to measure $E^\ominus_{cell}$ values for pairs of half-cells such as the zinc and copper electrochemical cell, rather than to use the standard hydrogen electrode. The $E^\ominus$ value for the zinc half-cell is -0.76 V and the $E^\ominus$ value for the copper half cell is $+0.34$ V. Here is how you calculate the $E^\ominus_{cell}$ value for the electrochemical cell:

$$E^\ominus_{cell} = E^\ominus_R - E^\ominus_L$$

$$E^\ominus_{cell} = +0.34 - (-0.76) = +1.10 \text{ V}$$

Remember that this $E^\ominus_{cell}$ value is for the cell:

$$Zn(s) \mid Zn^{2+}(aq) \parallel Cu^{2+}(aq) \mid Cu(s)$$

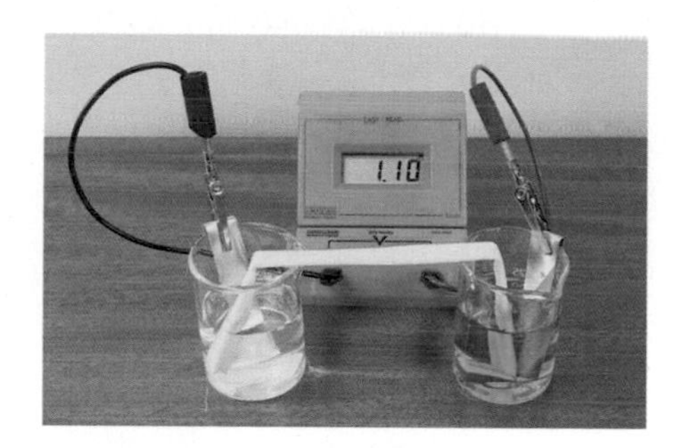

$E^\ominus_{cell}$ for the zinc and copper cell is $+1.10$ V. Note the simple salt bridge made from filter paper dipped in saturated aqueous potassium nitrate.

Changing the conditions

If the aqueous solutions in the two half-cells are at the same concentration, the E_{cell} value stays the same, whatever concentration is chosen. But the E_{cell} value decreases if the concentration of the reduced species decreases. Cu^{2+} ions are the reduced species in the zinc and copper cell. So the E_{cell} value goes down if the aqueous copper(II) sulfate is diluted. The E_{cell} value also goes down if the temperature is reduced.

16.05 The electrochemical series

OUTCOMES

already from A2 Level, you

- know the IUPAC convention for writing half-equations for electrode reactions
- know and can use the conventional representation of cells
- understand how cells are used to measure electrode potentials by reference to the standard hydrogen electrode
- know the importance of the conditions when measuring the electrode potential, E
- know that standard electrode potential, $E^{\ominus}$, refers to conditions of 298 K, 100 kPa, and 1.00 mol dm^{-3} solution of ions

and after this spread you should

- know that standard electrode potentials can be listed as an electrochemical series

Standard electrode potentials can be listed in order to produce an **electrochemical series**. This can be in order of increasing negative potential, as seen here, or in order of increasing positive potential. You should be prepared to use data given in either order.

electrode reaction	E (V)
$F_2(g) + 2e^- \rightleftharpoons 2F^-(aq)$	+2.87
$MnO_4^-(aq) + 8H^+(aq) + 5e^- \rightleftharpoons Mn^{2+}(aq) + 4H_2O(l)$	+1.51
$Cl_2(aq) + 2e^- \rightleftharpoons 2Cl^-(aq)$	+1.36
$Cr_2O_7^{2-}(aq) + 14H^+(aq) + 6e^- \rightleftharpoons 2Cr^{3+}(aq) + 7H_2O(l)$	+1.33
$Br_2(aq) + 2e^- \rightleftharpoons 2Br^-(aq)$	+1.09
$Ag^+(aq) + e^- \rightleftharpoons Ag(s)$	+0.80
$Fe^{3+}(aq) + e^- \rightleftharpoons Fe^{2+}(aq)$	+0.77
$I_2(aq) + 2e^- \rightleftharpoons 2I^-(aq)$	+0.54
$Cu^+(aq) + e^- \rightleftharpoons Cu(s)$	+0.52
$Cu^{2+}(aq) + 2e^- \rightleftharpoons Cu(s)$	+0.34
$Sn^{4+}(aq) + 2e^- \rightleftharpoons Sn^{2+}(aq)$	+0.15
$Cu^{2+}(aq) + e^- \rightleftharpoons Cu^+(aq)$	+0.15
$2H^+(aq) + 2e^- \rightleftharpoons H_2(g)$	0.00
$Pb^{2+}(aq) + 2e^- \rightleftharpoons Pb(s)$	−0.13
$Sn^{2+}(aq) + 2e^- \rightleftharpoons Sn(s)$	−0.14
$Ni^{2+}(aq) + 2e^- \rightleftharpoons Ni(s)$	−0.25
$Fe^{2+}(aq) + 2e^- \rightleftharpoons Fe(s)$	−0.44
$Cr^{3+}(aq) + 3e^- \rightleftharpoons Cr(s)$	−0.74
$Zn^{2+}(aq) + 2e^- \rightleftharpoons Zn(s)$	−0.76
$Mn^{2+}(aq) + 2e^- \rightleftharpoons Mn(s)$	−1.19
$Al^{3+}(aq) + 3e^- \rightleftharpoons Al(s)$	−1.66
$Mg^{2+}(aq) + 2e^- \rightleftharpoons Mg(s)$	−2.37
$Na^+(aq) + e^- \rightleftharpoons Na(s)$	−2.71
$Ca^{2+}(aq) + 2e^- \rightleftharpoons Ca(s)$	−2.87
$K^+(aq) + e^- \rightleftharpoons K(s)$	−2.92
$Rb^+(aq) + e^- \rightleftharpoons Rb(s)$	−2.93
$Li^+(aq) + e^- \rightleftharpoons Li(s)$	−3.03

strongest oxidizing agent ↑ weakest oxidizing agent

weakest reducing agent ↓ strongest reducing agent

Part of the electrochemical series, listed in order of increasing negative potential. The electrode reactions are shown in the IUPAC convention of reduction reactions.

Half-cell diagrams

Some tables of the electrochemical series show the cell diagrams for the half-cells instead of the electrode reactions. For example:

$F_2(g) \mid F^-(aq) \mid Pt(s)$ ... instead of ... $\frac{1}{2}F_2(g) + e^- \rightleftharpoons F^-(aq)$

$Li^+(aq) \mid Li(s)$ ... instead of ... $Li^+(aq) + e^- \rightleftharpoons Li(s)$

Features of the electrochemical series

Similarity to the reactivity series

The halogens appear from the top of the series in the same order as they do in the periodic table. This is also the order of their decreasing reactivity.

The metals appear from the bottom in almost the same order as the reactivity series of the metals. The match is not exact because their reactivity depends on other factors in addition to their position in the electrochemical series.

Reducing ability

The strongest reducing agents have the most negative $E^{\ominus}$ values. A species on the right hand side (as seen in the series opposite) can donate electrons to any of the species above and to the left, and so reduce them.

Oxidizing ability

The strongest oxidizing agents have the most positive $E^{\ominus}$ values. A species on the left hand side (as seen in the series opposite) can gain electrons from any of the species below and to the right, and so oxidize them.

The reactivity series of the metals

The reactivity series lists metals in order of their reactivity, based on observations of their reactions with other metals and their compounds. It is useful for predicting the likely outcome of many reactions.

potassium — most reactive
sodium
lithium
calcium
magnesium
aluminium
zinc
iron
tin
lead
hydrogen
copper
silver
gold
platinum — least reactive

The reactivity series of metals, with hydrogen for comparison.

Check your understanding

1 What is the electrochemical series?

2 For each of the pairs of substances below, state which one is the more powerful reducing agent.

a Na or Fe

b Br^- or Cl^-

3 For each of the pairs of substances below, state which one is the more powerful oxidizing agent.

a $Cr_2O_7^{2-}$ or MnO_4^-

b Fe^{2+} or Fe^{3+}

16.06 Redox reactions – which way?

OUTCOMES

already from A2 Level, you

- know the IUPAC convention for writing half-equations for electrode reactions
- know and can use the conventional representation of cells
- understand how cells are used to measure electrode potentials by reference to the standard hydrogen electrode
- know that standard electrode potentials can be listed as an electrochemical series

and after this spread you should

- be able to use $E^\ominus$ values to calculate the e.m.f. of a cell and to predict the direction of simple redox reactions

Standard electrode potentials can be used to calculate the e.m.f. of a cell and to predict the direction of simple redox reactions.

Calculating e.m.f.

You can calculate the e.m.f. of a cell if you are given the cell diagram and the appropriate standard electrode potential data. The cell diagram shows you which half-cell is on the right, and which one is on the left. The standard electrode potentials give you the values you need. Remember how to calculate $E^\ominus_{cell}$, the e.m.f. of a cell:

$$E^\ominus_{cell} = E^\ominus_{right} - E^\ominus_{left}$$

Worked example

Calculate the e.m.f. of this cell: Cu(s) |Cu(aq) || Zn^{2+}(aq) | Zn(s)

$Zn^{2+}(aq) + 2e^- \rightarrow Zn(s)$ $\quad E^\ominus = -0.76$ V

$Cu^{2+}(aq) + 2e^- \rightarrow Cu(s)$ $\quad E^\ominus = +0.34$ V

$E^\ominus_{cell} = E^\ominus_{right} - E^\ominus_{left}$

$E^\ominus_{cell} = -0.76 - (+0.34) = -1.10$ V

You might have noticed that the electrochemical cell in the example is the same as the one discussed in Spread 16.04, except that it is the other way around. In the earlier spread it was:

$$Zn(s) \mid Zn^{2+}(aq) \parallel Cu^{2+}(aq) \mid Cu(s) \qquad E^\ominus_{cell} = +1.10 \text{ V}$$

The sign of $E^\ominus_{cell}$ is reversed if the cell is reversed. It is important that you know which way round the two half-cells are arranged.

Predicting the direction of spontaneous change

There are several different ways to work out the direction of spontaneous change in an electrochemical cell. But it is easiest to remember that the spontaneous change is the one that produces a positive $E^\ominus_{cell}$ value. For this to happen, the half-cell with the more positive electrode potential must be on the right. Here is an easy way to predict the direction of spontaneous change.

Step 1 Draw two vertical lines a few centimetres apart. These represent the electrodes.

Step 2 Label the left hand line as negative ⊖ and the right hand line as positive ⊕.

Step 3 Draw a spiral line from the bottom of the left hand electrode to the bottom of the right hand electrode. Add an arrow pointing to the right and label it e^-.

Step 4 Write the equation for the left hand electrode next to the left hand electrode, and the equation for the right hand electrode next to the right hand electrode.

Step 5 Add the two $E^\ominus$ values to the two electrodes.

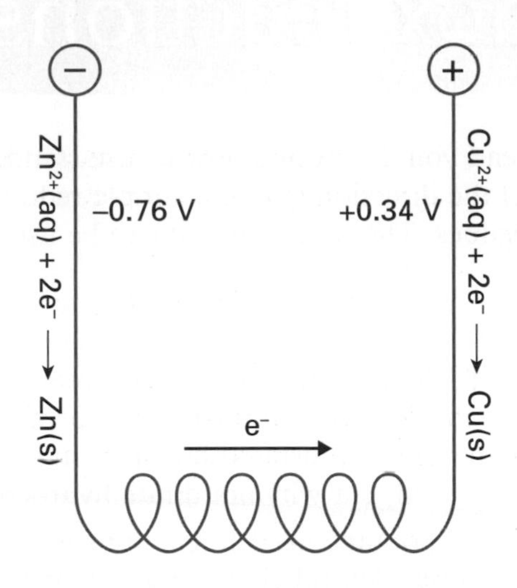

Drawing a diagram for the zinc and copper cell using the five steps.

Direction of electron flow

In the diagram, electrons flow from left to right, in the direction of the arrow. This makes it easy to work out where oxidation and reduction happen.

Oxidation and reduction

Oxidation is loss of electrons and reduction is gain of electrons. So

- oxidation happens on the left
- reduction happens on the right

The overall redox equation

By convention, electrode reactions are written as reduction reactions. This means that the reaction at the right hand electrode is already shown correctly. But the equation for the reaction at the left hand electrode must be reversed so that it shows an oxidation reaction involving electron loss. Once this is done, it is easy to write the overall redox equation for the spontaneous change. Taking the zinc and copper cell as an example:

Left hand electrode reaction: $Zn^{2+}(aq) + 2e^- \rightarrow Zn(s)$

This is reversed to give an oxidation reaction:

Left hand electrode reaction: $Zn(s) \rightarrow Zn^{2+}(aq) + 2e^-$

Right hand electrode reaction: $Cu^{2+}(aq) + 2e^- \rightarrow Cu(s)$

Add the two half-equations together, multiplying if necessary to get the same number of electrons in each one:

$$Zn(s) + Cu^{2+}(aq) + \cancel{2e^-} \rightarrow Zn^{2+}(aq) + \cancel{2e^-} + Cu(s)$$

$$Zn(s) + Cu^{2+}(aq) \rightarrow Zn^{2+}(aq) + Cu(s) \text{ ... the spontaneous cell reaction}$$

Calculating $E^\ominus_{cell}$

Use the standard electrode potentials written on the diagram to calculate $E^\ominus_{cell}$ in the usual way:

$$E^\ominus_{cell} = E^\ominus_{right} - E^\ominus_{left}$$

$$E^\ominus_{cell} = +0.34 - (-0.76) = +1.10 \text{ V}$$

How fast?

The value of $E^\ominus_{cell}$ for an electrochemical cell gives no indication of the likely rate of reaction. A large value does not necessarily mean that the rate will be high. Reactions with high activation energies may be feasible but they may not be spontaneous.

Check your understanding

1 For each of the following pairs of redox reactions, write the equation for the spontaneous redox reaction, identify the reduced species, and calculate $E^\ominus_{cell}$. Use data from the table on Spread 16.05 to help you answer these questions.

a Fe^{2+}/Fe and Pb^{2+}/Pb

b Cl_2/Cl^- and Br_2/Br^-

c MnO_4^-/Mn^{2+} and Fe^{3+}/Fe^{2+}

d Ag^+/Ag and Ni^{2+}/Ni

16.07 More redox reactions

OUTCOMES

already from A2 Level, you can

- use $E^{\ominus}$ values to calculate the e.m.f. of a cell and to predict the direction of simple redox reactions

and after this spread you should be able to

- use $E^{\ominus}$ values to predict the direction of redox reactions

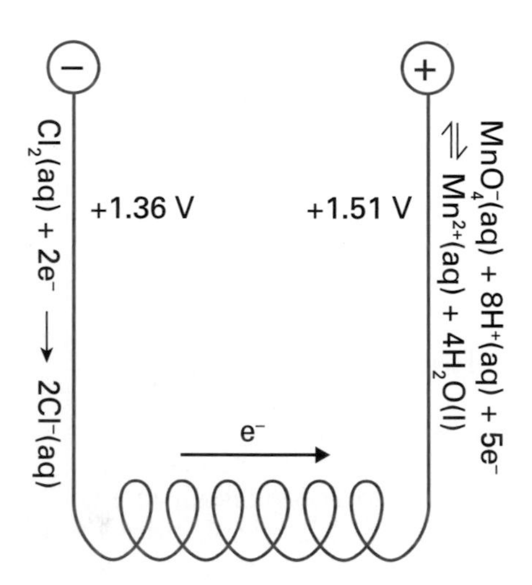

Working out the spontaneous cell reaction. Remember to put the most positive electrode reaction on the right.

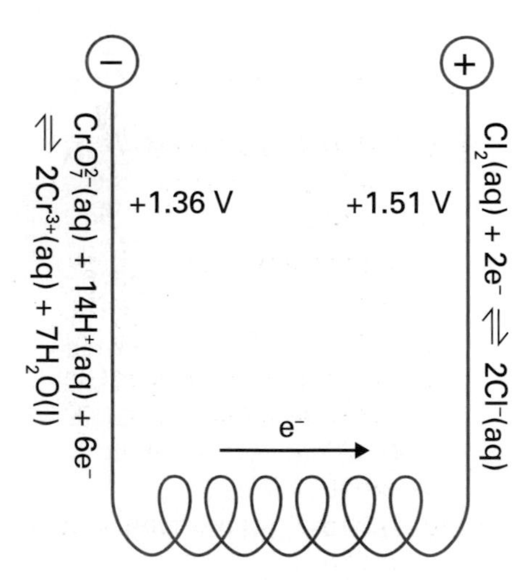

Working out the spontaneous cell reaction.

In the previous spread you discovered how to use standard electrode potentials to predict the direction of some simple redox reactions, and to calculate e.m.f. values. The same method can be used to answer more complex questions.

Which acid?

Potassium manganate(VII) and potassium dichromate(VI) are oxidizing agents used in redox titrations (see Spread 17.02 for more details). Hydrogen ions must be supplied using dilute acid in both cases. Dilute sulfuric acid is usually used, but why not dilute hydrochloric acid?

An examination of the spontaneous cell reactions explains why. Hydrochloric acid contains chloride ions, and one of the oxidizing agents mentioned can oxidize them to chlorine. This will increase the measured titre in a redox titration and give an invalid result.

Manganate(VII)

Here are the required standard electrode potentials:

$$MnO_4^-(aq) + 8H^+(aq) + 5e^- \rightleftharpoons Mn^{2+}(aq) + 4H_2O(l) \quad E^{\ominus} = +1.51\text{ V}$$

$$Cl_2(aq) + 2e^- \rightleftharpoons 2Cl^-(aq) \quad E^{\ominus} = +1.36\text{ V}$$

Construct a simple diagram as described in the previous spread.

Notice from the diagram that electrons flow from the Cl_2/Cl^- couple to the MnO_4^-/Mn^{2+} couple. Here is the overall redox equation for the spontaneous change (remembering to adjust both half-equations to get the same number of electrons in each one):

$$10Cl^-(aq) + 2MnO_4^-(aq) + 16H^+(aq) \rightleftharpoons 5Cl_2(aq) + 2Mn^{2+}(aq) + 8H_2O(l)$$

This means that manganate(VII) ions will oxidize the chloride ions in hydrochloric acid to chlorine. This will prevent the titration giving accurate results, so hydrochloric acid cannot be used.

Dichromate(VI)

As before, here are the required standard electrode potentials and the diagram:

$$Cr_2O_7^{2-}(aq) + 14H^+(aq) + 6e^- \rightleftharpoons 2Cr^{3+}(aq) + 7H_2O(l) \quad E^{\ominus} = +1.33\text{ V}$$

$$Cl_2(aq) + 2e^- \rightleftharpoons 2Cl^-(aq) \quad E^{\ominus} = +1.36\text{ V}$$

Notice from the diagram that electrons flow from the $Cr_2O_7^{2-}/Cr^{3+}$ couple to the Cl_2/Cl^- couple. This is in a different direction to the manganate(VII) example. Here is the overall redox equation for the spontaneous change:

$$6Cl_2(aq) + 2Cr^{3+}(aq) + 7H_2O(l) \rightleftharpoons 12Cl^-(aq) + Cr_2O_7^{2-}(aq) + 14H^+(aq)$$

This is the reverse of the reaction between chloride ions and dichromate(VI) ions. So dichromate(VI) ions cannot oxidize chloride ions, and hydrochloric acid can be used to acidify aqueous dichromate(VI) for redox titrations.

Predicting test tube reactions

The same method can be used to predict the outcome of other redox reactions. You can simplify the method to avoid drawing a diagram if you wish. The key is to identify the direction of electron flow in the spontaneous cell reaction. This is towards the more positive electrode reaction.

Worked example 1

Can aqueous Fe^{2+} ions reduce acidified aqueous VO_2^+ ions?

$Fe^{3+}(aq) + e^- \rightleftharpoons Fe^{2+}(aq)$ $E^\ominus = +0.77$ V

$VO_2^+(aq) + 2H^+(aq) + e^- \rightleftharpoons VO^{2+}(aq) + H_2O(l)$ $E^\ominus = +1.00$ V

The VO_2^+/VO^{2+} couple is more positive than the Fe^{3+}/Fe^{2+} couple, so electrons flow towards it. This means that VO_2^+ will be reduced to VO^{2+} (remember that reduction is the gain of electrons). So aqueous Fe^{2+} ions can reduce aqueous VO_2^+ ions.

Worked example 2

Can aqueous Fe^{2+} ions reduce acidified aqueous VO^{2+} ions?

$Fe^{3+}(aq) + e^- \rightleftharpoons Fe^{2+}(aq)$ $E^\ominus = +0.77$ V

$VO^{2+}(aq) + 2H^+(aq) + e^- \rightleftharpoons V^{3+}(aq) + H_2O(l)$ $E^\ominus = +0.34$ V

The Fe^{3+}/Fe^{2+} couple is more positive than the VO^{2+}/V^{3+} couple, so electrons flow towards it. This means that Fe^{3+} will be reduced to Fe^{2+}. This is the opposite of what is required for the reaction in the question, so aqueous Fe^{2+} ions cannot reduce aqueous VO^{2+} ions.

Worked example 3

Which metal ions will be left in solution if excess iron filings are added to acidified aqueous VO^{2+} ions?

$Fe^{2+}(aq) + 2e^- \rightleftharpoons Fe(s)$ $E^\ominus = -0.44$ V

$V^{3+}(aq) + e^- \rightleftharpoons V^{2+}(aq)$ $E^\ominus = -0.26$ V

$VO^{2+}(aq) + 2H^+(aq) + e^- \rightleftharpoons V^{3+}(aq) + H_2O(l)$ $E^\ominus = +0.34$ V

The VO^{2+}/V^{3+} couple is more positive than the Fe^{2+}/Fe couple, so electrons flow towards it. This means that the iron filings will be oxidized to Fe^{2+} ions and the VO^{2+} ions will be reduced to V^{3+} ions. But notice that the V^{3+}/V^{2+} couple is more positive than the Fe^{2+}/Fe couple, so electrons flow towards it. A further reaction will happen, reducing V^{3+} ions to V^{2+} ions. This means that Fe^{2+} ions and V^{2+} ions will be left in solution.

Check your understanding

1 Use data from the table on Spread 16.05 to answer these questions.

a Can magnesium reduce manganese(II) ions to manganese?

b What ions will remain in solution if excess nickel is added to aqueous Sn^{4+} ions?

c Is the disproportionation reaction:

$$2Cu^+(aq) \rightarrow Cu^{2+}(aq) + Cu(s)$$

possible?

Commercial cells

OUTCOMES

already from A2 Level, you can

- use $E^{\ominus}$ values to calculate the e.m.f. of a cell and to predict the direction of simple redox reactions

and after this spread you should

- appreciate that electrochemical cells can be used as a commercial source of electrical energy
- appreciate that cells can be non-rechargeable (irreversible), rechargeable, or fuel cells
- be able to use given electrode data to deduce the reactions occurring in non-rechargeable cells, and to deduce the cell's e.m.f.

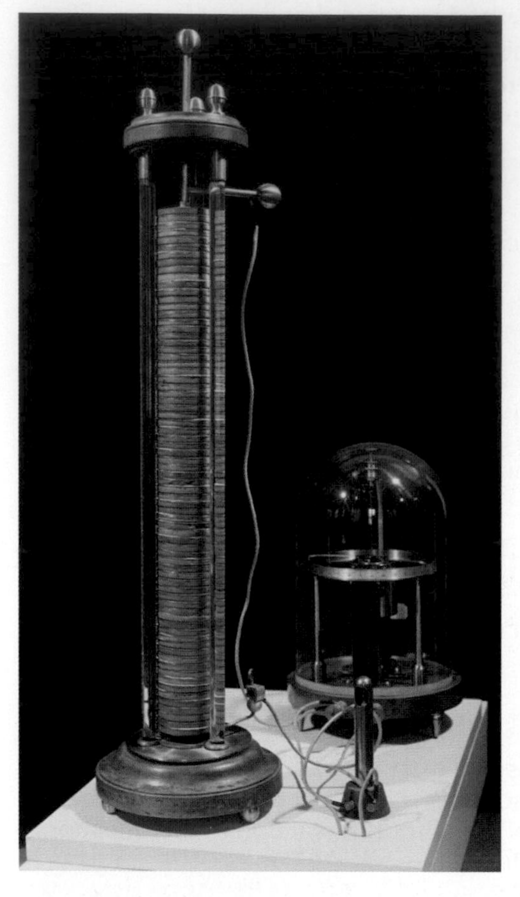

A voltaic pile. Copper disks were also used instead of silver disks.

In 1800 Alessandro Volta invented a primitive battery called a *voltaic pile*. This comprised several alternating layers of silver and zinc disks, separated by cardboard soaked in salt water. An electric current flowed when the two ends of the pile were connected by a wire. Volta's invention very quickly led to many new discoveries in chemistry, and to new designs for batteries.

A 1.5 V 'battery' and a 9 V battery.

Batteries and cells

Electrochemical cells are a convenient source of electrical energy. In everyday language the terms *battery* and *cell* tend to mean the same thing. But strictly-speaking a battery comprises two or more cells connected in series. The familiar cylindrical 1.5 V batteries are cells. The rectangular 9 V batteries really are batteries, as they contain six 1.5 V cells in series.

There are three main types of commercial electrochemical cell:

- **Primary cells** are not rechargeable and are thrown away after they run down.
- **Secondary cells** can be recharged after they run down.
- **Fuel cells** produce electricity from gaseous or liquid fuels.

Primary cells

Early designs of primary cells were *wet cells*. The Daniell cell was invented by John Daniell in 1836. It consists of a piece of zinc dipped into aqueous zinc sulfate in a porous pot. The pot itself is surrounded by aqueous copper sulfate in a copper can. The zinc acts as the **anode**. Electrons are lost and the zinc is oxidized:

$$Zn(s) \rightarrow Zn^{2+}(aq) + 2e^-$$

The copper acts as the **cathode**. Electrons are gained and copper(II) ions are reduced:

$$Cu^{2+}(aq) + 2e^- \rightarrow Cu(s)$$

The overall cell reaction is $Zn(s) + Cu^{2+}(aq) \rightarrow Zn^{2+}(aq) + Cu(s)$, with an E_{cell} of 1.1 V. This is a practical application of the electrochemical cell discussed in Spreads 16.04 and 16.05. But the Daniell cell is not very portable because of the liquid **electrolytes** it contains. Most modern primary cells are **dry cells** instead. Their electrolyte is present as a damp paste or gel, which does not spill if the cell is turned upside down.

Zinc–carbon cells

Zinc–carbon cells are cheap to make but easily run down under heavy use. They produce a potential difference of about 1.5 V, which gradually reduces to around 0.8 V with use. The anode is a zinc can, which contains a moist paste of ammonium chloride and zinc chloride. The cathode is a mixture of powdered manganese(IV) oxide and graphite surrounding a graphite rod.

At the anode, zinc is oxidized from an oxidation state of 0 to +2. Ammonia is produced when hydrogen ions are released from ammonium ions in the electrolyte. It reacts with the zinc ions:

$$Zn(s) + 2NH_3(aq) \rightarrow [Zn(NH_3)_2]^{2+}(aq) + 2e^-$$

At the cathode, manganese is reduced from an oxidation state of +4 to +3. The reaction involves hydrogen ions released from the ammonium ions:

$$2MnO_2(s) + 2H^+(aq) + 2e^- \rightarrow Mn_2O_3(s) + H_2O(l)$$

This is the overall cell reaction:

$$Zn(s) + 2MnO_2(s) + 2NH_4^+(aq) \rightarrow [Zn(NH_3)_2]^{2+}(aq) + Mn_2O_3(s) + H_2O(l)$$

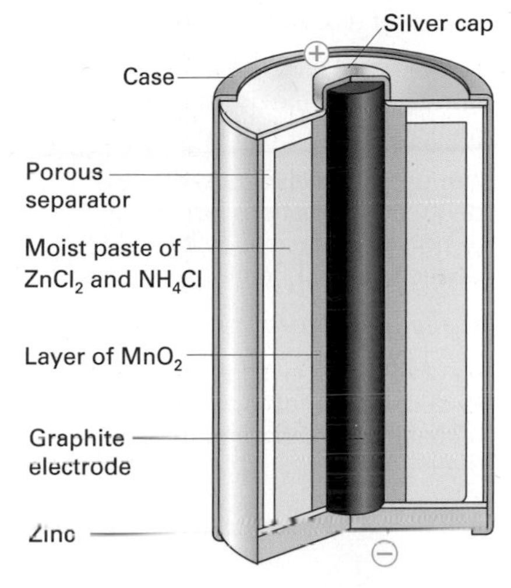

The construction of a typical zinc–carbon cell.

Alkaline dry cells

Alkaline cells produce the same potential difference as zinc–carbon cells but they last longer. They use potassium hydroxide as the electrolyte, rather than ammonium chloride or zinc chloride. The anode is powdered zinc alloy mixed with potassium hydroxide. The cathode is a mixture of potassium hydroxide, powdered manganese(IV) oxide, and graphite.

At the anode, zinc is oxidized from an oxidation state of 0 to +2. Hydroxide ions react with the zinc ions:

$$Zn(s) + 2OH^-(aq) \rightarrow ZnO(s) + H_2O(l) + 2e^-$$

At the cathode, manganese is reduced from an oxidation state of +4 to +3:

$$2MnO_2(s) + H_2O(l) + 2e^- \rightarrow Mn_2O_3(s) + 2OH^-(aq)$$

This is the overall cell reaction:

$$Zn(s) + 2MnO_2(s) \rightarrow ZnO(s) + 2Mn_2O_3(s)$$

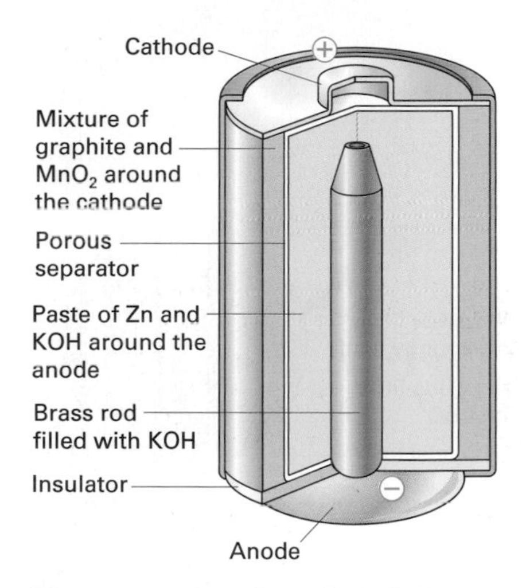

The construction of a typical alkaline dry cell.

Check your understanding

Use data from the table on Spread 16.05 to help you answer this question.

1 The small 'button' batteries used in watches are often zinc–silver oxide batteries. The anode is zinc and the cathode is silver oxide. The overall cell reaction can be written as:

$$Zn + 2Ag^+ \rightarrow Zn^{2+} + 2Ag.$$

a Write the half-equation for the reaction at
  i the anode **ii** the cathode

b Calculate $E^{\ominus}_{cell}$ for the zinc–silver oxide battery.

c Suggest why zinc–silver oxide batteries are usually small.

16.09 Rechargeable cells

OUTCOMES

already from A2 Level, you

- appreciate that electrochemical cells can be used as a commercial source of electrical energy
- appreciate that cells can be non-rechargeable (irreversible), rechargeable, or fuel cells
- can use given electrode data to deduce the reactions occurring in non-rechargeable cells, and to deduce the cell's e.m.f.

and after this spread you should

- be able to use given electrode data to deduce the reactions occurring in rechargeable cells, and to deduce the cell's e.m.f.

Rechargeable cells are secondary cells. They must be charged before use by connecting them to the electricity supply. The lead–acid battery, the type used in cars, is the oldest design.

The lead–acid battery

The lead-acid battery was invented in 1859 by Gaston Planté, a French scientist. A typical car battery comprises six cells in series, each producing 2 V, giving a total voltage of 12 V. The electrolyte is approximately 6 mol dm^{-3} sulfuric(VI) acid, H_2SO_4. When charged, the anode is spongy lead covering a grid made of a lead alloy, and the cathode is lead(IV) oxide in a similar grid. The grids provide a large surface area for electrode reactions to happen and improve the efficiency of the battery.

Anode reactions

This is the reaction that happens at the anode when the battery discharges:

$$Pb(s) + SO_4^{2-}(aq) \rightarrow PbSO_4(s) + 2e^-$$

The reaction is reversed during charging:

$$PbSO_4(s) + 2e^- \rightarrow Pb(s) + SO_4^{2-}(aq)$$

Cathode reactions

This is the reaction that happens at the cathode when the battery discharges:

$$PbO_2(s) + 4H^+(aq) + SO_4^{2-}(aq) + 2e^- \rightarrow PbSO_4(s) + 2H_2O(l)$$

The reaction is reversed during charging:

$$PbSO_4(s) + 2H_2O(l) \rightarrow PbO_2(s) + 4H^+(aq) + SO_4^{2-}(aq) + 2e^-$$

Notice that when the battery is discharged, both the anode and the cathode consist of lead(II) sulfate.

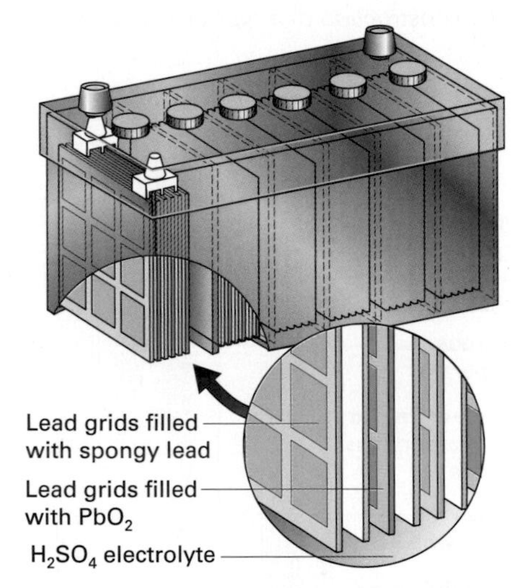

The construction of a lead–acid battery.

Nicads

The most common rechargeable cell in everyday use is the **nickel–cadmium cell**, often just called a **nicad**. Nicads produce a potential difference of 1.2 V, which is slightly less than the potential difference of zinc–carbon cells and alkaline dry cells. The anode is made from cadmium and the cathode from nickel(III) hydroxide. Potassium hydroxide is the electrolyte.

Anode reactions

This is the reaction that happens at the anode when the battery discharges:

$$Cd(s) + 2OH^-(aq) \rightarrow Cd(OH)_2(s) + 2e^-$$

The reaction is reversed during charging:

$$Cd(OH)_2(s) + 2e^- \rightarrow Cd(s) + 2OH^-(aq)$$

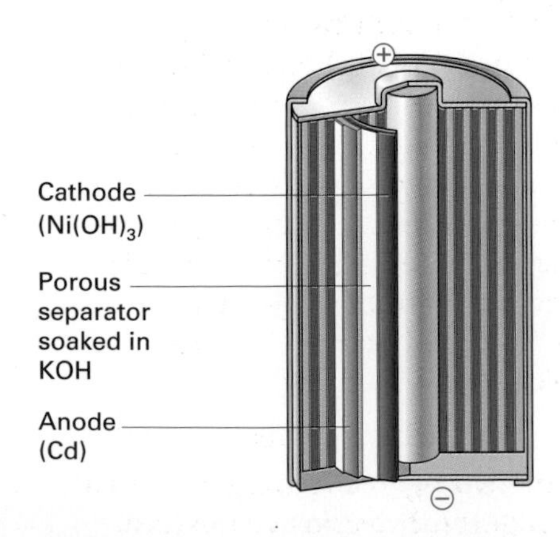

The construction of a nicad battery.

Cathode reactions

This is the reaction that happens at the cathode when the battery discharges:

$$Ni(OH)_3(s) + e^- \rightarrow Ni(OH)_2(s) + OH^-(aq)$$

The reaction is reversed during charging:

$$Ni(OH)_2(s) + OH^-(aq) \rightarrow Ni(OH)_3(s) + e^-$$

Lithium-ion batteries

Lithium-ion batteries are used in portable devices such as mobile phones and laptop computers. They are more complex than the other secondary cells described here, and include a computer chip to control charging and discharging. They produce a much higher potential difference, too, typically 3.7 V. The anode is made of graphite and the cathode is made of lithium cobalt oxide, $LiCoO_2$.

Anode reactions

Lithium ions are attracted to the graphite anode during charging, forming *lithiated graphite*. This is an example of the reaction that happens at the anode when the battery discharges:

$$LiC_6(s) \rightarrow Li^+(aq) + 6C(s) + e^-$$

The reaction is reversed during charging:

$$Li^+(aq) + 6C(s) + e^- \rightarrow LiC_6(s)$$

Lithium-ion battery packs like this one can power mobile phones for long periods of time between charges.

Cathode reactions

This is an example of the reaction that happens at the cathode when the battery discharges:

$$LiCoO_2(s) + Li^+(aq) + e^- \rightarrow Li_2CoO_2(s)$$

The reaction is reversed during charging:

$$Li_2CoO_2(s) \rightarrow LiCoO_2(s) + Li^+(aq) + e^-$$

Check your understanding

1 Write the overall cell reaction for the following situations:
  a discharging a lead–acid battery
  b charging a nickel–cadmium cell

2 These are the typical electrode potentials for the electrodes in a discharging lead–acid battery:
  Anode: $E = -0.35$ V
  Cathode: $E = +1.69$ V
  a Use these values to show why the potential difference of a single cell is about 2 V.
  b Suggest why the potential difference of a car battery is about 12 V.
  c Suggest why car batteries are usually recharged using about 14 V.

3 A reaction at the cathode in a lithium-ion cell can be written as:

$$Li_2CoO_2(s) \underset{\text{discharging}}{\overset{\text{charging}}{\rightleftharpoons}} LiCoO_2(s) + Li^+(aq) + e^-$$

Write a similar equation to show a reaction at the anode.

Explosive charging

Modern car batteries do not usually need recharging. This happens while the car is being driven. A type of dynamo, the alternator, generates electricity as the car moves. If a battery does need recharging separately, it is important not to overcharge it. When this happens, water in the electrolyte is electrolyzed:

$$2H_2O(l) \rightarrow 2H_2(g) + O_2(g)$$

The hydrogen and oxygen form a dangerously explosive mixture.

16.10 Fuel cells

OUTCOMES

already from A2 Level, you appreciate

- that cells can be non-rechargeable (irreversible), rechargeable, and fuel cells.

and after this spread you should

- understand the electrode reactions of a hydrogen–oxygen fuel cell and appreciate that a fuel cell does not need to be electrically recharged
- appreciate the benefits and risks to society associated with the use of electrochemical cells

A small demonstration fuel cell. The current is increased by increasing the surface area, and the voltage is increased by stacking several individual cells together.

In 1800, shortly after Volta's invention of the battery, the English chemist William Nicholson decomposed water by electrolysis. These are the reactions that happen at the two platinum electrodes:

At the positive electrode: $2H_2O(l) \rightarrow O_2(g) + 4H^+(aq) + 4e^-$

At the negative electrode: $4H^+(aq) + 4e^- \rightarrow 2H_2(g)$

The overall reaction is: $2H_2O(l) \rightarrow 2H_2(g) + O_2(g)$

Chemists reasoned that if water could be split into hydrogen and oxygen using electricity, it should be possible to produce electricity from the reaction of hydrogen with oxygen. The Welsh scientist William Grove invented a device in 1839 that did just that. He passed hydrogen and oxygen gases over platinum electrodes partly submerged in dilute sulfuric acid. An electric current flowed when the two electrodes were connected. It continued to flow as long as the two gases were present and the electrodes were damp with acid. Grove's *gaseous voltaic battery* was the first fuel cell.

The hydrogen–oxygen fuel cell

Fuel cells transform the chemical energy in a fuel such as hydrogen or methanol directly into electrical energy. The fuel is oxidized by oxygen from the air using electrochemical reactions in the fuel cell.

A typical hydrogen–oxygen fuel cell comprises two flat electrodes, each coated on one side by a thin layer of platinum catalyst. A *proton exchange membrane* is sandwiched between the two electrodes. Hydrogen gas flows to the anode and air flows to the cathode. Any unreacted hydrogen is recirculated. Water vapour, the reaction product, is pushed out by the stream of air.

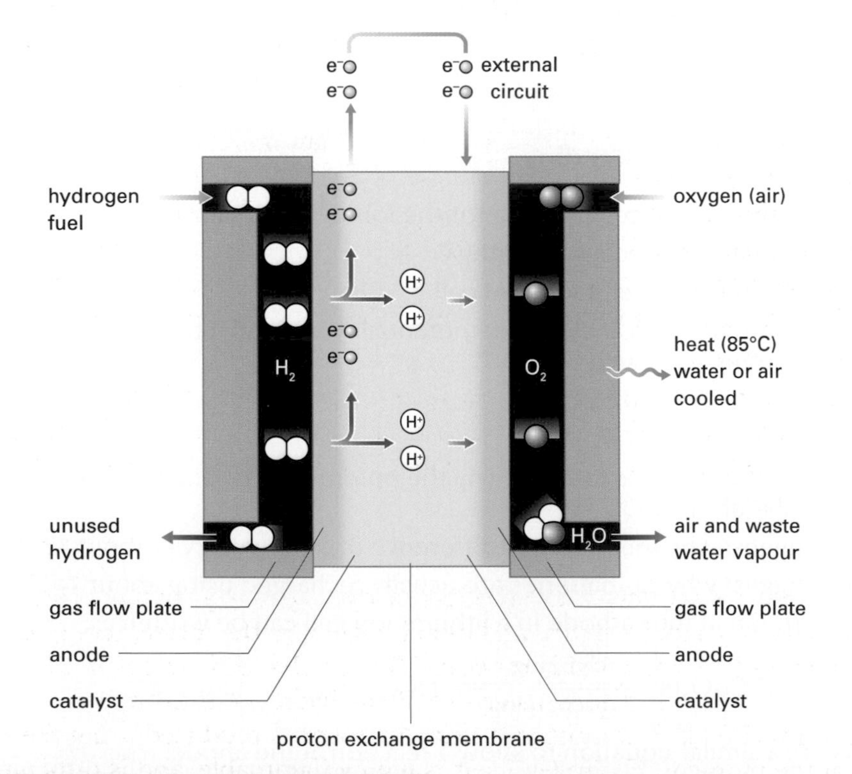

A cross-section through a typical fuel cell.

Anode reaction

The oxidation of hydrogen is catalysed by the layer of platinum:

$$H_2(g) \rightarrow 2H^+(aq) + 2e^-$$

The electrons released by this reaction flow through the external circuit. The hydrogen ions pass through the proton exchange membrane to the cathode.

This car is powered by a hydrogen fuel cell.

Cathode reaction

Oxygen is reduced to water vapour. It reacts with the hydrogen ions that pass through the proton exchange membrane and electrons from the external circuit:

$$4H^+(aq) + 4e^- + O_2(g) \rightarrow 2H_2O(g)$$

Overall reaction

The overall reaction is: $2H_2(g) + O_2(g) \rightarrow 2H_2O(g)$

The efficiency of a fuel cell is typically around 40%, depending on the design and the fuel used.

Benefits and risks of electrochemical cells

Commercial electrochemical cells are very convenient portable sources of electricity. They reduce the need for expensive cabling and bring electricity supplies to remote places. Spacecraft use hydrogen fuel cells to provide electricity and drinking water. But there are drawbacks, too.

Non-rechargeable cells

Non-rechargeable cells are cheap. They are manufactured in many different sizes to suit different applications. Tiny button batteries power watches while larger batteries power torches and other portable devices. But they gradually run down with use and then cannot be recharged. They are usually thrown away, which wastes the energy and resources needed to manufacture them.

Rechargeable cells

The use of rechargeable cells reduces the total number of batteries thrown away each year. Car batteries are vital to the normal running of a car. Without them we should have to start the engine by turning a starting handle! Rechargeable cells also improve the performance of solar cells. These convert sunlight directly into electricity but they do not work at night. Rechargeable cells store electricity generated during the day and release it at night.

The cadmium in nickel–cadmium cells is toxic. These cells should be recycled rather than disposed of in a landfill site. The lead and lead(II) sulfate in vehicle batteries is also toxic, but it is recycled to make new batteries.

Fuel cells

Water vapour is the only waste product of hydrogen–oxygen fuel cells. These cells may replace petrol and diesel engines in the future, reducing the amount of carbon dioxide produced by vehicles. This will help to reduce the release of greenhouse gases, if the hydrogen is produced without fossil fuels. But most industrial hydrogen is produced using these fuels at the moment. Hydrogen itself is highly flammable, and is difficult to store and handle safely.

Check your understanding

1 a Write the half-equations for the electrode reactions that happen in a hydrogen fuel cell.
  b Explain how hydrogen ions reach the cathode in a fuel cell.
  c What happens to the unreacted hydrogen and waste water vapour from a fuel cell?

2 Outline one benefit and one risk to society for each of the following commercial electrochemical cells:
  a non-rechargeable primary cells
  b rechargeable secondary cells
  c hydrogen fuel cells

Questions: Chs 13–16

1 a Define the term enthalpy of formation. [3]

b The diagram below shows the Born–Haber cycle for the formation of rubidium bromide from its elements.

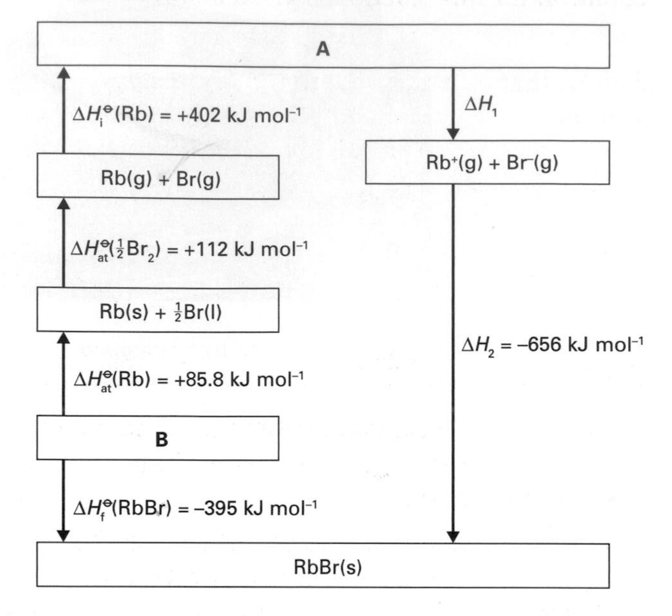

i Two species are missing from this cycle. State the formulas of the species that should be in Box A and Box B. [2]

ii Give the names of the enthalpy changes represented by ΔH_1 and ΔH_2. [2]

iii Calculate the value of the enthalpy change represented by ΔH_1. [2]

c Study the diagram below carefully. Use the diagram, and the following information, to answer question **c** parts **i** and **ii**.

Standard enthalpy change of solution is $-22.1\,kJ\,mol^{-1}$.

Lattice dissociation enthalpy is $-656\,kJ\,mol^{-1}$.

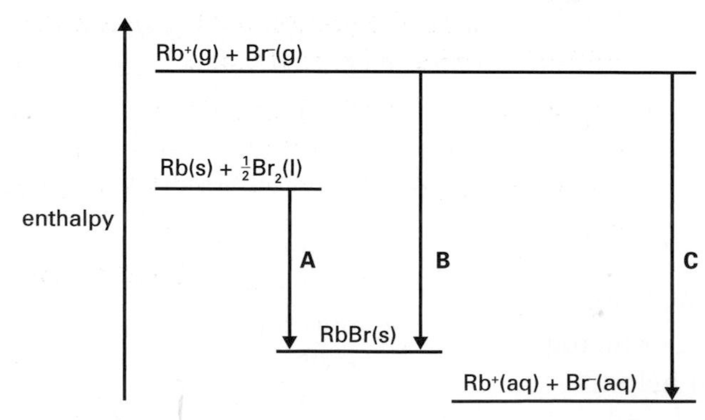

i Give the name of each of the changes **A** and **B**. [2]

ii State the name(s) of the enthalpy change(s) labelled **C**. Calculate the value of this enthalpy change. [3]

[Total 14 marks]

2 a Define the term *enthalpy of combustion*. [2]

b The enthalpies of combustion of butane and but-1-ene are $-2877\,kJ\,mol^{-1}$ and $-2717\,kJ\,mol^{-1}$ respectively.

Use this information to work out the value of the enthalpy change for the oxidation of 1 mol of butane to but-1-ene and steam:

$C_4H_{10}(g) + \frac{1}{2}O_2(g) \rightarrow C_4H_8(g) + H_2O(g)$. [3]

c Define the term *mean bond enthalpy*. [2]

d The table lists some mean bond enthalpy values.

bond	mean bond enthalpy ($kJ\,mol^{-1}$)
C–C	348
C=C	612
C–H	413
O–H	463

State the number and type of bonds made and broken in the oxidation of butane to but-1-ene and steam. Use the mean bond enthalpies in the table above, together with your answer to part **b**, to calculate the bond enthalpy of the O=O bond in the oxygen molecule. [4]

[Total 11 marks]

3 Use the values below to answer the questions.

electrode	electrode reaction	$E^{\ominus}$ (V)
A	$V^{2+}(aq) + 2e^- \rightleftharpoons V(s)$	−1.18
B	$Fe^{2+}(aq) + 2e^- \rightleftharpoons Fe(s)$	−0.44
C	$Ni^{2+}(aq) + 2e^- \rightleftharpoons Ni(s)$	−0.25
D	$Pb^{2+}(aq) + 2e^- \rightleftharpoons Pb(s)$	−0.13
E	$2H^{2+}(aq) + 2e^- \rightleftharpoons H_2(s)$	?

a i Give the name of electrode **E** and state its role in determining standard electrode potentials. [2]

ii What is the value of the standard electrode potential for electrode **E**? [1]

b The electrochemical cell set up between electrodes **C** and **D** can be represented by the cell diagram:

$Ni(s) \mid Ni^{2+}(aq) \parallel Pb^{2+}(aq) \mid Pb(s)$

i Calculate the e.m.f. of this cell. [1]

ii State which electrode would be positive. [1]

iii Write an equation to show the overall reaction in the cell. [1]

c Use the standard electrode potential data given in the table above to:

i Explain whether or not you would expect a reaction to occur if a piece of lead were to be added to a test tube containing aqueous vanadium(II) ions. [2]

ii Predict and explain two observations you would expect to make if a small piece of vanadium were to be added to a test tube containing 1.00 mol dm^{-3} hydrochloric acid. [4]

[Total 12 marks]

4 a Standard electrode potentials are measured relative to a reference electrode. Give the name of this reference electrode and state the value of its standard electrode potential. State the conditions to which the term *standard* refers. [5]

b The standard electrode potentials for two electrode reactions are given below:

$S_2O_8^{2-}(aq) + 2e^- \rightarrow 2SO_4^{2-}(aq) \qquad E^\ominus = +2.01\,V$

$I_2(aq) + 2e^- \rightarrow 2I^-(aq) \qquad E^\ominus = +0.54\,V$

i A cell is produced when these two half-cells are connected. Calculate the cell potential, $E^\ominus_{cell}$, for this cell, and write an equation for the spontaneous reaction. [3]

ii State how, if at all, the electrode potential of the $S_2O_8^{2-}/SO_4^{2-}$ equilibrium would change if the concentration of SO_4^{2-} ions were increased. Explain your answer. [3]

[Total 11 marks]

5 The metal tungsten, W, is found in a number of minerals including scheelite, $CaWO_4$. The extraction of tungsten from its ore has many stages but the ore is eventually converted into tungsten(VI) oxide, WO_3, and then heated with hydrogen to produce powdered tungsten. The equation for this reaction is given below:

$WO_3(s) + 3H_2(g) \rightarrow W(s) + 3H_2O(g)$

The table gives some data for this reaction:

	$\Delta H_f^\ominus$ (kJ mol^{-1})	$S^\ominus$ (J K^{-1} mol^{-1})
WO_3	−843	+75.9
H_2	0	+65.3
W	0	+32.6
H_2O	−242	+188.7

a Determine the standard enthalpy, the standard entropy, and standard free energy changes at 298 K for this reaction. [7]

b i The reaction is said to be not feasible. In terms of free energy change, explain the meaning of the term *feasible*. [1]

ii Calculate the temperature at which the reaction between tungsten(VI) oxide and hydrogen becomes feasible. [2]

[Total 10 marks]

6 Consider the following oxides:

$Na_2O \quad MgO \quad Al_2O_3 \quad SiO_2 \quad P_4O_{10} \quad SO_2 \quad SO_3$

a Indentify one of the oxides from the list above which can form:

i A solution with a pH less than 3.

ii A solution with a pH greater than 12. [2]

b i Write a balanced equation for the reaction of sodium oxide with cold water. [1]

ii Write a balanced equation for the reaction of sulfur(IV) oxide with cold water. [1]

iii Suggest why silicon(IV) oxide does not react with cold water. You should refer to the bonding and structure of silicon(IV) oxide in your answer. [2]

[Total 6 marks]

17.01 Transition metals and the d sub-level

OUTCOMES

already from AS Level, you

- know the electron configurations of atoms and ions up to $Z = 36$ in terms of levels and sub-levels (orbitals) s, p, and d
- can classify an element as being in the s, p, or d block according to its position in the periodic table

and after this spread you should know

- that transition metal characteristics of elements Ti–Cu arise from an incomplete d sub-level in atoms or ions
- that these characteristics include complex formation, formation of coloured ions, variable oxidation state, and catalytic activity

The d block in the periodic table contains transition metals. These are elements with an incomplete d sub-level that can form at least one stable ion with an incomplete d sub-level. According to this definition, the only elements in the first row of the d block that are transition metals are titanium to copper. Scandium has an incomplete d sub-level but it forms Sc^{3+} ions which have no electrons in the d sub-level. Zinc has a complete d sub-level and it forms Zn^{2+} ions with a complete d sub-level.

Scandium

IUPAC defines a transition metal as an element whose atom has an incomplete d sub-level, *or* which can give rise to cations with an incomplete d sub-level. According to this definition scandium *is* a transition metal, as it has an incomplete d sub-level: [Ar] $3d^1\ 4s^2$.

The elements titanium to copper

The properties of the elements titanium to copper are typical of transition metals. They include

- catalytic activity
- variable oxidation states
- the ability to form complexes
- the ability to form coloured ions

Transition metals also tend to be stronger, denser, and less reactive than the metals in groups 1 and 2. They are often used as catalysts. For example, vanadium(V) oxide is used in the contact process to manufacture sulfuric acid, and iron is used in the Haber Process to manufacture ammonia.

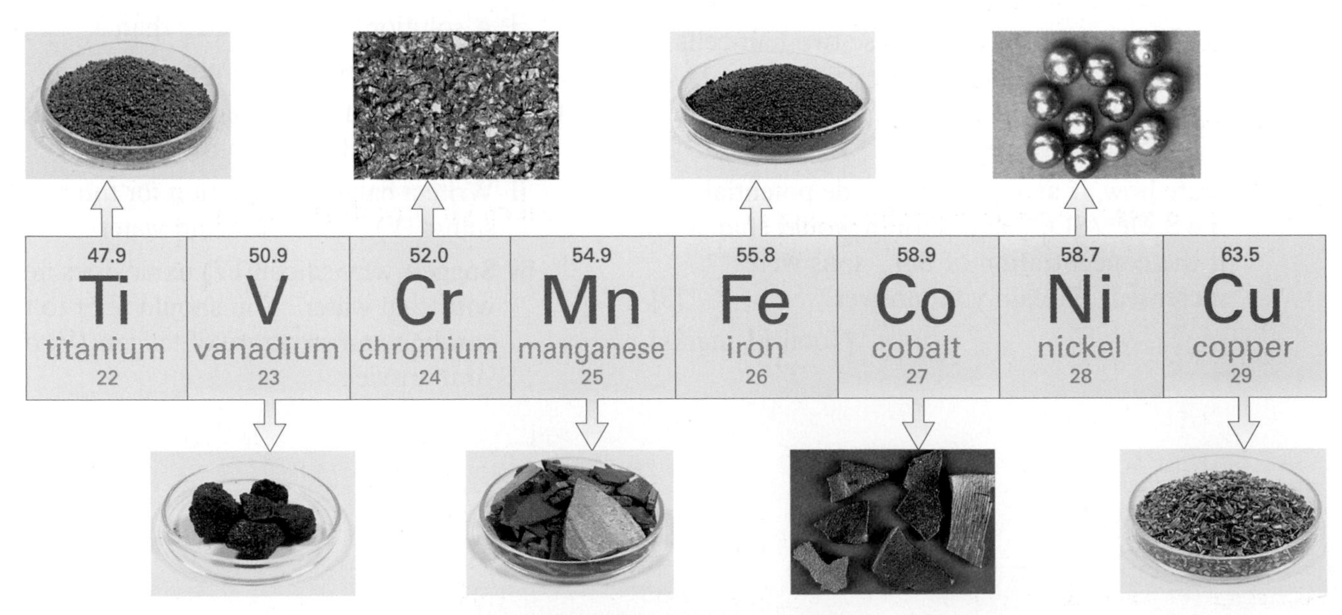

The transition elements titanium to copper

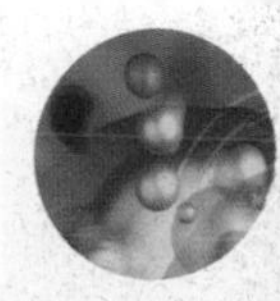

Electron configurations

The table summarizes the **electron configurations** of the elements titanium to copper.

element	symbol	electron configuration
titanium	Ti	[Ar] $3d^2$ $4s^2$
vanadium	V	[Ar] $3d^3$ $4s^2$
chromium	Cr	[Ar] $3d^5$ $4s^1$
manganese	Mn	[Ar] $3d^5$ $4s^2$
iron	Fe	[Ar] $3d^6$ $4s^2$
cobalt	Co	[Ar] $3d^7$ $4s^2$
nickel	Ni	[Ar] $3d^8$ $4s^2$
copper	Cu	[Ar] $3d^{10}$ $4s^1$

The common electron configuration $1s^2$ $2s^2$ $2p^6$ $3s^2$ $3p^6$ is shown as an argon core, [Ar]. This makes it easier to focus on the 3d and 4s sub-levels.

The 3d sub-level fills going from titanium to copper. It fills smoothly overall, but there are two places where this does not happen.

Chromium

You might expect the electron configuration of chromium to be [Ar] $3d^4$ $4s^2$. Instead it is [Ar] $3d^5$ $4s^1$. One of the 4s electrons is promoted to the 3d sub-level. As a result of this, the 3d and 4s sub-levels are both half-filled, and neither contain paired electrons.

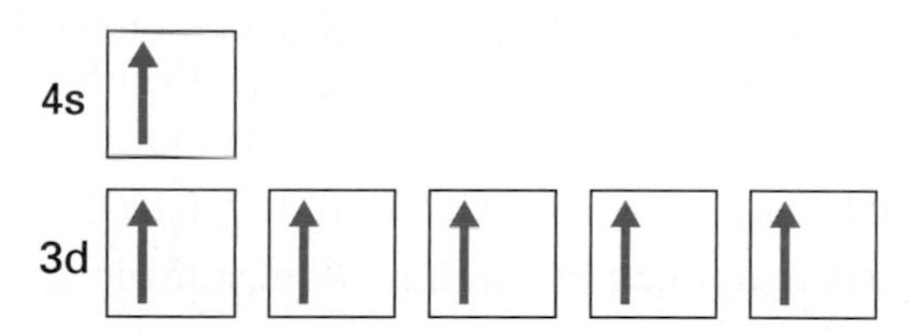

Electron spin diagrams for the 3d and 4s orbitals in chromium.

Copper

You might expect the electron configuration of copper to be [Ar] $3d^9$ $4s^2$. Instead it is [Ar] $3d^{10}$ $4s^1$. As in chromium, one of the 4s electrons is promoted to the 3d sub-level. As a result of this, the 3d sub-level is complete while the 4s sub-level is half-filled.

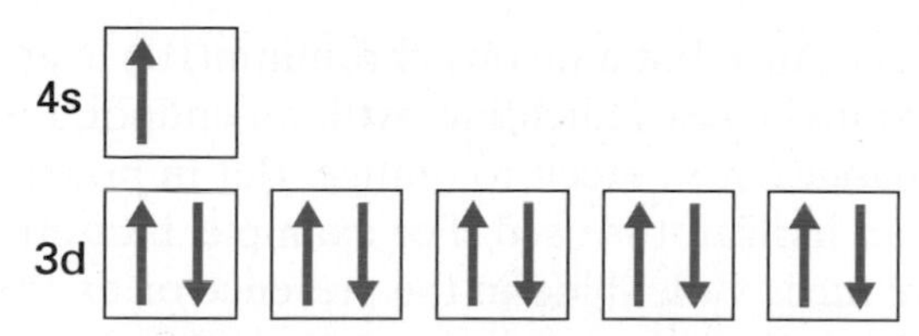

Electron spin diagrams for the 3d and 4s orbitals in copper.

Ion formation

When they form ions, the transition metals in period 4 lose electrons from their 4s sub-level first, then from their 3d sub-level. For example, the Ti^{2+} ion forms when two 4s electrons are lost. The electron configuration of Ti^{2+} is [Ar] $3d^2$ not [Ar] $4s^2$. The ions of transition metals are usually coloured, and they often exist as **complexes**. You can find out more about complexes in Chapter 18.

$$[H_3N: \longrightarrow Ag^+ \longleftarrow :NH_3]^+$$

This diagram represents a complex ion $[Ag(NH_3)_2]^+$ in which a silver ion forms co-ordinate bonds with ammonia molecules.

Variable oxidation state

Transition metals have variable oxidation states because the d and s sub-levels are at similar energy levels.

Ti	V	Cr	Mn	Fe	Co	Ni	Cu
			+7				
		+6	+6	+6			
	+5						
+4	+4						
+3	+3	+3	+3	+3	+3	+3	
+2	+2	+2	+2	+2	+2	+2	+2
							+1

The oxidation states of the elements titanium to copper. The most important ones are shown in red.

Check your understanding

1 State four common properties of the transition metals.

2 Write electron configurations for the following atoms and ions:

a V
b Cr
c Mn^{2+}
d Fe^{3+}
e Cu^+
f Cu^{2+}

17.02 Redox titrations

OUTCOMES

already from AS Level, you can

- calculate the concentrations and volumes for reactions in solutions, limited to titrations of monoprotic acids and bases, and examples for which the equations are given

already from A2 Level, you can

- perform calculations for the titrations of monoprotic and diprotic acids with sodium hydroxide, based on experimental results
- apply the electron transfer model of redox, including oxidation states and half-equations, to d block elements

and after this spread you should know

- the redox titration of Fe^{2+} with MnO_4^- and $Cr_2O_7^{2-}$ in acid solution

Which acid?

Sulfuric acid is suitable for both manganate(VII) titrations and dichromate(VI) titrations. But hydrochloric acid is only suitable for dichromate(VI) titrations (see Spread 16.07). Its chloride ions are oxidized to chlorine by manganate(VII). Weak acids such as ethanoic acid do not provide a sufficiently high concentration of hydrogen ions.

A dichromate(VI) titration. The orange colour in the burette and the green colour in the flask can be seen. Note that an indicator has not been added, and that the tip of the burette is further inside the flask than normal.

You will be familiar with acid–base titrations from your AS Chemistry studies and Unit 4. **Redox titrations** are similar in principle but rely on redox reactions rather than neutralization reactions. The manganate(VII) ion and the dichromate(VI) ion are common oxidizing agents used in redox titrations.

Manganate(VII) in titrations

Acidified aqueous potassium manganate(VII), $KMnO_4$, can act as an oxidizing agent in redox titrations. Here is the half-equation for the reduction of the manganate(VII) ion to the manganese(II) ion:

$$MnO_4^-(aq) + 8H^+(aq) + 5e^- \rightarrow Mn^{2+}(aq) + 4H_2O(l)$$

The hydrogen ions are usually supplied by adding dilute sulfuric acid.

Manganate(VII) is often used to find the concentration of aqueous iron(II) ions. It oxidizes iron(II) to iron(III):

$$Fe^{2+}(aq) \rightarrow Fe^{3+}(aq) + e^-$$

The two half-equations can be combined, after multiplying the iron half-equation by five so that the electrons will cancel out:

$$MnO_4^-(aq) + 8H^+(aq) + 5Fe^{2+}(aq) \rightarrow Mn^{2+}(aq) + 4H_2O(l) + 5Fe^{3+}(aq)$$

In a typical titration, aqueous potassium manganate(VII) is added from a burette to aqueous iron(II) ions in a conical flask. The reaction is *self-indicating*, so no indicator is needed. Aqueous manganate(VII) is dark purple but aqueous manganese(II) is very pale pink. It is essentially colourless in the concentrations used for titrations. Manganate(VII) loses its colour as it enters the iron(II) solution. The endpoint is when a permanent pink tinge caused by excess manganate(VII) first appears.

Dichromate(VI) in titrations

Acidified aqueous potassium dichromate(VI), $K_2Cr_2O_7$, can also act as an oxidizing agent in redox titrations. Here is the half-equation for the reduction of the dichromate(VI) ion to the chromium(III) ion:

$$Cr_2O_7^{2-}(aq) + 14H^+(aq) + 6e^- \rightarrow 2Cr^{3+}(aq) + 7H_2O(l)$$

The hydrogen ions are usually supplied by adding dilute sulfuric acid or dilute hydrochloric acid.

Dichromate(VII) is also used to find the concentration of aqueous iron(II) ions:

$$Cr_2O_7^{2-}(aq) + 14H^+(aq) + 6Fe^{2+}(aq) \rightarrow$$
$$2Cr^{3+}(aq) + 7H_2O(l) + 6Fe^{3+}(aq)$$

Notice that, in manganate(VII) titrations, the mole ratio for calculations is 5 mol Fe^{2+} to 1 mol MnO_4^-. But in dichromate(VI) titrations it is 6 mol Fe^{2+} to 1 mol $Cr_2O_7^{2-}$.

Aqueous dichromate(VI) is orange but aqueous chromium(III) is green. In principle the reaction could be self-indicating, with an endpoint when the mixture in the flask changes from green to orange. But in practice this is difficult to see and an indicator is used. For example, barium diphenylamine-4-sulfonate turns violet-blue in the presence of excess potassium dichromate(VI) solution.

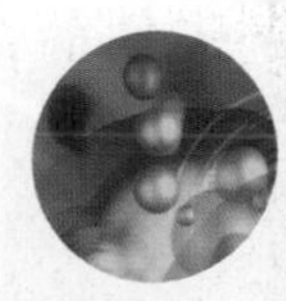

Thiosulfate in titrations – a reducing agent

Manganate(VII) and dichromate(VI) are oxidizing agents but the thiosulfate ion, $S_2O_3^{2-}$, is a reducing agent. Here is the half-equation for the reduction of the thiosulfate ion to the tetrathionate ion:

$$2S_2O_3^{2-}(aq) \rightarrow S_4O_6^{2-}(aq) + 2e^-$$

Thiosulfate is often used to find the concentration of aqueous iodine. It reduces iodine to iodide ions:

$$I_2(aq) + 2e^- \rightarrow 2I^-(aq)$$

The two half-equations are combined to give the overall redox equation:

$$2S_2O_3^{2-}(aq) + I_2(aq) \rightarrow S_4O_6^{2-}(aq) + 2I^-$$

Notice that the mole ratio for calculations is 0.5 mol I_2 to 1 mol $S_2O_3^{2-}$. In a typical titration, aqueous sodium thiosulfate, $Na_2S_2O_3$, is added from a burette to aqueous iodine in a conical flask. Starch suspension is added as an indicator near the endpoint. It forms a deep purple colour with iodine. The endpoint is when this purple colour first disappears.

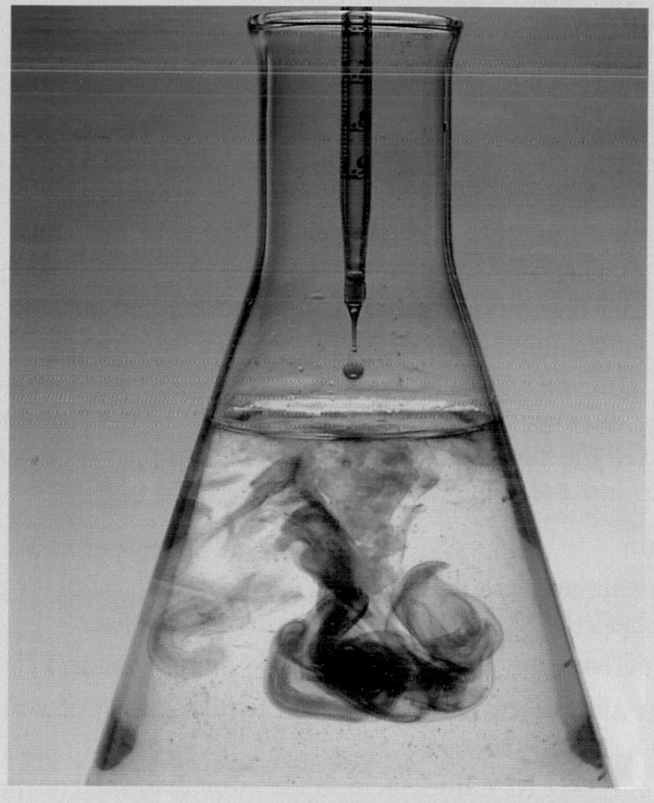

Starch forms a purple complex with iodine. Thiosulfate ions reduce the iodine in this complex to colourless iodide ions.

Check your understanding

1. For each of the following redox titrations, write the overall redox equation, state the mole ratio used in calculations, and describe the endpoint.
   - **a** manganate(VII) and iron(II)
   - **b** dichromate(VI) and iron(II)
   - **c** thiosulfate and iodine
2. Explain why the use of dilute sulfuric acid to provide hydrogen ions in redox titrations may be preferable to the use of hydrochloric acid.

17.03 Analysing iron

OUTCOMES

already from A2 Level, you

- can perform calculations for the titrations of monoprotic and diprotic acids with sodium hydroxide, based on experimental results
- know the redox titration of Fe^{2+} with MnO_4^- and $Cr_2O_7^{2-}$ in acid solution

and after this spread you should be able to

- carry out a redox titration
- perform calculations for the titration of Fe^{2+} with MnO_4^- or with $Cr_2O_7^{2-}$

A pestle and mortar. Remember you 'pound with the pestle' and 'mix in the mortar'.

The purple colour of potassium manganate(VII) disappears in excess aqueous iron(II) sulfate. It is important to swirl to mix the contents of the flask thoroughly.

One of the Investigative and Practical Skills tasks for A2 Inorganic Chemistry is to carry out a redox titration. A suitable experiment is the analysis of iron tablets by titration using acidified potassium manganate(VII). Iron tablets may be taken as a dietary supplement. A typical tablet contains several milligrams of iron, usually as iron(II) sulfate. The mass of iron in the tablet can be determined by titration against manganate(VII) or dichromate(VI).

An outline method

To carry out such an investigation successfully, you will need to use these techniques skilfully:

- weighing
- transferring known volumes of liquid
- redox titration

Iron(II) ions oxidize in air to iron(III) ions unless dilute sulfuric acid is present. This oxidation happens more quickly when the solution is warm, so it is best to keep the temperature below 50°C if any heating is needed to dissolve the iron tablet.

Preparing the sample

1. Weigh five iron tablets and record their total mass.
2. Crush them using a pestle and mortar, and transfer them to a beaker. Use 1 M sulfuric acid to rinse the pestle and mortar. Add the washings to the beaker and make up to about 100 cm^3 with the acid.
3. Stir to dissolve the crushed tablets. Filter the mixture using a clean, dry Buchner funnel and flask. Wash the beaker and add the washings to the Buchner funnel. Rinse the funnel with some de-ionized water.
4. Pour the contents of the Buchner flask into a 250 cm^3 standard flask. Wash the Buchner flask and add the washings to the standard flask. Make up to the mark with de-ionized water, stopper the flask, and swirl it to mix the contents.

The titration

1. Rinse the burette with de-ionized water, then with 0.001 M potassium manganate(VII). Make sure the tip is free of bubbles.
2. Fill the burette with the aqueous potassium manganate(VII).
3. Use a transfer pipette and pipette filler to add 25 cm^3 of your sample solution to a conical flask.
4. Add 25 cm^3 of 1 M sulfuric acid to the conical flask. This provides an excess of hydrogen ions, so a measuring cylinder may be used instead of a transfer pipette.
5. Add aqueous potassium manganate(VII) from the burette until the first appearance of a permanent pink tinge. Record the burette reading.
6. Repeat steps **2** to **5** until you get at least two concordant results. These are usually within 0.10 cm^3 of each other.

Titration precautions

Take care that you

- set the burette vertically with the tip just inside the conical flask
- remove the funnel before each run
- swirl the conical flask during the run
- add the titrant dropwise near the endpoint
- wash the inside of the flask with de-ionized water just before the endpoint
- record the initial and final readings to two decimal places, ending in 0 or 5
- read from the top of the meniscus if the bottom is difficult to see because aqueous potassium manganate(VII) is purple

The endpoint is the first appearance of a permanent pink tinge caused by excess potassium manganate(VII).

The calculation

Ignore your first rough run and tick the concordant titres from the accurate runs. Use these to calculate the mean titre.

Amount of manganate(VII)

The amount of manganate(VII) is calculated from the concentration of aqueous potassium manganate(VII) and the mean titre. For example, if 0.001 M potassium manganate(VII) is used and the mean titre is 23.50 cm^3:

$$\text{amount of } MnO_4^- = \frac{23.50}{1000} \times 0.001 = 2.35 \times 10^{-5} \text{ mol}$$

Amount of iron

The overall redox equation for the titration between manganate(VII) and iron(II) is:

$$MnO_4^-(aq) + 8H^+(aq) + 5Fe^{2+}(aq) \rightarrow Mn^{2+}(aq) + 4H_2O(l) + 5Fe^{3+}(aq)$$

So 1 mol of MnO_4^- ions react with 5 mol of Fe^{2+} ions. Using the amount of MnO_4^- ions from above:

$$\text{amount of } Fe^{2+} = 5 \times 2.35 \times 10^{-5} = 1.175 \times 10^{-4} \text{ mol}$$

Note that if you used acidified potassium dichromate instead, the amount of Fe^{2+} would be six times the amount of $Cr_2O_7^{2-}$, not five times.

Remember that you titrated 25 cm^3 from a 250 cm^3 sample:

total amount of $Fe^{2+} = 10 \times 1.175 \times 10^{-4} = 1.175 \times 10^{-3}$ mol

Mass of iron in each tablet

Five tablets were used to make the sample:

amount of iron in one tablet $= 1.175 \times 10^{-3} \div 5 = 2.35 \times 10^{-4}$ mol

The relative atomic mass of iron is 55.8:

mass of iron in one tablet $= A_r \times \text{mole} = 55.8 \times 2.35 \times 10^{-4} =$ **0.0131 g**

	Run 1	Run 2	Run 3
Final volume (cm^3)	23.85	47.15	23.55
Initial volume (cm^3)	0.15	23.70	0.00
Titre (cm^3)	23.70	23.45 ✓	23.55 ✓
Mean titre (cm^3)	23.50		

Record your results in a table like this one.

Science @ Work

Iron in iron tablets

The RNI for a nutrient is its Reference Nutrient Intake. This is the amount of the nutrient which is enough for at least 97% of the population. The RNI for iron varies, depending on age and gender.

men	RNI (mg)	women	RNI (mg)
11–18 years	11.3	11–49 years	14.8
19 and older	8.7	50 and older	8.7

A typical iron tablet may contain 15 mg of iron (0.015 g).

17.04 Heterogeneous catalysts

OUTCOMES

already from AS Level, you

- know the meaning of the term *catalyst*
- understand that catalysts work by providing an alternative reaction route of lower activation energy

already from A2 Level, you

- know that transition metal characteristics include variable oxidation state and catalytic activity

and after this spread you should

- know that transition metals and their compounds can act as heterogeneous catalysts
- know that a heterogeneous catalyst is in a different phase from the reactants and that the reaction occurs at the surface
- understand how V_2O_5 acts as a catalyst in the contact process
- understand the importance of variable oxidation states in heterogeneous catalysis

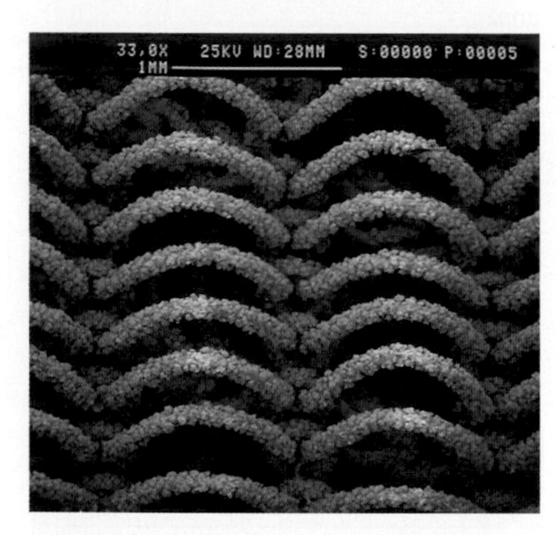

Ammonia is converted into nitric oxide in several stages in the Ostwald Process. Gauze made from platinum and rhodium is used in the first stage. The gauze has a large surface area and allows the reacting gases to flow through it.

A **catalyst** is a substance that speeds up chemical reactions by providing an alternative reaction route with a lower activation energy. Catalysts are important in industrial chemistry, where they improve the efficiency of manufacturing processes. Transition metals and their compounds can act as catalysts. For example, iron is used in the manufacture of ammonia by the Haber Process, and vanadium(V) oxide is used in the manufacture of sulfuric acid by the contact process.

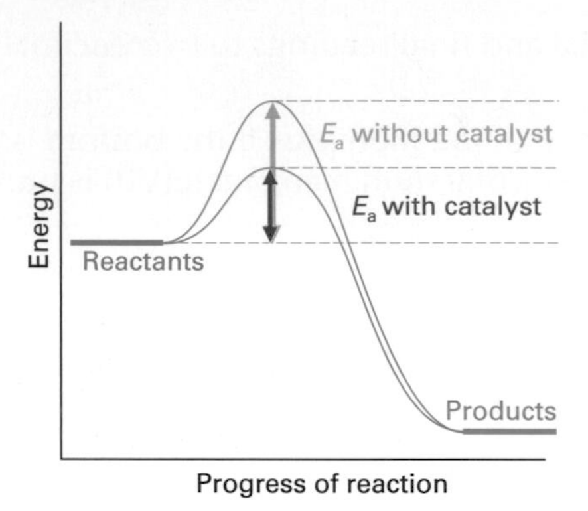

The energy level diagrams for a reaction with and without a catalyst.

Common features of catalysts

Catalysts have some common features. For example, they can be very specific and may only catalyse a certain reaction or a class of similar reactions.

Equilibrium position

Catalysts increase the rates of the forward reaction and the reverse reaction by the same ratio. So they do not alter the position of the equilibrium or the value of the equilibrium constant K_c. Instead, they reduce the time needed to reach equilibrium.

Size and surface area

Catalysts are often only needed in small amounts. For example, 2 g of platinum can catalyse the decomposition of 1000 m^3 of hydrogen peroxide. Solid catalysts are more effective when they are present as a thin layer or as a powder. This is because their surface area to volume ratio is then very large.

Heterogeneous catalysts

Heterogeneous catalysts exist in a different phase or state to the reactants in the reaction. They are the type of catalysts you met in your GCSE studies and in AS Chemistry Unit 2. The reaction happens at the surface of the catalyst in heterogeneous catalysis. The places where the reaction happens are called **active sites**. There are three main processes involved:

- **adsorption** of the reactants to the active sites
- the reaction itself
- **desorption** of the products from the active sites

For example, in the Haber Process, nitrogen and hydrogen react together in the presence of an iron catalyst to produce ammonia. The nitrogen and hydrogen molecules are adsorbed onto the surface of the iron. They react together and ammonia then desorbs from the surface.

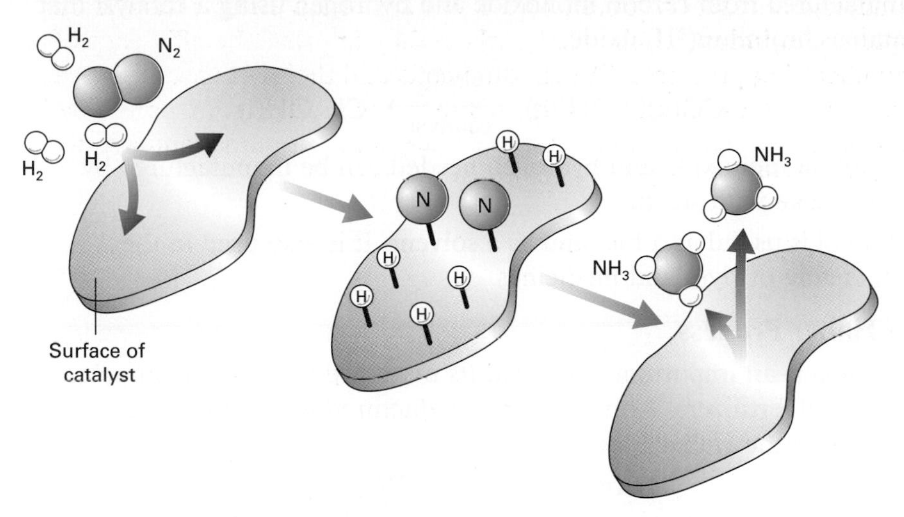

The three stages in the iron-catalysed reaction between nitrogen and hydrogen to produce ammonia.

Importance of variable oxidation states

Variable oxidation states are important in heterogeneous catalysis. The use of vanadium(V) oxide in the contact process is a good example of how this works. In the second stage of sulfuric acid manufacture, sulfur(IV) oxide is oxidized to sulfur(VI) oxide. The reaction mechanism happens in two steps:

Step 1 The vanadium(V) oxide catalyst oxidizes sulfur(IV) oxide to sulfur(VI) oxide. The oxidation state of vanadium decreases from +5 to +4:

$$SO_2(g) + \overset{+5}{V}_2O_5(s) \rightarrow SO_3(g) + \overset{+4}{V}_2O_4(s)$$

Step 2 The vanadium(IV) oxide reacts with oxygen, and vanadium(V) oxide is regenerated:

$$\overset{+4}{V}_2O_4(s) + \tfrac{1}{2}O_2(g) \rightarrow \overset{+5}{V}_2O_5(s)$$

The two equations may be added to together to give the overall equation:

$$SO_2(g) + \tfrac{1}{2}O_2(g) \xrightleftharpoons{V_2O_5 \text{ catalyst}} SO_3(g)$$

More about adsorption

Adsorption of a reactant onto the surface of a catalyst by van der Waals' forces is called *physisorption*. Adsorption because of the formation of covalent bonds is called *chemisorption*. Chemisorption to transition metal catalysts often involves the adsorbed reactants donating electrons to vacant d sub-levels.

The bonds between reactant and catalyst must be strong enough to adsorb the reactant effectively, but weak enough to release the product easily. Silver forms bonds that are too weak, and tungsten forms bonds that are too strong.

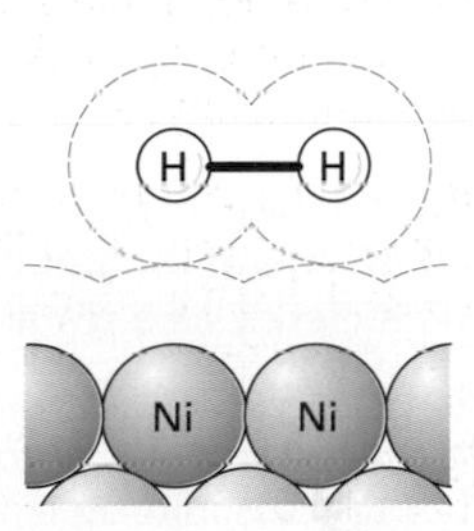

The physisorption of hydrogen onto a nickel surface. Nickel is used as a catalyst in the hydrogenation of vegetable oils for margarine manufacture.

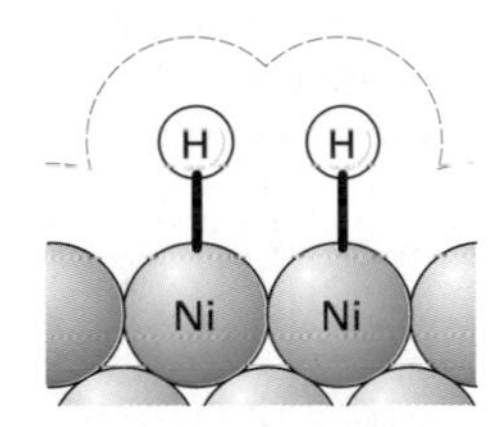

The chemisorption of hydrogen onto a nickel surface.

Check your understanding

1. What is a *heterogeneous catalyst*?
2. With reference to the Haber Process, describe the three main stages involved in heterogeneous catalysis.
3. With reference to the role of vanadium(V) oxide in the synthesis of sulfur(VI) oxide, explain the importance of variable oxidation states in heterogeneous catalysis.

17.05 Industrial catalysts

OUTCOMES

already from A2 Level, you know

- that transition metals and their compounds can act as heterogeneous catalysts
- that a heterogeneous catalyst is in a different phase from the reactants and that the reaction occurs at the surface

and after this spread you should

- know that a chromium(III) oxide catalyst is used in the manufacture of methanol from carbon monoxide and hydrogen
- know that iron is used as a catalyst in the Haber Process
- know that catalysts can become poisoned by impurities and have a reduced efficiency as a result
- know that catalyst poisoning has a cost implication, e.g. by sulfur in the Haber Process and by lead in catalytic converters
- understand the use of a support medium to maximize the surface area and minimize the cost, e.g. rhodium on a ceramic support in catalytic converters

Catalysts are important in industrial processes. For example, methanol is manufactured from carbon monoxide and hydrogen using a catalyst that contains chromium(III) oxide:

$$CO(g) + 2H_2(g) \xrightarrow[\text{catalyst}]{Cr_2O_3} CH_3OH(l)$$

The carbon monoxide and hydrogen needed can be manufactured by reacting coke with steam.

Methanol is useful as a fuel and as a solvent. It is also used in the manufacture of dyes and medicines.

The Haber Process

Ammonia is an important chemical. Its main use is in the manufacture of artificial fertilizers. Ammonia is manufactured from nitrogen and hydrogen in the Haber Process:

$$N_2(g) + 3H_2(g) \rightleftharpoons 2NH_3(g)$$

Iron is the catalyst in the process. The reaction proceeds in several steps:

- Nitrogen and hydrogen molecules adsorb onto the surface of the iron catalyst.
- N≡N and H—H bonds break, and N—H bonds form between individual atoms.
- Ammonia molecules desorb from the surface.

Various types of catalyst beads are used in the Haber process. Each one is designed to provide a large surface area.

The hydrogen needed can be manufactured by the reaction of methane with steam:

$$CH_4(g) + 2H_2O(g) \rightarrow CO_2(g) + 4H_2(g)$$

Natural gas is the source of the methane needed. It often contains hydrogen sulfide, H_2S, when it is first extracted from the ground. This is a problem because sulfur will **poison** the iron catalyst. The sulfur adsorbs strongly to the active sites, which blocks the adsorption of nitrogen and hydrogen. So the natural gas must be treated before use to remove the hydrogen sulfide.

Science @ Work

Sour gas

Natural gas that contains hydrogen sulfide is called *sour gas* because it smells of rotten eggs. The hydrogen sulfide also makes the gas corrosive, so it must be removed at the processing plant. It is then reduced to sulfur by the *Claus Process*, and converted into useful products such as sulfuric acid.

Catalyst supports

The iron used in the Haber Process is relatively cheap, but many other heterogeneous catalysts are expensive. For example, platinum and rhodium are used in catalytic converters. They are more expensive than gold, so they must be used in an efficient and cost-effective way. This is achieved by coating a thin layer of the catalyst on an inexpensive solid support, such as porous ceramic beads or a ceramic honeycomb. The support has a microscopically rough surface, producing a huge surface area using only a few grams of catalyst. In this way, the effectiveness of the catalyst is maximized and its cost is minimized.

In a catalytic converter, the platinum and rhodium catalyst is coated onto a ceramic support.

Leaded petrol and catalysts

Petrol with a low *octane rating* can ignite incorrectly in the cylinders of an engine. This is called knocking or pinking, and it produces a characteristic metallic sound. The shock waves from knocking can damage the engine, so petrol with the correct octane rating must be used. *Anti-knocking agents* are added to petrol to increase its octane rating. In the past, a substance called tetraethyl lead was used.

Lead poisons the platinum and rhodium used in catalytic converters. Cars with catalytic converters must not use leaded petrol, otherwise the catalyst will be poisoned and harmful gases will leave the exhaust pipe. Leaded petrol was banned in the UK in 2000.

Lead is not just a problem for cars. Scientific studies in the twentieth century showed that children living near motorways had lower IQs than children living in areas with less lead pollution. It was concluded that the lead from car exhausts could reduce the intelligence of children, which was another good reason for supplying unleaded petrol only.

People living near motorways and other busy roads need not worry about lead from car exhausts affecting their health today.

Check your understanding

1. Write a balanced equation for the manufacture of methanol from carbon monoxide and hydrogen, and give the formula of the catalyst used.
2. With reference to the Haber Process, what is a meant by the term *catalyst poisoning*?
3. With reference to catalytic converters, explain how the surface area of the catalyst is maximized and its cost is minimized.

17.06 Homogeneous catalysts

OUTCOMES

already from A2 Level, you

- know that transition metals and their compounds can act as heterogeneous catalysts
- know that a heterogeneous catalyst is in a different phase from the reactants and that the reaction occurs at the surface
- understand the importance of variable oxidation states in heterogeneous catalysis

and after this spread you should

- know that transition metals and their compounds can act as homogeneous catalysts
- know that a homogeneous catalyst is in the same phase as the reactants
- understand the importance of variable oxidation states in homogeneous catalysis
- know that in homogeneous catalysis the reaction proceeds through an intermediate species, e.g. the reaction between I^- and $S_2O_8^{2-}$ catalysed by Fe^{2+}, and autocatalysis by Mn^{2+} in reactions of $C_2O_4^{2-}$ with MnO_4^-

Homogeneous catalysts are in the same phase or state as the reactants. Just like heterogeneous catalysis, homogeneous catalysis involves variable oxidation states. Hydrogen peroxide decomposes slowly to form water and oxygen. Manganese(IV) oxide is a heterogeneous catalyst for the reaction, and bromine is a homogeneous catalyst for the reaction. There are two steps in the reaction mechanism when bromine is used:

Step 1 $\mathbf{H_2O_2(aq)} + Br_2(aq) \rightarrow 2H^+(aq) + 2Br^-(aq) + \mathbf{O_2(g)}$

Step 2 $\mathbf{H_2O_2(aq)} + 2H^+(aq) + 2Br^-(aq) \rightarrow \mathbf{2H_2O(l)} + Br_2(aq)$

Overall: $2H_2O_2(aq) \xrightarrow[\text{catalyst}]{Br_2(aq)} 2H_2O(l) + O_2(g)$

Bromine acts as a catalyst because, although it is involved in the reaction mechanism, it is unchanged at the end of the two steps. It is reduced from bromine to bromide ions in step 1, then oxidized from bromide ions to bromine again in step 2.

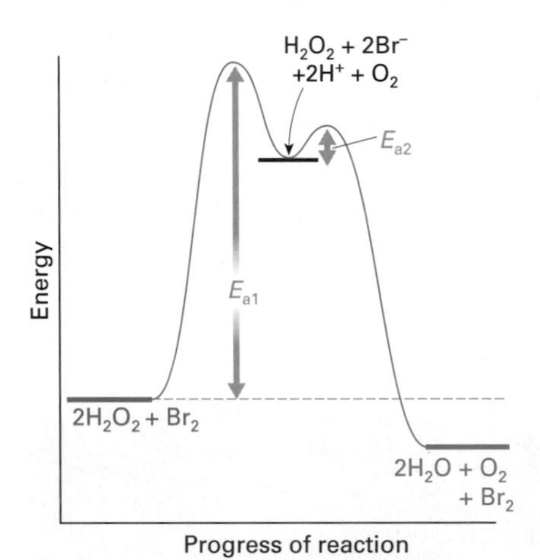

The reaction profile for the decomposition of hydrogen peroxide catalysed by bromine. There are two separate activation energies, E_{a1} and E_{a2}, representing the two steps.

Iodide ions and peroxodisulfate ions

Aqueous peroxodisulfate ions can oxidize aqueous iodide ions to iodine:

$$S_2O_8^{2-}(aq) + 2I^-(aq) \rightarrow 2SO_4^{2-}(aq) + I_2(aq)$$

The reaction is energetically favourable but slow because both reactants are negatively charged. They tend to repel each other rather than colliding. Aqueous iron(II) ions catalyze the reaction. They act as an intermediate in the transfer of electrons from iodide ions to peroxodisulfate ions:

Step 1 $S_2O_8^{2-}(aq) + 2Fe^{2+}(aq) \rightarrow 2SO_4^{2-}(aq) + 2Fe^{3+}(aq)$

Step 2 $2Fe^{3+}(aq) + 2I^-(aq) \rightarrow 2Fe^{2+}(aq) + I_2(aq)$

Iron(II) ions act as a catalyst because, although they are involved in the reaction mechanism, they are unchanged at the end of the two steps. Iron(II) is oxidized to iron(III) in step 1, then reduced to iron(II) again in step 2.

Check your understanding

1. With reference to the reaction between peroxodisulfate ions and iodide ions, explain the importance of variable oxidation states in homogeneous catalysis.
2. Draw a diagram, similar to the one above but without the $E^\ominus$ values, to show how manganese(II) ions act as an intermediate in the reaction between manganate(VII) ions and ethanedioate ions.

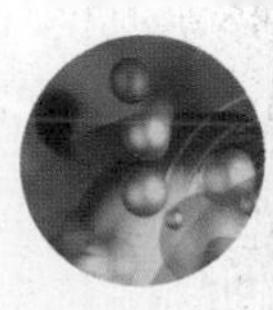

Redox potentials and the iodide/peroxodisulfate reaction

Both steps in the iron(II) catalyzed reaction between peroxodisulfate ions and iodide ions have positive $E^\ominus_{cell}$ values. The table shows the relevant standard electrode potentials.

half-equation	$E^\ominus$ (V)
$S_2O_8^{2-}(aq) + 2e^- \rightleftharpoons 2SO_4^{2-}(aq)$	+2.01
$Fe^{3+}(aq) + e^- \rightleftharpoons Fe^{2+}(aq)$	+0.77
$I_2(aq) + 2e^- \rightleftharpoons 2I^-(aq)$	+0.54

Step 1 Oxidation of iron(II) by peroxodisulfate

The two half-equations are:

$$2Fe^{2+}(aq) \rightarrow 2Fe^{3+}(aq) + 2e^- \text{ and}$$
$$S_2O_8^{2-}(aq) + 2e^- \rightarrow 2SO_4^{2-}(aq)$$

Reduction happens on the right:

$$E^\ominus_{cell} = E^\ominus_R - E^\ominus_L = +2.01 - (+0.77) = +1.24\text{ V}$$

So the spontaneous cell reaction is the oxidation of iron(II) ions by peroxodisulfate ions.

Step 2 Reduction of iron(III) by iodide

The two half-equations are:

$$2I^-(aq) \rightarrow 2I_2(aq) + 2e^- \text{ and}$$
$$2Fe^{2+}(aq) \rightarrow 2Fe^{3+}(aq) + 2e^-$$

Reduction happens on the right:

$$E^\ominus_{cell} = E^\ominus_R - E^\ominus_L = +0.77 - (+0.54) = +0.23\text{ V}$$

So the spontaneous cell reaction is the reduction of iron(III) ions by iodide ions.

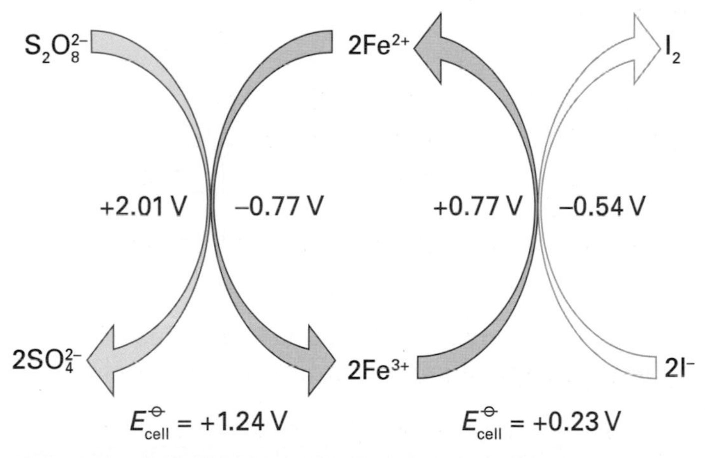

The reaction between peroxodisulfate ions and iodide ions can also be catalysed by Fe^{3+} ions.

Manganate(VII) ions and ethanedioate ions

An interesting thing happens when aqueous manganate(VII) is titrated with acidified aqueous ethanedioate. The purple colour decolorizes slowly at first, but then decolorizes immediately as more manganate(VII) is added until the endpoint. It happens because manganese(II) is formed in the reaction, which then catalyzes the reaction. This is called **autocatalysis**.

Here is the equation for the overall reaction:

$$2MnO_4^-(aq) + 8H^+(aq) + 5C_2O_4^{2-}(aq) \rightarrow$$
$$2Mn^{2+}(aq) + 10CO_2(g) + 4H_2O(l)$$

The reaction is energetically favourable but slow because both reactants are negatively charged. But once manganese(II) ions have formed, they can act as an intermediate:

Step 1 $MnO_4^-(aq) + 8H^+(aq) + 4Mn^{2+}(aq) \rightarrow 5Mn^{3+}(aq) + 4H_2O(l)$

Step 2 $5C_2O_4^{2-}(aq) + 5Mn^{3+}(aq) \rightarrow 10CO_2(g) + 5Mn^{2+}(aq)$

Manganese(II) is oxidized to manganese(III) in step 1, then reduced to manganese(II) again in step 2.

Aqueous manganese(II) ions autocatalyse the reaction between aqueous manganate(VII) ions and aqueous ethanedioate ions.

18.01 Ligands

OUTCOMES

already from A2 Level, you know

- that transition metals form complexes
- that aldehydes are readily oxidized to carboxylic acids, and that this forms the basis of the simple chemical test using Tollens' reagent to distinguish between aldehydes and ketones

and after this spread you should

- understand that a complex is a central metal ion surrounded by ligands
- be able to define the term *ligand*
- know that co-ordinate bonding is involved in complex formation
- understand the importance of lone pairs of electrons in co-ordinate bond formation
- know the meaning of co-ordination number
- know that Ag^+ forms the linear complex $[Ag(NH_3)_2]^+$ used in Tollens' reagent to distinguish between aldehydes and ketones

$\ddot{N}H_2CH_2CH_2\ddot{N}H_2$

1,2-diaminoethane is a bidentate ligand.

A **complex** comprises a central metal ion surrounded by **ligands**. A ligand is an ion or molecule that can donate a pair of electrons to the central metal ion, forming a co-ordinate bond. Remember that a co-ordinate bond is a covalent bond in which one atom donates both electrons.

Ligands

Ligands can be negative ions. These include

- halide ions, such as chloride ions, Cl^-
- hydroxide ions, OH^-
- cyanide ions, CN^-

Ligands can also be uncharged molecules that have one or more lone pairs of electrons to donate. These include

- ammonia, NH_3
- water, H_2O

In the ammonia molecule, the nitrogen atom has one lone pair of electrons. In the water molecule, the oxygen atom has two lone pairs of electrons.

A ligand, such as ammonia, that can donate one pair of electrons is a **unidentate** ligand. A ligand that can donate two pairs of electrons is a **bidentate** ligand, while **multidentate** ligands can donate more than two pairs of electrons. Note that water is a unidentate ligand. Even though the oxygen atom in the water molecule has two lone pairs of electrons, each water molecule can only donate one pair of electrons when it acts as a ligand.

Dot and cross diagrams for ammonia and water.

EDTA⁴⁻ is a multidentate ligand.

Co-ordination number

The **co-ordination number** of a complex is the number of pairs of electrons donated to the central metal ion. Co-ordination numbers are usually 2, 4, or 6.

Co-ordination number	2	4	6
Examples	$[Ag(NH_3)_2]^+$	$[CuCl_4]^{2-}$	$[Cu(H_2O)_6]^{2+}$
	$[AgCl_2]^-$	$[FeCl_4]^{2-}$	$[V(H_2O)_6]^{3+}$
	$[Ag(H_2O)_2]^+$	$[Ni(CN)_4]^{2-}$	$[Cu(H_2O)_4(OH)_2]$

Some complexes and their co-ordination numbers.

Oxidation state of the central metal ion

The overall charge on the complex depends upon the charge on the central metal ion, and the charge on its ligands. Metal ions are positively charged, so neutral ligands such as water and ammonia produce positively charged complexes. For example, the $[Cu(H_2O)_6]^{2+}$ complex contains six neutral ligands. The overall charge on the complex is 2+, so the charge on the copper ion must be 2+ too. It has an oxidation state of +2.

Negatively charged ligands produce complexes that are either negatively charged or neutral, depending on the number of ligands. For example, the $[Ni(CN)_4]^{2-}$ complex contains four cyanide ions, each with a single negative charge. The overall charge on the complex is 2−, so the charge on the nickel ion must be 2+, with an oxidation state of +2.

The $[Cu(H_2O)_4(OH)_2]$ complex contains four neutral water molecules and two hydroxide ions, each with a single negative charge. There is no overall charge on the complex, so the charge on the copper ion must be 2+, with an oxidation state of +2.

Silver complexes

Silver forms complexes with a co-ordination number of 2. For example, the $[Ag(NH_3)_2]^+$ ion contains two ammonia ligands, each donating one pair of electrons to the central Ag^+ ion. This is the soluble ion formed in Tollens' reagent, used to test for aldehydes.

$$\left[H_3N\!: \longrightarrow Ag^+ \longleftarrow :\!NH_3\right]^+$$

The diamminesilver(I) complex ion $[Ag(NH_3)_2]^+$ is linear.

When Tollens' reagent is warmed with an aldehyde, the silver(I) ion is reduced to silver. It forms a silver mirror on the inside surface of the container. Ketones do not reduce the silver(I) ion, so they do not form a silver mirror with Tollens' reagent.

Testing for halides

The halide ions Cl^-, Br^-, and I^- form precipitates with aqueous silver nitrate. The AgCl precipitate dissolves in dilute aqueous ammonia, the AgBr precipitate dissolves in concentrated aqueous ammonia, and AgI does not dissolve. These observations are used to confirm the identity of the halide ion present. The two precipitates dissolve because soluble $[Ag(NH_3)_2]^+$ forms.

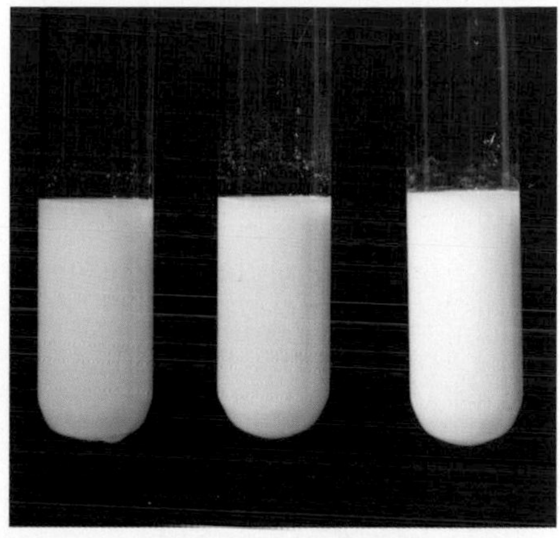

From left to right, precipitates of: silver chloride, silver bromide, and silver iodide.

Check your understanding

1 Explain the terms *complex, ligand,* and *co-ordination number.*

2 The complex $[Ag(NH_3)_2]^+$ forms in Tollens' reagent.
  - **a** What is its co-ordination number and shape?
  - **b** How does Tollens' reagent distinguish between aldehydes and ketones?

3 State the co-ordination number and oxidation states of the central metal ion in **a** $[AgCl_2]^-$, **b** $[CuCl_4]^{2-}$, **c** $[V(H_2O)_6]^{3+}$.

18.02 Naming complexes

OUTCOMES

already from A2 Level, you understand

- that a complex is a central metal ion surrounded by ligands

and after this spread you should know

- how to name complexes

The *formula* of a complex gives you information about the number and type of ligands present, the identity of the central metal ion, and the overall charge on the complex. For example, here is the formula of the hexaaquacopper(II) ion:

$$[Cu(H_2O)_6]^{2+}$$

It shows that the overall charge is 2+, and that the complex contains one central copper ion and six water molecules as ligands. Here is the formula of another complex that contains copper, the tetrachlorocuprate(II) ion:

$$[CuCl_4]^{2-}$$

It shows that the overall charge is 2− and that four chloride ions are present as ligands. Notice how two different types of brackets are used:

- Round brackets are used around ligands comprising more than one element, such as water, ammonia, hydroxide ions, and cyanide ions. These are not needed if the ligand is an ion formed from one element, such as chloride ions.
- Square brackets are used to enclose the complex, with the overall charge written outside.

The *name* of a complex ion also gives information about the complex.

Naming complexes

The name of a complex gives you information about

- the number of each type of ligand
- the identity of each ligand
- the identity of the central metal ion and its oxidation state

Number of ligands

When there are two or more ligands around the central metal ion, the number is shown by a prefix, as shown in the table.

number of ligands	prefix
2	di
3	tri
4	tetra
5	penta
6	hexa

Names of ligands

The identity of each type of ligand is shown by a name, but this is not identical to the name of the ligand itself. For example, water is shown by the name *aqua* when it is in a complex. The table shows some ligands and their names in complexes.

ligand	name in complex
water	aqua
ammonia	ammine
chloride	chloro
hydroxide	hydroxo
cyanide	cyano

Names of metals

The name of the metal changes, depending upon whether the complex is positively charged or negatively charged. If the complex is positively charged, the metal is identified by its normal name, with the oxidation state shown in brackets. For example, in the hexaaquacopper(II) ion, copper is present with an oxidation state of +2.

If the complex is negatively charged, the oxidation state of the metal is still shown in brackets. But the ending of the metal's name has the suffix *ate* added to it, and sometimes its Latin name is used instead. For example, in the tetrachlorocuprate(II) ion, copper is also present with an oxidation state of +2, but the complex is negatively charged.

metal	name in anionic complexes
chromium	chromate
cobalt	cobaltate
copper	cuprate
iron	ferrate
manganese	manganate
nickel	nickelate
silver	argentate
vanadium	vanadate

Aluminium is not a transition metal but it can form complexes. It is named aluminate in anionic (negatively charged) complexes.

Some examples

The table shows some simple examples of the formulae and names of complexes. Note that it is acceptable to have letter a's next to each other, as in hexaaqua and hexaammine. Take care not to confuse *diammine* with *diamine* from organic chemistry.

formula	name
$[Ag(NH_3)_2]^+$	diamminesilver(I)
$[Co(H_2O)_6]^{2+}$	hexaaquacobalt(II)
$[Co(NH_3)_6]^{2+}$	hexaamminecobalt(II)
$[CoCl_4]^{2-}$	tetrachlorocobaltate(II)
$[AgCl_2]^-$	dichloroargentate(I)

If two or more different ligands are present in a complex, they are named in alphabetical order. For example:

- $[Cu(H_2O)_5(OH)]^+$ is pentaaquahydroxocopper(II)
- $[Cu(H_2O)_2(NH_3)_4]^{2+}$ is tetraamminediaquacopper(II)

Note that it is the name of the ligand itself which is written in alphabetical order, not its numbering prefix.

Writing formulae

When you write the formula for a complex, you put the central atom first, then the ligands. These are written in alphabetical order by the first symbol in their formulae, rather than by their names. For example, the tetraaquadichlorochromium(III) ion is shown as $[CrCl_2(H_2O)_4]^+$ and not as $[Cr(H_2O)_4Cl_2]^+$.

IUPAC recommend that the formula for each ligand should be written so its donor atom symbol is nearest the central metal ion symbol. This means that aqua should be written as (OH_2) rather than (H_2O), and the tetraamminediaquacopper(II) ion would be written as $[Cu(NH_3)_4(OH_2)_2]^{2+}$ rather than as $[Cu(H_2O)_2(NH_3)_4]^{2+}$. We show aqua as (H_2O) in this book to avoid confusion with the hydroxo ligand (OH), other books, and past examination papers.

Aqueous hexaaquacopper(II) ions give a blue colour and aqueous tetrachlorocuprate(II) ions give a green colour.

Oxidation state not charge

In the name of a complex, the number in brackets shows the oxidation state of the central metal ion, not the charge on the complex. Copper has an oxidation state of +2 in the hexaaquacopper(II) ion, $[Cu(H_2O)_6]^{2+}$, and in the tetrachlorocuprate(II) ion, $[CuCl_4]^{2-}$.

Check your understanding

1 Name the following complexes:

a $[Fe(H_2O)_6]^{2+}$

b $[Cr(NH_3)_6]^{3+}$

c $[Ag(CN)_2]^-$

d $[MnCl_4]^{2-}$

e $[Fe(H_2O)_4(OH)_2]^+$

2 Write the formulae of the following complexes:

a diaquasilver(I)

b hexaaquavanadium(III)

c tetracyanonickelate(II)

d tetrachloroferrate(II)

e diaquatetrahydroxo-aluminate(III)

18.03 Shapes of complexes

OUTCOMES

already from AS Level, you can

- predict the shapes of simple molecules and ions in terms of electron pair repulsion

already from A2 Level, you

- understand that a complex is a central metal ion surrounded by ligands
- know the meaning of co-ordination number
- know that Ag^+ forms a linear complex $[Ag(NH_3)_2]^+$
- know how to name complexes

and after this spread you should know

- that transition metal ions commonly form octahedral complexes with small ligands (e.g. H_2O and NH_3)
- that transition metal ions commonly form tetrahedral complexes with larger ligands (e.g. Cl^-)

You already know how to work out the shape of a simple molecule by counting the number of electron pairs around the central atom. Pairs of electrons repel each other so that they are as far apart as possible, keeping the repulsive forces to a minimum and giving each molecule a characteristic shape.

number of electron pairs around central atom	name of shape	bond angle(s) (°)	example
2	linear	180	$BeCl_2$
3	trigonal planar	120	BCl_3
4	tetrahedral	109.5	CH_4
5	trigonal bipyramidal	90 and 120	PF_5
6	octahedral	90	SF_6

The main shapes of molecules with some examples.

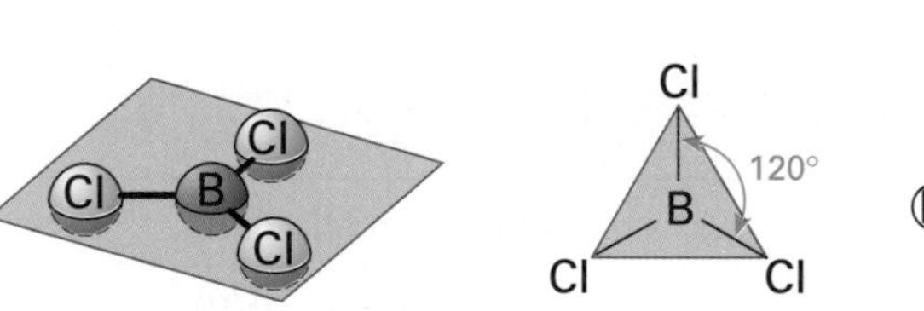

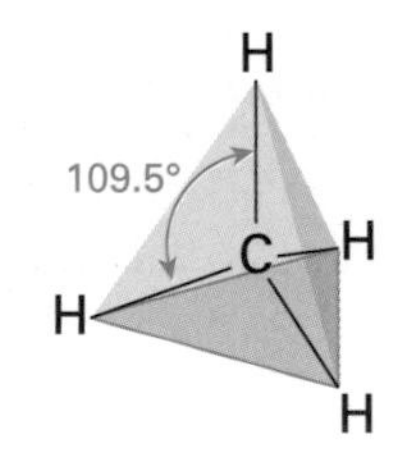

The boron trichloride molecule is trigonal planar.

The methane molecule is tetrahedral.

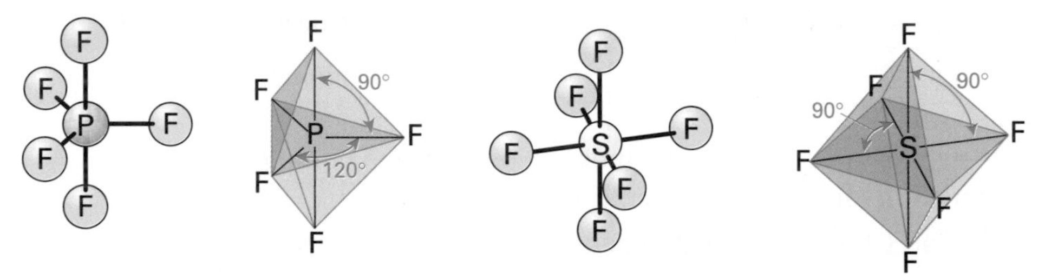

The phosphorus pentafluoride molecule is trigonal bypyramidal.

The sulfur hexafluoride molecule is octahedral.

The shapes of complexes can be predicted in a similar way.

Shapes of complexes

You need to know the shapes commonly formed by complexes with co-ordination numbers of 2, 4, or 6. In general, for monodentate ligands (those that can donate one pair of electrons):

- two ligands give a linear shape
- four ligands give a tetrahedral shape
- six ligands give an octahedral shape

Square planar complexes

Some complexes with four ligands are square planar, rather than tetrahedral. For example, the tetracyanonickelate(II) ion $[Ni(CN)_4]^{2-}$ is square planar. You can find out more about such complexes on Spread 18.04.

The shape of the tetracyanonickelate(II) ion.

You already know that the diamminesilver(I) ion is linear. Large, charged, ligands such as the chloride ion commonly form tetrahedral complexes.

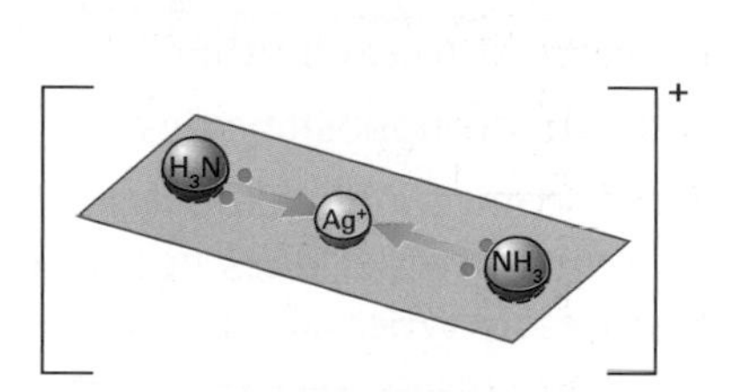

The shape of the diamminesilver(I) ion.

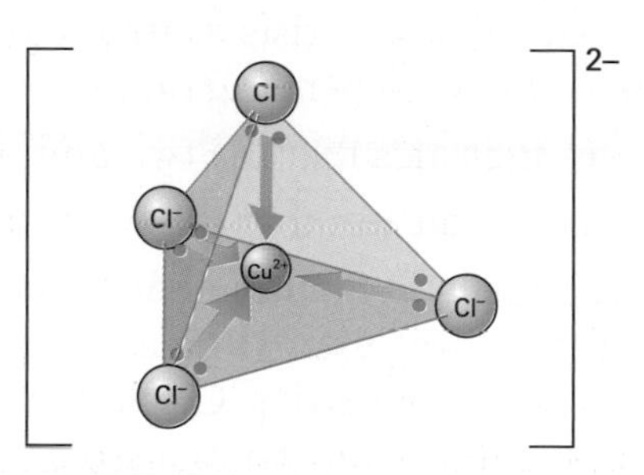

The shape of the tetrachlorocuprate(II) ion.

Small uncharged ligands, such as water and ammonia, commonly form octahedral complexes. For example, the hexaaquacopper(II) ion has an octahedral shape.

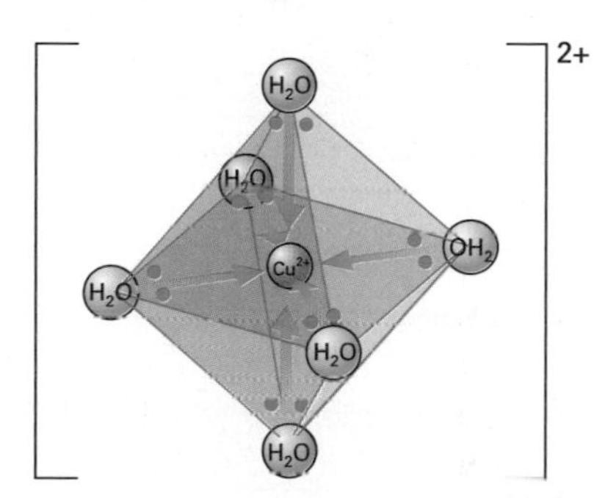

The shape of the hexaaquacopper(II) ion.

Check your understanding

1 Predict the shapes of the following complexes:

a $[Fe(H_2O)_6]^{2+}$
b $[Cr(NH_3)_6]^{3+}$
c $[Ag(CN)_2]^-$
d $[Cu(H_2O)_2(NH_3)_4]^{2+}$
e $[CrCl_2(H_2O)_4]^+$
f $[CoCl_4]^{2-}$
g $[Ni(CN)_4]^{2-}$

18.04 Cisplatin

OUTCOMES

already from A2 Level, you know

- that transition metal ions commonly form octahedral complexes with small ligands (e.g. H_2O and NH_3)
- that transition metal ions commonly form tetrahedral complexes with larger ligands (e.g. Cl^-)
- that square planar complexes are also formed
- that *E-Z* isomerism and optical isomerism are forms of stereoisomerism

and after this spread you should

- know that the Pt(II) complex *cisplatin* is used as an anticancer drug
- appreciate the benefits and risks associated with cisplatin
- know that cisplatin is a square planar complex

Complexes with a co-ordination number of 4, such as the tetrachlorocuprate(II) ion $[CuCl_4]^{2-}$, usually have a tetrahedral shape. Some complexes with a co-ordination number of 4, such as the tetracyanonickelate(II) ion $[Ni(CN)_4]^{2-}$, may have a square planar shape instead.

Cisplatin

The complex diamminedichloroplatinum(II), $[PtCl_2(NH_3)_2]$, has a square planar shape. It exists as two stereoisomers. These are given the prefix *cis-* or *trans-*, depending on the arrangement of the two different ligands:

- *cis-* indicates that the two ammine ligands are next to each other
- *trans-* indicates that the two ammine ligands are on opposite sides

The isomer *cis*-diamminedichloroplatinum(II) is commonly called **cisplatin**. It is used as an anticancer drug. The *trans* isomer does not have anticancer properties. Cisplatin is used to treat many types of cancer, but it is most commonly used against cancer of the testes, ovaries, bladder, stomach, and lungs. Like most drugs, it has side-effects. These include nausea and vomiting, allergic reactions, hearing loss, and kidney problems. The risk of these side-effects is greater at higher doses, and it increases as the total amount of cisplatin used increases over time.

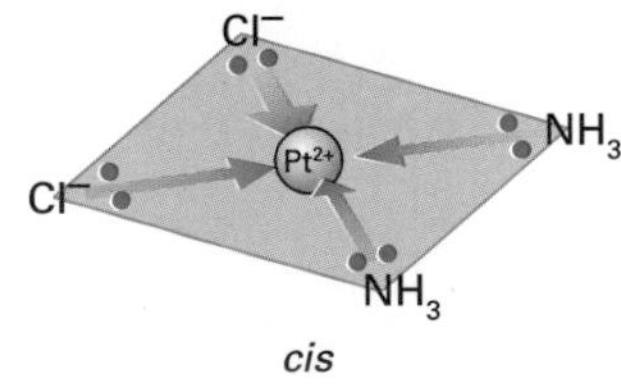

cis

cis-*[PtCl₂(NH₃)₂] or cisplatin*

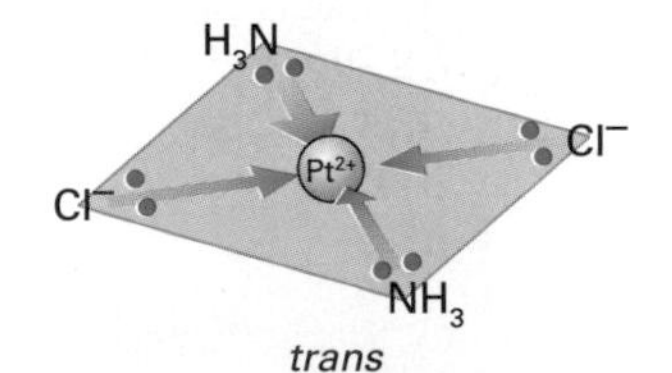

trans

trans-*[PtCl₂(NH₃)₂]*

Cis–trans isomers of alkenes

Alkenes such as but-2-ene have *E-Z* isomers. In (*E*)-but-2-ene, the methyl groups attached to the C=C functional group are on opposite sides. In (*Z*)-but-2-ene they are on the same side. Such isomers were also described as *cis–trans* isomers, and you will still see these names used. *cis*-But-2-ene is (*Z*)-but-2-ene and *trans*-but-2-ene is (*E*)-but-2-ene.

(*E*)-but-2-ene (the *trans* isomer)

(*Z*)-but-2-ene (the *cis* isomer)

E-Z *isomers of but-2-ene*

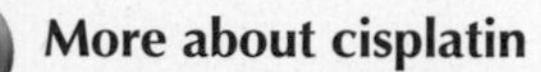

More about cisplatin

Cisplatin was first synthesized in 1845 and its structure was known by 1893. At that time its anticancer properties were unknown. In the 1960s, an American scientist called Barnett Rosenberg carried out a series of experiments to study the effects of electric currents on the growth of bacteria. He discovered that he could produce unusually long *E. coli* cells. Something was interfering with cell division in these cells. This turned out not to be the electric currents themselves, but cisplatin formed from the platinum electrodes used in the experiments.

Rosenberg's group of scientists then tested the effectiveness of cisplatin against cancer in mice. Its success was reported in 1965 and clinical trials in people began in 1971. Cisplatin was eventually licensed for use in 1977. Patients are given an intravenous drip containing a solution of cisplatin. Scientists are not quite sure exactly how it works, but it seems to be converted in cells into a reactive ion that binds to DNA. This stops cancer cells dividing, so treating the tumour.

Computer artwork of a section of DNA with two cisplatin molecules attached. The two DNA strands are coloured grey and blue, and each cisplatin molecule comprises silver and dark blue spheres. The cisplatin molecules bind to the areas coloured red on the DNA strands.

Isomers of octahedral complexes

Many octahedral complexes also have *cis–trans* isomers. For example, the tetraamminedichloro-cobalt(III) ion $[CoCl_2(NH_3)_4]^+$ has *cis* and *trans* isomers.

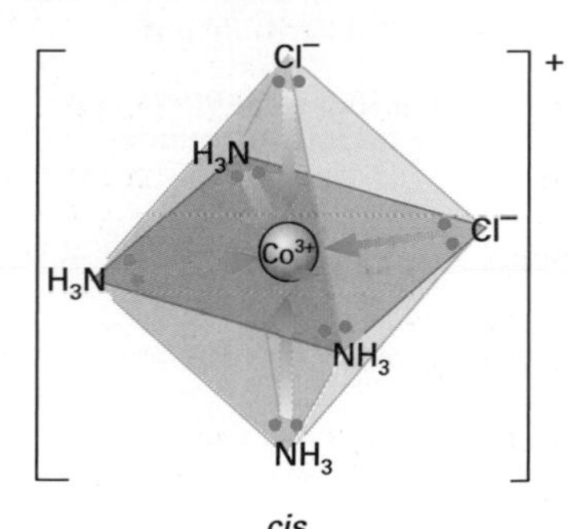

The cis-*tetraamminedichlorocobalt(III) ion*

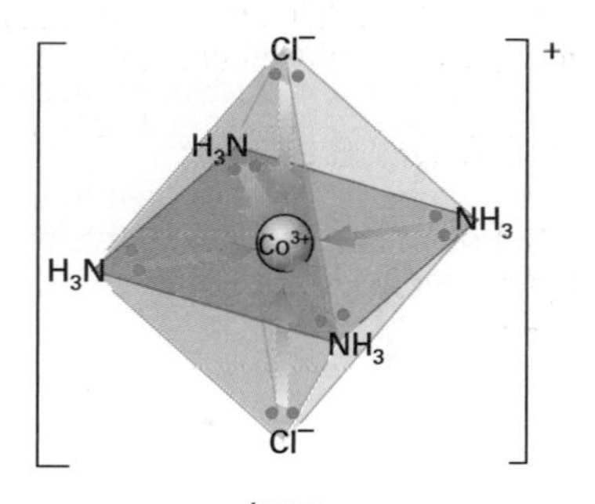

The trans-*tetraamminedichlorocobalt(III) ion*

Check your understanding

1 a Write the formula of cisplatin.

b Explain why diamminedichloroplatinum(II) exists as *cis* and *trans* isomers.

c State a medical use of cisplatin.

18.05 More ligands

OUTCOMES

already from A2 Level, you

- can define the term *ligand*
- understand that a complex is a central metal ion surrounded by ligands
- know that co-ordinate bonding is involved in complex formation
- know the meaning of co-ordination number

and after this spread you should

- understand that ligands can be unidentate, bidentate, or multidentate
- know that haem is an iron(II) complex with a multidentate ligand
- understand that Fe(II) in haemoglobin enables oxygen to be transported in the blood, and why carbon monoxide is toxic

Unidentate ligands

Unidentate ligands can donate one pair of electrons to the central metal ion in a complex. Common unidentate ligands include chloride ions, cyanide ions, ammonia, and water. In diagrams, you can show each co-ordinate bond as an arrow pointing from the ligand to the metal ion.

$[H_3N \rightarrow Ag \leftarrow NH_3]^+$

Co-ordinate bonding in the diamminesilver(I) ion, $[Ag(NH_3)_2]^+$.

Co-ordinate bonding in the tetrachlorocobaltate(II) ion, $[CoCl_4]^{2-}$.

Co-ordinate bonding in the hexaaquacopper(II) ion, $[Cu(H_2O)_6]^{2+}$.

Bidentate ligands

Bidentate ligands can donate two pairs of electrons to the central metal ion in a complex. Common bidentate ligands include 1,2-diaminoethane and the ethanedioate ion.

1,2-diaminoethane

The organic compound 1,2-diaminoethane, $H_2NCH_2CH_2NH_2$, has two amino groups. The nitrogen atom in each amino group has one lone pair of electrons. So the molecule can donate two pairs of electrons to the central metal ion in a complex.

Three 1,2-diaminoethane molecules can form co-ordinate bonds with a copper(II) ion.

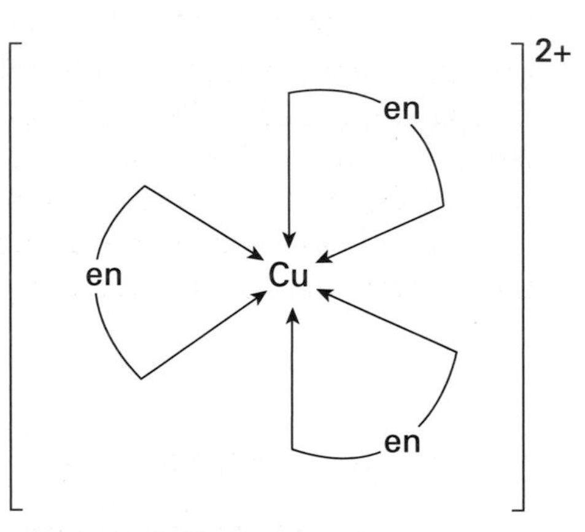

The abbreviation 'en' may be used to show 1,2-diaminoethane in simplified diagrams of complexes.

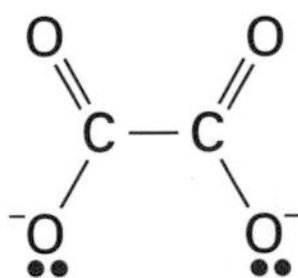

The ethanedioate ion can act as a bidentate ligand.

The ethanedioate ion

The ethanedioate ion, $C_2O_4^{2-}$, has two carboxylate groups. Each one can donate a pair of electrons to the central metal ion in a complex.

Multidentate ligands

Multidentate ligands can donate more than two pairs of electrons to the central metal ion in a complex.

$EDTA^{4-}$

The ligand $EDTA^{4-}$ can donate six pairs of electrons to the metal ion. Two of these are on its nitrogen atoms, and four are on oxygen atoms in the carboxylate groups.

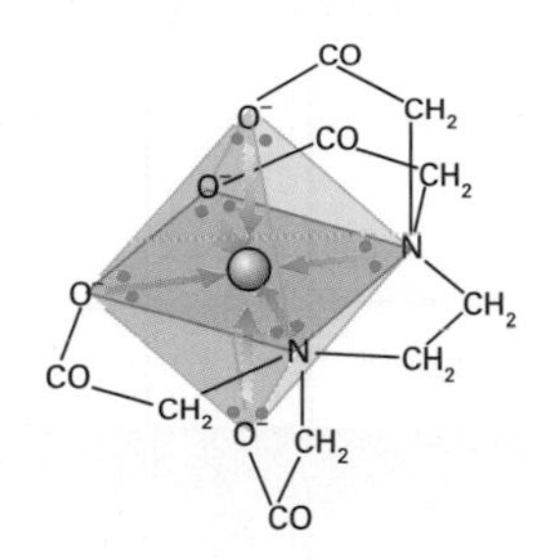

The structure of an octahedral $EDTA^{4-}$ complex.

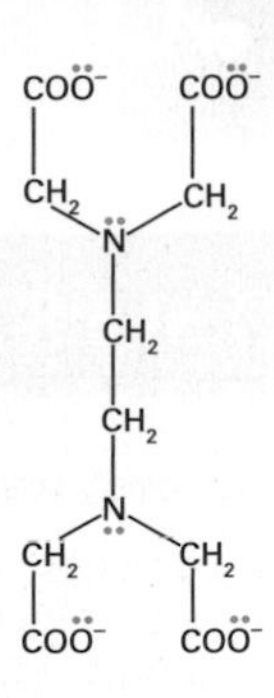

EDTA stands for ethylenediaminetetraacetic acid. The $EDTA^{4-}$ ion can act as a multidentate ligand.

Haem

Haemoglobin is a protein found in red blood cells. It binds to oxygen, allowing oxygen to be carried around the body in the bloodstream. Each haemoglobin molecule comprises four smaller sub-units, each of which contains a **haem** group. This is a complex involving an iron(II) ion and a multidentate ligand. This ligand forms four co-ordinate bonds with the central metal ion.

Haem has a square planar shape and further co-ordination can happen. A histidine amino acid residue in the protein sub-unit forms a fifth co-ordinate bond with the iron(II) ion, and oxygen can form a sixth co-ordinate bond. Haemoglobin binds to oxygen in regions of high oxygen concentration, forming oxyhaemoglobin. This releases oxygen in regions of low oxygen concentration.

Oxygen is not the only ligand that can form co-ordinate bonds with the iron(II) ion in haem. Carbon monoxide can, too. It forms bonds that are less easily broken than those formed by oxygen, so it reduces the capacity of the blood to carry oxygen. This makes carbon monoxide toxic. As it is formed during incomplete combustion of fuels, it is important the gas fires and boilers have adequate ventilation to ensure complete combustion.

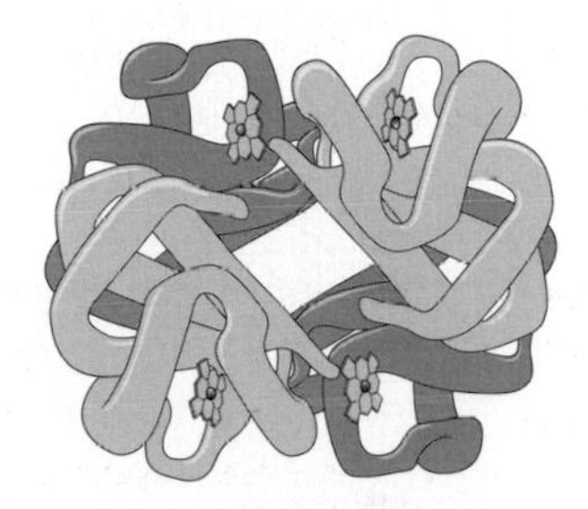

Haemoglobin is made from four sub-units, each containing a haem group.

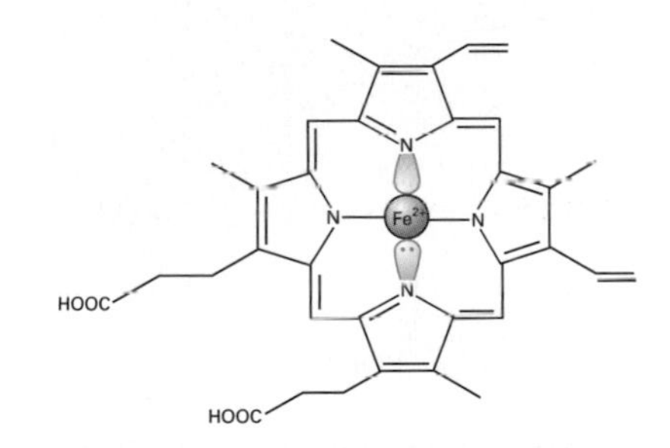

Four nitrogen atoms form co-ordinate bonds with the Fe^{2+} ion in a haem group.

Check your understanding

1 a Explain the meaning of the term *unidentate ligand*.
b Give the formula of the tetrahedral complex formed by chloride ions and copper(II) ions.

2 a With reference to 1,2-diaminoethane, explain the meaning of the term *bidentate ligand*.
b Sketch the structure of the octahedral complex formed by ethanedioate ions and iron(III) ions.

3 a Explain why $EDTA^{4-}$ can act as a multidentate ligand.
b Explain why carbon monoxide is toxic.

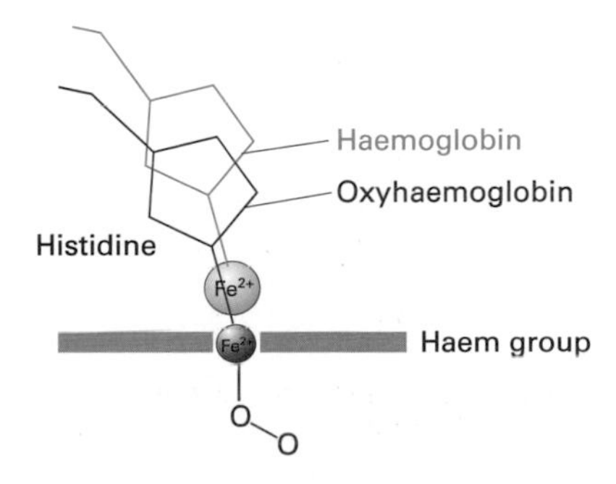

Oxygen binds reversibly to the Fe^{2+} ion in a haem group. Cyanide ions and carbon monoxide bind more strongly.

19.01 Complexes and colour

OUTCOMES

already from A2 Level, you know

- that the characteristics of transition metals arise from an incomplete d sub-level in atoms or ions, and include the formation of coloured ions

and after this spread you should know

- that transition metal ions can be identified by their colour
- that colour arises from electronic transitions from the ground state to excited states: $\Delta E = h\nu$
- that colour changes arise from changes in oxidation state, co-ordination number and ligand

The visible spectrum is formed by electromagnetic radiation in a range of wavelengths between about 380 nm at the blue end, and 760 nm at the red end. White light contains all the colours of the spectrum. A substance appears coloured if it absorbs some of the colours in white light. The colours we see are *complementary colours*, the colours remaining after absorption of the other colours by a substance.

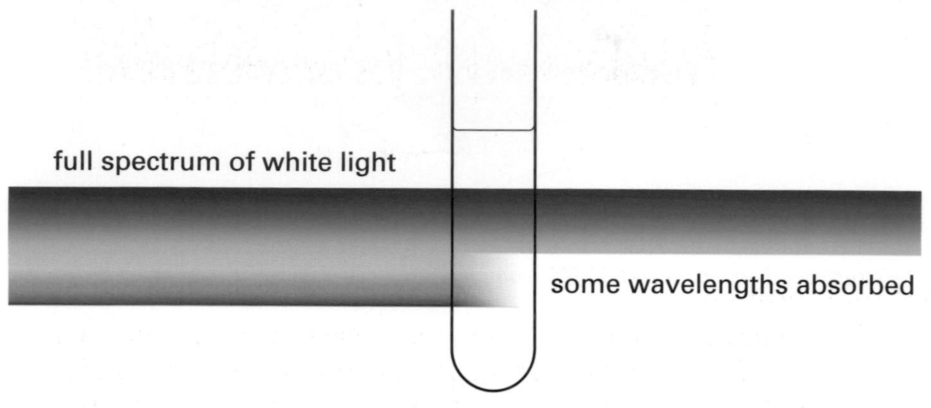

A solution of a substance that absorbs red and orange light will appear blue-green.

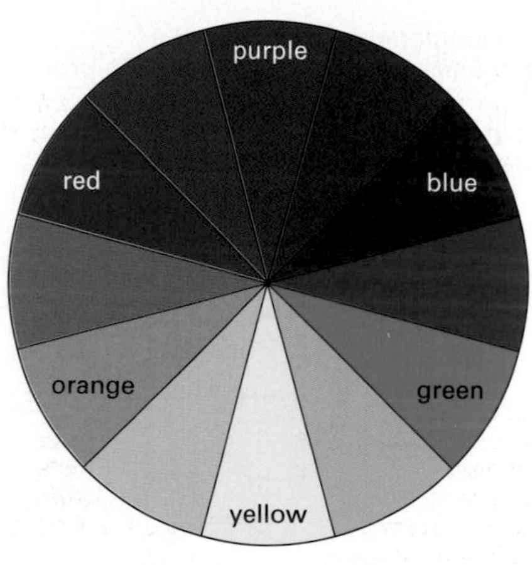

Colour wheels are often used by artists, decorators, and website designers to select their colour schemes. Complementary colours are on opposite sides of the wheel.

Transition metal ions can be identified by their colour. You need to know how colour arises, and you will need to know the colours of the ions studied in the Specification.

Electronic transitions

Electrons occupy particular energy levels in atoms. When an atom absorbs energy, an electron can be promoted from its normal **ground state** to a higher energy level. The electron is then in an **excited state**. It emits energy as electromagnetic radiation when it returns to its ground state. Such changes from one energy level to another are called **electronic transitions**.

The frequency of electromagnetic radiation absorbed or emitted depends upon the difference in energy between the ground state and the excited state. Planck's equation, named after the German physicist Max Planck, links the two quantities:

$\Delta E = h\nu$ ΔE is the difference in energy in joule, J

h is **Planck's constant** (6.626×10^{-34} J s)

ν is the frequency of the electromagnetic radiation in hertz, Hz

Planck's equation shows that the frequency of the electromagnetic radiation absorbed, or emitted, is directly proportional to the energy of the electronic transition. Since wavelength is inversely proportional to frequency:

- A large energy transition involves high frequency, short wavelength radiation in the blue end of the spectrum.
- A small energy transition involves low frequency, long wavelength radiation in the red end of the spectrum.

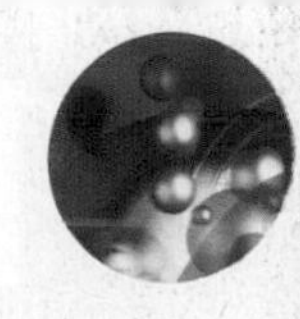

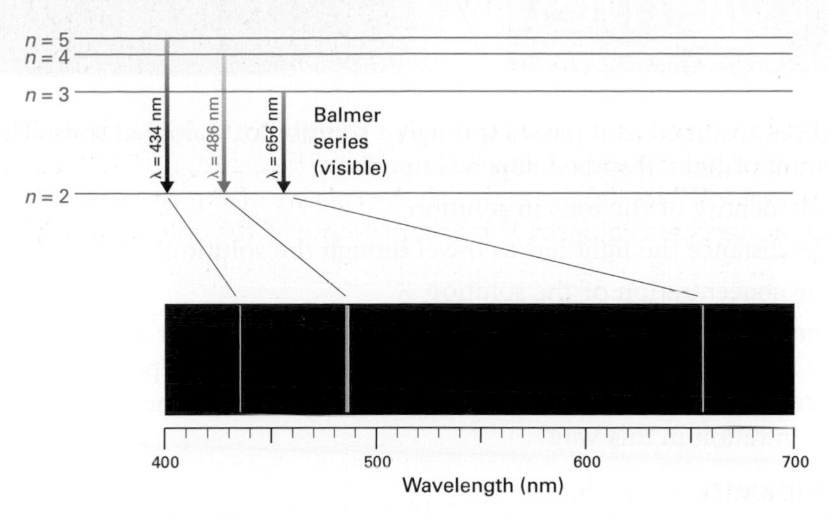

Electronic transitions in hydrogen atoms produce the Balmer series, which comprises characteristic emissions in the visible range.

d-to-d transitions

The five d orbitals in an isolated transition metal or ion are at the same energy level. The presence of ligands causes the d sub-level to split into two slightly different energy levels. An electron in a lower-energy d orbital can absorb energy and be promoted into a higher-energy d orbital. A complex will be coloured if the energy absorbed corresponds to a frequency of visible light.

Such d-to-d transitions can only happen if there is a d electron to promote, and a space in a higher-energy d orbital to accept it. So an ion must have a partially filled d sub-level to be coloured.

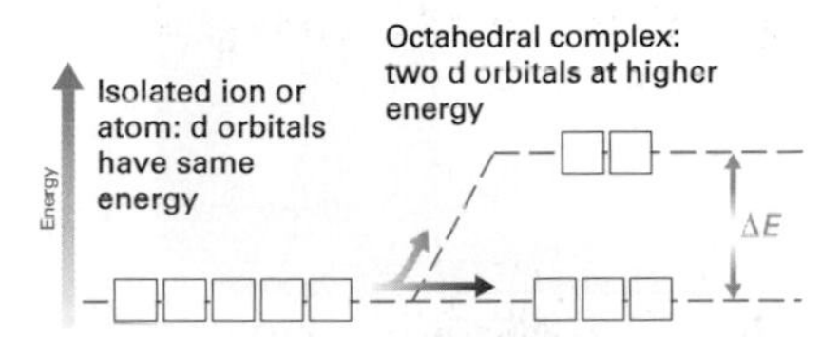

The splitting of d orbitals in an octahedral complex. In a tetrahedral complex, three d orbitals occupy a higher energy level.

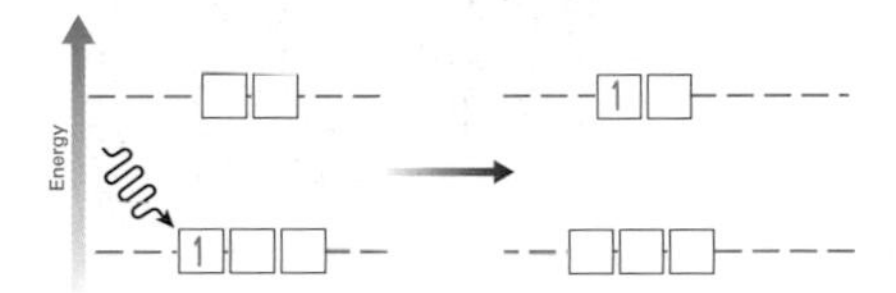

An electron in a lower d orbital is promoted into a higher-energy d orbital when it absorbs light of the appropriate frequency. It returns to its original orbital by an alternative route that does not emit visible light.

Factors affecting colour

The colour of a transition metal's complexes depends its oxidation state, the ligand involved, and the co-ordination number.

Oxidation state

For a given metal, the higher the oxidation state of its ions the greater the amount of d sub-level splitting. This increases the energy level difference and so changes the colour of light absorbed. For example, $[Fe(H_2O)_6]^{2+}$(aq) is pale green but $[Fe(H_2O)_6]^{3+}$ is yellow.

Ligand

Different ligands cause different amounts of d sub-level splitting. In general, splitting increases in the order $Cl^- < OH^- < H_2O < NH_3 < CN^-$. This increases the energy level difference and so changes the colour of light absorbed. For example, $[Co(H_2O)_6]^{2+}$(aq) is pink but $[Co(NH_3)_6]^{2+}$ is pale straw-coloured.

Co-ordination number

The amount of d sub-level splitting is greater in an octahedral complex than in a tetrahedral complex. So a change in co-ordination number changes the colour of the complex. But any such change will also involve a change in ligand, and any colour differences will involve a combination of both factors.

Check your understanding

1 a State what is meant by the symbols ΔE, h, and ν in Planck's equation.

b Outline how colour arises in transition metal complexes.

c State three changes that can alter the colour of a transition metal complex.

d Explain why Sc^{3+}(aq) and Zn^{2+}(aq) ions are not coloured.

19.02 Spectrometry

OUTCOMES

already from A2 Level, you know

- why transition metal ions are coloured
- that transition metal ions can be identified by their colour

and after this spread you should appreciate

- that the absorption of visible light is used in spectrometry to determine the concentration of coloured ions

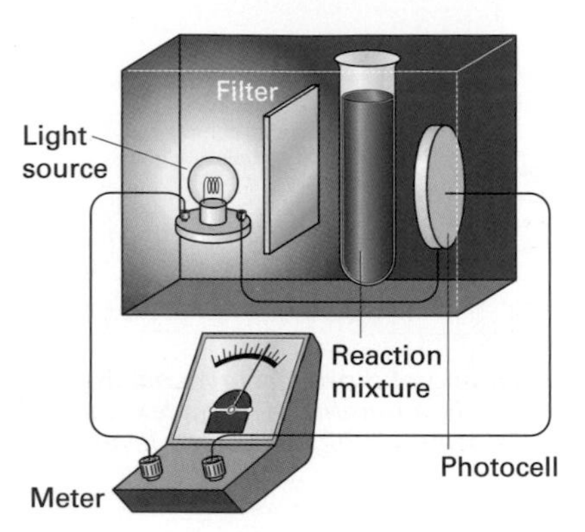

A simple colorimeter

Light is absorbed as it passes through a solution of coloured ions. The amount of light absorbed depends upon

- the identity of the ions in solution
- the distance the light has to travel though the solution
- the concentration of the solution

If the distance the light has to travel is fixed, the amount of light absorbed by an aqueous solution of ions depends upon the concentration. A **colorimeter** is an instrument used to measure concentration in this way.

Colorimetry

The colorimeter

A colorimeter comprises a white light source, a filter or diffraction grating, and a photocell connected to a meter. The filter or diffraction grating allows coloured light to pass through the sample, which is contained in a small cylindrical glass or plastic container called a **cuvette**. The photocell detects the amount of light transmitted through the sample, and this is displayed on the meter. Colorimeters usually contain electronics that will convert the transmittance values to absorbance values, as absorbance is directly proportional to the concentration of the ions in the sample.

Percentage transmittance and absorbance

The percentage transmittance, %*T*, is the percentage of light that passes through a sample. If no light gets through, %*T* is zero. If all the light gets through, %*T* is 100. A graph of %*T* against concentration produces a curve, but a graph of absorbance *A* against concentration produces a straight line. Absorbance and % transmittance are related by this equation:

$$A = 2 - \log_{10}(\%T)$$

If no light is absorbed and it is all transmitted, the absorbance is zero. If all the light is absorbed and none is transmitted, the absorbance is infinitely high. The absorbance scale on colorimeters usually goes up to 2 (equivalent to 1% transmittance).

The filter is chosen so that it allows light through that is the complementary colour to the colour of the ions in the sample. This makes the measurements more sensitive as more absorption can happen. For example, a red filter might be used when determining the concentration of blue hexaaquacopper(II) ions. A diffraction grating does the same job as the filter, but it can be adjusted to give various precise wavelengths of light in the visible spectrum.

The determination

A calibration graph or standard curve is produced using solutions with known concentrations. A graph of absorbance against concentration will begin to level off at high concentrations, so the range of concentrations must be chosen carefully. The absorbance of the sample with the unknown concentration is measured, and the equivalent concentration is read off the graph.

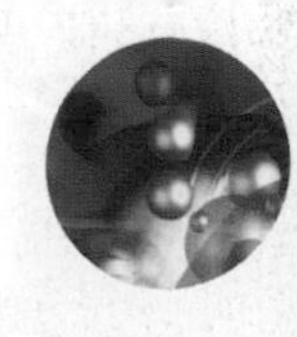

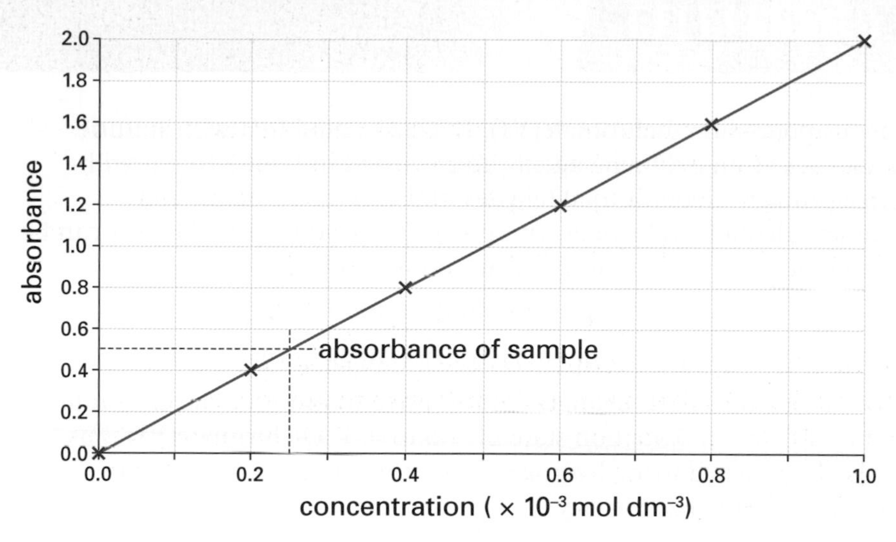

A typical calibration graph. In this example, the concentration of the sample is 0.25×10^{-3} mol dm^{-3}.

The colorimeter may be designed to carry two cuvettes at once. One contains the sample or standard solution, and the other contains a *blank* containing just the solvent. The machine subtracts the absorbance of the solvent from the absorbance of each sample or standard solution.

Absorption spectra

The UV/vis spectrometer is a more complex instrument than the colorimeter. It can measure absorbance in the visible range, just like the colorimeter. But in addition, it can measure absorbance in the ultraviolet range. This makes it useful for analysing colourless substances, as they may absorb in the ultraviolet range instead. The spectrometer can also determine the absorption spectrum of a substance. This is a graph of absorbance against wavelength.

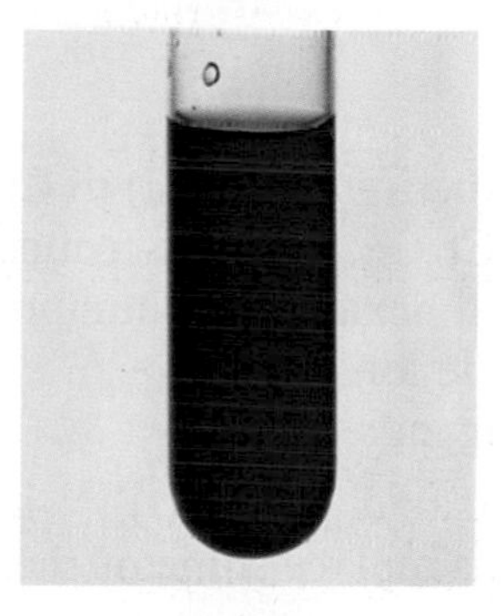

A solution containing the complex ion $[Ti(H_2O)_6]^{3+}$(aq).

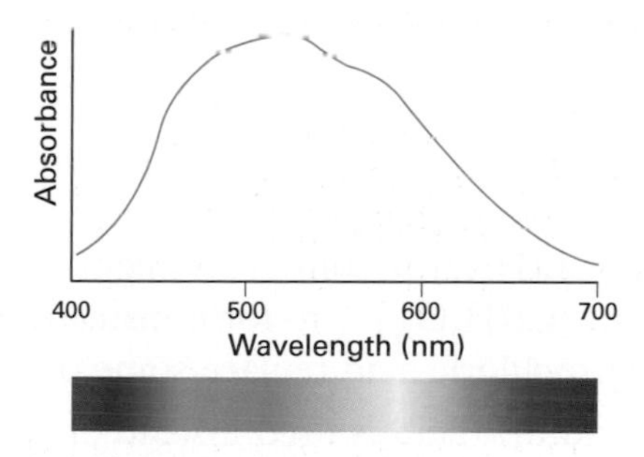

The absorption spectrum of $[Ti(H_2O)_6]^{3+}$(aq). Note that $[Ti(H_2O)_6]^{3+}$(aq) absorbs green and yellow the most, so it appears purple coloured.

Colour chemists are interested in absorption spectra because they provide information about potential pigments and dyes. A narrow peak of absorbance indicates a bright, pure colour. The greater the absorption by a substance, the less of it is needed to colour fabrics and other materials. The wavelength at which most light is absorbed is often chosen for the filter, or diffraction grating setting, for the determination of the concentration of the substance.

Check your understanding

1. a Explain why the dimensions of the cuvettes used in a colorimetry experiment must be kept the same.
   b Suggest an advantage of measuring absorbance rather than percentage transmittance.
   c Use the calibration graph on this spread to determine the absorbance of a 0.7×10^{-3} mol dm^{-3} sample of the same substance.
   d Suggest a suitable wavelength for measuring the concentration of a sample of $[Ti(H_2O)_6]^{3+}$(aq) by colorimetry.

19.03 Some colour reactions of chromium

OUTCOMES

already from AS Level, you understand

- that primary alcohols can be oxidized to aldehydes and carboxylic acids, and that secondary alcohols can be oxidized to ketones, by acidified potassium dichromate(VI)

already from A2 Level, you know

- that transition metal ions can be identified by their colour
- the redox titration of iron(II) with dichromate(VI) in acid solution

and after this spread you should know

- that Cr^{3+} and Cr^{2+} are formed by reduction of $Cr_2O_7^{2-}$ by zinc in acid solution
- the oxidation in alkaline solution of Cr^{3+} by H_2O_2

Summary

The table summarizes the complexes and their colours.

oxidation state	complex	colour
+6	$Cr_2O_7^{2-}(aq)$	orange
+3	$[Cr(H_2O)_4Cl_2]^{+}(aq)$	green
+2	$[Cr(H_2O)_6]_2{}^{+}(aq)$	blue

Aqueous potassium dichromate(VI), $K_2Cr_2O_7$, acidified with sulfuric acid can act as an oxidizing agent. You learned in Unit 2 that it will oxidize primary alcohols to aldehydes and carboxylic acids, and secondary alcohols to ketones. In Spread 17.02 you learned that it can be used to oxidize iron(II) to iron(III) in redox titrations:

$$Cr_2O_7^{2-}(aq) + 14H^+(aq) + 6Fe^{2+}(aq) \rightarrow$$
$$2Cr^{3+}(aq) + 7H_2O(l) + 6Fe^{3+}(aq)$$

In each case, the colour changes from orange to green as the chromium is reduced from an oxidation state of +6 to +3. Dichromate(VI) ions can also be reduced using hydrogen, and it is possible to reach an oxidation state of +2 this way.

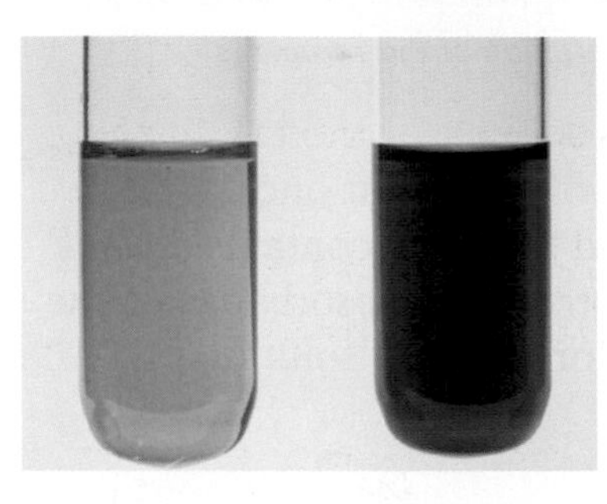

Orange dichromate(VI) is reduced to green chromium(III) by primary and secondary alcohols.

Reduction of dichromate(VI)

Dichromate(VI) ions can be reduced using a mixture of zinc and hydrochloric acid. The reaction between the metal and the acid produces hydrogen gas, which acts as the reducing agent. It is important to keep air out of the reaction mixture, otherwise the chromium(II) stage will not be reached. This can be achieved by carrying the reaction out in a conical flask stoppered by cotton wool.

Oxidation state +6 to +3

The first stage in the reaction involves the reduction of Cr(VI) to Cr(III). The colour changes from orange to green as $Cr_2O_7^{2-}$ ions are reduced to $[CrCl_2(H_2O)_4]^+$ ions. You might have expected hexaaquachromium(III) ions $[Cr(H_2O)_6]^{3+}$ to form instead. But chloride ions from the hydrochloric acid replace some of the aqua ligands.

If sulfuric acid is used instead of hydrochloric acid, $[Cr(H_2O)_6]^{3+}$ ions still do not form, because this time sulfate ions replace some of the aqua ligands.

Oxidation state +3 to +2

The second stage in the reaction involves the reduction of Cr(III) to Cr(II). The colour changes from green to blue as $[CrCl_2(H_2O)_4]^+$ ions are reduced to $[Cr(H_2O)_6]^{2+}$ ions.

Oxidation state +2 to +3

The hexaaquachromium(II) ion is a powerful reducing agent and only exists in the absence of air. If air is let into the reaction mixture, the colour rapidly changes from blue to green as the $[Cr(H_2O)_6]^{2+}$ ions are oxidized to $[CrCl_2(H_2O)_4]^+$ ions.

The Bunsen valve

An effective way to keep air out involves a *Bunsen valve*. This is a short length of rubber tubing with a slit up the side. One end is attached to a glass tube running through a bung, and the other end to a short length of glass rod. Hydrogen escapes under pressure through the slit but air cannot get in.

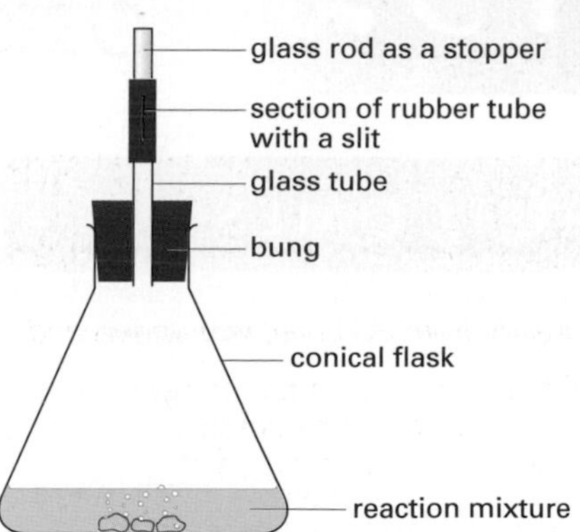

The reaction can be carried out in a conical flask fitted with a Bunsen valve.

Oxidation in alkaline solution

Cr(III) can be oxidized to Cr(VI) in alkaline solution using hydrogen peroxide, H_2O_2. This time Cr(VI) is present as the yellow tetraoxochromate(VI) ion CrO_4^{2-}, instead of the orange dichromate(VI) ion $Cr_2O_7^{2-}$. There are three steps.

Aqueous sodium hydroxide is added, forming a grey-green precipitate:

$$[Cr(H_2O)_6]^{3+}(aq) + 3OH^-(aq) \rightarrow [Cr(H_2O)_3(OH)_3](s) + 3H_2O(l)$$

The precipitate redissolves in excess aqueous sodium hydroxide, forming a dark green solution:

$$[Cr(H_2O)_3(OH)_3](s) + 3OH^-(aq) \rightarrow [Cr(OH)_6]^{3-}(aq) + 3H_2O(l)$$

The solution is then warmed with aqueous hydrogen peroxide, forming a yellow solution:

$$2[Cr(OH)_6]^{3-}(aq) + 3H_2O_2(aq) \rightarrow 2CrO_4^{2-}(aq) + 2OH^-(aq) + 8H_2O(l)$$

Solutions of orange dichromate(VI) and yellow chromate(VI).

Summary

The table summarizes the complexes and their colours. Note that you may not see the ruby colour of $[Cr(H_2O)_6]^{3+}(aq)$, as chloride ions and other ligands may replace water molecules to give green complexes.

oxidation state	complex	colour
+3	$[Cr(H_2O)_6]^{3+}(aq)$	ruby
+3	$[Cr(H_2O)_3(OH)_3](s)$	grey-green
+3	$[Cr(OH)_6]^{3-}(aq)$	dark green
+6	$CrO_4^{2-}(aq)$	yellow

Check your understanding

1 An orange solution of a chromium complex **A** turns to a green solution of complex **B** in the presence of zinc and hydrochloric acid. This then forms a blue solution of complex **C**.

a Identify complexes **A**, **B**, and **C**.

b What will happen to complex **C** in the presence of air?

2 A green solution of a chromium complex **X** turns to a dark green solution of complex **Y** when excess aqueous sodium hydroxide is added. A yellow solution of complex **Z** is formed when **Y** is warmed with aqueous hydrogen peroxide.

a Identify complexes **X**, **Y**, and **Z**.

b What is the role of the hydrogen peroxide?

19.04 Some colour reactions of cobalt

OUTCOMES

already from A2 Level, you know

- that transition metal ions can be identified by their colour
- the oxidation in alkaline solution of Cr^{3+} by H_2O_2

and after this spread you should know

- the oxidation in alkaline solution of Co^{2+} by H_2O_2
- the oxidation of Co^{2+} by air in ammoniacal solution

A precipitate of cobalt(II) hydroxide formed by adding aqueous sodium hydroxide to aqueous cobalt(II) ions.

In the previous spread, you discovered how Cr(III) can be oxidized to Cr(VI) in alkaline solution using hydrogen peroxide. In excess aqueous sodium hydroxide, blue hexaaquachromium(III) ions are converted to green hexahydroxochromate(III) ions:

$$[Cr(H_2O)_6]^{3+}(aq) + 6OH^-(aq) \rightarrow [Cr(OH)_6]^{3-}(aq) + 6H_2O(l)$$

When the solution is warmed with aqueous hydrogen peroxide, yellow chromate(VI) ions form:

$$2[Cr(OH)_6]^{3-}(aq) + 3H_2O_2(aq) \rightarrow 2CrO_4^{2-}(aq) + 2OH^-(aq) + 8H_2O(l)$$

In a similar way, cobalt(II) complexes can be oxidized to cobalt(III) complexes.

Oxidation in alkaline solution

Co(II) is oxidized in two steps to Co(III) in alkaline solution using hydrogen peroxide, H_2O_2:

Aqueous sodium hydroxide is added, forming a blue-green precipitate:

$$[Co(H_2O)_6]^{2+}(aq) + 2OH^-(aq) \rightarrow [Co(H_2O)_4(OH)_2](s) + 2H_2O(l)$$

The mixture is then warmed with aqueous hydrogen peroxide, forming a dark brown precipitate:

$$2[Co(H_2O)_4(OH)_2](s) + H_2O_2(aq) \rightarrow 2[Co(H_2O)_3(OH)_3](s) + 2H_2O(l)$$

Summary

The table summarizes the complexes and their colours. Note that the cobalt(III) hydroxide precipitate is also described as hydrated cobalt(III) oxide, $Cr_2O_3.xH_2O$.

oxidation state	complex	colour
+2	$[Co(H_2O)_6]^{2+}(aq)$	pink
+2	$[Co(H_2O)_4(OH)_2](s)$	blue-green
+3	$[Co(H_2O)_3(OH)_3](s)$	dark brown

Oxidation by air

Co(II) can be oxidized by air in ammoniacal solution. It happens in three steps:

Aqueous ammonia is added to aqueous Co^{2+} ions, forming a blue-green precipitate:

$$[Co(H_2O)_6]^{2+}(aq) + 2NH_3(aq) \rightarrow [Co(H_2O)_4(OH)_2](s) + 2NH_4^+(aq)$$

The precipitate redissolves in excess ammonia to form a pale brown, straw-coloured solution:

$$[Co(H_2O)_4(OH)_2](s) + 6NH_3(aq) \rightarrow$$
$$[Co(NH_3)_6]^{2+}(aq) + 4H_2O(l) + 2OH^-(aq)$$

The aqueous hexaamminecobalt(II) complex then rapidly oxidizes in air to form a dark brown mixture containing aqueous hexaamminecobalt(III) ions:

$$[Co(NH_3)_6]^{2+}(aq) \rightarrow [Co(NH_3)_6]^{3+}(aq) + e^-$$

Note that the $[Co(NH_3)_6]^{3+}$ ion is actually yellow. But other complexes form, such as purple $[Co(H_2O)(NH_3)_5]^{3+}$ ions, turning the mixture brown.

Ease of oxidation

Transition metal ions in low oxidation states are easily oxidized to higher oxidation states in alkaline solution. They are less easily oxidized in acidic or neutral solutions. Aqueous hydrogen peroxide will oxidize chromium(III) hydroxide to chromate(VI) in the presence of an alkali such as aqueous sodium hydroxide. It will also oxidize cobalt(II) hydroxide to cobalt(III) hydroxide, but oxygen in air is a strong enough oxidizing agent to do this, too.

Oxygen in air can also oxidize alkaline solutions of iron(II) hydroxide to iron(III) hydroxide. This means that green precipitates of iron(II) hydroxide, formed by adding aqueous sodium hydroxide to aqueous iron(II) ions, rapidly oxidize to an orange-brown precipitate of iron(III) hydroxide.

Solutions of iron(II) ions will oxidize during storage unless they are in acidic conditions. For example, iron(II) sulfate solutions are usually prepared by dissolving iron(II) sulfate in dilute sulfuric acid (see Spread 17.03).

A green precipitate of iron(II) hydroxide formed by adding aqueous sodium hydroxide to aqueous iron(II) ions. Note that it is beginning to turn orange-brown.

Aqueous iron(II) sulfate is green but will oxidize rapidly in air unless dilute sulfuric acid is present.

Summary

The table summarizes the complexes and their colours.

oxidation state	complex	colour
+2	$[Co(H_2O)_6]^{2+}(aq)$	pink
+2	$[Co(H_2O)_4(OH)_2](s)$	blue-green
+2	$[Co(NH_3)_6]^{2+}(aq)$	pale brown
+3	$[Co(NH_3)_6]^{3+}(aq)$	yellow

Check your understanding

1 A pink solution of a cobalt complex **A** forms a blue-green precipitate **B** when excess aqueous sodium hydroxide is added. A dark brown precipitate of complex **C** is formed when **B** is warmed with aqueous hydrogen peroxide.

  a Write the names and formulae of complexes **A**, **B**, and **C**.

  b What is the role of the hydrogen peroxide?

2 Pink aqueous cobalt chloride forms a blue-green precipitate when aqueous ammonia is added to it.

  a Identify the two complexes present.

  b What happens when excess aqueous ammonia is added to the precipitate?

  c Write the name and formula of the complex formed in part **b**.

  d The complex formed in part **b** oxidizes in air to form a brown mixture of complexes. Identify the yellow Co(III) complex in this mixture.

20.01 Metal aqua ions

OUTCOMES

already from A2 Level, you

- know the definitions of a Brønsted–Lowry acid and a Brønsted–Lowry base
- understand that a complex is a central metal ion surrounded by ligands
- can define the term ligand
- know that co-ordinate bonding is involved in complex formation

and after this spread you should know

- the definitions of a Lewis acid and a Lewis base
- that metal–aqua ions are formed in aqueous solution:
- $[M(H_2O)_6]^{2+}$, limited to copper, cobalt, and iron
- $[M(H_2O)_6]^{3+}$, limited to aluminium, chromium, and iron

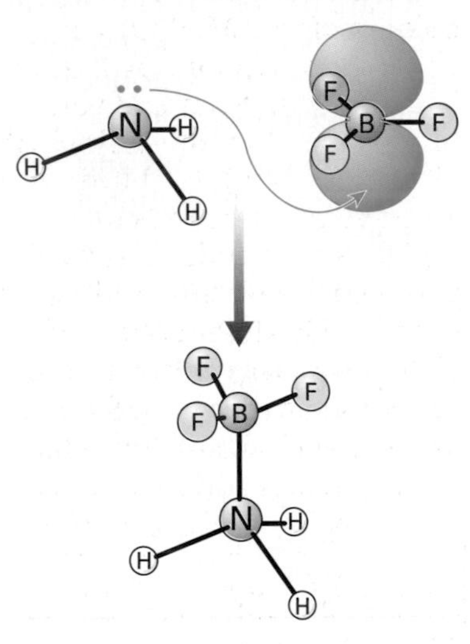

Ammonia forms a co-ordinate bond with boron trifluoride.

Acid–base reactions can be described in terms of the transfer of protons or hydrogen ions, H^+:

- a Brønsted–Lowry acid is a proton donor
- a Brønsted–Lowry base is a proton acceptor

For example, benzene is nitrated using a mixture of concentrated sulfuric acid and concentrated nitric acid (Spread 7.03). They react according to this equation:

$$H_2SO_4 + HNO_3 \rightarrow H_2NO_3^+ + HSO_4^-$$

Sulfuric acid donates protons to nitric acid. So in this example, sulfuric acid acts as a Brønsted–Lowry acid, and nitric acid acts as a Brønsted–Lowry base because it accepts protons. Some reactions appear to be acid–base reactions but do not involve transfer of protons. For example, calcium oxide reacts with sulfur(VI) oxide to produce a salt, calcium sulfate:

$$CaO(s) + SO_3(g) \rightarrow CaSO_4(s)$$

The theory of Lewis acids and bases gets around this difficulty.

Lewis acid–base theory

The American chemist Gilbert Lewis introduced his theory of acids and bases in 1923. It involves the donation of a pair of electrons by one atom to another:

- a **Lewis acid** is an electron pair acceptor
- a **Lewis base** is an electron pair donor

A co-ordinate bond forms between the acid and base. For example, ammonia and boron trifluoride react together to form a solid product:

$$NH_3(g) + BF_3(g) \rightarrow NH_3BF_3(s)$$

Ammonia acts as a Lewis base because it donates a pair of electrons to the vacant orbital in the boron atom, and boron trifluoride acts as a Lewis acid because it accepts a pair of electrons. The co-ordinate bond is shown by an arrow pointing from the donor atom to the accepting atom: $H_3N \rightarrow BF_3$.

Ligands donate pairs of electrons to a central metal ion in complexes. This means that ligands act as Lewis bases, and the metal ions act as Lewis acids.

Metal–aqua ions

Transition metal ions such as Fe^{2+} and Fe^{3+} have a high charge density. In aqueous solutions, water molecules are attracted very strongly to these ions. The oxygen atom in a water molecule donates a lone pair of electrons to a vacant orbital in the metal ion, forming a co-ordinate bond. Metal ions form hexaaqua complexes, $[M(H_2O)_6]^{2+}$ and $[M(H_2O)_6]^{3+}$, which are octahedral and have a co-ordination number of 6.

$[M(H_2O)_6]^{2+}$ ions

Copper, cobalt, and iron(II) ions form hexaaqua complexes in aqueous solution. For example, anhydrous copper(II) sulfate dissolves in water to produce the hexaaquacopper(II) complex:

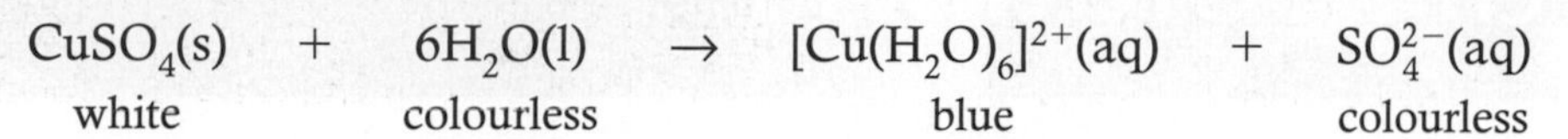

$$CuSO_4(s) + 6H_2O(l) \rightarrow [Cu(H_2O)_6]^{2+}(aq) + SO_4^{2-}(aq)$$

white — colourless — blue — colourless

In a similar way, blue anhydrous cobalt(II) chloride dissolves in water to form the hexaaquacobalt(II) complex, which is pink:

$$CoCl_2(s) + 6H_2O(l) \rightarrow [Co(H_2O)_6]^{2+}(aq) + 2Cl^-(aq)$$

Pale green iron(II) sulfate dissolves in water to form the hexaaquairon(II) complex, which is also pale green:

$$FeSO_4(s) + 6H_2O(l) \rightarrow [Fe(H_2O)_6]^{2+}(aq) + SO_4^{2-}(aq)$$

Water reacts with anhydrous copper(II) sulfate to produce blue hydrated copper(II) sulfate, containing the hexaaquacopper(II) complex. The reaction is vigorously exothermic.

Summary

The table summarizes the colours of the $[M(H_2O)_6]^{2+}$ complexes required by the Specification.

complex	colour
$[Cu(H_2O)_6]^{2+}$ (aq)	blue
$[Co(H_2O)_6]^{2+}$ (aq)	pink
$[Fe(H_2O)_6]^{2+}$ (aq)	pale green

$[M(H_2O)_6]^{3+}$ ions

Aluminium is not a transition metal, but its Al^{3+} ion has a high charge density and forms hexaaqua complexes in aqueous solution:

$$AlCl_3(s) + 6H_2O(l) \rightarrow [Al(H_2O)_6]^{3+}(aq) + 3Cl^-(aq)$$

The reaction is violently exothermic. The solution of the hexaaquaaluminium complex is colourless, since aluminium is not a transition metal and its ions do not undergo d-to-d transitions.

Chromium(III) compounds dissolve in water to form the hexaaquachromium(III) complex, which is ruby coloured:

$$CrCl_3(s) + 6H_2O(l) \rightarrow [Cr(H_2O)_6]^{2+}(aq) + 3Cl^-(aq)$$

Iron(III) compounds dissolve in water to form the hexaaquairon(III) complex, which is pale violet:

$$FeCl_3(s) + 6H_2O(l) \rightarrow [Fe(H_2O)_6]^{3+}(aq) + 3Cl^-(aq)$$

Summary

The table summarizes the colours of the $[M(H_2O)_6]^{3+}$ complexes required by the Specification. Note that you may not see the ruby colour of $[Cr(H_2O)_6]^{3+}$(aq), as chloride ions and other ligands may replace water molecules to give green complexes. You will not see the pale violet colour of $[Fe(H_2O)_6]^{3+}$(aq) either, as the complex hydrolyses to form brown $[Fe(H_2O)_5(OH)]^{2+}$(aq).

complex	colour
$[Al(H_2O)_6]^{3+}$(aq)	colourless
$[Cr(H_2O)_6]^{3+}$(aq)	ruby
$[Fe(H_2O)_6]^{3+}$(aq)	pale violet

Check your understanding

1 a Define the terms *Lewis acid* and *Lewis base*.

b Identify the Lewis acid and Lewis base in this reaction:

$$H_2O(l) + H^+(aq) \rightarrow H_3O^+(aq).$$

2 State the names, formulae, and colours of the aqua complexes formed by Cu(II), Co(II), Fe(II), Al, Cr(III), and Fe(III).

20.02 Acidic solutions

OUTCOMES

already from A2 Level, you know

- that metal–aqua ions are formed in aqueous solution:

and after this spread you should understand

- the equilibrium $[M(H_2O)_6]^{3+} + H_2O \rightleftharpoons [M(H_2O)_5(OH)]^{2+} + H_3O^+$ to show generation of weakly acidic solutions with M^{3+}
- the equilibrium $[M(H_2O)_6]^{2+} + H_2O \rightleftharpoons M(H_2O)_5(OH)]^+ + H_3O^+$ to show generation of very weakly acidic solutions with M^{2+}
- that the acidity of $[M(H_2O)_6]^{3+}$ is greater than that of $M(H_2O)_6]^{2+}$ in terms of the charge/size ratio of the metal ion

Metal–aqua ions can undergo two types of reaction, depending upon which bonds in the complex are broken. If a co-ordinate bond between an aqua ligand and the central metal ion breaks, the ligand can be replaced by a different ligand in a *substitution reaction.* If the O—H bond in an aqua ligand breaks, a hydrogen ion is released in a *hydrolysis reaction.* You are going to study hydrolysis reactions on this spread, and substitution reactions in the following spreads.

Hydrolysis reactions in complexes

Aluminium is not a transition metal but it does form hexaaquaaluminium complexes in aqueous solution:

$$AlCl_3(s) + 6H_2O(l) \rightarrow [Al(H_2O)_6]^{3+}(aq) + 3Cl^-(aq)$$

The resulting solution is acidic, with a typical pH of about 3. This is because the $[Al(H_2O)_6]^{3+}$ complex reacts with water in a hydrolysis reaction to produce oxonium ions H_3O^+:

$$[Al(H_2O)_6]^{3+}(aq) + H_2O(l) \rightleftharpoons [Al(H_2O)_5(OH)]^{2+}(aq) + H_3O^+(aq)$$

The $[Al(H_2O)_6]^{3+}$ complex acts as a Brønsted–Lowry acid because it donates a proton to a water molecule in the solution. The position of equilibrium lies far to the left, which is why solution is weakly acidic. The reaction is also called an **acidity reaction** because it produces an acidic solution. Transition metal hexaaqua complexes where the central metal ion has a 2+ or 3+ charge also undergo hydrolysis reactions.

Breaking bonds

Metal ions with high charge/size ratios are very polarizing. Aluminium ions and many transition metal ions are like this. They have a high positive charge and a relatively small size. As a result, these ions can polarize their aqua ligands and so weaken the O—H bonds. A water molecule in the solvent then acts as a Brønsted–Lowry base and removes a hydrogen ion from an aqua ligand, forming a hydroxo ligand in the complex.

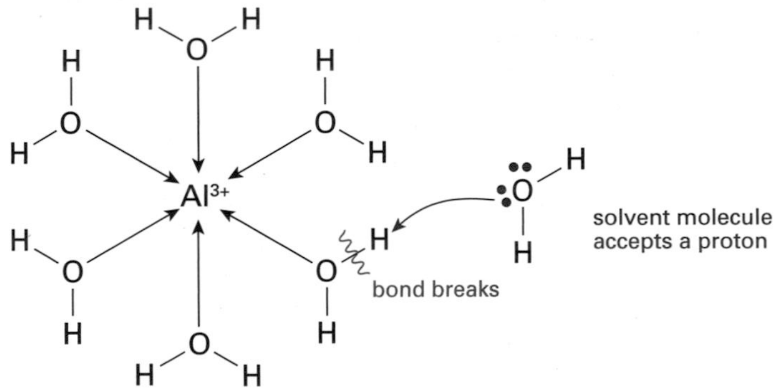

In a hydrolysis reaction, an O—H bond breaks in an aqua ligand and a proton is accepted by a solvent molecule.

A different equation

You may sometimes see the equilibrium equations written without water. In general you may see:

$$[M(H_2O)_6]^{3+}(aq) \rightleftharpoons [M(H_2O)_5(OH)]^{2+}(aq) + H^+(aq)$$

$$[M(H_2O)_6]^{2+}(aq) \rightleftharpoons [M(H_2O)_5(OH)]^{+}(aq) + H^+(aq)$$

This makes it easy to see that a hydrogen ion is released from an aqua ligand to form a hydroxo ligand, but it ignores the role of the solvent water molecule as a Brønsted–Lowry base.

M^{3+} transition metal ions

Chromium(III) forms the hexaaquachromium(III) complex in water. This hydrolyses to form the pentaaquahydroxochromium(III) complex:

$$[Cr(H_2O)_6]^{3+}(aq) + H_2O(l) \rightleftharpoons [Cr(H_2O)_5(OH)]^{2+}(aq) + H_3O^+(aq)$$

Iron(III) forms the hexaaquairon(III) complex in water. This hydrolyses to form the pentaaquahydroxoiron(III) complex:

$$[Fe(H_2O)_6]^{3+}(aq) + H_2O(l) \rightleftharpoons [Fe(H_2O)_5(OH)]^{2+}(aq) + H_3O^+(aq)$$

M^{2+} transition metal ions

Copper(II) forms the hexaaquacopper(II) complex in water. This hydrolyses to form the pentaaquahydroxocopper(II) complex:

$$[Cu(H_2O)_6]^{2+}(aq) + H_2O(l) \rightleftharpoons [Cu(H_2O)_5(OH)]^{+}(aq) + H_3O^+(aq)$$

The position of equilibrium lies very far to the left, so the solution is only very weakly acidic, with a pH of around 6.

Cobalt(II) forms the hexaaquacobalt(II) complex in water. This hydrolyses to form the pentaaquahydroxocobalt(II) complex:

$$[Co(H_2O)_6]^{2+}(aq) + H_2O(l) \rightleftharpoons [Co(H_2O)_5(OH)]^{+}(aq) + H_3O^+(aq)$$

Iron(II) forms the hexaaquairon(II) complex in water. This hydrolyses to form the pentaaquahydroxoiron(II) complex:

$$[Fe(H_2O)_6]^{2+}(aq) + H_2O(l) \rightleftharpoons [Fe(H_2O)_5(OH)]^{+}(aq) + H_3O^+(aq)$$

Factors affecting acidity

In general, acid strength increases as

- the ionic radius decreases
- the size of the charge increases

Highly charged, relatively small metal ions such as Al^{3+}, Cr^{3+}, and Fe^{3+} have a higher charge/size ratio than metal ions such as Cu^{2+}, Co^{2+}, and Fe^{2+}. They have a greater polarizing power, so they more strongly attract electron density from the oxygen atoms in aqua ligands. This makes the O—H bond weaker, so less energy is needed to break it to release a hydrogen ion. As a result, the acidity of $[M(H_2O)_6]^{3+}$ complexes is greater than that of $M(H_2O)_6]^{2+}$ complexes.

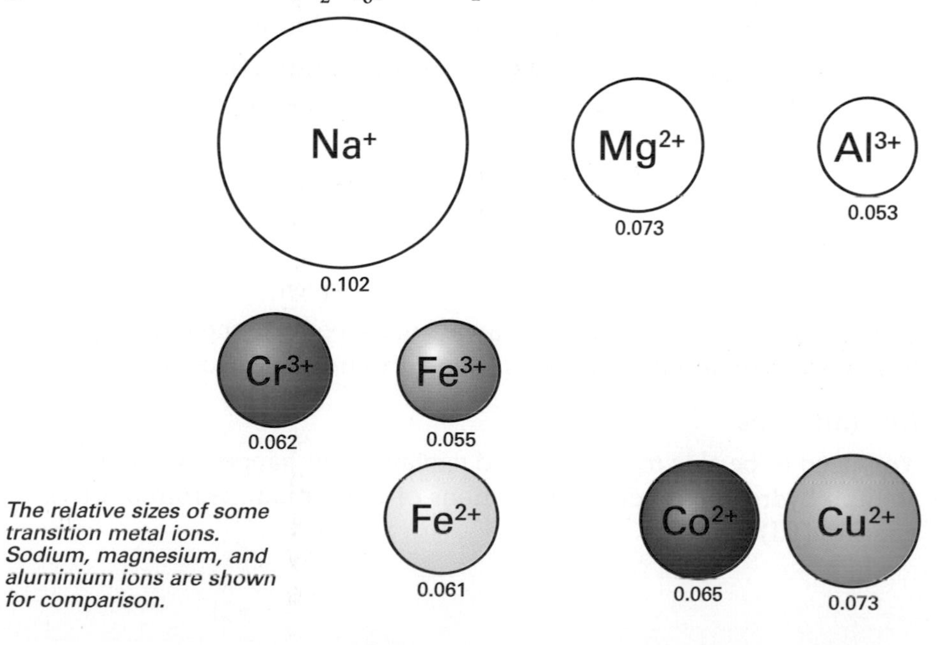

The relative sizes of some transition metal ions. Sodium, magnesium, and aluminium ions are shown for comparison.

The colour of $[Fe(H_2O)_6]^{3+}$

The $[Fe(H_2O)_6]^{3+}$ complex has a pale violet colour. This is seen in solids such as iron(III) nitrate-9-water, $Fe(NO_3)_3.9H_2O$. But when this is dissolved, the $[Fe(H_2O)_6]^{3+}$ complex hydrolyses to form brown aqueous $[Fe(H_2O)_5(OH)]^{2+}$. Even though the position of equilibrium lies far to the left, the brown colour of the $[Fe(H_2O)_5(OH)]^{2+}$ complexes hides the pale violet colour of the more numerous $[Fe(H_2O)_6]^{3+}$ complexes.

Check your understanding

1 a Explain why metal aqua ions produce weakly acidic solutions.

b Explain why a solution of iron(III) nitrate is likely to be more acidic than a solution of iron(II) nitrate.

2 Explain why $[Fe(H_2O)_6]^{3+}$ has a pale violet colour but appears brown in aqueous solution.

20.03 Reactions with hydroxide ions

OUTCOMES

already from AS Level, you can

- use Le Chatelier's principle to predict the effects of changes in concentration on the position of equilibrium

already from A2 Level, you understand

- the equilibria $[M(H_2O)_6]^{2+} + H_2O \rightleftharpoons M(H_2O)_5(OH)]^+ + H_3O^+$ and $[M(H_2O)_6]^{3+} + H_2O \rightleftharpoons [M(H_2O)_5(OH)]^{2+} + H_3O^+$

and after this spread you should be able to

- describe and explain the simple test-tube reactions of $M^{2+}(aq)$ ions with hydroxide ions, limited to M = Cu, Co, and Fe
- describe and explain the simple test-tube reactions of $M^{3+}(aq)$ ions with hydroxide ions, limited to M = Al, Cr, and Fe

Introducing metal aqua ions

Metal–aqua ions can undergo hydrolysis reactions in which O—H bonds in aqua ligands are broken. These are reversible reactions, and in aqueous solution the position of equilibrium lies far to the left. The addition of a base such as aqueous sodium hydroxide alters the position of equilibrium.

Summary M^{2+}

The table summarizes the precipitates and their colours.

precipitate	colour
$[Cu(H_2O)_4(OH)_2](s)$	blue
$[Co(H_2O)_4(OH)_2](s)$	blue-green
$[Fe(H_2O)_4(OH)_2](s)$	green

$M^{2+}(aq)$ ions

You need to be able to describe and explain what happens when aqueous sodium hydroxide is added to solutions of copper(II), cobalt(II), and iron(II) ions.

Background

An aqueous solution of M^{2+} transition metal ions contains the metal hexaaqua ion $[M(H_2O)_6]^{2+}(aq)$, and a small proportion of the $[M(H_2O)_5(OH)]^+(aq)$ ion:

$$[M(H_2O)_6]^{2+}(aq) + H_2O(l) \rightleftharpoons [M(H_2O)_5(OH)]^+(aq) + H_3O^+(aq)$$

When hydroxide ions are added, they act as a base and react with the oxonium ions H_3O^+:

$$OH^-(aq) + H_3O^+(aq) \rightarrow 2H_2O(l)$$

This moves the position of equilibrium to the right, and a new equilibrium is set up:

$$[M(H_2O)_5(OH)]^+(aq) + H_2O(l) \rightleftharpoons [M(H_2O)_4(OH)_2](s) + H_3O^+(aq)$$

The reaction between hydroxide ions and oxonium ions causes the position of this equilibrium to move to the right, and an insoluble metal(II) hydroxide precipitate forms. The overall general equation for the reaction is:

$$[M(H_2O)_6]^{2+}(aq) + 2OH^-(aq) \rightleftharpoons [M(H_2O)_4(OH)_2](s) + 2H_2O(l)$$

Aqueous copper(II), cobalt(II), and iron(II) react in this way.

Copper(II)

A blue precipitate of copper(II) hydroxide forms when aqueous sodium hydroxide is added to aqueous copper(II) ions:

$$\underset{\text{blue}}{[Cu(H_2O)_6]^{2+}(aq)} + 2OH^-(aq) \rightleftharpoons \underset{\text{blue}}{[Cu(H_2O)_4(OH)_2](s)} + 2H_2O(l)$$

Cobalt(II)

A blue-green precipitate of cobalt(II) hydroxide forms:

$$\underset{\text{pink}}{[Co(H_2O)_6]^{2+}(aq)} + 2OH^-(aq) \rightleftharpoons \underset{\text{blue-green}}{[Co(H_2O)_4(OH)_2](s)} + 2H_2O(l)$$

Iron(II)

A green precipitate of iron(II) hydroxide forms:

$$\underset{\text{pale green}}{[Fe(H_2O)_6]^{2+}(aq)} + 2OH^-(aq) \rightleftharpoons \underset{\text{green}}{[Fe(H_2O)_4(OH)_2](s)} + 2H_2O(l)$$

The precipitate gradually turns orange-brown because the iron(II) hydroxide is oxidized by oxygen in the air to iron(III) hydroxide.

$M^{3+}(aq)$ ions

You need to be able to describe and explain what happens when aqueous sodium hydroxide is added to solutions of aluminium, chromium(III), and iron(III) ions.

Background

An aqueous solution of M^{3+} metal ions contains the metal hexaaqua ion $[M(H_2O)_6]^{3+}(aq)$, and a small proportion of the $[M(H_2O)_5(OH)]^{2+}(aq)$ ion:

$$[M(H_2O)_6]^{3+}(aq) + H_2O(l) \rightleftharpoons [M(H_2O)_5(OH)]^{2+}(aq) + H_3O^+(aq)$$

The reaction between hydroxide ions and oxonium ions causes the position of this equilibrium to move to the right:

$$[M(H_2O)_5(OH)]^{2+}(aq) + H_2O(l) \rightleftharpoons [M(H_2O)_4(OH)_2]^+(aq) + H_3O^+(aq)$$

Further reaction between hydroxide ions and oxonium ions moves the position of this equilibrium to the right, and a new equilibrium is set up:

$$[M(H_2O)_5(OH)_2]^+(aq) + H_2O(l) \rightleftharpoons [M(H_2O)_3(OH)_3](s) + H_3O^+(aq)$$

Further reaction between hydroxide ions and oxonium ions moves the position of this equilibrium to the right, and an insoluble metal(III) hydroxide precipitate forms. Here is the overall general equation for the reaction:

$$[M(H_2O)_6]^{3+}(aq) + 3OH^-(aq) \rightleftharpoons [M(H_2O)_3(OH)_3](s) + 3H_2O(l)$$

Aluminium

A white precipitate of aluminium hydroxide forms when aqueous sodium hydroxide is added to aqueous aluminium ions:

$$\underset{\text{colourless}}{[Al(H_2O)_6]^{3+}(aq)} + 3OH^-(aq) \rightleftharpoons \underset{\text{white}}{[Al(H_2O)_3(OH)_3](s)} + 3H_2O(l)$$

Metal(II) hydroxides tend not to redissolve in excess aqueous sodium hydroxide but metal(III) hydroxides often do. Aluminium hydroxide redissolves to form the tetrahydroxoaluminate complex:

$$\underset{\text{white}}{[Al(H_2O)_3(OH)_3](s)} + OH^-(aq) \rightleftharpoons \underset{\text{colourless}}{[Al(OH)_4]^-(aq)} + 3H_2O(l)$$

Chromium(III)

A grey-green precipitate of chromium(III) hydroxide forms when aqueous sodium hydroxide is added to aqueous chromium(III) ions:

$$\underset{\text{ruby}}{[Cr(H_2O)_6]^{3+}(aq)} + 3OH^-(aq) \rightleftharpoons \underset{\text{grey-green}}{[Cr(H_2O)_3(OH)_3](s)} + 3H_2O(l)$$

You may not see a ruby colour at the start, as chloride ions and other ligands may replace water molecules to give green complexes. Chromium(III) hydroxide redissolves in excess aqueous sodium hydroxide to form the hexahydroxochromate(III) complex:

$$\underset{\text{grey-green}}{[Cr(H_2O)_3(OH)_3](s)} + 3OH^-(aq) \rightleftharpoons \underset{\text{dark green}}{[Cr(OH)_6]^{3-}(aq)} + 3H_2O(l)$$

Iron(III)

A brown precipitate of iron(III) hydroxide forms when aqueous sodium hydroxide is added to aqueous solutions of iron(III) ions:

$$\underset{\text{violet}}{[Fe(H_2O)_6]^{3+}(aq)} + 3OH^-(aq) \rightleftharpoons \underset{\text{brown}}{[Fe(H_2O)_3(OH)_3](s)} + 3H_2O(l)$$

Remember that the presence of brown $[Fe(H_2O)_5(OH)]^{2+}(aq)$ will mask the violet colour at the start.

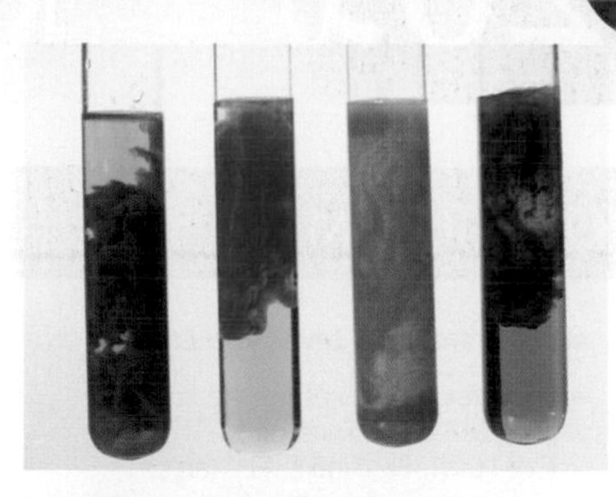

Precipitates of iron(III) hydroxide, copper(II) hydroxide, chromium(III) hydroxide, and cobalt(II) hydroxide.

Summary M^{3+}

The table summarizes the precipitates and their colours.

precipitate	colour
$[Al(H_2O)_3(OH)_3](s)$	white
$[Cr(H_2O)_3(OH)_3](s)$	grey-green
$[Fe(H_2O)_3(OH)_3](s)$	brown

Reversible reactions

The position of all these equilibria can be moved towards the left by adding a strong acid such as nitric acid.

Check your understanding

1 Describe the colour changes observed when aqueous sodium hydroxide is added to solutions of:
   a cobalt(II) chloride
   b iron(II) sulfate
   c iron(III) sulfate
2 Explain these observations:
   a Aqueous copper(II) sulfate reacts with aqueous sodium hydroxide to form a blue precipitate.
   b Aqueous aluminium sulfate reacts with aqueous sodium hydroxide to form a white precipitate that redissolves in excess alkali to form a colourless solution.

20.04 Reactions with ammonia

OUTCOMES

already from A2 Level, you can

- describe and explain the simple test-tube reactions of $Cu^{2+}(aq)$, $Co^{2+}(aq)$, $Fe^{2+}(aq)$, $Al^{3+}(aq)$, $Cr^{3+}(aq)$, and $Fe^{3+}(aq)$ ions with hydroxide ions

and after this spread you should be able to

- describe and explain the simple test-tube reactions of $M^{2+}(aq)$ ions with ammonia, limited to M = Cu, Co, and Fe
- describe and explain the simple test-tube reactions of $M^{3+}(aq)$ ions with ammonia, limited to M = Al, Cr, and Fe

Precipitates of copper(II) hydroxide and iron(II) hydroxide. The deep blue solution above the copper(II) hydroxide precipitate is due to ligand substitution by ammonia. The iron(II) hydroxide precipitate has begun to oxidize.

Summary M^{2+}

The table summarizes the precipitates and their colours.

precipitate	colour
$[Cu(H_2O)_4(OH)_2](s)$	blue
$[Co(H_2O)_4(OH)_2](s)$	blue-green
$[Fe(H_2O)_4(OH)_2](s)$	green

Aqueous Cu^{2+}, Co^{2+}, Fe^{2+}, Al^{3+}, Cr^{3+}, and Fe^{3+} all produce metal hydroxide precipitates when aqueous sodium hydroxide is added to them. These precipitates are all coloured, apart from aluminium hydroxide. Similar reactions also happen when ammonia is added.

Metal–aqua ions can also undergo ligand substitution reactions in the presence of ammonia, not just hydrolysis reactions. Reactions due to ammonia acting as a Brønsted–Lowry base are discussed on this spread, and ligand substitution reactions involving ammonia are discussed on Spread 21.01.

$M^{2+}(aq)$ ions

You need to be able to describe and explain what happens when a small amount of aqueous ammonia is added to solutions of copper(II), cobalt(II), and iron(II) ions.

Background

An aqueous solution of M^{2+} transition metal ions contains the metal hexaaqua ion $[M(H_2O)_6]^{2+}(aq)$, and a small proportion of the $[M(H_2O)_5(OH)]^{+}(aq)$ ion:

$$[M(H_2O)_6]^{2+}(aq) + H_2O(l) \rightleftharpoons [M(H_2O)_5(OH)]^{+}(aq) + H_3O^{+}(aq)$$

When aqueous ammonia is added, it acts as a base and reacts with the oxonium ions H_3O^+:

$$NH_3(aq) + H_3O^{+}(aq) \rightarrow NH_4^{+}(aq) + H_2O(l)$$

This moves the position of equilibrium to the right, and a new equilibrium is set up:

$$[M(H_2O)_5(OH)]^{+}(aq) + H_2O(l) \rightleftharpoons [M(H_2O)_4(OH)_2](s) + H_3O^{+}(aq)$$

The reaction between ammonia and oxonium ions causes the position of this equilibrium to move to the right, and an insoluble metal(II) hydroxide precipitate forms. Here is the overall general equation for the reaction:

$$[M(H_2O)_6]^{2+}(aq) + 2NH_3(aq) \rightleftharpoons [M(H_2O)_4(OH)_2](s) + 2NH_4^{+}(aq)$$

Copper(II), cobalt(II), and iron(II) react in this way. Their precipitates are the same as those observed when aqueous sodium hydroxide is added, although there might be differences in the shade of the colour.

Copper(II)

A blue precipitate of copper(II) hydroxide forms:

$$\underset{\text{blue}}{[Cu(H_2O)_6]^{2+}(aq)} + 2NH_3(aq) \rightleftharpoons \underset{\text{blue}}{[Cu(H_2O)_4(OH)_2](s)} + 2NH_4^{+}(aq)$$

Cobalt(II)

A blue-green precipitate of cobalt(II) hydroxide forms:

$$\underset{\text{pink}}{[Co(H_2O)_6]^{2+}(aq)} + 2NH_3(aq) \rightleftharpoons \underset{\text{blue-green}}{[Co(H_2O)_4(OH)_2](s)} + 2NH_4^{+}(aq)$$

Iron(II)

A green precipitate of iron(II) hydroxide forms:

$$[Fe(H_2O)_6]^{2+}(aq) + 2NH_3(aq) \rightleftharpoons [Fe(H_2O)_4(OH)_2](s) + 2NH_4^+(aq)$$

pale green — green

The precipitate gradually turns orange-brown because the iron(II) hydroxide is oxidized by oxygen in the air to iron(III) hydroxide.

M^{3+}(aq) ions

Background

An aqueous solution of M^{3+} metal ions contains the metal hexaaqua ion $[M(H_2O)_6]^{3+}(aq)$, and a small proportion of the $[M(H_2O)_5(OH)]^{2+}(aq)$ ion:

$$[M(H_2O)_6]^{3+}(aq) + H_2O(l) \rightleftharpoons [M(H_2O)_5(OH)]^{2+}(aq) + H_3O^+(aq)$$

The reaction between ammonia and oxonium ions causes the position of this equilibrium to move to the right:

$$[M(H_2O)_5(OH)]^{2+}(aq) + H_2O(l) \rightleftharpoons [M(H_2O)_4(OH)_2]^+(aq) + H_3O^+(aq)$$

Further reaction between ammonia and oxonium ions moves the position of this equilibrium to the right, and a new equilibrium is set up:

$$[M(H_2O)_5(OH)_2]^+(aq) + H_2O(l) \rightleftharpoons [M(H_2O)_3(OH)_3](s) + H_3O^+(aq)$$

The reaction between ammonia and oxonium ions moves the position of this equilibrium to the right, and an insoluble metal(III) hydroxide precipitate forms. Here is the overall general equation for the reaction:

$$[M(H_2O)_6]^{3+}(aq) + 3NH_3(aq) \rightleftharpoons [M(H_2O)_3(OH)_3](s) + 3NH_4^+(aq)$$

Aluminium, chromium(III), and iron(III) react in this way. Their precipitates are the same as those observed when aqueous sodium hydroxide is added, although there might be differences in the shade of the colour.

Aluminium

A white precipitate of aluminium hydroxide forms:

$$[Al(H_2O)_6]^{3+}(aq) + 3NH_3(aq) \rightleftharpoons [Al(H_2O)_3(OH)_3](s) + 3NH_4^+(aq)$$

colourless — white

Chromium(III)

A grey-green precipitate of chromium(III) hydroxide forms:

$$[Cr(H_2O)_6]^{3+}(aq) + 3NH_3(aq) \rightleftharpoons [Cr(H_2O)_3(OH)_3](s) + 3NH_4^+(aq)$$

ruby — grey-green

You may not see a ruby colour at the start, as chloride ions and other ligands may replace water molecules to give green complexes.

Iron(III)

A brown precipitate of iron(III) hydroxide forms:

$$[Fe(H_2O)_6]^{3+}(aq) + 3NH_3(aq) \rightleftharpoons [Fe(H_2O)_3(OH)_3](s) + 3NH_4^+(aq)$$

violet — brown

The presence of brown $[Fe(H_2O)_5(OH)]^{2+}(aq)$ will mask the violet colour at the start.

Check your understanding

1 Describe the colour changes observed when a small amount of aqueous ammonia is added to solutions of:
  a copper(II) sulfate
  c iron(II) nitrate
  b chromium(III) nitrate
2 Explain why aqueous iron(III) sulfate reacts with a small amount of aqueous ammonia to form a brown precipitate.

Aluminium hydroxide

Aluminium hydroxide redissolves in excess aqueous sodium hydroxide to form the tetrahydroxoaluminate complex:

$$[Al(H_2O)_3(OH)_3](s) + OH^-(aq) \rightleftharpoons [Al(OH)_4]^-(aq) + 3H_2O(l)$$

white — colourless

It does not redissolve in excess aqueous ammonia, but the transition metal hydroxides described on this spread do redissolve in excess ammonia (see Spread 21.01).

Summary M^{3+}

The table summarizes the precipitates and their colours.

precipitate	colour
$[Al(H_2O)_3(OH)_3](s)$	white
$[Cr(H_2O)_3(OH)_3](s)$	grey-green
$[Fe(H_2O)_3(OH)_3](s)$	brown

20.05 Reactions with carbonate ions

OUTCOMES

already from A2 Level, you can

- describe and explain the simple test-tube reactions of Cu^{2+}(aq), Co^{2+}(aq), Fe^{2+}(aq), Al^{3+}(aq), Cr^{3+}(aq), and Fe^{3+}(aq) ions with hydroxide ions and with ammonia

and after this spread you should

- be able to describe and explain the simple test-tube reactions of M^{2+}(aq) ions with carbonate ions, limited to M = Cu, Co, and Fe
- be able to describe and explain the simple test-tube reactions of M^{3+}(aq) ions with carbonate ions, limited to M = Al, Cr, and Fe
- know that MCO_3 is formed but that $M_2(CO_3)_3$ is not formed

Aqueous sodium carbonate on the left reacts with aqueous iron(II) sulfate on the right to produce a green precipitate of iron(II) carbonate. Note the brown iron(III) product near the top due to oxidation by air.

Summary

The table summarizes the precipitates and their colours.

precipitate	colour
$CuCO_3$(s)	green-blue
$CoCO_3$(s)	pink
$FeCO_3$	green

Sodium carbonate and other carbonates will react with metal–aqua ions in solution. The carbonate ion CO_3^{2-} can act as a Brønsted–Lowry base. It reacts with acids to produce the hydrogen carbonate ion HCO_3^-:

$$CO_3^{2-}(aq) + H_3O^+(aq) \rightleftharpoons HCO_3^-(aq) + H_2O(l)$$

The hydrogen carbonate ion reacts further to produce carbon dioxide:

$$HCO_3^-(aq) + H_3O^+(aq) \rightleftharpoons CO_2(g) + 2H_2O(l)$$

This is the overall reaction, provided the concentration of H_3O^+(aq) ions is high enough:

$$CO_3^{2-}(aq) + 2H_3O^+(aq) \rightarrow CO_2(g) + 3H_2O(l)$$

Metal–aqua ions react with carbonate ions in different ways, depending upon whether they are M^{2+}(aq) ions or M^{3+}(aq) ions.

M^{2+}(aq) ions

M^{2+} transition metal compounds produce the metal hexaaqua ion $[M(H_2O)_6]^{2+}$(aq) in aqueous solution. This hydrolyses to form the $[M(H_2O)_5(OH)]^+$(aq) complex:

$$[M(H_2O)_6]^{2+}(aq) + H_2O(l) \rightleftharpoons [M(H_2O)_5(OH)]^+(aq) + H_3O^+(aq)$$

The position of equilibrium lies very far to the left, so the solution is only very weakly acidic. So when carbonate ions are added, the reaction that produces carbon dioxide gas does not happen. Instead, metal(II) carbonates MCO_3 are produced. These are insoluble in water and they form precipitates:

$$[M(H_2O)_6]^{2+}(aq) + CO_3^{2-}(aq) \rightarrow MCO_3(s) + 6H_2O(l)$$

You need to be able to describe and explain the reactions between carbonate ions and Cu^{2+}(aq), Co^{2+}(aq), and Fe^{2+}(aq).

Copper(II)

A green-blue precipitate of copper(II) carbonate forms:

$$\underset{\text{blue}}{[Cu(H_2O)_6]^{2+}(aq)} + CO_3^{2-}(aq) \rightarrow \underset{\text{green-blue}}{CuCO_3(s)} + 6H_2O(l)$$

Cobalt(II)

A pink precipitate of cobalt(II) carbonate forms:

$$\underset{\text{pink}}{[Co(H_2O)_6]^{2+}(aq)} + CO_3^{2-}(aq) \rightarrow \underset{\text{pink}}{CoCO_3(s)} + 6H_2O(l)$$

Iron(II)

A green precipitate of iron(II) carbonate forms:

$$\underset{\text{pale green}}{[Fe(H_2O)_6]^{2+}(aq)} + CO_3^{2-}(aq) \rightarrow \underset{\text{green}}{FeCO_3(s)} + 6H_2O(l)$$

M^{3+}(aq) ions

M^{3+} transition metal compounds produce the metal hexaaqua ion $[M(H_2O)_6]^{3+}$(aq) in aqueous solution. This hydrolyses to form the $[M(H_2O)_5(OH)]^{2+}$(aq) complex:

$$[M(H_2O)_6]^{3+}(aq) + H_2O(l) \rightleftharpoons [M(H_2O)_5(OH)]^{2+}(aq) + H_3O^+(aq)$$

Bases react with the oxonium ions $H_3O^+(aq)$ and the position of equilibrium moves to the right. A new equilibrium is set up as a result:

$$[M(H_2O)_5(OH)]^{2+}(aq) + H_2O(l) \rightleftharpoons [M(H_2O)_4(OH)_2]^+(aq) + H_3O^+(aq)$$

The base reacts with more oxonium ions, and the position of this equilibrium moves to the right. Another new equilibrium is set up as a result:

$$[M(H_2O)_4(OH)_2]^+(aq) + H_2O(l) \rightleftharpoons [M(H_2O)_3(OH)_3](s) + H_3O^+(aq)$$

The position of this equilibrium moves to the right and an insoluble metal(III) hydroxide precipitate forms. This happens when aqueous hydroxide ions or aqueous ammonia is added. It also happens when carbonate ions are added. Here is the overall general equation for the reaction:

$$2[M(H_2O)_6]^{3+}(aq) + 3CO_3^{2-}(aq) \rightleftharpoons 2[M(H_2O)_3(OH)_3](s) + 3H_2O(l) + 3CO_2(g)$$

Note that metal(III) carbonates are not produced. It is not possible to make them from aqueous solutions, and $Al_2(CO_3)_3$, $Cr_2(CO_3)_3$, and $Fe_2(CO_3)_3$ do not exist.

Aqueous sodium carbonate and aqueous aluminium chloride react vigorously together, producing bubbles of carbon dioxide and a white precipitate of aluminium hydroxide.

Aluminium

A white precipitate of aluminium hydroxide forms:

$$\underset{\text{colourless}}{2[Al(H_2O)_6]^{3+}(aq)} + 3CO_3^{2-}(aq) \rightleftharpoons \underset{\text{white}}{2[Al(H_2O)_3(OH)_3](s)} + 3H_2O(l) + 3CO_2(g)$$

Chromium(III)

A grey-green precipitate of chromium(III) hydroxide forms:

$$\underset{\text{ruby}}{2[Cr(H_2O)_6]^{3+}(aq)} + 3CO_3^{2-}(aq) \rightleftharpoons \underset{\text{grey-green}}{2[Cr(H_2O)_3(OH)_3](s)} + 3H_2O(l) + 3CO_2(g)$$

You may not see a ruby colour at the start, as chloride ions and other ligands may replace water molecules to give green complexes.

Iron(III)

A brown precipitate of iron(III) hydroxide forms:

$$\underset{\text{violet}}{2[Fe(H_2O)_6]^{3+}(aq)} + 3CO_3^{2-}(aq) \rightleftharpoons \underset{\text{brown}}{2[Fe(H_2O)_3(OH)_3](s)} + 3H_2O(l) + 3CO_2(g)$$

The presence of brown $[Fe(H_2O)_5(OH)]^{2+}(aq)$ will mask the violet colour at the start.

Summary

The table summarizes the precipitates and their colours. Remember that you will see effervescence in each reaction as the carbon dioxide bubbles off.

precipitate	colour
$[Al(H_2O)_3(OH)_3](s)$	white
$[Cr(H_2O)_3(OH)_3](s)$	grey-green
$[Fe(H_2O)_3(OH)_3](s)$	brown

Check your understanding

1 State what you would observe when sodium carbonate is added to solutions of iron(II) sulfate and iron(III) sulfate. Give the formula of each iron-containing compound formed.

2 A green precipitate forms when sodium carbonate is added to aqueous chromium(III) chloride.
   a Name the precipitate.
   b Write an equation for the reaction.
   c What is the role of the sodium carbonate?

20.06 Amphoteric metal hydroxides

OUTCOMES

already from A2 Level, you

- can describe and explain the simple test-tube reactions of $Al^{3+}(aq)$ and $Cr^{3+}(aq)$ with bases
- know that Cr^{3+} is formed by reduction of $Cr_2O_7^{2-}$ by zinc in acid solution
- know the oxidation in alkaline solution of Cr^{3+} by H_2O_2

and after this spread you should know

- that some metal hydroxides show amphoteric character by dissolving in both acids and bases (e.g. hydroxides of Al^{3+} and Cr^{3+})
- the equilibrium reaction $2CrO_4^{2-} + 2H^+ \rightleftharpoons Cr_2O_7^{2-} + H_2O$

Acids, acids

It is best to use sulfuric acid or nitric acid in test tube reactions. If hydrochloric acid is used, chloride ions may substitute for ligands in the complexes, changing the colours observed.

Summary

The table summarizes the complexes and their colours.

precipitate	colour
$[Al(H_2O)_6]^{3+}(aq)$	colourless
$[Al(H_2O)_3(OH)_3](s)$	white
$[Al(OH)_4]^-(aq)$	colourless

Amphoteric substances can react with both acids and bases. Some metal(III) hydroxides are amphoteric, including aluminium hydroxide and chromium(III) hydroxide.

Metal(III) hydroxides in strong acid

An aqueous solution of M^{3+} metal ions contains the metal hexaaqua ion $[M(H_2O)_6]^{3+}(aq)$, and a small proportion of the $[M(H_2O)_5(OH)]^{2+}(aq)$ ion:

$$[M(H_2O)_6]^{3+}(aq) + H_2O(l) \rightleftharpoons [M(H_2O)_5(OH)]^{2+}(aq) + H_3O^+(aq)$$

The reactions between oxonium ions and a strong base such as hydroxide ions causes the position of this equilibrium to move to the right, eventually forming an insoluble metal(III) hydroxide precipitate. The overall general equation for the reaction is:

$$[M(H_2O)_6]^{3+}(aq) + 3OH^-(aq) \rightarrow [M(H_2O)_3(OH)_3](s) + 3H_2O(l)$$

This reaction can be reversed using strong acids:

$$[M(H_2O)_3(OH)_3](s) + 3H_3O^+(aq) \rightarrow [M(H_2O)_6]^{3+}(aq) + 3H_2O(l)$$

This means that metal(III) hydroxides will dissolve in excess strong acid. But some will also dissolve in excess strong base.

Metal(III) hydroxides in strong base

Metal(III) hydroxides are not hydrolysed by water but they may be attacked by strong bases such as hydroxide ions. The reactions produce negatively charged metal hydroxo complexes that dissolve in water. The ability of some metal(III) hydroxides to dissolve in both acid and base means that they amphoteric.

Amphoteric behaviour

Aluminium hydroxide and chromium(III) hydroxide are amphoteric.

Aluminium hydroxide

A white precipitate of aluminium hydroxide forms when aqueous sodium hydroxide is added to aqueous aluminium ions:

$$\underset{\text{colourless}}{[Al(H_2O)_6]^{3+}(aq)} + 3OH^-(aq) \rightleftharpoons \underset{\text{white}}{[Al(H_2O)_3(OH)_3](s)} + 3H_2O(l)$$

Aluminium hydroxide reacts with excess aqueous sodium hydroxide to form the soluble tetrahydroxoaluminate complex:

$$\underset{\text{white}}{[Al(H_2O)_3(OH)_3](s)} + OH^-(aq) \rightleftharpoons \underset{\text{colourless}}{[Al(OH)_4]^-(aq)} + 3H_2O(l)$$

Aluminium hydroxide also reacts with excess acid to form the soluble hexaaquaaluminium complex:

$$\underset{\text{white}}{[Al(H_2O)_3(OH)_3](s)} + 3H_3O^+(aq) \rightleftharpoons \underset{\text{colourless}}{[Al(H_2O)_6]^{3+}(aq)} + 3H_2O(l)$$

Chromium(III)

A grey-green precipitate of chromium(III) hydroxide forms when aqueous sodium hydroxide is added to aqueous chromium(III) ions:

$$\underset{\text{ruby}}{[Cr(H_2O)_6]^{3+}(aq)} + 3OH^-(aq) \rightleftharpoons \underset{\text{grey-green}}{[Cr(H_2O)_3(OH)_3](s)} + 3H_2O(l)$$

Chromium(III) hydroxide reacts with excess aqueous sodium hydroxide to form the soluble hexahydroxochromate(III) complex:

$$[Cr(H_2O)_3(OH)_3](s) + 3OH^-(aq) \rightleftharpoons [Cr(OH)_6]^{3-}(aq) + 3H_2O(l)$$

grey-green → dark green

Chromium(III) hydroxide also reacts with excess acid to form the soluble hexaaquachromium(III) complex:

$$[Cr(H_2O)_3(OH)_3](s) + 3H_3O^+(aq) \rightleftharpoons [Cr(H_2O)_6]^{3+}(aq) + 3H_2O(l)$$

grey-green → ruby

Chromium(III) hydroxide forms a grey-green precipitate.

Chromate(VI) and dichromate(VI)

Cr(III) can be oxidized to Cr(VI) using hydrogen peroxide (see Spread 19.03). When aqueous hexahydrochromate(III) is warmed with aqueous hydrogen peroxide, a yellow solution forms containing chromate(VI) ions, CrO_4^{2-}:

$$2[Cr(OH)_6]^{3-}(aq) + 3H_2O_2(aq) \rightarrow 2CrO_4^{2-}(aq) + 2OH^-(aq) + 8H_2O(l)$$

dark green → yellow

If the hydrogen peroxide is decomposed by boiling the mixture, an equilibrium forms between chromate(VI) ions and dichromate(VI) ions, $Cr_2O_7^{2-}$:

$$2CrO_4^{2-}(aq) + 2H^+(aq) \rightleftharpoons Cr_2O_7^{2-}(aq) + H_2O(l)$$

yellow → orange

The position of equilibrium moves to the right if dilute sulfuric acid is added, producing a higher concentration of orange $Cr_2O_7^{2-}(aq)$. The position of equilibrium moves to the left if aqueous sodium hydroxide is added, producing a higher concentration of yellow $CrO_4^{2-}(aq)$.

Summary

The table summarizes the complexes and their colours. You may not see a ruby colour at the start, as chloride ions and other ligands may replace water molecules to give green complexes.

precipitate	colour
$[Cr(H_2O)_6]^{3+}(aq)$	ruby
$[Cr(H_2O)_3(OH)_3](s)$	grey-green
$[Cr(OH)_6]^{3-}(aq)$	dark green

Check your understanding

1. a What is meant by the term *amphoteric*?
   b Write equations to explain why aluminium hydroxide and chromium(III) hydroxide are amphoteric.
2. Explain why an aqueous mixture of chromate(VI) ions and dichromate(VI) ions turns yellow when aqueous sodium hydroxide is added, and orange when dilute sulfuric acid is added.

21.01 Ligand exchange (1)

OUTCOMES

already from A2 Level, you

- understand that a complex is a central metal ion surrounded by ligands
- can define the term *ligand*
- know the meaning of *co-ordination number*
- know that water and ammonia can act as unidentate ligands
- know that transition metal ions commonly form octahedral complexes with small ligands such as water and ammonia
- know that metal–aqua ions are formed in aqueous solution
- know that colour changes can arise from changes in oxidation state, co-ordination number, and ligand
- can describe and explain the simple test-tube reactions of $Cr^{3+}(aq)$, $Co^{2+}(aq)$, and $Cu^{2+}(aq)$ ions with small amounts of ammonia

and after this spread you should

- understand that the ligands NH_3 and H_2O are similar in size and are uncharged, and that ligand exchange occurs without change of co-ordination number (e.g. Cr^{3+} and Co^{2+})
- know that substitution may be incomplete (e.g. the formation of $[Cu(NH_3)_4(H_2O)_2]^{2+}$)

Summary

The table summarizes the colours of the ammine complexes.

complex	colour
$[Cr(NH_3)_6]^{3+}(aq)$	purple
$[Co(NH_3)_6]^{2+}(aq)$	straw-coloured
$[Cu(NH_3)_4(H_2O)_2]^{2+}(aq)$	deep blue

Transition metal ions have a high charge/size ratio. In aqueous solutions, water molecules are attracted very strongly to them. The metal ions can act as Lewis acids and the water as a Lewis base. The oxygen atom in a water molecule donates a lone pair of electrons to a vacant orbital in the metal ion, forming a co-ordinate bond. Hexaaqua complexes form, with the general formulae $[M(H_2O)_6]^{2+}$ and $[M(H_2O)_6]^{3+}$. They have an octahedral shape and a co-ordination number of 6.

Metal–aqua ions can undergo two types of reaction, depending upon which bonds in the complex are broken. If the O—H bond in an aqua ligand breaks, a hydrogen ion is released in an *acidity* or *hydrolysis reaction*. If a co-ordinate bond between an aqua ligand and the central metal ion breaks, the ligand can be replaced by a different ligand in a *substitution reaction*. The spreads in Chapter 20 deal with hydrolysis reactions. You are going to study substitution reactions by ammonia in this spread, and by chloride ions in Spread 21.02.

Substitution by ammonia

Water and ammonia can both act as neutral ligands. They are unidentate ligands because they can each donate one lone pair of electrons to the central metal ion. Ammonia molecules can replace water molecules in complexes. This is an example of a **ligand substitution** reaction. Aqua ligands are replaced, one by one, in the presence of aqueous ammonia. Here are the general equations when a metal(III) aqua complex is involved:

$$[M(H_2O)_6]^{3+}(aq) + NH_3(aq) \rightleftharpoons [M(NH_3)(H_2O)_5]^{3+}(aq) + H_2O(l)$$
$$[M(NH_3)(H_2O)_5]^{3+}(aq) + NH_3(aq) \rightleftharpoons [M(NH_3)_2(H_2O)_4]^{3+}(aq) + H_2O(l)$$
$$[M(NH_3)_2(H_2O)_4]^{3+}(aq) + NH_3(aq) \rightleftharpoons [M(NH_3)_3(H_2O)_3]^{3+}(aq) + H_2O(l)$$
$$[M(NH_3)_3(H_2O)_3]^{3+}(aq) + NH_3(aq) \rightleftharpoons [M(NH_3)_4(H_2O)_2]^{3+}(aq) + H_2O(l)$$
$$[M(NH_3)_4(H_2O)_2]^{3+}(aq) + NH_3(aq) \rightleftharpoons [M(NH_3)_5(H_2O)]^{3+}(aq) + H_2O(l)$$
$$[M(NH_3)_5(H_2O)]^{3+}(aq) + NH_3(aq) \rightleftharpoons [M(NH_3)_6]^{3+}(aq) + H_2O(l)$$

Notice that the charge on the complex stays the same, because one neutral ligand is exchanged for another neutral ligand. Water and ammonia molecules have a similar size, and the co-ordination number stays at 6 in all the complexes. This also means that the complexes are octahedral in shape. The general equation for the overall reaction is:

$$\underset{\text{hexaaqua complex}}{[M(H_2O)_6]^{3+}(aq)} + 6NH_3(aq) \rightarrow \underset{\text{hexaammine complex}}{[M(NH_3)_6]^{3+}(aq)} + 6H_2O(l)$$

Chromium(III)

When a small amount of aqueous ammonia is added to aqueous chromium(III) ions, a grey-green precipitate of chromium(III) hydroxide forms:

$$\underset{\text{ruby}}{[Cr(H_2O)_6]^{3+}(aq)} + 3NH_3(aq) \rightleftharpoons \underset{\text{grey-green}}{[Cr(H_2O)_3(OH)_3](s)} + 3NH_4^+(aq)$$

If excess aqueous ammonia is added, a purple solution of hexaamminechromium(III) forms:

$$\underset{\text{ruby}}{[Cr(H_2O)_6]^{3+}(aq)} + 6NH_3(aq) \rightarrow \underset{\text{purple}}{[Cr(NH_3)_6]^{3+}(aq)} + 6H_2O(l)$$

Cobalt(II)

When a small amount of aqueous ammonia is added to aqueous cobalt(II) ions, a blue-green precipitate of cobalt(II) hydroxide forms:

$$\underset{\text{pink}}{[Co(H_2O)_6]^{2+}(aq)} + 2NH_3(aq) \rightleftharpoons \underset{\text{blue-green}}{[Co(H_2O)_4(OH)_2](s)} + 2NH_4^+(aq)$$

If excess aqueous ammonia is added, a pale brown, straw-coloured solution of hexaamminecobalt(II) forms:

$$\underset{\text{pink}}{[Co(H_2O)_6](s)} + 6NH_3(aq) \rightarrow \underset{\text{straw-coloured}}{[Co(NH_3)_6]^{2+}(aq)} + 6H_2O(l)$$

Copper(II)

When a small amount of aqueous ammonia is added to aqueous copper(II) ions, a blue precipitate of copper(II) hydroxide forms:

$$\underset{\text{blue}}{[Cu(H_2O)_6]^{2+}(aq)} + 2NH_3(aq) \rightleftharpoons \underset{\text{blue}}{[Cu(H_2O)_4(OH)_2](s)} + 2NH_4^+(aq)$$

A deep blue solution forms if excess aqueous ammonia is added. But this does not contain hexaamminecopper(II). Instead, just four of the six aqua ligands are replaced by ammine ligands, and the tetraamminediaqu acopper(II) complex forms:

$$\underset{\text{blue}}{[Cu(H_2O)_6](s)} + 4NH_3(aq) \rightarrow \underset{\text{deep blue}}{[Cu(NH_3)_4(H_2O)_2]^{2+}(aq)} + 4H_2O(l)$$

This complex has four ammine ligands in a square planar arrangement around the central copper(II) ion. The two aqua ligands occupy positions above and below the plane, completing the octahedral complex.

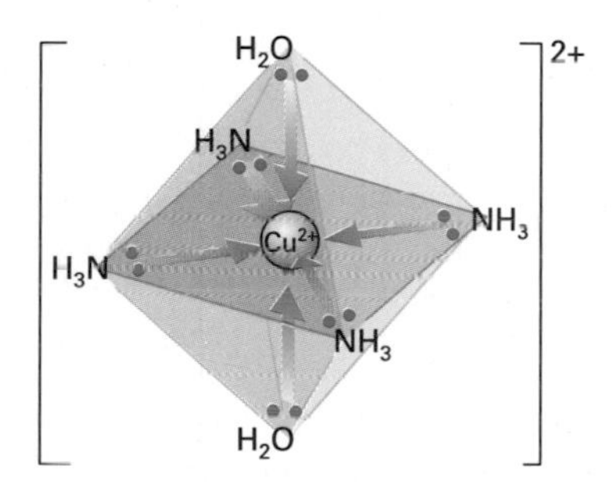

The structure of the octahedral tetraamminediaquacopper(II) complex.

Aerial oxidation

Oxygen in the air rapidly oxidizes the aqueous hexaamminecobalt(II) complex, forming a dark brown mixture containing aqueous hexaamminecobalt(III):

$$[Co(NH_3)_6]^{2+}(aq) \rightarrow [Co(NH_3)_6]^{3+}(aq) + e^-$$

Aqueous ammonia reacts with aqueous hexaaquacopper(II) to produce a deep blue solution of tetraamminediaquacopper(II).

Hexaamminecopper(II)

Further substitution by ammonia will happen if liquid ammonia is used, or the mixture is chilled and ammonia gas is added.

Check your understanding

1 Describe, and explain using suitable equations, what happens when excess aqueous ammonia is added to the following solutions.

- a $Cr^{3+}(aq)$
- b $Co^{2+}(aq)$
- c $Cu^{2+}(aq)$

21.02 Ligand exchange (2)

OUTCOMES

already from A2 Level, you

- understand that the ligands NH_3 and H_2O are similar in size and are uncharged, and that ligand exchange occurs without change of co-ordination number (e.g. Cr^{3+} and Co^{2+})
- know that chloride ions can act as unidentate ligands
- know that transition metal ions commonly form octahedral complexes with small ligands such as water
- know that transition metal ions commonly form tetrahedral complexes with large ligands such as chloride ions

and after this spread you should understand

- that the chloride ligand is larger than water and ammonia, and that ligand exchange can involve a change of co-ordination number (e.g. Co^{2+} and Cu^{2+})

Ligand substitution involves breaking the co-ordinate bond between a ligand and the central metal ion, and replacing one ligand by another ligand. Water and ammonia are neutral molecules with a similar size. Ligand substitution reactions involving these ligands produce a change of colour, but not a change of co-ordination number or shape. Chloride ions are charged, and larger than these ligands. Ligand substitution by chloride ions can involve a change in co-ordination number and shape, as well as colour.

Substitution by chloride ions

The addition of concentrated hydrochloric acid to aqueous transition metal ions causes a substitution reaction, and aqua ligands are replaced by chloro ligands. In general for metal(II) ions:

$$[M(H_2O)_6]^{2+}(aq) + 4Cl^-(aq) \rightleftharpoons [MCl_4]^{2-}(aq) + 6H_2O(l)$$

Notice that not only has the ligand changed, but the co-ordination number has changed from 6 to 4. The colour of tetrachloro complexes differ from the original hexaaqua complexes. In addition they have a tetrahedral shape, rather than an octahedral shape. This happens because chloride ions are larger than water molecules. The repulsive forces between chloride ions are too large for an octahedral arrangement with bond angles of 90° to be as stable as a tetrahedral arrangement with bond angles of 109.5°.

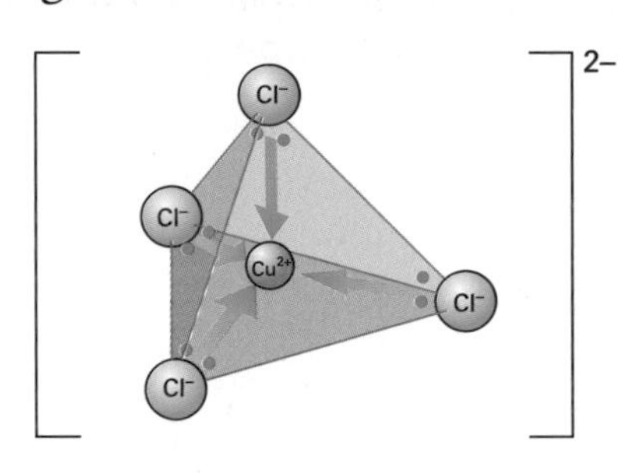

The $[CuCl_4]^{2-}$ complex has a tetrahedral shape.

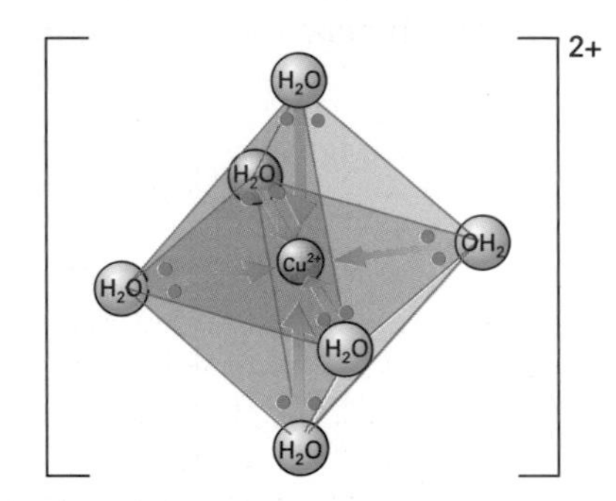

The $[Cu(H_2O)_6]^{2+}$ complex has an octahedral shape.

Cobalt(II)

When concentrated hydrochloric acid is added to aqueous cobalt(II) ions, the colour changes from pink to blue as the tetrachlorocobaltate(II) complex forms:

$$\underset{\text{pink}}{[Co(H_2O)_6]^{2+}(aq)} + 4Cl^-(aq) \rightarrow \underset{\text{blue}}{[CoCl_4]^{2-}(aq)} + 6H_2O(l)$$

The reaction is reversible. The colour becomes pink again if excess water is added:

$$\underset{\text{blue}}{[CoCl_4]^{2-}(aq)} + 6H_2O(l) \rightarrow \underset{\text{pink}}{[Co(H_2O)_6]^{2+}(aq)} + 4Cl^-(aq)$$

Cobalt(II) sulfate solution before (left) and after the addition of concentrated hydrochloric acid (right).

Copper(II)

When concentrated hydrochloric acid is added to aqueous copper(II) ions, the colour changes from blue to olive-green as the tetrachlorocuprate(II) complex forms:

$$\underset{\text{blue}}{[Cu(H_2O)_6]^{2+}(aq)} + 4Cl^-(aq) \rightarrow \underset{\text{olive-green}}{[CuCl_4]^{2-}(aq)} + 6H_2O(l)$$

The reaction is reversible. The colour becomes blue again if excess water is added:

$$\underset{\text{olive-green}}{[CuCl_4]^{2-}(aq)} + 6H_2O(l) \rightarrow \underset{\text{blue}}{[Cu(H_2O)_6]^{2+}(aq)} + 4Cl^-(aq)$$

Copper(II) sulfate solution before (left) and after the addition of concentrated hydrochloric acid (right).

Using hydrochloric acid

Transition metal ions form hexaaqua complexes in aqueous solution. These complexes hydrolyse to form the pentaaquahydroxo complexes. For example, here is the general equation for metal(II) ions:

$$[M(H_2O)_6]^{2+}(aq) + H_2O(l) \rightleftharpoons [M(H_2O)_5(OH)]^+(aq) + H_3O^+(aq)$$

The position of equilibrium lies very far to the left, so the solution is only very weakly acidic, with a pH of around 6. A similar situation exists for metal(III) ions, but the position of equilibrium does not lie so far to the left, so their solutions have a pH of around 3. The position of equilibrium moves even further to the left when hydrochloric acid is added, so metal aqua ions are essentially the only complexes present. Concentrated hydrochloric acid contains a high concentration of chloride ions, and these are able to substitute for water as ligands in the complexes.

Summary

The table summarizes the colours of the complexes involved.

complex	colour
$[Co(H_2O)_6]^{2+}(aq)$	pink
$[CoCl_4]^{2-}(aq)$	blue
$[Cu(H_2O)_6]^{2+}(aq)$	blue
$[CuCl_4]^{2-}(aq)$	olive-green

Hydroxide precipitates

The position of this equilibrium moves to the right if bases such as hydroxide ions or ammonia are added. This causes the formation of metal hydroxide precipitates (see Spreads 20.03 and 20.04). The addition of carbonate ions to solutions containing metal(III) aqua complexes causes metal(III) hydroxides to form. But metal(II) carbonate precipitates form when carbonate ions are added to metal(II) aqua complexes (see Spread 20.05).

Check your understanding

1 Describe, and explain using suitable equations, what happens when concentrated hydrochloric acid is added to the following solutions.

  a $Co^{2+}(aq)$

  b $Cu^{2+}(aq)$

2 Explain the changes in co-ordination number and shape that occur as a result of ligand substitution by chloride ions in metal aqua complexes.

21.03 Preparing inorganic compounds

OUTCOMES

already from A2 Level, you can

- describe and explain simple test-tube reactions involving Cu^{2+}, Co^{2+}, Fe^{2+}, Cr^{3+}, and Fe^{3+} with the bases OH^-, NH_3, and CO_3^{2-}, and with Cl^- ions

and after this spread you should know

- how to prepare an inorganic complex

Tetraamminecopper(II) sulfate-1-water is dark blue.

Water baths

You may have access to a thermostatically controlled water bath. If not, you can make a simple hot water bath by pouring hot water from a kettle into a beaker. If you need to use a Bunsen burner to heat the water, make sure the flame is extinguished before using ethanol, which is flammable.

One of the Investigative and Practical Skills tasks for A2 Inorganic Chemistry is to prepare an inorganic complex. There are many possible complexes you might be asked to prepare, but the preparation is likely to involve these steps:

- weighing the solid reactant at the start
- dissolving the solid
- adding another reactant to produce the desired product
- separation and purification of the product
- drying the product
- weighing the product to determine the yield

Outline method

An outline method for making tetraamminecopper(II) sulfate-1-water using copper(II) sulfate, concentrated ammonia, and ethanol is described on this spread:

$$CuSO_4.5H_2O + 4NH_3 \rightarrow Cu(NH_3)_4SO_4.H_2O + 4H_2O$$

It is given as an example only, and you may be asked to prepare a different inorganic complex.

The starting mass of copper(II) sulfate-5-water

Before starting, leave some ethanol on ice to chill. You will need it later. If you forget to do this, you will have to wait and may run out of time.

Weigh between 1.4 g and 1.6 g of copper(II) sulfate-5-water using a ±0.1 g balance. Weigh a test tube using a ±0.01 g balance and record its mass. Add the solid, re-weigh using the ±0.01 g balance, and record the mass of the test tube and solid. The difference between the two masses is the precise mass of solid used.

The reaction

Use a graduated pipette and pipette filler to add 4 cm^3 of de-ionized water to the test tube. Place the test tube in a hot water bath and stir the contents gently to dissolve the solid.

The next step will need to be carried out while wearing gloves and using a fume cupboard. This is because concentrated ammonia is involved. Add 2 cm^3 of concentrated ammonia to the test tube and stir gently as you add it.

Separation and purification

Use a graduated pipette and pipette filler to add 6 cm^3 of ethanol to a beaker. Carefully pour the contents of the test tube into the ethanol. Shake the test tube and then cool it on ice.

Filter the crystals using a Buchner funnel and flask. Use cold ethanol to wash out the test tube, and add the washings to the Buchner funnel. Rinse the crystals with cold ethanol.

Carefully scrape the crystals off the filter paper and onto an unused piece of filter paper. Take care not to scrape bits of paper away as you do this. Pat the crystals dry with another piece of filter paper. You may need to put the crystals in a desiccator to dry completely.

Determining the yield

Weigh a dry sample bottle using a ±0.01 g balance and record its mass. With care, add your crystals to the sample bottle. Weigh the sample bottle again and record its mass. The mass of product is the difference between the two masses.

Analysis of the experiment

Here is the formula needed to calculate the percentage yield of product:

$$\% \text{ yield} = \frac{\text{actual mass of product obtained}}{\text{theoretical mass of product}} \times 100$$

You will need to calculate the relative formula masses of the starting reagent, $CuSO_4.5H_2O$, and the product $Cu(NH_3)_4SO_4.H_2O$.

From the mass of the starting reagent and the relative formula mass, calculate the number of moles you used. Use the balanced equation to calculate the number of moles of product you should have made, and then the theoretical mass of product (the maximum possible). Finally calculate the percentage yield. A yield of around 65% is typical for the preparation of tetraamminecopper(II) sulfate-1-water using copper(II) sulfate.

Typical marking points

To attain the maximum marks, you will be expected to

- carry out all experiments competently and safely
- use appropriate quantities of the reagents
- set up your apparatus correctly
- produce a sufficient amount of a good quality product

You will lose marks if you carry out parts of the investigation poorly. For example, you might work carelessly or use inappropriate quantities of reagents. Your product may be poor quality. It may be the wrong colour or be contaminated with bits of filter paper. You will also lose marks if your yield is poor. This can happen if you spill some of your reaction mixture or work in a rush.

21.04 Chelation

OUTCOMES

already from A2 Level, you

- understand the concept of increasing disorder (entropy change ΔS)
- can calculate entropy changes from absolute entropy values
- understand that ligands can be unidentate (e.g. H_2O, NH_3, and Cl^-), bidentate (e.g. $NH_2CH_2CH_2NH_2$ and $C_2O_4^{2-}$), or multidentate (e.g. $EDTA^{4-}$)

and after this spread you should

- know that substitution of unidentate ligand with a bidentate or a multidentate ligand leads to a more stable complex
- understand the effect of chelation in terms of a positive entropy change

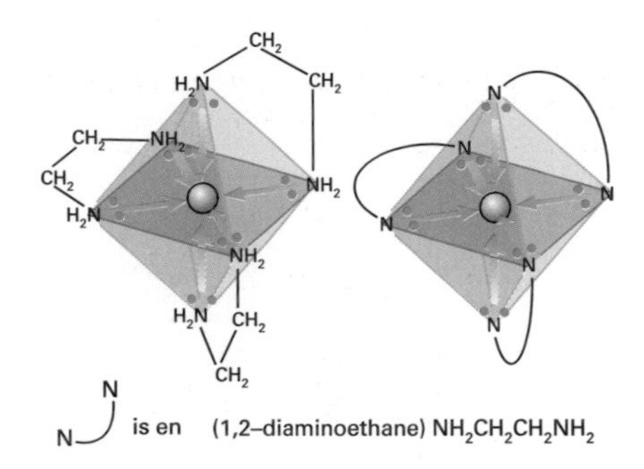

Three 1,2-diaminoethane molecules can form co-ordinate bonds with a central metal ion, as seen in these two diagrams, one of which is simplified to emphasize the co-ordinate bonds.

Unidentate ligands such as water, ammonia, and chloride ions can each donate one lone pair of electrons to the central metal ion in a complex. Bidentate ligands such as 1,2-diaminoethane and the ethanedioate ion can each donate two lone pairs of electrons. $EDTA^{4-}$ is a multidentate ligand, and each ion can donate six lone pairs of electrons. One unidentate ligand can substitute for another. For example, ammonia and chloride ions can substitute for water (see Spreads 21.01 and 21.02). In addition, bidentate or multidentate ligands can substitute for unidentate ligands, forming complexes that are more stable than the original ones.

Chelation

Complexes containing bidentate or multidentate ligands are called **chelates**, and the process of forming chelates is called **chelation**. The words come from the Greek word for *claw*, as it seems as if the central metal ion is gripped by claws – the ligands.

Bidentate ligands

1,2-diaminoethane, $H_2NCH_2CH_2NH_2$, has two amino groups. The nitrogen atom in each amino group has one lone pair of electrons that can be donated to the central metal ion in a complex. The ethanedioate ion, $C_2O_4^{2-}$, has two carboxylate groups. Each one can donate a pair of electrons to the central metal ion in a complex.

1,2-diaminoethane and the ethanedioate ion

Up to three bidentate ligands can replace unidentate ligands in complexes. This happens one by one:

$$[M(H_2O)_6]^{2+} + H_2NCH_2CH_2NH_2 \rightleftharpoons [M(H_2O)_4(H_2NCH_2CH_2NH_2)]^{2+} + 2H_2O$$

$$[M(H_2O)_4(H_2NCH_2CH_2NH_2)]^{2+} + H_2NCH_2CH_2NH_2 \rightleftharpoons [M(H_2O)_2(H_2NCH_2CH_2NH_2)_2]^{2+} + 2H_2O$$

$$[M(H_2O)_2(H_2NCH_2CH_2NH_2)_2]^{2+} + H_2NCH_2CH_2NH_2 \rightleftharpoons [M(H_2NCH_2CH_2NH_2)_3]^{2+} + 2H_2O$$

Chelates are very stable and the positions of these equilibria lie far to the right.

$$[Cu(H_2O)_6]^{2+} + 3H_2NCH_2CH_2NH_2 \rightleftharpoons [Cu(H_2NCH_2CH_2NH_2)_3]^{2+} + 6H_2O$$

$$[Cr(H_2O)_6]^{3+} + 3H_2NCH_2CH_2NH_2 \rightleftharpoons [Cr(H_2NCH_2CH_2NH_2)_3]^{3+} + 6H_2O$$

Metal(III) chelates also form, and are more stable than metal(II) chelates. Here are the overall equations for the reaction of 1,2-diaminoethane with the copper(II) hexaaqua complex and with the chromium(III) hexaaqua complex:

Note that there are no changes in oxidation state or co-ordination number in these reactions.

1,2-diaminoethane can also replace ammine ligands. For example:

$$[Co(NH_3)_6]^{2+} + 3H_2NCH_2CH_2NH_2 \rightleftharpoons [Co(H_2NCH_2CH_2NH_2)_3]^{2+} + 6NH_3$$

The enthalpy change ΔH in this reaction is almost zero, because the same number and type of bonds are being broken and made (N→metal bonds). But there is an increase in entropy, so the entropy change ΔS is positive. There are four molecules on the left hand side of the equation but seven on the right. This means that the free energy change ΔG is negative, so the reaction is feasible. It proceeds to completion and stable chelates form.

$EDTA^{4-}$

$EDTA^{4-}$ can donate six pairs of electrons to the metal ion. Two of these are on its nitrogen atoms, and four are on oxygen atoms in the carboxylate groups.

$EDTA^{4-}$ can replace all six unidentate ligands in a complex. For example:

$$[Cu(H_2O)_6]^{2+} + EDTA^{4-} \rightleftharpoons [Cu(EDTA)]^{2-} + 6H_2O$$
$$[Cr(H_2O)_6]^{3+} + EDTA^{4-} \rightleftharpoons [Cr(EDTA)]^{-} + 6H_2O$$

Note that the oxidation state and co-ordination number stays the same. But the overall charge of the complex changes (as it also does when substitution reactions involving the ethanedioate ion happen). Again, the enthalpy change is almost zero, because the same number of similar bonds is being broken and made. There is a positive entropy change, as there are two molecules on the left hand side of the equation and seven on the right. The free energy change is negative and the reaction proceeds to completion to form very stable chelates.

EDTA stands for ethylenediaminetetraacetic acid. The $EDTA^{4-}$ ion can act as a multidentate ligand.

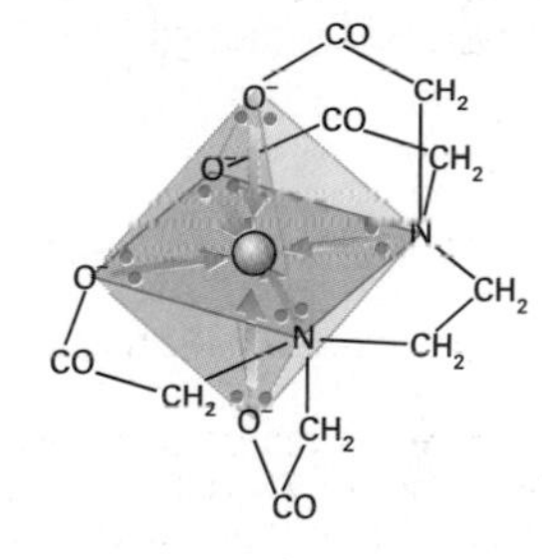

The structure of an octahedral $EDTA^{4-}$ complex.

$EDTA^{4-}$ at work

$EDTA^{4-}$ is used in medicine to remove toxic heavy metal ions, such as cadmium and lead, from the body. It is also used as an *anticoagulant* to stop blood samples clotting. Hard water contains magnesium and calcium salts that make it difficult to get a lather. $EDTA^{4-}$ is used as a water softener in detergents and shampoos, as it forms chelates with the magnesium and calcium ions.

Check your understanding

1 a Explain why diaminoethane and the ethanedioate ion are bidentate ligands.

b Explain why $EDTA^{4-}$ is a multidentate ligand.

2 a What is meant by the term *chelation*?

b Write an equation for the formation of a chelate from hexaamminechromium(III) ions and $EDTA^{4-}$.

c Explain why the chelate in part **b** is more stable than the complex from which it was formed.

22.01 Cobalt

OUTCOMES

already from A2 Level, you

- know that $[Co(H_2O)_6]^{2+}$ ions are formed in aqueous solution
- understand that aqueous solutions of Co^{2+} are very weakly acidic because of the equilibrium $[Co(H_2O)_6]^{2+} + H_2O \rightleftharpoons [Co(H_2O)_5(OH)]^{+} + H_3O^{+}$
- can describe and explain the simple test-tube reactions of $Co^{2+}(aq)$ ions with the bases OH^-, NH_3, and CO_3^{2-}
- understand that ligand exchanges between NH_3 and H_2O occur without change of co-ordination number, and that exchange involving Cl^- can involve a change of co-ordination number
- know the oxidation in alkaline solution of Co^{2+} by H_2O_2
- know the oxidation of Co^{2+} by air in ammoniacal solution

and after this spread you should

- have a summary of the reactions of Co^{2+} required by the Specification

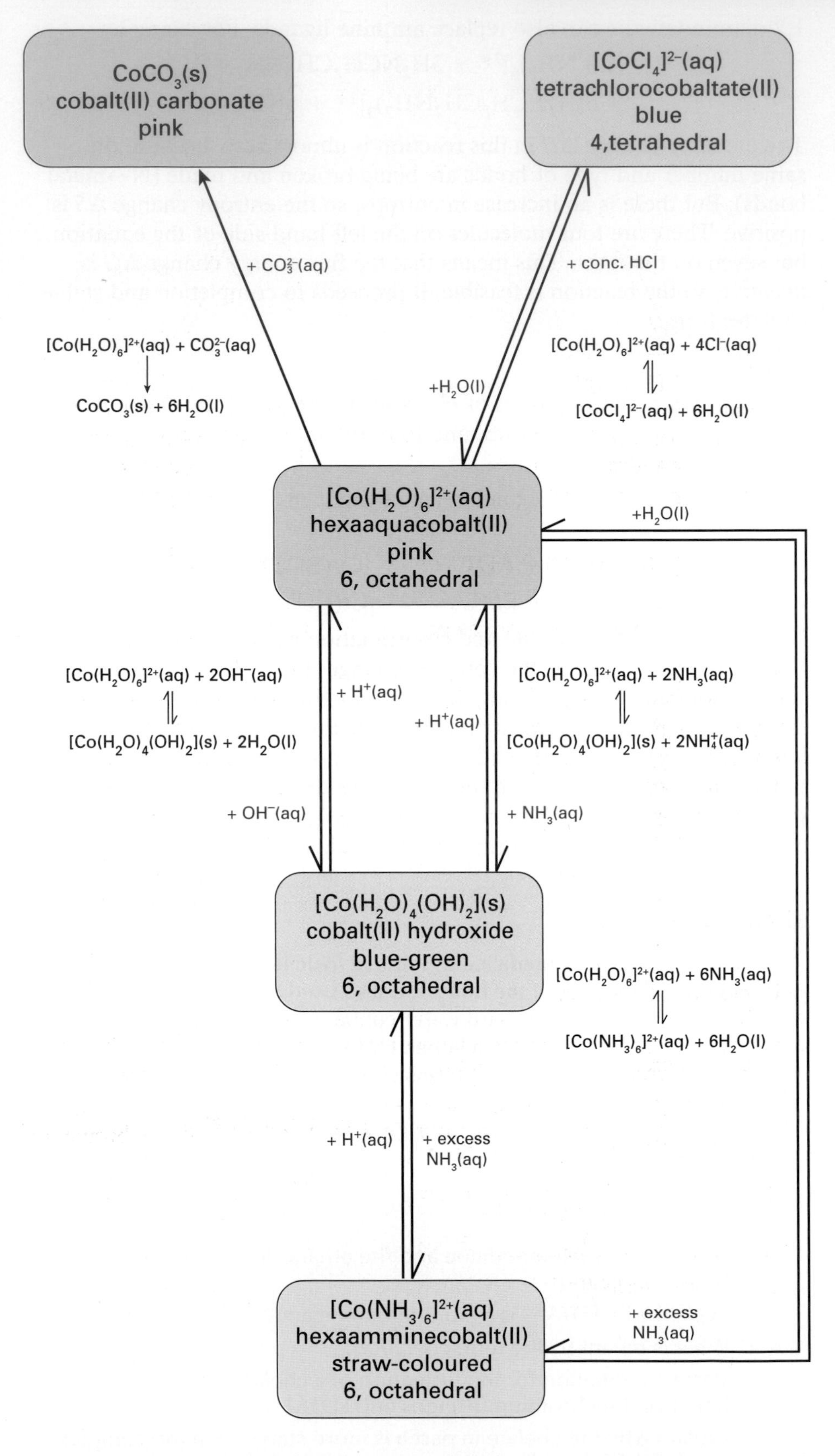

Hydrolysis, ligand substitution, and carbonate formation. Note the change in co-ordination number as the $[CoCl_4]^{2-}$ complex forms.

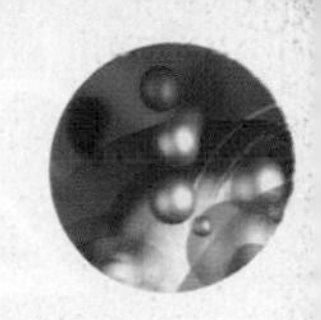

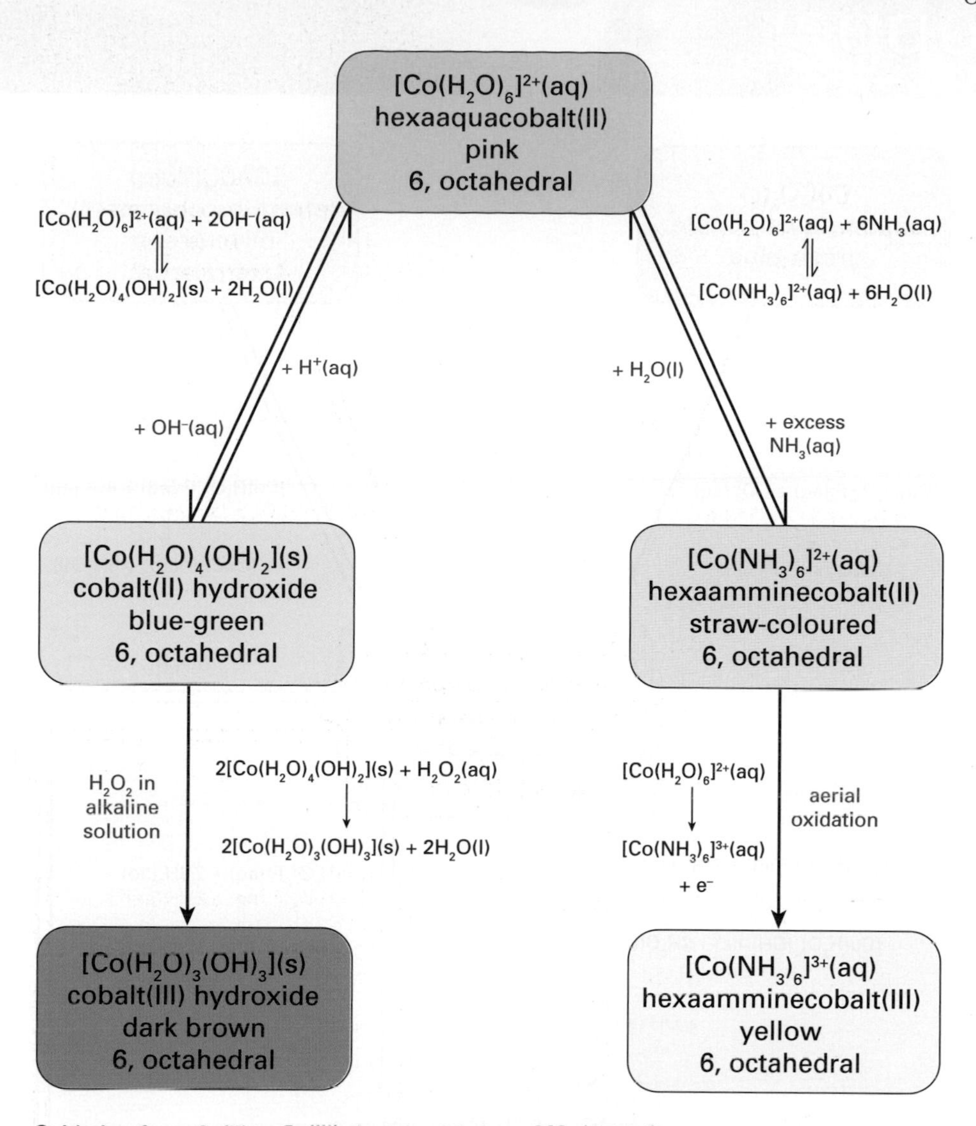

Oxidation from Co(II) to Co(III). Aerial oxidation of $[Co(NH_3)_6]^{2+}$ produces a brown mixture. The $[Co(NH_3)_6]^{3+}$ complex is yellow, but other complexes form too, such as the purple $[Co(NH_3)_5(H_2O)]^{3+}$ complex.

Summary of colours

Co(II) complex	colour
$[Co(H_2O)_4(OH)_2](s)$	blue-green
$[Co(H_2O)_6]^{2+}(aq)$	pink
$[Co(NH_3)_6]^{2+}(aq)$	straw-coloured
$[CoCl_4]^{2-}(aq)$	blue

Co(III) complex	colour
$[Co(H_2O)_3(OH)_3](s)$	dark brown
$[Co(NH_3)_6]^{3+}(aq)$	yellow

Cobalt carbonate, $CoCO_3(s)$ is pink.

22.02 Copper

OUTCOMES

already from A2 Level, you

- know that $[Cu(H_2O)_6]^{2+}$ ions are formed in aqueous solution
- understand that aqueous solutions of Cu^{2+} are very weakly acidic because of the equilibrium $[Cu(H_2O)_6]^{2+} + H_2O \rightleftharpoons [Cu(H_2O)_5(OH)]^+ + H_3O^+$
- can describe and explain the simple test-tube reactions of $Cu^{2+}(aq)$ ions with the bases OH^-, NH_3, and CO_3^{2-}
- understand that ligand exchanges between NH_3 and H_2O occur without change of co-ordination number, and that ligand exchange involving Cl^- ions can involve a change of co-ordination number

and after this spread you should

- have a summary of the reactions of Cu^{2+} required by the Specification
- know how to investigate the chemistry of transition metal compounds in a series of experiments, using copper(II) complexes as an example

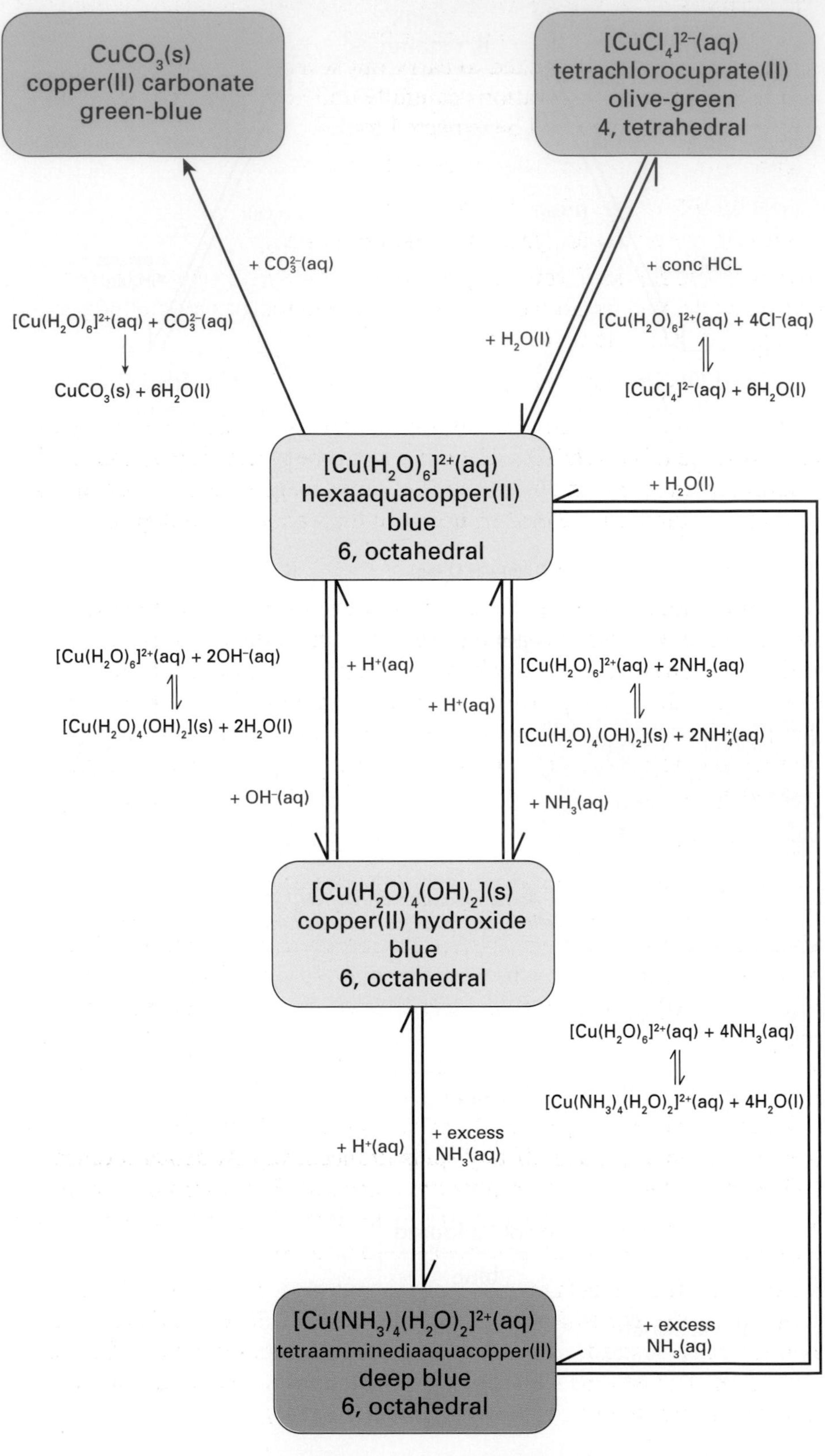

Hydrolysis, ligand substitution, and formation of copper(II) carbonate. Note that the formation of the $[CuCl_4]^{2-}$ complex from the $[Cu(H_2O)_6]^{2+}$ complex involves a change in co-ordination number from 6 to 4. Also note that ammonia does not completely substitute for water.

Summary of copper(II) colours

complex	colour
$[Cu(H_2O)_4(OH)_2](s)$	blue
$[Cu(H_2O)_6]^{2+}(aq)$	blue
$[Cu(NH_3)_4(H_2O)_2]^{2+}(aq)$	deep blue
$[CuCl_4]^{2-}(aq)$	olive-green

Copper carbonate, $CuCO_3(s)$ is green-blue.

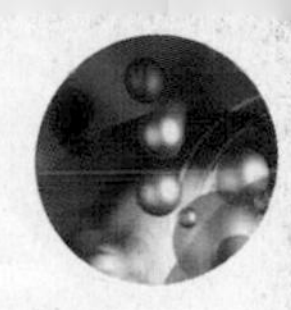

Investigating the chemistry of copper compounds

One of the Investigative and Practical Skills tasks for A2 Inorganic Chemistry is to investigate the chemistry of transition metal compounds in a series of experiments. The chemistry of copper(II) compounds is often used. You may be asked to carry out several test-tube reactions, and to record your observations carefully and accurately. To attain the maximum marks, you will be expected to:

- carry out all experiments competently, safely, and with care
- use appropriate quantities of the reagents
- have all, or nearly all, your observations correct

You will lose marks if you carry out parts of the investigation p orly. For example, you might work carelessly, use inappropriate quantities of reagents, or make some incorrect observations.

Appropriate quantities

It is important to use modest volumes of test solutions. For example, a large volume of copper(II) sulfate in a test tube is not only wasteful of resources, but makes it difficult to mix other reagents with it adequately. Keep the volume to 1–2 cm^3 or no more than about 1 cm depth.

Patience

It is important to work carefully and with patience. For example, if you squirt a large volume of aqueous ammonia into aqueous copper(II) sulfate, you may miss the precipitate of copper(II) hydroxide and go straight to a deep blue solution of the tetraamminediaquacopper(II) complex. It is wise to add reagents dropwise until you see a change. Be prepared to repeat any experiments where you are not sure of your observations.

Observations

Record your observations as you work. Remember to record the colours you see, and whether you have a solution or a precipitate. Take care to distinguish between the words *clear* and *colourless*. Solutions are clear because you can see through them. But they may be coloured, such as blue copper(II) sulfate solution, or colourless, such as aqueous sodium hydroxide.

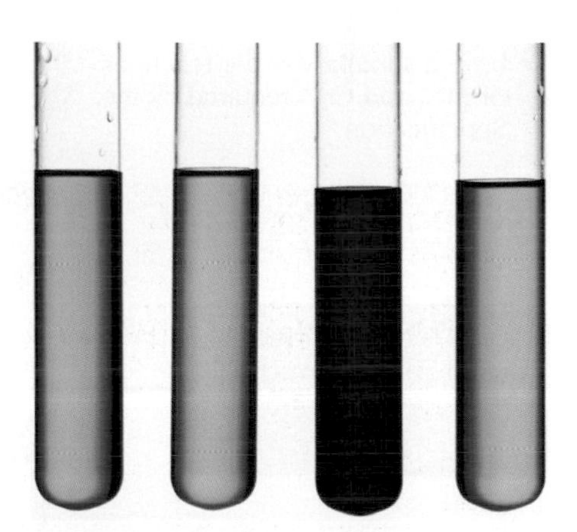

From left to right, solutions of hexaaquacopper(II), tetrachlorocuprate(II), and tetraamminediaquacopper(II). The test tube on the far right contains the $(Cu(EDTA))^{2-}(aq)$ complex.

Safety

Take appropriate safety precautions as you work. You will need to wear eye protection and wash off any spills immediately. Aqueous sodium hydroxide and aqueous ammonia are corrosive. But ammonia also has an irritating vapour, so avoid breathing its fumes. Copper compounds are harmful if swallowed and irritate the skin and eyes.

When you mix reagents together, avoid putting a thumb over the mouth of the test tube and shaking up and down. You will contaminate your skin this way. Instead, hold the test tube near its mouth and shake the bottom of the test tube from side to side. This will ensure thorough mixing without the risk of spraying the laboratory.

22.03 Aluminium and chromium

OUTCOMES

already from A2 Level, you

- know that $[Al(H_2O)_6]^{3+}$ and $[Cr(H_2O)_6]^{3+}$ ions are formed in aqueous solution
- understand that aqueous solutions of Al^{3+} and Cr^{3+} are weakly acidic because of the equilibrium $[M(H_2O)_6]^{3+} + H_2O \rightleftharpoons [M(H_2O)_5(OH)]^{2+} + H_3O^+$
- can describe and explain the simple test-tube reactions of Al^{3+}(aq) ions and Cr^{3+}(aq) ions with the bases OH^-, NH_3, and CO_3^{2-}
- understand that ligand exchanges between NH_3 and H_2O occur without change of co-ordination number, and that exchange involving Cl^- can involve a change of co-ordination number
- know that aluminium hydroxide and chromium(III) hydroxide are amphoteric
- know the equilibrium reaction $2CrO_4^{2-} + 2H^+ \rightleftharpoons Cr_2O_7^{2-} + H_2O$
- know that Cr^{3+} and Cr^{2+} are formed by reduction of $Cr_2O_7^{2-}$ by zinc in acid solution
- know the oxidation in alkaline solution of Cr^{3+} by H_2O_2

and after this spread you should

- have a summary of the reactions of Al^{3+} and Cr^{3+} required by the Specification

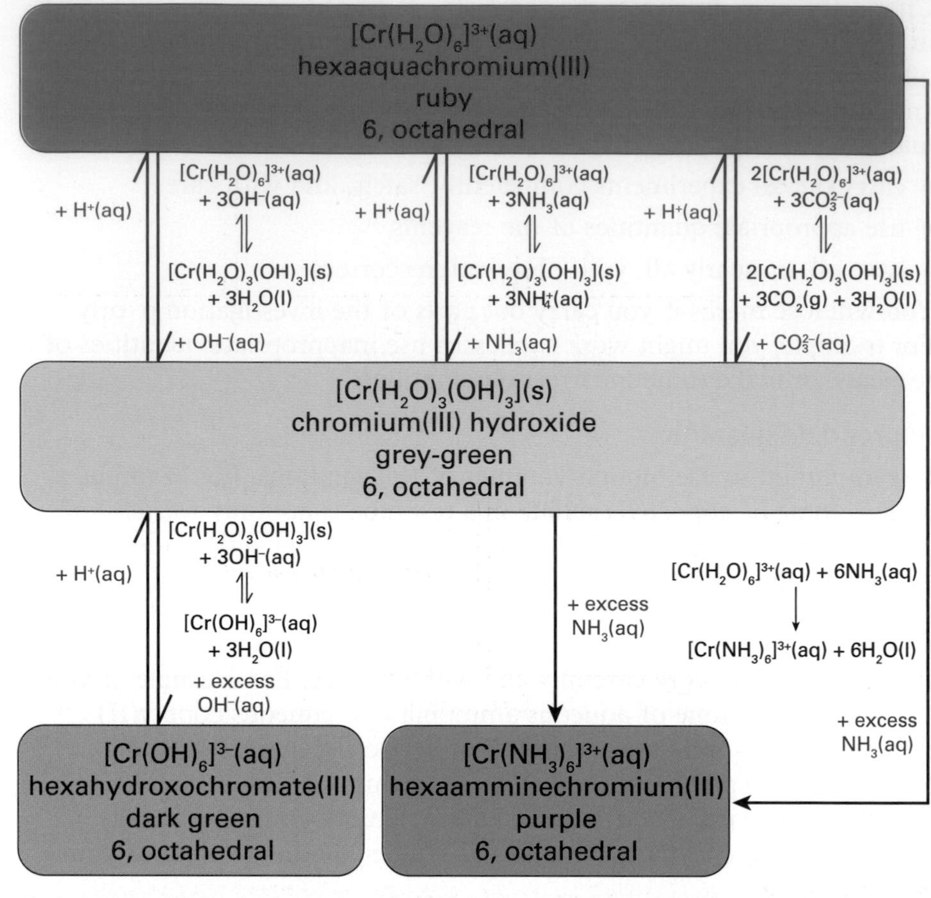

Hydrolysis and ligand substitution. Note that chromium(III) carbonate does not form. You may not see the ruby colour of $[Cr(H_2O)_6]^{3+}$ (aq), as chloride ions and other ligands may replace water molecules to give green complexes. Chromium(III) hydroxide is amphoteric, as it dissolves in both acids and bases.

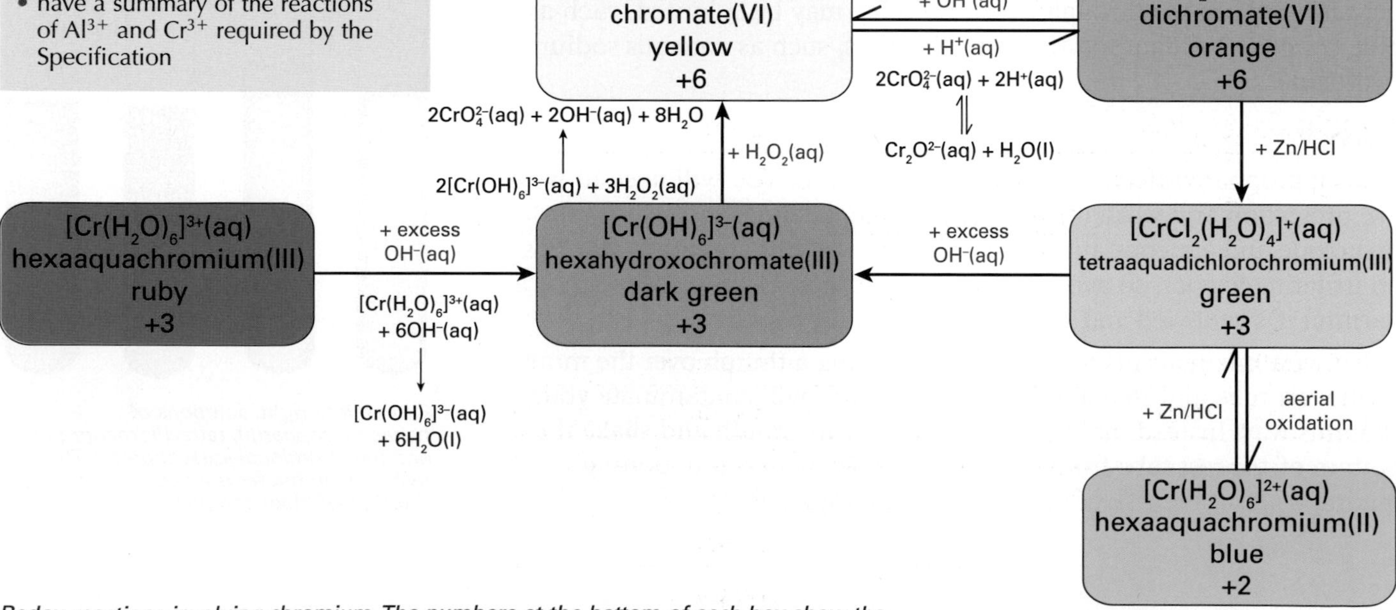

Redox reactions involving chromium. The numbers at the bottom of each box show the oxidation state of chromium in the complex.

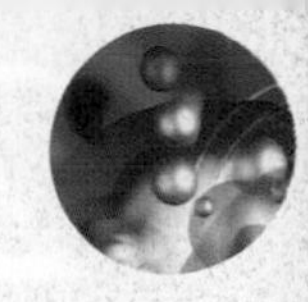

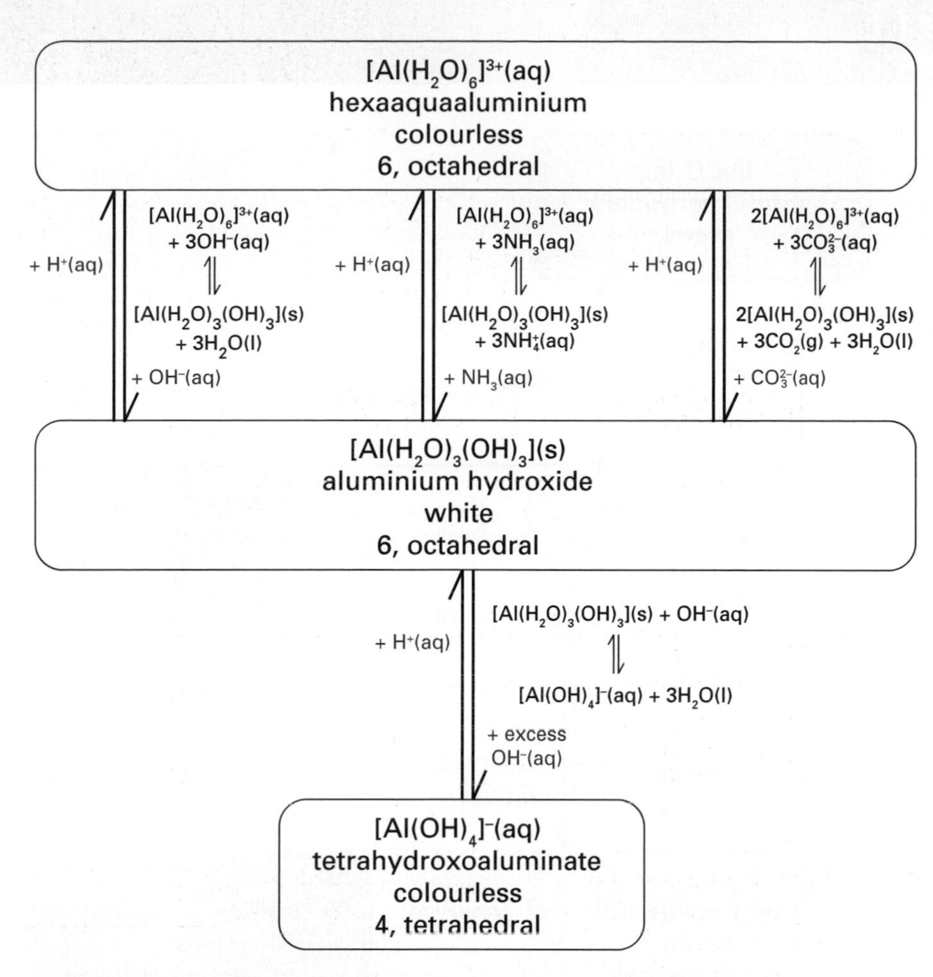

Hydrolysis and ligand substitution. Aluminium is not a transition metal so its complexes are colourless or white. Note that aluminium carbonate does not form. Aluminium hydroxide is amphoteric, as it dissolves in both acids and bases.

Summary of chromium colours

Cr(II) complex	colour
$[Cr(H_2O)_6]^{2+}(aq)$	blue
$[Cr(H_2O)_4Cl_2]^+(aq)$	green

Cr(III) complex	colour
$[Cr(H_2O)_6]^{3+}(aq)$	ruby
$[Cr(H_2O)_3(OH)_3](s)$	grey-green
$[Cr(OH)_6]^{3-}(aq)$	dark green
$[Cr(NH_3)_6]^{3+}(aq)$	purple

Cr(VI) complex	colour
$Cr_2O_7^{2-}(aq)$	orange
$CrO_4^{2-}(aq)$	yellow

22.04 Iron

OUTCOMES

already from A2 Level, you

- know that $[Fe(H_2O)_6]^{2+}$ and $[Fe(H_2O)_6]^{3+}$ ions are formed in aqueous solution
- understand that aqueous solutions of Fe^{2+} are very weakly acidic because of the equilibrium $[Fe(H_2O)_6]^{2+} + H_2O \rightleftharpoons [Fe(H_2O)_5(OH)]^{+} + H_3O^{+}$
- understand that aqueous solutions of Fe^{3+} are weakly acidic because of the equilibrium $[Fe(H_2O)_6]^{3+} + H_2O \rightleftharpoons [Fe(H_2O)_5(OH)]^{2+} + H_3O^{+}$
- can describe and explain the simple test-tube reactions of $Fe^{2+}(aq)$ and $Fe^{3+}(aq)$ ions with the bases OH^-, NH_3, and CO_3^{2-}
- understand that the ligand exchanges between NH_3 and H_2O occur without change of co-ordination number

and after this spread you should

- have a summary of the reactions of Fe^{2+} and Fe^{3+} required by the Specification

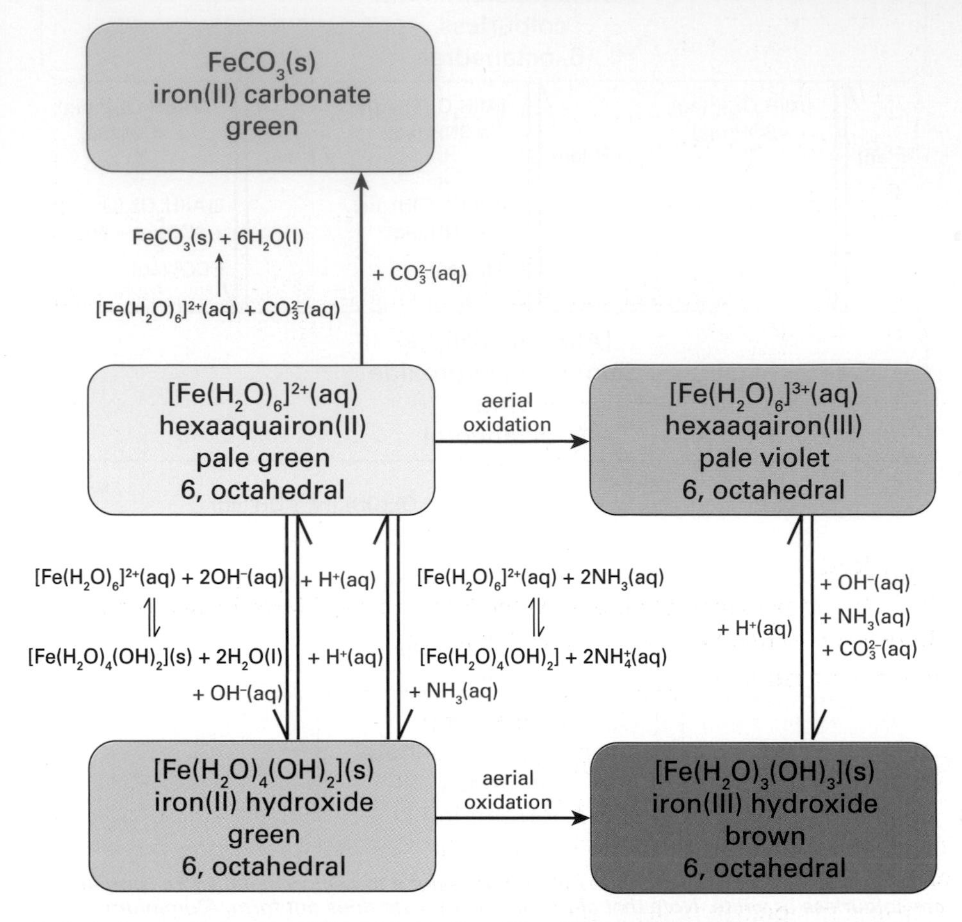

Hydrolysis reactions of iron(II) and formation of iron(II) carbonate. Note that $[Fe(H_2O)_6]^{2+}(aq)$ oxidizes in air to form $[Fe(H_2O)_6]^{3+}(aq)$. This hydrolyses to form brown $[Fe(H_2O)_5(OH)]^{2+}(aq)$, which masks the pale violet colour. Iron(II) hydroxide also oxidizes in air, forming brown iron(III) hydroxide.

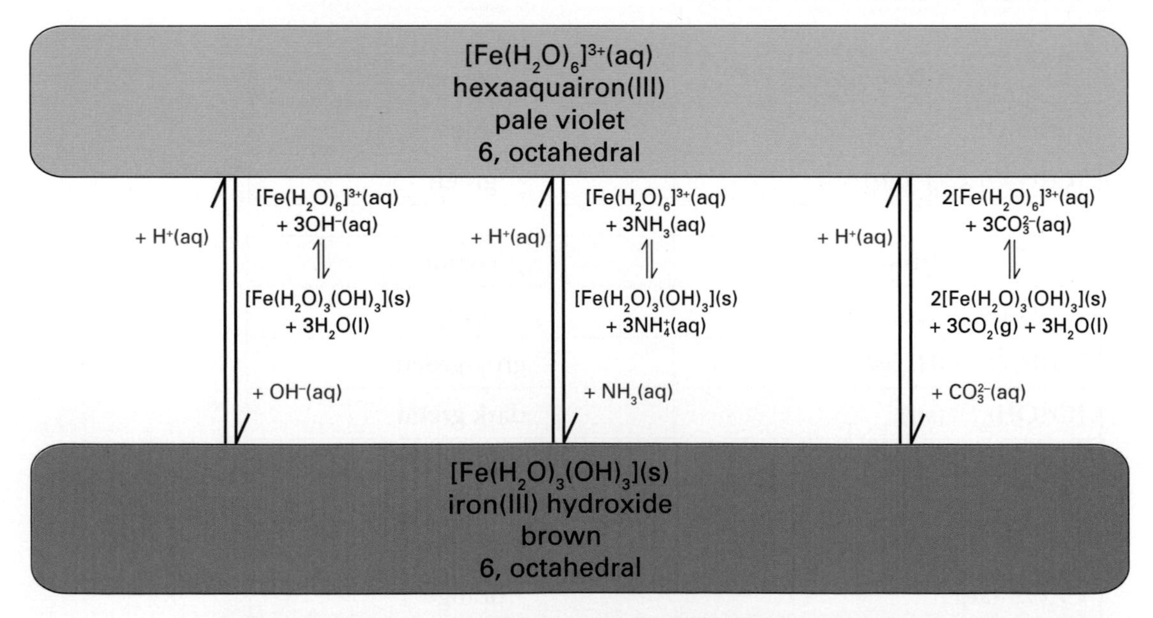

Hydrolysis reactions of iron(III). Note that iron(III) carbonate does not form. Remember that hexaaquairon(II) oxidizes in air to form hexaaquairon(III), and iron(II) hydroxide oxidizes in air to form iron(III) hydroxide.

Summary of iron colours

Fe(II) complex	colour
$[Fe(H_2O)_6]^{2+}(aq)$	pale green
$[Fe(H_2O)_4(OH)_2](s)$	green
$FeCO_3(s)$	green

Fe(III) complex	colour
$[Fe(H_2O)_6]^{3+}(aq)$	pale violet
$[Fe(H_2O)_3(OH)_3](s)$	brown

The brown $[Fe(H_2O)_5(OH)]^{2+}(aq)$ complex forms in aqueous solutions of iron(III) compounds, so the pale violet colour is hidden.

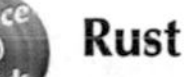

Rust

In the presence of air and water, iron and steel form brown rust. There are two main process involved in rusting.

Iron is oxidized to iron(II):

$$Fe(s) \rightarrow Fe^{2+}(aq) + 2e^-$$

Oxygen is reduced to hydroxide ions:

$$O_2(g) + 2H_2O(l) + 4e^- \rightarrow 4OH^-(aq)$$

The two ions combine to form iron(II) hydroxide, $Fe(OH)_2$. This oxidizes rapidly to form brown iron(III) hydroxide, $Fe(OH)_3$. This is also described as hydrated iron(III) oxide, $Fe_2O_3.xH_2O$.

Questions: Chs 17–22

1 a Define the term *transition element.* [1]

b State three characteristics of a transition element. [3]

c Give the oxidation state of manganese in each of the following species:

i MnO_4^-

ii MnO_2

iii Mn^{2+} [3]

d Write a half-equation to show the reduction of MnO_4^- to Mn^{2+}. [1]

e Write a half-equation to show the oxidation of Fe^{2+} to Fe^{3+}. [1]

f Combine the half-equations from parts **d** and **e** to produce an overall redox equation. [2]

g The concentration of iron(II) ions in aqueous solution can be found by titrating the solution, after acidification, with a standard solution of potassium manganate(VII). Work out the concentration of iron(II) ions in 25 cm^3 of a solution which needed 32.8 cm^3 of 0.02 $mol\,dm^{-3}$ $KMnO_4$ for complete oxidation. [5]

[Total 16 marks]

2 The rates of chemical reactions can be altered by homogeneous or heterogeneous catalysts.

a Explain what is meant by the term *heterogeneous catalyst.* [1]

b Finely divided iron can be used as a catalyst in the Haber Process. Give another example of a transition metal or transition metal compound used as a heterogeneous catalyst, and name the process in which it is used. [2]

c Give two reasons why an inert medium may be used as a support for a heterogeneous catalyst. [2]

d Small amounts of an impurity in the reaction mixture can reduce the efficiency of a catalyst. Give one example of this effect, and explain why it happens. [3]

[Total 8 marks]

3 Methanol is formed when a mixture of hydrogen and carbon monoxide, in the ratio 2:1, is passed over a catalyst under high pressure and at high temperature.

$$CO(g) + 2H_2(g) \xrightarrow{400°C} CH_3OH(g)$$

a Give the name or formula of the catalyst used in this reaction. [1]

b Define the term *activation energy.* [1]

c Explain how a catalyst can increase the rate of a chemical reaction. [2]

d Catalysts can be classified as homogeneous or heterogeneous. State the difference between these two types of catalyst. [1]

e Write half- equations to represent:

i The oxidation of $I^-(aq)$ to $I_2(aq)$

ii The reduction of $S_2O_8^{2-}(aq)$ to $SO_4^{2-}(aq)$ [2]

f Combine the two half-equations in part e) to give the overall redox equation for the reaction between $I^-(aq)$ and $S_2O_8^{2-}(aq)$. [2]

g Explain why the reaction described by the equation in part **f** is likely to be slow in the absence of a catalyst. [1]

[Total 10 marks]

4. a i Define the term *ligand.* [1]

ii Define the term *coordinate bond.* [1]

iii Explain why co-ordinate bonds can be formed between transition metal ions and water molecules. [1]

iv What name is given to any ligand that can form two co-ordinate bonds to one metal ion? Give an example of such a ligand. [2]

b $[Cu(H_2O)_6]^{2+}$ is a complex ion. State the shape of this ion, the $H_2O–Cu–H_2O$ bond angle, and the coordination number of copper. [3]

c The size and charge of a ligand can affect the shape of a transition metal complex ion. Give an example of a complex ion that is:

i tetrahedral

ii square planar

iii linear. [3]

[Total 11 marks]

5 a Transition metal complex ions can be identified by their colour. State the colours of the following complex ions:

i $[Co(H_2O)_6]^{2+}$

ii $[Fe(H_2O)_6]^{3+}$

iii $[CuCl_4]^{2-}$ [3]

b Write out the electron configurations for Co, the cobalt atom, and Co^{2+}, the cobalt(II) ion. Use [Ar] to represent the electron configuration of the argon atom. [2]

c **i** A precipitate is formed when aqueous sodium hydroxide is added to an aqueous solution containing cobalt(II) ions. Give the colour and name of this precipitate. Write an ionic equation for its formation. [3]

ii Describe what would be observed if excess aqueous ammonia were to be added to the precipitate formed in **ci**. [2]

[Total 10 marks]

6 **a** Explain why transition metal complexes are coloured. [2]

b State three reasons why a transition metal complex would change colour. [3]

c Describe how you could find the concentration of cobalt(II) ions in a solution of unknown concentration. You are provided with a 1.00 mol dm^{-3} solution of cobalt(II) ions and a visible-light spectrophotometer (colorimeter). [4]

[Total 9 marks]

7 **a** Iron is a transition element. State three characteristic features of transition elements and their compounds. [3]

b Iron(III) chloride is a Lewis acid.

i Define the term Lewis acid. [1]

ii Explain why iron(III) chloride will react with methanol but not with cyclohexane. [3]

c When sodium carbonate solution is added to an aqueous solution of iron(III) chloride, effervescence occurs and a precipitate forms. State the formula of the gas and of the precipitate. [2]

[Total 9 marks]

8 **a** Write out the electronic configurations of the:

i Cu atom

ii Cu^{+} ion

iii Cu^{2+} ion [3]

b Explain why a copper(I) salt appears white whereas a copper(II) salt is blue. [1]

c **i** A pale blue precipitate forms when a few drops of dilute aqueous ammonia are added to an aqueous solution of copper(II) sulfate. State the colour of the copper(II) sulfate solution and give the formula of the ion responsible for this colour. Give the name and formula of the pale blue precipitate. [4]

ii Aqueous copper(II) sulfate solution turns yellow when an excess of concentrated hydrochloric acid is added to it. State the type of reaction that has taken place. Give the formula and shape of the final copper-containing species in the solution. [3]

[Total 11 marks]

9 Crystals of iron(III) ammonium alum are pale violet in colour because they contain the ion $[Fe(H_2O)_6]^{3+}$. These crystals dissolve in water to form a brown solution. When an acid is added to this solution, the brown colour disappears and a very pale violet solution forms.

a Give the shape of the $[Fe(H_2O)_6]^{3+}$ ion. [1]

b Write an equation to show the $[Fe(H_2O)_6]^{3+}$ ion behaving as an acid in aqueous solution. [1]

c Give the formula of the species responsible for the brown colour of the solution. [1]

d Explain why adding acid causes the colour of the solution to change from brown to very pale violet. [2]

e The molecule 1,2-diaminoethane, $NH_2CH_2CH_2NH_2$, acts as a bidentate ligand.

i Define the term bidentate ligand.

ii This ligand forms an octahedral complex with iron(III) ions. State the formula of this complex, and draw its structure showing all the atoms present. [5]

[Total 10 marks]

Synoptic questions

1 Oil of Wintergreen is an essential oil used to flavour sweets and chewing gum. It contains an ester, methyl salicylate, whose structure is shown below:

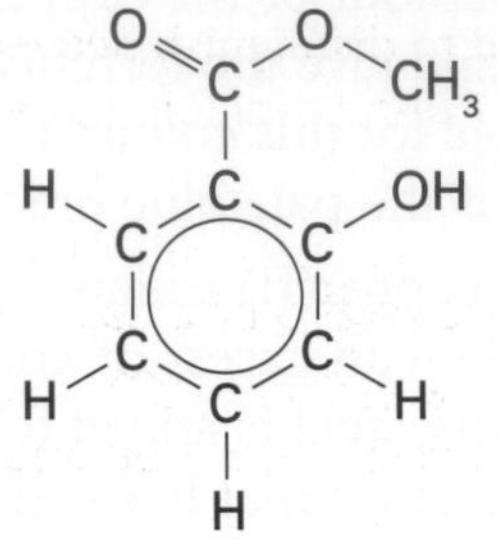

Because of the high demand in the food industry for methyl salicylate most of it is made synthetically, which is both cheaper and easier than extracting it from the natural sources.

a **i** Name the carboxylic acid and alcohol that would react together to produce methyl salicylate. [2]

ii Write an equation for this reaction. [1]

b State two uses of esters other than those given above. [2]

c Esters can be hydrolysed by heating with dilute hydrochloric acid to form an alcohol and a carboxylic acid. The rate equation for the hydrolysis of this ester is:

$$\text{rate} = k[C_6H_4(HO)COOCH_3][H^+]$$

When the initial concentration of the ester is $1.0\,mol\,dm^{-3}$ and that of hydrochloric acid is $2.0\,mol\,dm^{-3}$, the initial rate of the reaction is $9.60 \times 10^{-3}\,mol\,dm^{-3}\,s^{-1}$ at 323 K.

i Calculate the value of the rate constant at this temperature and give its units. [2]

ii Calculate the initial rate of the reaction if the concentration of hydrochloric acid is increased to $3.0\,mol\,dm^{-3}$ but all other conditions remain unchanged. [1]

iii Calculate the initial rate of the reaction at 323 K if more solvent is added to the original mixture so that the total volume is doubled. [4]

d **i** Explain how you would find the relative molecular mass of an ester from its mass spectrum. [1]

ii An ester formed by a reaction with butanol has a relative molecular mass of 130. Deduce the molecular formula of this ester. [3]

[Total 12 marks]

2 **a** **i** Define the term *Brønsted–Lowry base.*

ii State the essential feature of an acid-base reaction in aqueous solution, and write an ionic equation to illustrate your answer. [3]

b Define the term *weak* when used to describe acids and bases. [1]

c In aqueous solution, the weak acid butanoic acid, $CH_3CH_2CH_2COOH(aq)$, produces butanoate ions $CH_3CH_2CH_2CO_2^-(aq)$. Write an expression for the acid dissociation constant, K_a, of butanoic acid and state its units. [2]

d **i** Define the term *buffer solution.*

ii Give the names or formulas of two components that could be used to make an acidic buffer solution.

iii Give an example of the use of a buffer solution. [4]

[Total 10 marks]

3 **a** When concentrated sulfuric acid is added to solid potassium iodide, hydrogen iodide is evolved.

i Write a balanced symbol equation for this reaction.

ii State the role of sulfuric acid in this reaction. [2]

b In a further reaction, iodine and sulfur dioxide are evolved when concentrated sulfuric acid is warmed with solid potassium iodide.

i Deduce the oxidation state of sulphur in H_2SO_4. [1]

ii State the role of sulfuric acid in the conversion of I^- into I_2. [1]

iii Write a half equation for the conversion of Br^- into Br_2. [1]

iv Write a half equation for the conversion of H_2SO_4 into SO_2. [1]

v Use your answers to **iii** and **iv** above to deduce the overall equation for this reaction. [1]

c In a titration, $25.00\,cm^3$ of an aqueous solution of iodine required $17.40\,cm^3$ of aqueous sodium thiosulfate, of concentration $0.125\,mol\,dm^{-3}$, for complete reaction.

$$2S_2O_3^{2-}(aq) + I_2(aq) \rightarrow S_4O_6^{2-}(aq) + 2I^-(aq)$$

Calculate the mass of iodine present in $25.00\,cm^3$ of the original aqueous solution of iodine. Give your answer to 2 decimal places. [4]

[Total 11 marks]

4 An analytical chemist was given three aqueous solutions of metal sulfates and asked to identify the positive ion present in each of the solutions. The table below shows the results of some tests that were carried out.

Solution	Colour of solution	Result of reacting the solution with aqueous sodium hydroxide	Result of reacting the solution with an excess of aqueous sodium hydroxide
A	blue	pale blue gelatinous precipitate	no change
B	pale pink	blue-green precipitate	no change
C	colourless	white precipitate	precipitate re-dissolves to give a colourless solution

a Give the formula of the hydrated ion in:

i Solution A [1]

ii Solution C. [1]

b Write a balanced equation in each case to show the reaction of:

i Solution A with aqueous sodium hydroxide [1]

ii Solution C with excess aqueous sodium hydroxide. [1]

c Write out the electronic configuration of:

i Mn [1]

ii Mn^{2+} [1]

d A standard solution of the purple ion MnO_4^- obtained can be used to determine the concentration of a solution of ethanedioate ions, $C_2O_4^{2-}$.

i Describe the experimental procedure for carrying out this determination. [5]

ii Write a balanced ionic equation to represent the reaction taking place. [2]

[Total 13 marks]

5 This question is about four compounds whose structural formulae are shown below.

A	B	C	D
$CH_3(CH_2)_2CHO$	$(CH_3)_2CHCHO$	$CH_3COC_2H_5$	$CH_3CH_2COOCH_2CH_3$

Simple organic compounds, like the ones shown in the table, can be distinguished from each other by chemical tests, or by spectroscopic methods, or by a combination of both.

a Name and draw out graphical (displayed) formulae for **A**, **B**, **C**, and **D**. [8]

b Compound **D** can be made by esterification. State the names of two compounds needed for the reaction, and give a condition used in the reaction. [3]

c Give the name and draw the graphical formula of an isomer of compound **D** that is not an ester. [2]

d Describe one chemical test that would distinguish between **B** and **C**, and state what you would observe in each case. [3]

e The low resolution ^1H n.m.r. spectra of **A**, **B**, **C**, and **D** can be used to distinguish between some of the structures, by considering the numbers of peaks and the ratios of the areas under them.

For each compound, state how many peaks would be seen, and give the ratio of areas under the peaks. [8]

f Compound **C** reacts with HCN. Give the name of the type of mechanism, and an outline of the mechanism for this reaction. [4]

[Total 30 marks]

6 Copper is a constituent of all living tissues and is essential for the normal growth of plants and animals. Copper compounds have many uses in agriculture, as well as a variety of specialized industrial uses. Copper sulfate is particularly useful and, with sodium carbonate, is used to form *Burgundy mixture*. This is used to stop mildew forming on grapes.

a **i** Copper is a transition element. Define the term *transition element*. [1]

ii Give the formula of the copper-containing complex ion formed when copper(II) sulfate is added to water. [1]

iii State the shape of, and bond angle(s) in, the complex ion formed in part **ii**. [2]

b Write the electronic configuration of:

i Cu [1]

ii Cu^{2+} [1]

c Write an ionic equation to show the reaction between the complex ion formed in part **a ii** and sodium carbonate. [2]

d When CuCl is dissolved in an excess of concentrated hydrochloric acid, a colourless solution containing the complex ion, $[CuCl_2]^-$, is formed. When hydrogen peroxide, H_2O_2, is added to this acidified solution, a green solution containing a copper complex ion, X, and water are formed.

i State the oxidation number of copper in $[CuCl_2]^-$. Give the electronic configuration of copper in this species and explain why it is colourless. [3]

ii State the role of hydrogen peroxide in the formation of X. [1]

iii Write a half-equation for the formation of water, as the only product, from hydrogen peroxide in acid solution. [1]

[Total 13 marks]

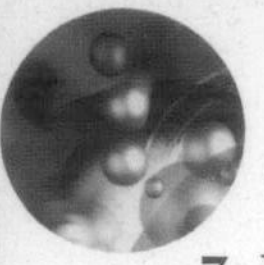

7 Lawnsand contains two products that work together to condition lawns: a nitrogen fertilizer that feeds the grass, encouraging growth, and giving a green healthy lawn; and ferrous sulphate, which kills moss causing it to blacken and die. As part of a manufacturer's quality control process, an industrial chemist was asked to analyze samples of lawnsand. It contained ammonium iron(II) sulfate-6-water, $Fe(NH_4)_2(SO_4)_2.6H_2O$, by titrating weighed samples against 0.0250 M $KMnO_4$.

a The chemist dissolved the weighed sample in water, and added one further reagent before titrating against 0.0250 M $KMnO_4$. Give the name of a suitable reagent, and state the colour change at the end-point of the titration. [2]

b Write the half-equations for the two reactions that take place in the redox reaction of iron(II) with manganate(VII) ions. Give an overall equation for this reaction. [3]

c Calculate the mass of pure $Fe(NH_4)_2(SO_4)_2.6H_2O$ which reacts exactly with 25.0 cm^3 of 0.0250 M $KMnO_4$. [4]

[Total 9 marks]

8 Analytical chemists may need to work out the names of the metal ion and non-metal ion in an unknown salt. Salt **A** is a white, crystalline solid. The following tests were carried out to establish the identity of salt **A**.

- An aqueous solution of **A** was divided into two equal portions. The first portion produced a white precipitate of compound **B** when treated with dilute sulfuric acid.
- The second portion of solution of **A** gave a solution of compound **C**, and a white precipitate of compound **D**, when treated with aqueous sodium carbonate.
- Compound **D** was removed from the mixture by filtration and, when treated with dilute hydrochloric acid, produced a gas. This gas gave a white precipitate when bubbled through limewater.
- Half of the filtrate, containing compound C, was treated with dilute nitric acid followed by aqueous silver nitrate. A cream precipitate of compound **E** was formed. This precipitate of **E** did not dissolve in either aqueous or concentrated ammonia.
- When chlorine water was added to the remaining portion of the filtrate, containing compound **C**, a brown solution of substance **F** was formed.

a Identify the metal ion present in **A**. [1]

b Give the name or formula of the substances **A** to **F**. [6]

c Write an equation for the reaction between **A** and aqueous sodium carbonate. [1]

d Give the name or formula of the gas produced when dilute hydrochloric acid is added to solid **D**. [1]

e When chlorine water is added to an aqueous solution of **C**, a brown solution is formed. Write an ionic equation for this reaction, and state the name of the type of reaction involved. [2]

[Total 11 marks]

9 Before the introduction of anaesthesia, surgery was agony for the patient and a terrifying last resort. Surgeons would only operate in a final attempt to save life and were judged by their speed. Some surgeons tried alcohol, morphine and other sedatives to dull the pain of surgery. But most patients were held or strapped down. Many died in agony before operations could be completed. The development of anaesthetics allowed surgeons to take more time and refine their techniques.

Early anaesthetics included nitrous oxide, N_2O, ether (ethoxyethane, $CH_3CH_2OCH_2CH_3$) and chloroform (trichloromethane, $CHCl_3$). Chloroform became popular because it worked well and was easier to use than ether, but it could have severe side effects such as sudden death. Chloroform can be made from chloromethane in a free radical substitution reaction which has several propagation steps.

a **i** Write an overall equation for this reaction.

ii Define the term *propagation step*.

iii Give an equation for a propagation step in which chloromethane is a reactant.

iv Write an equation for a propagation step which has trichloromethane as one of its products. [4]

b Write a balanced symbol equation, and draw out a mechanism, for the reaction of chloromethane with a solution of sodium ethoxide, $Na^{+-}OCH_2CH_3$. [5]

c Write a balanced symbol equation and draw out a mechanism for the reaction of chloromethane with an ethanolic solution of sodium ethoxide, $Na^{+-}OCH_2CH_3$. What is the role of the ethoxide ion in this reaction? [6]

[Total 15 marks]

Periodic table and data sheet

You will be given a Data Sheet in the examinations, similar to this one.

Proton n.m.r. chemical shift data

Type of proton	δ (ppm)
RCH_3	0.7–1.1
R_2CH_2	1.2–1.4
R_3CH	1.4–1.6
$RCOCH_3$	2.1–2.6
$ROCH_3$	3.1–3.9
$RCOOCH_3$	3.7–4.1
ROH	0.5–5.0

Infrared absorption data

Bond	Wavenumber (cm^{-1})
C–H	2850–3300
C–C	750–1100
C=C	1620–1680
C=O	1680–1750
C–O	1000–1300
O–H (alcohols)	3230–3550
O–H (acids)	2500–3000

Times of discovery

- before 1800
- 1800–1849
- 1849-1899
- 1900-1949
- 1949-1999

Key: relative atomic mass / atomic number / Name / atomic (proton) number

1.0 H Hydrogen 1

(IUPAC recommended group numbers)

Period	Group 1 (1)	2 (2)	(3)	(4)	(5)	(6)	(7)	(8)	(9)	(10)	(11)	(12)	3 (13)	4 (14)	5 (15)	6 (16)	7 (17)	8 (18)
1																		4.0 He Helium 2
2	6.9 Li Lithium 3	9.0 Be Beryllium 4											10.8 B Boron 5	12.0 C Carbon 6	14.0 N Nitrogen 7	16.0 O Oxygen 8	19.0 F Fluorine 9	20.2 Ne Neon 10
3	23.0 Na Sodium 11	24.3 Mg Magnesium 12											27.0 Al Aluminium 13	28.1 Si Silicon 14	31.0 P Phosphorus 15	32.1 S Sulfur 16	35.5 Cl Chlorine 17	39.9 Ar Argon 18
4	39.1 K Potassium 19	40.1 Ca Calcium 20	45.0 Sc Scandium 21	47.9 Ti Titanium 22	50.9 V Vanadium 23	52.0 Cr Chromium 24	54.9 Mn Manganese 25	55.8 Fe Iron 26	58.9 Co Cobalt 27	58.7 Ni Nickel 28	63.5 Cu Copper 29	65.4 Zn Zinc 30	69.7 Ga Gallium 31	72.6 Ge Germanium 32	74.9 As Arsenic 33	79.0 Se Selenium 34	79.9 Br Bromine 35	83.8 Kr Krypton 36
5	85.5 Rb Rubidium 37	87.6 Sr Strontium 38	88.9 Y Yttrium 39	91.2 Zr Zirconium 40	92.9 Nb Niobium 41	95.9 Mo Molybdenum 42	(98) Tc Technetium 43	101.1 Ru Ruthenium 44	102.9 Rh Rhodium 45	106.4 Pd Palladium 46	107.9 Ag Silver 47	112.4 Cd Cadmium 48	114.8 In Indium 49	118.7 Sn Tin 50	121.8 Sb Antimony 51	127.6 Te Tellurium 52	126.9 I Iodine 53	131.3 Xe Xenon 54
6	132.9 Cs Caesium 55	137.3 Ba Barium 56	138.9 * La Lanthanum 57	178.5 Hf Hafnium 72	180.9 Ta Tantalum 73	183.9 W Tungsten 74	186.2 Re Rhenium 75	190.2 Os Osmium 76	192.2 Ir Iridium 77	195.1 Pt Platinum 78	197.0 Au Gold 79	200.6 Hg Mercury 80	204.4 Tl Thallium 81	207.2 Pb Lead 82	209.0 Bi Bismuth 83	210.0 Po Polonium 84	210.0 At Astatine 85	222.0 Rn Radon 86
7	(223.0) Fr Francium 87	(226.0) Ra Radium 88	(227) # Ac Actinium 89	(261) Rf Rutherfordium 104	(262) Db Dubnium 105	(266) Sg Seaborgium 106	(264) Bh Bohrium 107	(277) Hs Hassium 108	(268) Mt Meitnerium 109	(271) Ds Darmstadtium 110	(272) Rg Roentgenium 111	Elements with atomic numbers 112-116 have been reported but not fully authenticated						

*58 – 71 Lanthanides	140.1 Ce Cerium 58	140.9 Pr Praseodymium 59	144.2 Nd Neodymium 60	(145) Pm Promethium 61	150.4 Sm Samarium 62	152.0 Eu Europium 63	157.3 Gd Gadolinium 64	158.9 Tb Terbium 65	162.5 Dy Dysprosium 66	164.9 Ho Holmium 67	167.3 Er Erbium 68	168.9 Tm Thulium 69	173.0 Yb Ytterbium 70	175.0 Lu Lutetium 71
#90 – 103 Actinides	232.0 Th Thorium 90	231.0 Pa Protactinium 91	238.0 U Uranium 92	237.0 Np Neptunium 93	239.1 Pu Plutonium 94	243.1 Am Americium 95	247.1 Cm Curium 96	247.1 Bk Berkelium 97	252.1 Cf Californium 98	(252) Es Einsteinium 99	(257) Fm Fermium 100	(258) Md Mendelevium 101	(259) No Nobelium 102	(260) Lr Lawrencium 103

The periodic table of the elements (photocopiable in the purchaser's institute).

Answers to calculations

1.02 The rate equation p18

1 b $10\,mol^{-2}\,dm^6\,s^{-1}$
c $9.6 \times 10^{-6}\,mol\,dm^{-3}\,s^{-1}$
2 a $1.0 \times 10^{-3}\,mol^{-1}\,dm^3\,s^{-1}$
b $2.5 \times 10^{-5}\,mol\,dm^{-3}\,s^{-1}$

2.02 The equilibrium constant p28

2 800 (no units)
3 $1.0\,mol^{-2}\,dm^6$

3.02 The pH of strong acids p36

2 a 1.3
b −0.81
c 3.7
3 a $3.16 \times 10^{-3}\,mol\,dm^{-3}$
b $1.00 \times 10^{-7}\,mol\,dm^{-3}$
c $5.01 \times 10^{-12}\,mol\,dm^{-3}$
4 a 1.00
b −0.176
c 0.921

3.03 The pH of strong bases p38

2 a 13.0
b 13.9
c 14.3
3 a 6.77
b 12.5

3.04 Weak acids p40

2 c 9.3

3.05 The pH of weak acids p42

2 a pH 2.1
b pH 1.0
3 a 4.77
b pH 2.7
4 pH 4.8

3.09 Titration calculations p50

1 a $0.300\,mol\,dm^{-3}$
b $31.25\,cm^3$
2 a 1.60
b 1.30
c 12.7
3 a 3.03
b 12.7

3.10 Buffers p52

3 4.88

Answers to *Questions: Ch 1–3* p54

1 b iv $k = \frac{rate}{[A]^2} = \frac{3.5 \times 10^{-4}}{(0.2)^2}$
$= 8.75 \times 10^{-3}\,dm^3\,mol^{-1}\,s^{-1}$

2 b Initial rate is 2.8×10^{-5} ($mol\,dm^{-3}\,s^{-1}$)
c i $k = \frac{7.5 \times 10^{-3}}{(0.25)^2} \times (0.50)^2$
$= 0.48\,mol^{-3}\,dm^9\,s^{-1}$

3 a i moles of C_2F_2 = 0.40
moles of HCl = 0.80
iii $k_c = \frac{(0.40/18.5)(0.80/18.5)^2}{(0.20/18.5)^2}$
$= 0.35\,mol\,dm^{-3}$

4 c pH = $-\log_{10}(2.00) = -0.30$
d $[H^+] = 4.0 \times 10^{-15}\,mol\,dm^{-3}$
pH = 14.4
e mol of H^+ = 19.0 × 2.0/1000 = 0.038
mol of OH^- = 16.0 × 2.50/1000
= 0.040
total volume = 19.0 + 16.0 = 35.0 cm^3
excess $[OH^-]$ = (0.040 − 0.038) ×
1000/35.0 = $0.0571\,mol\,dm^{-3}$
$[H^+] = K_w/[OH^-] = 1.0 \times 10^{-14}/0.0571$
$= 1.75 \times 10^{-13}\,mol\,dm^{-3}$
pH = $-\log_{10}(1.75 \times 10^{-13}) = 12.76$

5 b $[H^+] = \sqrt{(1.45 \times 10^{-4}) \times 0.45}$
$= 8.08 \times 10^{-3}$
pH = 2.09
c ii $[H^+] = 10^{-3.59} = 2.57 \times$
$10^{-4}\,mol\,dm^{-3}$
$[A^-] = K_a[HA]/[H^+]$
$[A^-] = (1.45 \times 10^{-4}) \times 0.45$ /
(2.57×10^{-4})
$[A^-] = 0.254\,mol\,dm^{-3}$

6 c i pH = $-\log_{10}[H^+]$
ii $[H^+] = 10^{-11.90} = 1.259 \times$
$10^{-12}\,mol\,dm^{-3}$
$[OH^-] = Kw/[H^+] = (1.0 \times$
$10^{-14})/(1.259 \times 10^{-12})$
$[NaOH] = 7.94 \times 10^{-3}\,mol\,dm^{-3}$
d i $K_a = [H^+]^2/[CH_3COOH]$
$[H^+] = \sqrt{(1.70 \times 10^{-5}) \times 0.117}$
$= 1.41 \times 10^{-3}$
pH = $-\log_{10}(1.41 \times 10^{-3}) = 2.85$
ii At half-equivalence, $[H^+] = K_a$
pH = $-\log_{10}(1.70 \times 10^{-5}) = 4.77$

12.03 Nuclear magnetic resonance spectroscopy p138

2 b δ = 900 ÷ 50 = 18 ppm

13.01 Enthalpy cycles p152

2 a −1358.6 −(−1410.8) = $+52.2\,kJ\,mol^{-1}$
b −(−84.7) −1614.4 = $-1559.7\,kJ\,mol^{-1}$

13.02 Mean bond enthalpies p154

2 a 685 − 926 = $-241\,kJ\,mol^{-1}$
b 2700 − 2826 = $-126\,kJ\,mol^{-1}$
c 4000 − 3740 = $+260\,kJ\,mol^{-1}$

13.04 Enthalpy of solution p158

1 b $\Delta H^\ominus_{soln}(NH_4NO_3) = +26\,kJ\,mol^{-1}$
2 a $\Delta H^\ominus_L(CaCl_2) = +2232\,kJ\,mol^{-1}$

13.06 Born–Haber cycles 2 p162

1 b $\Delta H^\ominus_L(KCl) = +718\,kJ\,mol^{-1}$
c $-718\,kJ\,mol^{-1}$

13.07 Born–Haber cycles 3 p166

1 b $\Delta H^\ominus_L(MgO) = +3845\,kJ\,mol^{-1}$
So lattice formation enthalpy
$= -3845\,kJ\,mol^{-1}$
2 $\Delta H^\ominus_f(MgCl_2) = +3949\,kJ\,mol^{-1}$

14.02 Entropy in chemical changes p172

2 a 229.5 − 284.8 = $-55.3\,J\,K^{-1}\,mol^{-1}$
b 42.6 − 84.4 = $-41.8\,J\,K^{-1}\,mol^{-1}$
c 474.8 − 563.6 = $-88.8\,J\,K^{-1}\,mol^{-1}$
d 3809.4 − 1827.3 = $+1982.1\,J\,K^{-1}\,mol^{-1}$
e 2.4 − 5.7 = $-3.3\,J\,K^{-1}\,mol^{-1}$

14.03 Spontaneous reactions p174

3 a $T\Delta S^\ominus = +27.7\,kJ\,mol^{-1}$, which is greater than $\Delta H^\ominus$, so the reaction is feasible under standard conditions.
b $T\Delta S^\ominus = +49.9\,kJ\,mol^{-1}$, which is less than $\Delta H^\ominus$, so the reaction is not feasible under standard conditions.
c $T\Delta S^\ominus = -0.98\,kJ\,mol^{-1}$, which is less than $\Delta H^\ominus$, so the reaction is not feasible under standard conditions.

14.04 Free energy and feasibility p176

2 b T = −356.3 ÷ (−0.5581) = 638 K

16.08 Commercial cells p200

1 b $E^\ominus_{cell}$ = +0.80 −(−0.76) = +1.56 V

16.09 Rechargeable cells p202

2 a E_{cell} = +1.69 −(−0.35) = +2.04 V (which is about 2 V)
b Car batteries contain six cells in series, so 6 × 2 = 12 V

Answers to *Questions: Ch 13–16* p206

1 b iii $\Delta H_1 = -338.8\,kJ\,mol^{-1}$
c ii Value = $-678.1\,kJ\,mol^{-1}$
2 b $\Delta H_r = -160\,kJ\,mol^{-1}$
d bond enthalpy = $+408\,kJ\,mol^{-1}$
5 a $\Delta H^\ominus = +117\,kJ\,mol^{-1}$
$\Delta S^\ominus = +326.9\,J\,K^{-1}\,mol^{-1}$
$\Delta G^\ominus = +195.8\,kJ\,mol^{-1}$
b 358 K

19.02 Spectrometry p230

1 c 1.4
d Answer in the range 500–540 nm.

Answers to *Questions: Ch 17–22* p266

1 g Mol of $MnO^{-4} = 6.56 \times 10^{-4}$
Mol of $Fe^{2+} = 3.28 \times 10^{-3}$
Concentration of $Fe^{2+} = 0.131\,mol\,dm^{-3}$

Answers to *Synoptic questions* p268

1 c i $k = 4.8 \times 10^{-3}\,mol^{-1}\,dm^3\,s^{-1}$
ii $1.44 \times 10^{-2}\,mol\,dm^{-3}\,s^{-1}$
iii $2.40 \times 10^{-4}\,mol\,dm^{-3}\,s^{-1}$
3 c mass of iodine = 0.28 g
7 c mass = 1.225 g

Terms first introduced in *AS Chemistry for AQA* are presented in grey text.

absolute temperature: The temperature measured in kelvin, K.

absolute zero: The lowest possible temperature theoretically achievable, −273.15 °C or 0 K.

absorption spectrum: A graph of absorbance against wavelength absorbed for a substance.

acceleration: The stage in mass spectrometry where positive ions are speeded up by an electric field.

acid amide: See amide

acid anhydride: An organic compound containing two acyl (RCO) groups bonded to the same oxygen atom.

acid chloride: See acyl chloride

acid dissociation constant: For a weak acid in solution, the ratio of the concentrations of the conjugate base and the acid present.

acid rain: Rain that has been acidified by pollutants such as sulfur dioxide.

acidic: Having the properties of an acid.

acidity reaction: Also called a hydrolysis reaction, a reaction in which a metal aqua ion loses a hydrogen ion.

activation energy: The minimum energy with which particles must collide for a reaction to happen.

active site: The place on the surface of a catalyst where the reactants are adsorbed.

actual yield: The mass of product obtained in a reaction.

acyl chloride: An organic compound containing the –COCl functional group.

acyl group: The –COR functional group where R is an alkyl or an aryl group.

acylation: The reaction between arenes and acid chlorides to prepare aromatic ketones.

acylium ion: The electrophile in a Friedel–Crafts acylation.

addition polymer: Polymer formed when many small unsaturated molecules join together.

addition reaction: The adding together of two or more molecules to form one larger molecule.

addition reaction: The adding together of two or more molecules to form one larger molecule.

adsorption: The process of bonding a substance to the active site of a catalyst.

aldehyde: An organic compound containing the –CHO functional group.

alkaline earth metal: An element from group 2 of the periodic table.

alkane: A saturated hydrocarbon with the general molecular formula C_nH_{2n+2}.

alkene: An unsaturated hydrocarbon with the general molecular formula C_nH_{2n}.

allotrope: Allotropes are different forms of the same element that exist in the same physical state. Diamond and graphite are allotropes of carbon.

alloy: A mixture of two or more metals, or a mixture of a metal and a non-metal.

alpha particle: A particle consisting of two protons and two neutrons, ejected from a nucleus.

amide: An organic compound containing the $-CONH_2$ functional group.

amide link: Identical to the peptide link but formed in a synthetic condensation polymer.

amine: An organic compound that contains a nitrogen atom joined to one or more carbon atoms, and two or less hydrogen atoms.

amino acid: An organic compound containing a carboxylic acid group and an amino group in the same molecule.

amino group: An $-NH_2$ group which is attached to a carbon atom.

amount of substance: The number of particles present, symbol n. It is measured in mole, mol.

amphoteric: Having the properties of an acid and a base.

analgesic: Painkiller

anion: A negatively charged ion, attracted to the anode during electrolysis.

anode: In a cell, the terminal where electrons are lost and oxidation happens.

antacid: A base taken as a medicine to neutralize excess stomach acid.

aramid: A polyamide in which the amide links are directly attached to benzene rings.

arene: See aromatic compound.

aromatic compound: Containing one or more benzene rings.

aromatic: Containing one or more benzene rings.

asymmetric carbon: A carbon atom with four different atoms or groups attached to it.

atom economy: The proportion of reactants that are converted into useful products rather than waste products.

atom: The smallest particle of an element that has the properties of that element.

atomic number: The number of protons in the nucleus of an atom, symbol Z.

atomic radius: The distance from the centre of the nucleus to the outer electrons of an atom.

Aufbau principle: The building up process that describes the filling of atomic orbitals in order of increasing energy.

autocatalysis: The phenomenon where one of the products in a reaction can catalyse the reaction.

Avogadro constant: The number of particles in one mole of a substance, 6.022×10^{23}.

Avogadro's principle: The idea that equal volumes of gases contain the same number of molecules, under the same conditions of temperature and pressure.

axial atom: Atom positioned at the top or bottom of a trigonal bipyramidal molecule.

barometer: An instrument for measuring atmospheric pressure.

base peak: The tallest peak in a mass spectrum.

base-catalyzed transesterification: The process of exchanging the alkoxy group of an ester compound with another alcohol in the presence of a base.

basic: Having the properties of a base.

batch process: An industrial process that is started and stopped at intervals. Production of ethanol by fermentation is a batch process.

bent line: The shape of a molecule that contains two bonding pairs and two lone pairs of electrons.

bidentate: A ligand that can donate two pairs of electrons.

biodegradable: Can be broken down by microorganisms.

biodiesel: Liquid fuel consisting of a mixture of methyl esters of long-chain carboxylic acids from vegetable oils.

biofuel: A fuel made from the products of living things.

bond angle: The angle between two adjacent bonds on the same atom.

bond: Attractive force between two atoms, ions or molecules.

Born–Haber cycle: An enthalpy cycle for the formation of an ionic compound.

Boyle's law: The volume of a fixed mass of gas (at a constant temperature) is inversely proportional to its pressure.

Brønsted–Lowry acid: A proton donor.

Brønsted–Lowry base: A proton acceptor.

Buchner flask: Thick-walled side arm flask, used for vacuum filtration.

Buchner funnel: Filter funnel with a perforated base, used for vacuum filtration.

buffer solution: A mixture that resists changes in pH when small amounts of acid or base are added to it, or when it is diluted.

burette: Laboratory apparatus used to add precise volumes of liquid during a titration.

carbocation: Ion with a positively charged carbon atom.

carbonyl: An organic compound containing the >C=O functional group, found in aldehydes and ketones.

carboxyl: An organic compound containing the –COOH functional group.

carboxylate ion: Anion formed from a carboxylic acid.

carboxylic acid: Organic acids with the general formula RCOOH. Their names end in -oic acid.

carrier gas: An inert gas such as nitrogen used to move the sample through a GLC column.

catalyst: A substance that speeds up chemical reactions by providing an alternative reaction route with a lower activation energy.

cathode: In a cell, the terminal where electrons are gained and reduction happens.

cation: A positively charged ion, attracted to the cathode during electrolysis.

cationic surfactant: A positively charged surface-active agent that can reduce the surface tension of liquids.

CFC: Abbreviation for chlorofluorocarbon, a hydrocarbon in which some or all the hydrogen atoms are replaced by chlorine and fluorine atoms.

chain isomer: See chain isomerism

chain isomerism: A type of structural isomerism in which compounds have identical molecular formulae but their carbon atoms are joined together in different arrangements. Chain isomers involve branched and unbranched carbon chains

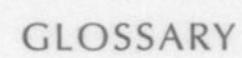

charge density: The charge:size ratio of an ion. Small ions with a high charges have large charge densities.

Charles's law: The volume of a fixed mass of gas (at a constant pressure) is proportional to its absolute temperature.

chelate: A complex formed by a metal ion, and bidentate or multidentate ligands.

chelation: The process of forming a chelate.

chemical environment: The particular situation of an atom in a molecule, determined by the atoms covalently bonded to it, and its neighbouring atoms or groups of atoms.

chemical shift: The position of an n.m.r. absorption peak relative to the position of tetramethylsilane, measured in parts per million.

chiral centre: See asymmetric carbon

chlorofluorocarbon: See CFC.

chromatography: A separation technique involving a stationary phase and a mobile phase.

cisplatin: The diamminedichloroplatinum(II) complex, $[PtCl_2(NH_3)_2]$, used as an anti-cancer drug.

coke: Solid produced by heating coal in the absence of air, almost pure carbon.

colorimeter: An instrument used to determine the amount of visible light absorbed by a coloured solution.

column chromatography: A separation technique where a mixture of solutes is passed through a tube packed with a stationary phase of powdered material or beads.

complete combustion: Burning a fuel in excess oxygen. Carbon dioxide and water vapour are produced from the complete combustion of hydrocarbons.

complex: A (central) metal ion surrounded by ligands.

concordant results: Titres that are in agreement, usually within 0.10 cm^3 of each other.

condensation polymer: Polymers formed by the reaction between molecules with two different functional groups, involving the loss of small molecules such as water or hydrogen chloride.

condensation reaction: A reaction in which two molecules join together with the elimination of a small molecule, often water.

conjugate acid: The species produced when a base gains a proton.

conjugate acid–base pair: Two species related to one another by the presence of a hydrogen ion (in the acid part of the pair) or its absence (in the base part of the pair).

conjugate base: The species produced when an acid loses a proton.

constituent amino acid: One of the amino acids that make a particular protein.

continuous process: An industrial process in which products are made all the time without any break. Production of iron in the blast furnace is a continuous process.

co-ordinate bond: A covalent bond in which the shared pair of electrons is provided by only one of the bonded atoms. In the bond X-Y, X provides both electrons.

co-ordination number: The number of pairs of electrons donated by ligands to the central metal ion in a complex.

correlation chart: A table of functional groups and their typical associated wavenumbers in infrared spectroscopy.

covalent bond: A shared pair of electrons.

cracking: A process used by the petroleum industry to produce shorter alkanes and alkenes from longer alkanes.

cuvette: A small cylindrical transparent container used to hold samples in the colorimeter.

cyclic: Hydrocarbons in which there are closed rings of carbon atoms are described as cyclic.

d block: The central section of the periodic table between groups 2 and 3, containing the transition metals.

dative covalent bond: Another name for a co-ordinate bond.

decomposition reaction: A reaction where one substance is broken down into two or more different substances.

deflection: The stage in mass spectrometry where positive ions are moved from their original path by a magnetic field.

dehydration reaction: A reaction where the elements hydrogen and oxygen are removed from a reactant in the ratio of 2:1, effectively the removal of water.

delocalization enthalpy: The increase in stability associated with electron delocalization.

delocalized: Electrons that are free to move between all atoms in a structure are delocalized. Delocalized electrons are found in metals and graphite.

delocalized electron: An electron in a molecule or metal that is not associated with a single atom or a covalent bond.

denatured: A protein that loses its function because of extreme conditions is said to be denatured.

deshielded: In n.m.r. the nucleus of a proton is deshielded when the electron density surrounding the nucleus is reduced due to the presence of an electron-withdrawing group.

desorption: The process of losing a substance from the active site of a catalyst.

detection: The stage in mass spectrometry where positive ions reach a detector and produce an electrical signal.

deuterated solvent: A solvent in which the hydrogen atoms are all ^2H atoms rather than ^1H atoms.

dextrorotatory: A substance that can rotate the plane of polarized light to the right is dextrorotatory.

diamine: A compound containing two amino groups.

diastereoisomers: Stereoisomers that are not mirror-images of each other, sometimes called geometrical isomers.

diatomic: A molecule containing just two atoms.

dicarboxylic acid: A compound containing two carboxyl groups.

difunctional compound: A compound containing two different functional groups, such as amino acids.

dimer: A molecule consisting of two monomer molecules joined together.

diol: A compound containing two hydroxyl groups.

dipeptide: An organic compound formed by the reaction of two amino acids.

dipole: Opposite charges separated by a short distance in a molecule or ion.

diprotic acid: An acid that can produce two hydrogen ions per molecule.

displace: To replace an atom or ion in a compound in a chemical reaction. For example, chlorine displaces iodine in sodium iodide.

displayed formula: A chemical formula showing all the atoms in a compound and their bonds.

disproportionation: The simultaneous oxidation and reduction of a species.

dot and cross diagram: A diagram showing all of the bonding electrons in a molecule. The electrons in one atom are shown as dots and the electrons in the other atom are shown as crosses.

double covalent bond: A bond in which two atoms are joined by two shared pairs of electrons.

doublet: A cluster of two peaks in the ratio 1:1 in a high-resolution n.m.r. spectrum.

dry cell: A type of electrical cell containing no free liquid.

ductile: Easily pulled into a thin wire.

dynamic equilibrium: A reaction in which the concentrations of reactants and products remain constant under fixed conditions, but the forward and backward reactions still continue.

dynamic: In a dynamic equilibrium, the reactions are still continuing.

dynamite: An explosive invented by Alfred Nobel, comprising nitroglycerine and silica.

e.m.f.: Electromotive force, the maximum potential difference an electrochemical cell can develop, measured using a high-resistance voltmeter.

effervescence: Fizzing or bubbling due to a chemical reaction.

electrochemical cell: A device that produces an electric current from a redox reaction.

electrochemical series: A list of electrode reactions and the corresponding standard electrode potentials in order of voltage.

electrode: The solid electrical conductor in a half-cell.

electrolysis: The decomposition of a compound into simpler substances using an electric current.

electrolyte: A chemical compound that can conduct an electric current.

electron configuration: The arrangement of electrons in an atom or ion.

electron gun: The source of high-energy electrons used to ionize the sample in mass spectrometry.

electron: Sub-atomic particle with a negative electric charge.

electron-deficient: An atom with a vacant orbital.

electronegativity: Electronegativity is the power of an atom to withdraw electron density from a covalent bond.

electronegativity: The power of an atom to withdraw electron density from a covalent bond.

electronic transitions: Changes by electrons from one energy level to another.

electrophile: A species that can accept a pair of electrons.

electrophilic addition: A reaction in which an electrophile is attracted to a region of high electron density, such as a carbon-carbon double bond, and adds on to an atom or group.

electrophilic substitution: A reaction in which an electrophile is attracted to a region of high electron density, such as delocalized electrons in a benzene ring, and replaces an existing atom or group.

electrostatic: Involving opposite charges.

element: A substance containing atoms which all have the same atomic number.

elimination reaction: A reaction in which a small, simple molecule is removed from a compound, forming a double covalent bond.

eluate: The liquid emerging from the end of the column in column chromatography.

eluted: Washed through the column in column chromatography.

empirical formula: A formula that gives the simplest whole number ratio of atoms of each element in a compound.

enantiomer: Optical isomer.

endothermic: A reaction in which heat energy is absorbed from the surroundings.

endpoint: Where the indicator just changes colour in a titration.

energy level: A certain fixed amount of energy that electrons in an atom can have, also called a shell.

enthalpy change: The heat energy change measured under conditions of constant pressure.

enthalpy cycle diagram: A chart that shows the relationship between various reactants and products, and the enthalpy changes between them.

enthalpy level diagram: A chart showing the enthalpies of the reactants and products in a reaction.

enthalpy of atomization: The enthalpy change when one mole of gaseous atoms is formed from an element or compound in its standard state.

entropy: A measure of the disorder of a system.

enzyme: A biological catalyst.

equatorial atom: Atom positioned around the middle of a trigonal bipyramidal molecule.

equilibrium: A reaction in which the concentrations of reactants and products remain constant under fixed conditions.

equilibrium constant: A numerical measure of the position of equilibrium in a reversible reaction.

equimolar: Containing equal numbers of moles.

equivalence point: In an acid–base titration, the point where equal numbers of moles of hydrogen ions and hydroxide ions have reacted.

equivalent atoms: Atoms in the same chemical environment.

ester: An organic compound containing –COO– functional group.

ester hydrolysis: The process in which an ester molecule is broken down by water.

esterification: The process by which an alcohol and a carboxylic acid are converted into an ester and water.

excess: More than the amount of reactant needed in a reaction.

excited state: An electron in a higher energy level than normal is said to be in an excited state.

exothermic: A reaction in which heat energy is released to the surroundings.

***E–Z* isomer:** A stereoisomer due to the presence of two different groups attached to the carbon atoms involved in the double bond in an alkene.

feasible reaction: A reaction which is energetically possible.

feedstock: Raw material used in a manufacturing process.

Fehling's solution: A reagent used to distinguish between aldehydes and ketones. It turns from blue to brick red when aldehydes are warmed with it.

fermentation: The process for making ethanol from sugar using yeast.

fingerprint region: The part of an infrared spectrum that is unique to a particular compound.

first electron affinity: The enthalpy change when electrons are gained by a mole of gaseous atoms or molecules to form a mole of gaseous ions, each with a single negative charge.

first ionization energy: The energy needed to remove one mole of electrons from one mole of gaseous atoms, forming one mole of ions with a single positive charge.

first order: A reaction in which the rate is directly proportional to the concentration of a substance is said to be first order with respect to that substance.

flue gas desulfurization: Removing sulfur compounds from waste gases produced by combustion.

flue gas: Waste gases from the combustion of a fuel, for example in a power station.

forward reaction: In a reversible reaction, the reaction in the direction from left to right when looking at the equation.

fraction: A part of a mixture collected at a particular temperature range by fractional distillation. A crude oil fraction contains hydrocarbons with a similar chain length.

fractional distillation: A method of separating mixtures of liquids or gases according to their boiling temperatures.

fragmentation pattern: The characteristic mass spectrum produced when molecules break up in the mass spectrometer.

free energy change: A measure of the enthalpy change and entropy change in a reaction at a given temperature.

free radical: A species that contains an unpaired electron, produced by the homolytic fission of a covalent bond.

frequency: The number of waves per second, measured in hertz, Hz.

Friedel–Crafts acylation: Electrophilic substitution reaction using aluminium chloride as a catalyst.

fuel cell: An electrochemical cell in which electricity is produced by the reaction of a fuel with oxygen.

full equation: A chemical equation showing the formulae and correct amounts of all the reactants and products in a reaction.

functional group: An atom, or group of atoms, in a molecule which determines its chemical properties.

gas constant: The constant used in the ideal gas equation. It has the symbol R and is approximately 8.31 $J K^{-1} mol^{-1}$.

gas syringe: A glass syringe used for collecting gases and measuring their volumes.

gas-liquid chromatography: see GLC

giant covalent: A structure in which very many atoms are joined by covalent bonds to form a regular structure. Diamond and graphite have giant covalent structures.

GLC: Gas-liquid chromatography – a type of chromatography where the mobile phase is a gas and the stationary phase is a liquid.

global warming: Increasing worldwide average temperatures.

greenhouse effect: Absorption of thermal energy by certain gases in the atmosphere, keeping the planet warmer than it would otherwise be.

greenhouse gas: An atmospheric gas that traps infrared radiation that would otherwise be radiated from the Earth's surface into space.

ground state: The lowest energy level normally occupied by an electron.

haem: A complex found in haemoglobin involving an iron(II) ion and a multidentate ligand

half-cell: A typical half-cell comprises a piece of metal dipped in an aqueous solution of its ions.

halide ion: A negatively charged ion formed when a halogen atom gains an electron.

haloalkane: A saturated organic compound in which at least one hydrogen atom has been replaced by a halogen atom.

haloarene: An aromatic compound in which at least one hydrogen atom has been replaced by a halogen atom.

Hess's law: If a reaction can occur by more than one route, the overall enthalpy change is independent of the route taken.

heterogeneous catalyst: A catalyst that exists in a different phase or state to the reactants in the reaction.

high resolution mass spectrometer: Device capable of measuring relative atomic masses and relative molecular masses to a high degree of precision.

homogeneous catalyst: A catalyst in a different phase or state from the reactants, such as nickel in the hydrogenation of vegetable oils.

homologous series: A series of compounds with the same general formula and functional group. Each member differs from the next by the presence of one more $–CH^2$ group.

homolytic fission: Breaking of a covalent bond so that each atom takes one electron from the shared pair, becoming a radical.

Hund's rule: Only when all the orbitals in a particular sub-level contain an electron do electrons begin to occupy the orbitals in pairs.

hydration: The addition of water across a double bond.

hydrocarbon: A compound containing hydrogen and carbon atoms only.

hydrogen bond: An intermolecular force between a lone pair of electrons on an N, O, or F atom in one molecule, and an H atom joined to an N, O, or F atom in another molecule.

hydrolysis: The splitting of a compound by reaction with water.

hydrolysed: When a chemical compound is broken down by reacting it with water, the compound is said to have been hydrolysed.

hydroxyl group: An –OH group, the functional group found in alcohols.

hydroxynitrile: An organic compound containing an –OH group and a –CN group attached to the same carbon atom.

ideal gas equation: The equation that describes the relationship between pressure, volume, amount of substance, and absolute temperature of a gas: $pV = nRT$.

incineration: The destruction of solid, liquid, or gaseous wastes by controlled burning at high temperatures.

incomplete combustion: Burning of a fuel in a restricted amount of oxygen.

indicator: A substance that changes colour according to the pH of a solution.

induced dipole: An uneven distribution of charge in a molecule or atom, caused by a charge in an adjacent particle.

infrared absorption spectrum: Spectrum produced when infrared radiation of various frequencies is absorbed by covalent bonds in a molecule.

infrared radiation: Electromagnetic radiation with a lower frequency than visible light, felt as heat.

infrared spectrometer: Device used in infrared spectroscopy.

infrared spectroscopy: Method used to analyze compounds by their absorption of infrared radiation.

initiation: The first stage in a free radical reaction.

integral: The area under the curve in a graph.

integrated spectra: A spectrum that includes a line whose height is proportional to the area under the peaks.

intermediate: Unstable species produced during a reaction before the final product is made.

intermolecular force: Weak attractive force between molecules.

ion: A charged particle formed when an atom or molecule gains or loses one or more electrons.

ionic bond: Electrostatic force of attraction between oppositely charged ions.

ionic compound: A compound made up of oppositely charged ions.

ionic equation: Chemical equation showing the separate ions in a chemical reaction and any essential reactant or product.

ionic product of water: The value obtained when the concentration of hydrogen ions in water is multiplied by its hydroxide ion concentration.

ionic radius: The distance from the centre of the nucleus to the outer electrons of an ion.

ionization: Producing an electrically charged particle by adding or removing electrons from an atom or molecule.

ionized: An atom or molecule that has gained or lost electrons is said to be ionized.

isoelectric point: The pH at which an amino acid has no overall charge.

isoelectronic: Having the same electron configuration as another species.

isomer: Compounds with the same molecular formula but different structural formulae.

isotope: Atoms with the same number of protons but different numbers of neutrons in their nuclei.

Kekulé structure: A structure proposed for benzene in which six carbon atoms are joined in a ring by alternating single and double carbon–carbon bonds.

kelvin: The unit of absolute temperature, symbol K.

ketone: An organic compound containing the carbonyl functional group >C=O, and with the general formula R_1COR_2.

laevorotatory: A substance that can rotate the plane of polarized light to the left is laevorotatory.

landfill: The controlled deposit of waste into holes in the ground which are then filled in with soil.

lattice: A regular arrangement of atoms, ions or molecules in a structure.

lattice dissociation enthalpy: The enthalpy change when one mole of an ionic solid is separated into its gaseous ions.

lattice formation enthalpy: The enthalpy change when one mole of an ionic solid is formed from its gaseous ions.

Lewis acid: An electron pair acceptor

Lewis base: An electron pair donor

ligand substitution: The replacement of one ligand in a complex by another ligand.

ligands: An ion or molecule that can donate a pair of electrons to the central metal ion in a complex.

limiting reactant: A reactant that is completely used up before the other reactants are converted into products.

linear molecule: A molecule with all of its atoms in a straight line.

lone pair: A pair of electrons in the highest occupied energy level that are not used in bonding.

macromolecular: See giant covalent.

main chain: The longest chain of carbon atoms in an organic compound.

main chain: The longest chain of carbon atoms in an organic compound.

malleable: Can be bent or hammered into shape without breaking.

mass number: The total number of protons and neutrons in the nucleus of an atom, symbol A.

mass spectrometer: An instrument used to determine the relative atomic mass of an element or relative molecular mass of a compound. The structure of a complex molecule can be worked out by analysis of a mass spectrum.

mass spectrum: The output from a mass spectrometer, plotting relative abundance against mass to charge ratio, m/z.

mass to charge ratio: The mass of an ion divided by its charge, symbol m/z.

mean bond enthalpy: Average enthalpy change when the same type of bond is broken in many similar substances.

melting point determination: A method used to identify a substance by its melting point, and to determine its purity.

metallic bond: The electrostatic force of attraction between metal ions and the delocalized electrons in a metallic lattice.

metalloid: Element with properties that are intermediate between the properties of a metal and the properties of a non-metal.

mobile phase: The phase that moves in column chromatography, usually the solvent.

mol: The symbol for amount of substance measured in moles.

molar volume: The volume occupied by one mole of gas at a specified temperature and pressure.

molarity: The concentration of a solution measured in moles of solvent per cubic decimetre of solution, mol dm^{-3}.

mole: The amount of substance that contains as many particles as there are atoms in exactly 12 g of ^{12}C.

molecular crystal: Covalent molecules held together in a regular arrangement by intermolecular forces.

molecular formula: A formula that gives the actual number of atoms of each element in a molecule.

molecular ion: In mass spectrometry, the ion that produces a peak in the mass spectrum at the highest m/z value.

molecular ion peak: Peak in a mass spectrum due to the molecular ion.

molecular sieve: Porous materials, such as zeolites, that let some molecules pass through but not others.

molecule: A particle containing two or more atoms joined together.

monomer: A small molecule that can join together to make a polymer.

monoprotic: An acid containing one replaceable hydrogen ion, such as HCl and HNO_3.

multidentate: A ligand that can donate more than two pairs of electrons.

***n* + 1 rule:** In high-resolution n.m.r. spectroscopy, the peak produced by a nucleus in a chemical environment is split into a number of peaks equal to the number of adjacent non-equivalent nuclei plus one.

n.m.r. spectrometer: A device used to analyse substances by their nuclear magnetic resonance.

n.m.r. spectroscopy: An analytical method used to determine the structures of substances by their nuclear magnetic resonance patterns.

n.m.r. spectrum: The characteristic pattern of peaks produced by a substance in nuclear magnetic resonance spectroscopy.

neurotransmitter: Biological substance involved in the passage of nerve impulses from one nerve cell to another.

neutral: In terms of acids and bases, a solution containing equal concentrations of hydrogen ions and hydroxide ions.

neutron: A neutral sub-atomic particle found in the nucleus of an atom.

nicad: A nickel–cadmium rechargeable battery.

nickel–cadmium cell: See nicad

nitrating mixture: A mixture of concentrated nitric acid and concentrated sulfuric acid used to add a nitro group to an arene.

nitration: Adding a nitro group to an arene.

nitrile: An organic compound containing the –C≡N functional group.

nitro group: The $-NO_2$ functional group

nitronium ion: The NO_2^+ ion

non-equivalent atoms: Atoms in different chemical environments.

non-polar: Having no dipole. A molecule with polar bonds may be non-polar if its shape is such that the dipoles cancel each other out.

non-renewable resource: Resource that cannot be replaced once it has all been used up. Fossil fuels and metal ores are non-renewable resources.

non-superimposable: Mirror-image objects are non-superimposable.

N-substituted amide: A compound based on an amide but with one of the hydrogen atoms in the $CONH_2$ group replaced by an alkyl group.

nuclear magnetic resonance spectroscopy: See n.m.r. spectroscopy

nucleophile: A species with a lone pair of electrons that is available to form a co-ordinate bond.

nucleophilic addition–elimination: A reaction mechanism in which a nucleophile adds onto a compound and a small molecule is removed.

nucleophilic substitution: A chemical reaction in which one nucleophile replaces another in a molecule.

nucleus: The central part of the atom, made from protons and neutrons.

nylon 6,6: A polyamide in which the two monomers each contain six carbon atoms.

octahedral: The shape of a molecule containing six bonding pairs of electrons.

optical isomer: A molecule which can exist in two mirror image forms because it has a chiral centre.

orbital: The volume of space in an atom where one or two electrons are most likely to be found.

ore: A mineral from which metals can be extracted and purified.

overall order of reaction: The sum of the individual orders of reaction for a certain reaction.

oxidant: See oxidizing agent

oxidation state: The number of electrons that would have to be added to an atom in a compound to make a neutral atom of that element.

oxidation: A loss of electrons or hydrogen, or gain of oxygen by a chemical species.

oxidized: A species that has lost electron(s), gained oxygen, lost hydrogen, or increased in oxidation number.

oxidizing agent: A substance that can oxidize another substance; an electron acceptor.

ozone layer: The part of the atmosphere with the greatest concentration of ozone. Found in the stratosphere, it absorbs harmful ultraviolet radiation from the Sun.

ozone: An allotrope of oxygen with the formula O_3.

p block: The part of the periodic table containing groups 3 to 0.

Pauli exclusion principle: The idea that an orbital cannot hold more than two electrons.

Pauling electronegativity scale: A scale showing the ability of elements to withdraw electron density from a covalent bond. The larger the number, the more electronegative the element.

peptide link: The –CONH– linkage formed when the carboxyl group of one amino acid reacts with the amino group of another amino acid.

percentage composition: The percentage mass of a compound due to a particular element.

percentage difference: The difference between an experimental result and an expected result, shown as a percentage of the expected result.

percentage yield: The actual yield of a product shown as a percentage of the expected yield.

permanent dipole–dipole forces: Attractive forces that exist between polar molecules.

pH curve: A graph of pH against volume of titrant added.

pH meter: An instrument used to determine the pH of a solution.

pH scale: A scale of acidity based on the concentration of hydrogen ions.

phenylamine: An organic compound with the formula $C_6H_5NH_2$, also called aniline or aminobenzene.

planar: A flat arrangement of atoms or ions.

Planck's constant: The constant of proportionality between change in energy and frequency of electromagnetic radiation, $h = 6.626 \times 10^{-34}$ J s.

plane-polarized light: Light which vibrates in one plane only.

plasticizer: A substance added to a polymer to alter its properties.

plum pudding model: A disproved model of the atom in which electrons move in sea of positive charge.

polar bond: A covalent bond between atoms with different electronegativities.

polar: A covalent bond between atoms with different electronegativities is said to be polar.

polarimeter: An instrument for measuring the degree of rotation of plane polarized light.

polarized: Having opposite charges, separated by a small distance.

polyamide: A condensation polymer containing amide links, –CONH–

polyatomic ion: An ion containing more than one atom.

polyester: A condensation polymer containing ester links, –COO–.

polymer: A large molecule made up of many repeating units or monomers.

polypeptide: Condensation polymer formed by amino acids.

position isomerism: A type of isomerism where the functional group can be joined at different places on the carbon skeleton.

position isomerism: A type of isomerism where the functional group can be joined at different places on the carbon skeleton.

position of equilibrium: If a reaction's position of equilibrium lies to the right, the ratio of products to reactants will be high. If it lies to the left, the ratio of products to reactants will be low.

potential difference: The difference in electrical potential, measured in volts.

precipitate: An insoluble solid formed when two solutions are mixed.

primary alcohol: An alcohol in which the hydroxyl group is attached to a carbon atom that is directly attached to no more than one other carbon atom.

primary amine: An organic compound derived from ammonia in which one of the three hydrogen atoms has been replaced by an alkyl or aryl group.

primary cell: A non-rechargeable cell.

primary haloalkane: An haloalkane with one carbon atom, or where the carbon atom carrying the halogen atom is directly attached to just one other carbon atom.

primary structure: The sequence of amino acid residues in a polypeptide or protein.

product: A substance made in a chemical reaction.

propagation: The stage in a free radical mechanism where a particular radical is used in one reaction then produced again in a subsequent reaction.

proton n.m.r.: The analysis of hydrogen environments in a compound by nuclear magnetic resonance spectroscopy.

proton: A positively charged sub-atomic particle found in the nucleus of an atom.

quartet: A cluster of four peaks in the ratio 1:3:3:1 in a high-resolution n.m.r. spectrum.

quaternary ammonium ion: An organic ion derived from ammonia in which four alkyl or aryl groups are attached to the same nitrogen atom.

quaternary structure: The arrangement of sub-units in a protein.

racemate: See racemic mixture

racemic mixture: In optical isomerism this is a mixture of equimolar amounts of both enantiomers.

radical: Very reactive species formed by homolytic fission, and containing an excited unpaired electron.

radioactive: A substance that produces radiation is said to be radioactive.

rate coefficient: See rate constant

rate constant: The constant of proportionality in the rate expression.

rate determining step: The slowest step in a reaction which determines the rate of the whole reaction.

rate equation: An equation that shows the relationship between the concentrations of the reactants and the rate of the reaction.

rate limiting step: See rate determining step

raw material: Substance in its natural state intended for use in a chemical process.

reactant: A substance used in a chemical reaction.

reaction mechanism: A step by step description of how a reaction happens.

recrystallization: A method used to purify a substance by dissolving its crystals in a minimum amount of hot solvent, filtering, and then letting crystals form again as the mixture cools.

redox titration: Method to find the concentration of a sample using a reactant of known concentration by a redox reaction rather than a neutralization reaction.

reducing agent: A substance that can reduce another substance; an electron donor.

reductant: See reducing agent

reduction: A gain of electrons or hydrogen, or loss of oxygen by a chemical species.

reference compound: A substance against which other substances are compared in analyses such as n.m.r. spectroscopy.

refluxing: A laboratory technique that allows a reaction mixture to be heated for a long time without loss by evaporation.

relative abundance: The proportion of a particular species in a sample.

relative atomic mass: The mean mass of an atom of an element compared to one-twelfth the mass of a ^{12}C atom, symbol *Ar*.

relative charge: The charge of a sub-atomic particle compared to the charge on a proton, taken as +1.

relative formula mass: The mean mass of a unit of a compound compared to one-twelfth the mass of a ^{12}C atom, symbol *Mr* .

relative isotopic mass: The mass of an atom of a particular isotope compared to one-twelfth the mass of a ^{12}C atom.

relative mass: The mass of a sub-atomic particle compared to the mass of a proton, taken as 1.

relative molecular mass: The mean mass of a molecule compared to one-twelfth the mass of a ^{12}C atom, symbol *Mr* .

repeating unit: The short sequence of atoms that is repeated many times in a polymer.

residue: The part of an amino acid that remains in a peptide or protein after polymerization.

resonance frequency: The frequency of electromagnetic radiation absorbed by an atom in n.m.r. spectroscopy.

retention factor: see R_f value

retention time: The time taken for a component of a mixture to pass through a chromatographic column.

reverse reaction: In a reversible reaction, the reaction in the direction from right to left when looking at the equation.

reversible reaction: A reaction in which the products are able to decompose to form the original reactants again.

R_f value: In thin-layer chromatography, the distance travelled by a solute divided by the distance travelled by the solvent.

roast: Heating strongly in a stream of air, usually applied to metal ores.

s block: The part of the periodic table containing groups 1 and 2.

salt bridge: Filter paper, typically soaked in saturated potassium nitrate, that allows two half-cells to be in electrical contact without the solutions mixing in bulk.

saponification: Converting oils and fats into soap by hydrolysing them with a base.

saturated: A compound containing only single covalent bonds between carbon atoms.

second electron affinity: The enthalpy change when electrons are gained by a mole of gaseous ions with single negative charges to form a mole of gaseous ions, each with two negative charges.

second ionization enthalpy: The enthalpy change when one mole of electrons is removed from one mole of gaseous ions, each with a single positive charge, forming one mole of ions with two positive charges.

second order: A reaction in which the rate is directly proportional to the square of the concentration of a substance is said to be second order with respect to that substance.

secondary alcohol: An alcohol in which the carbon atom to which the hydroxyl group is attached is directly attached to two carbon atoms.

secondary amine: An organic compound derived from ammonia in which two of the three hydrogen atoms have been replaced by alkyl or aryl groups.

secondary cell: A rechargeable cell.

secondary structure: The folding of a polypeptide chain maintained by hydrogen bonds.

semiconductor: A substance that is an electrical insulator at room temperature, but a conductor when warmed or when other elements are added to it.

shield: In an atom with more than one occupied energy level, a decrease in the force of attraction between an electron and the nucleus because of electrons in lower energy levels.

shielded: In n.m.r. the nucleus of a proton is shielded when the electron density surrounding the nucleus is increased due to the presence of an electron-donating group.

shortened structural formula: Abbreviated structural formula in which the arrangement of atoms and group is shown without drawing bonds. For example, hexane would be $CH_3\ (CH_2)_4\ CH_3$.

side chain: A shorter chain of carbon atoms attached to a longer main chain, a branch.

simple covalent molecule: A molecule containing just a few atoms covalently bonded together, such as O_2 and NH_3.

simple molecule: A molecule containing just a few atoms, such as O_2 and NH_3.

singlet: A single peak in a high-resolution n.m.r. spectrum.

skeletal formula: A type of displayed formula in which the symbols for carbon atoms are left out.

solubility: The extent to which one substance dissolves in another.

solute: The substance that will dissolve in a solvent.

solvent: The substance in which a solute will dissolve.

sparingly soluble: Almost insoluble but a very small amount will dissolve.

species: An atom, molecule or ion.

spectator ion: An ion that appears on both sides of an equation but does not take part in the reaction.

spin–spin coupling: In n.m.r spectroscopy, the interaction between the nuclear spins of adjacent non-equivalent nuclei that cause characteristic splitting of peaks in the spectrum.

spontaneous change: A reaction that happens without needing an input of energy to start it.

square pyramidal: The shape of a molecule that contains five bonding pairs and one lone pair of electrons.

standard bond dissociation enthalpy: The enthalpy change when one mole of bonds of the same type in gaseous molecules is broken under standard conditions, producing gaseous fragments.

standard conditions: 298 K and 100 kPa

standard enthalpy change: An enthalpy change measured under standard conditions.

standard entropy: A measure of the disorder of a system under standard conditions.

standard solution: A solution whose exact molarity is known.

standard state: The state (solid, liquid, or gas) of a substance at 298 K and 100 kPa.

standard temperature and pressure: 273 K and 100 kPa.

state symbol: Symbols used in chemical equations to show the state of the substance: solid (s), liquid (l), gas (g), aqueous or dissolved in water (aq).

states of matter: Solid, liquid, and gas.

stationary phase: In chromatography, the component that does not move and through which the mobile phase moves.

stereoisomer: Molecules with the same structural formula but a different arrangement of their bonds in space.

stratosphere: Upper part of the atmosphere, approximately 17 km to 50 km high, containing the ozone layer.

strong acid: An acid that is fully ionized or dissociated in aqueous solution.

strong base: A base that is fully ionized or dissociated in aqueous solution.

structural formula: A formula showing the atoms present and the bonds between them.

structural isomerism: When two or more compounds have the same molecular formulae, but different structures.

sub-atomic particle: A particle found within an atom, for example the proton, neutron, or electron.

sub-level: Part of an energy level in an atom, containing pairs of electrons: s sub-levels contain up to one electron pair, p sub-levels contain up tothree electron pairs, and d sub-levels contain up to five electron pairs.

sublimation: Passing directly from the solid state to the gas state.

sublime: To pass directly from the solid state to the gas state.

substitution: The replacement of one atom or group of atoms in a molecule by another atom or group of atoms.

surfactant: See cationic surfactant

sustainable development: Living in such a way that we meet our needs without damaging the ability of future generations to meet their own needs.

temporary dipole: The asymmetrical distribution of the electron pair in a covalent bond.

termination: The final stage in a free radical substitution reaction.

tertiary amine: An organic compound derived from ammonia in which all three hydrogen atoms have been replaced by three alkyl or aryl groups.

tertiary structure: The three-dimensional structure of a protein maintained by covalent bonds and hydrogen bonds.

tetrahedral: The shape of a molecule with 4 bonding pairs of electrons.

tetramethylsilane: The internal standard used in n.m.r. spectroscopy.

theoretical yield: The maximum mass of product possible, calculated using the mass of reactantsand the balanced equation.

thermal cracking: The thermal decomposition of hydrocarbons to produce shorter alkanes and alkenes.

thin-layer chromatography: A type of chromatography where the mobile phase is liquid and the stationary phase is a thin coating of solid

titrant: The solution added from the burette in a titration.

titration: Method used to find the concentration of a sample using a reactant of known concentration.

titre: The volume of titrant added to reach the end-point in a titration.

TLC: See thin-layer chromatography

TMS: Tetramethylsilane

TNT: Trinitrotoluene, an explosive.

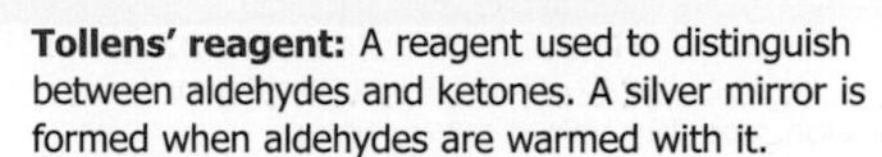

Tollens' reagent: A reagent used to distinguish between aldehydes and ketones. A silver mirror is formed when aldehydes are warmed with it.

trans fat: An unsaturated fat produced as a by-product of hydrogenating vegetable oils.

transfer pipette: A piece of glassware used to add an accurate volume of liquid.

transition metal: A d block element that can form at least one stable ion with a partially filled d sub-level.

triacylglycerol: An organic compound formed by the reaction of three fatty acid molecules with a molecule of propane-1,2,3-triol (glycerol).

triester: An organic compound containing three ester groups, such as triglycerides

triglyceride: See triacylglycerol

trigonal bipyramidal: The shape of a molecule with 5 bonding pairs of electrons.

trigonal planar: The shape of a molecule with 3 bonding pairs of electrons.

trigonal pyramidal: The shape of a molecule with 3 bonding pairs and 1 lone pair of electrons.

triple covalent bond: A bond in which two atoms are joined by three shared pairs of electrons.

triplet: A cluster of three peaks in the ratio 1:2:1 in a high-resolution n.m.r. spectrum.

unburned hydrocarbons: Pollutants in the exhaust from car engines, due to incomplete combustion of the fuel.

unidentate: A ligand that can donate one pair of electrons.

unsaturated: Containing at least one carbon–carbon double bond.

UV/vis spectrometer: An instrument that measures the absorbance by a substance of ultraviolet and visible light at different wavelengths.

vacant orbital: An orbital that can accept a pair of electrons.

vacuum filtration: Filtration under reduced pressure, such as using a Buchner funnel and flask.

valence shell: The energy level in an atom that is involved in forming bonds.

van der Waals' forces: Temporary, induced dipole–dipole attractions between covalent molecules.

volatile: A liquid that easily vaporizes is said to be very volatile.

volumetric flask: An item of glassware used to make up a standard solution. Also called a standard flask or graduated flask.

VSEPR: **V**alence **S**hell **E**lectron **P**air **R**epulsion theory. The theory used to predict the shape of a covalent molecule using the idea of repulsion by pairs of electrons.

wavenumber: A measure of frequency used in infrared spectroscopy. It has units of cm-1.

weak acid: An acid that is only partially dissociated in aqueous solution.

weak base: A base that is only partially dissociated in aqueous solution.

yield: The mass of product formed in a reaction.

zeolite: Compounds of aluminium, silicon, and oxygen with microscopic pores. Zeolites are used as catalysts and molecular sieves in the petrochemical industry.

zero order: A reaction in which the rate is independent of the concentration of a substance is said to be zero order with respect to that substance.

Ziegler–Natta catalyst: Catalysts used to produce polymers with particular arrangements of side groups.

zwitterion: A chemical compound that is electrically neutral overall but has positive and negative charges on different atoms.

Index

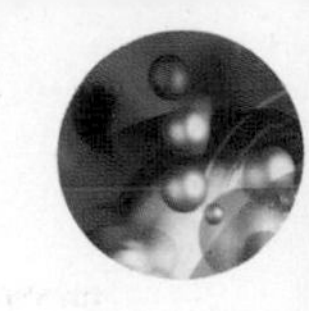

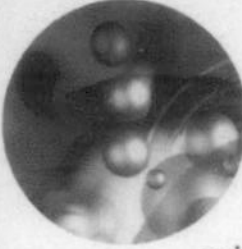